Sandlot Stats

Sandlot Stats

Learning Statistics with Baseball

STANLEY ROTHMAN

The Johns Hopkins University Press • *Baltimore*

Printed in the United States of America on acid-free paper
9 8 7 6 5 4 3 2 1

The Johns Hopkins University Press
2715 North Charles Street
Baltimore, Maryland 21218-4363
www.press.jhu.edu

Library of Congress Cataloging-in-Publication Data

Rothman, Stanley, 1944–
Sandlot stats : learning statistics with baseball / Stanley Rothman.
p. cm.
Includes index.
ISBN 978-1-4214-0602-2 (hdbk. : alk. paper) — ISBN 978-1-4214-0867-5 (electronic)—
ISBN 1-4214-0602-0 (hdbk. : alk. paper) — ISBN 1-4214-0867-8 (electronic)
1. Statistics—Study and teaching. 2. Baseball—Statistics. 3. Baseball—
Mathematics. 4. Mathematics—Study and teaching. 5. Probabilities—Study
and teaching. I. Title.
QA276.R685 2012
519.5—dc23 2011048238

A catalog record for this book is available from the British Library.

The Johns Hopkins University Press uses environmentally friendly book materials, including recycled text paper that is composed of at least 30 percent post-consumer waste, whenever possible.

For Eva, Yana, Cameron, and all future grandchildren
in hopes that they inherit not only Grandpa's love of baseball,
but also his love of mathematics

Contents

Acknowledgments

When my colleague Larry Levine decided to teach a course on the history of baseball, he approached me with the idea of a course involving learning statistics with baseball. Because I am a baseball fanatic, the idea immediately appealed to me, and thus, the course Baseball and Statistics 101 was born. A course needs a textbook. My next step was to find a book that could be used to teach an introductory course in statistics, based solely on baseball data. There are books on the market that use statistics to analyze baseball records. Unfortunately, I found that these books were not appropriate for teaching statistics. So I started preparing for the course writing my own lecture notes. Trevor Lipscombe, who was then the editor-in-chief at Johns Hopkins University Press, found my course online and contacted me to see if I would be interested in writing a textbook using baseball examples to teach statistics. I told him I would be interested, and so the process of formalizing my work began. If it hadn't been for Trevor's encouragement, enthusiasm for the subject, and guidance, this book would not be a reality. He also added baseball trivia to my text. When Trevor left Johns Hopkins, I was devastated. I figured that was the end of my textbook. However, I began to work with executive editor Vincent Burke. Not only did I acquire a new editor, but also a new friend. He got me to streamline the text to make it more manageable, and thus I also owe him a great deal of gratitude for getting me to finish what I started. I've never considered myself a writer, and working with my copy editor Jeremy Horsefield was like being back in grade school having my papers corrected. I can only imagine the hours he put into editing! I want to take this opportunity to personally thank him for his excellent comments and corrections.

Now I would like to take this opportunity to thank all the students in my Baseball and Statistics classes for helping to proofread the text, and in particular Quoc Le and Kevin Faggella, for helping with the research that is included in the last three chapters of the book.

I would be remiss if I didn't thank everyone from my math colleagues, all my friends, my sons' friends, my sisters-in-law and my brothers-in-law (Ira and Sue Blieden, Avra and Michael Tietze, Mavra and Jon McCann), my nieces and nephews (Marissa Blieden and Ben Blieden, Maxwell Tietze and Harrison Tietze, Trenton McCann and Madison McCann), my daughters-in-law (Victoria and Christina), and other family members, especially Edgar and Linda Ednalino, for their constant encouragement and genuine interest in this book. One friend in particular, Marty Cobern, constantly sent me articles of interest on the subject of baseball and statistics and also debated with me many of the finer points about these articles. Even the Tech Support department at QU (thank you Billy Murphy) kept me going when I had computer issues and almost missed deadlines. Also, special thanks to Dr. Michael Nabel for his advice on statistical topics used in the book, and to Dr. Vincent Celeste for his help with Excel.

Quinnipiac University, my home for 42 years, supported my efforts by giving me a sabbatical so that I could spend an entire semester writing this book.

Lastly, I would like to thank my wife, Tara, and sons, Bradley and Matthew, for their endless patience and encouragement. Tara, who knows nothing about baseball, constantly sent me articles, asked questions about my course, and made sure I stayed focused. She also created my webpage and set up my Facebook page and blog. I had fun discussing topics with my sons and listening to their opinions and suggestions. Without them, I never would have had the stamina to turn my course notes into a full-fledged textbook.

Abbreviations

1B	single
2B	double
3B	triple
AB	at bat
Adj_BA	adjusted batting average
Adj_HRA	adjusted home run average
Adj_OBP	adjusted on-base percentage
Adj_SLG	adjusted slugging average
AL	American League
AVG	batting average
BA	batting average (alternate abbreviation)
BB	base on balls
BBA	base-on-balls average
BF	Ballpark Factor
BOP	bases over plate appearances
BRA	batter's run average
CI	confidence interval
COPS	collection, organization, presentation, and summarization
CV	coefficient of variation
cv	critical value
df	degrees of freedom
EO	extended out
H	hit
HBP	hit by pitch
HR	home run
HRA	home run average
HRR	home run rate
HT	hypothesis testing

IBB	intentional base on balls
IPAB	in-play at bat
IPBA	in-play batting average
IPHR	in-play home run average
IPHRA	alternate abbreviation for IPHR
IPO	in-play out
ISO	isolated power
ME	mutually exclusive
MVP	most valuable player
NL	National League
Not H	not a hit
O	out
OB	on base
OBA	on-base average
OBP	on-base percentage
OPS	on-base plus slugging
PA	plate appearance
PCL	Pacific Coast League
pdf	probability distribution function (or probability density function)
Pr	probability
R	runs scored
R/27	runs scored by a player per 27 outs
RBI	runs batted in
RC	runs created
RC/27	runs created by a player per 27 outs
R/G	runs per game
RPR	run production
SB	stolen base
SE	safe on error
SF	sacrifice fly
SH	sacrifice hit
SLG	slugging percentage
SO	strikeout
SOA	strikeout average
sts	standard test statistic
TB	total bases
TPQ	total power quotient
ts	test statistic

Sandlot Stats

Introduction

Baseball is by far the most statistically oriented sport. While watching a game, one becomes exposed to all sorts of statistics. As soon as a player approaches home plate, his batting average, on-base percentage, number of home runs, and number of runs batted in are displayed and talked about. The delay of approximately 45 seconds between pitches gives the commentators an opportunity to bombard you with statistics. We learn a player's batting average against this pitcher, his batting average against this team, his batting average during the last week, during the day versus at night, and so on.

Statistical techniques are used to collect, organize, summarize, graphically present, and derive conclusions from baseball data and a host of other sources. This book teaches you about statistics in the context of baseball, but the lessons are transferable to health professions, accounting, marketing, educational research, sociology, and almost every other profession. Beyond using statistics at work, the statistical techniques learned from this book will help you make important consumer decisions and let you more accurately judge the validity of polls. If you are already interested in baseball, you will be familiar with such descriptive measures as batting average, on-base percentage, and slugging percentage. In the pages that follow you learn that these terms are related to important statistical concepts such as mean, proportion, and weighted mean.

The underlying data for the study of statistics in this book come from the well-archived records of professional baseball. We will revisit some of baseball's greats, such as Joe DiMaggio, Babe Ruth, and Henry Aaron, as well as many lesser known but statistically important players.

Baseball fans debate many issues in baseball. Some of the debated issues that will provide statistical fodder for learning in this book are as follows:

- Which player was the best hitter of all time?

- Which baseball team was the best team of all time?
- Which league was better in a given year?
- At what age does a Major League player reach his peak as a hitter?
- Will a player ever break Joe DiMaggio's 56-game hitting streak?
- What does it take for a player to hit .400 in today's game?
- Which baseball strategies do not work?
- Does a certain player belong in the Hall of Fame?
- Is there a home-field advantage in baseball?
- How do we go about comparing two players?
- Which batting feat from the past would be the toughest to duplicate today?
- Which batting statistics can be used to best predict a win–loss record for a team?

Two Branches of Statistics

The study of statistics is divided into two areas, *descriptive statistics* and *inferential statistics*, both of which attempt to make sense out of data. *Descriptive statistics* deals with the collection, organization, summarization, and graphical presentation of data. After applying the techniques of descriptive statistics to the collected data, we are in the position to draw *subjective* conclusions about the data. The conclusions made are limited to only the collected data.

Inferential statistics attempts to use the collected observed data to predict (infer) results for a much larger group of data. The collected data is called the *sample data*, and the larger group of data is called the *population data*. Applying the results of a small sample to make decisions about a population is a method used in many disciplines. Some examples are the testing of a new drug on a small sample before approving it for use; the use of surveys, administered to a small number of households, to estimate public opinion; and the use of exit polls to predict the winner of an election.

With the help of a calculator, you will perform each of the statistical processes in the book. By doing so, you will gain an understanding and appreciation of statistical concepts. You will also use Microsoft Excel as a statistical software package. It is not the strongest statistical package, but it is available on almost all computers and has enough statistical features to handle the tasks in this book.

The following five websites serve as major sources for baseball data; these sites are used throughout this book, and you will need to become familiar with them:

www.baseball-reference.com
www.mlb.com
www.baseball-almanac.com
www.retrosheet.org
www.baseballhalloffame.org

Each student is asked to choose two Major League players. The first player must already be a Hall of Famer. The second player is not a member of the Hall of Fame; however, the student should believe that this player may deserve induction. The website www.baseballhalloffame.org contains all the Hall of Fame players. In selecting the two players, the following guidelines can be followed:

- The careers of the two players either overlap or are close to overlapping.
- The two players must have played most of their careers at the same position. The possible positions are catcher (C), first baseman (1B), second baseman (2B), shortstop (SS), third baseman (3B), corner outfielder, or center fielder.
- The player not in the Hall of Fame must have played at least 10 years in the Major Leagues.

Of course, these guidelines are just guidelines, and an instructor may change some of them.

Beginning in Chapter 4, students will be asked to apply the same statistical methods used to compare the batting performance of Henry Aaron with the batting performance of Barry Bonds to compare the batting performances of their two chosen players. This process of comparing batting performances will continue in subsequent chapters.

The students will create a model in the form of a spinner. The spinner can be assembled by physically making a cardboard disk with a spinning pointer for each of the chosen players or, more easily, by using one of the many available online spinners. A reliable online spinner is available from the National Council of Teachers of Mathematics (http://illuminations.nctm.org/activitydetail.aspx?ID=79). The purpose of the spinner disk is to mimic a player's batting performance. The disk is composed of sectors. Each sector corresponds to an outcome of a plate appearance. The size of each of the sectors corresponds to the probability of that player having that outcome. Spinning the pointer will correspond to a plate appearance (the disk is similar to that found in the classic game All Star Baseball by Cadaco).

At the end of Chapter 10, the students will be asked to give a preliminary argument on whether the potential Hall of Fame player they chose should be admitted to the Hall of Fame. This argument will be based on the descriptive statistics covered in Chapters 1–10.

At the end of Chapter 18, the students will again be asked to answer the same question. They may or may not choose to change their mind, based on the new statistical techniques learned in Chapters 11–18.

Throughout the chapters of this book, the following baseball questions will be addressed:

- What is the difference between a player's batting performance and batting ability?
- How can a player's batting performance be used to estimate a player's batting ability?
- How can statistics be used to compare two or more baseball players?
- Which baseball statistics are most important in assessing the batting ability of a baseball player?
- Which hitting feat will be the hardest to duplicate now and in the future?
- Which baseball statistics are the best indicators for determining the number of runs a player contributes to his team?
- What properties must a player have to duplicate Joe DiMaggio's 56-game hitting streak?
- Will we ever have another .400 hitter?
- What are the greatest hitting feats of all time?
- Who are the top 10 hitters of all time?

Overview of the Book

Chapter 1 introduces the terms and definitions for the concepts presented in any statistics course. An actual "hot dog study" done by the Los Angeles Dodgers is used to illustrate these concepts. Examples, in and out of the world of baseball, are provided for each of the concepts introduced.

Chapter 2 introduces the standard techniques used in descriptive statistics. These techniques are applied to one quantitative variable. Standard statistical measures are used to find the middle of the data and how spread out the data are. Graphs are used to give a shape to the data. The normal curve is the most common shape. The properties common to all normal curves are presented.

Chapter 3 introduces the descriptive measures specific to baseball such as batting average, on-base percentage, slugging percentage, and home run rate.

Chapter 4 uses the techniques of descriptive statistics to compare two quantitative data sets. In particular, a comparison of the batting performances of Henry Aaron and Barry Bonds is made.

Chapter 5 introduces regression and correlation analysis. Relationships between various baseball descriptive measures are explored. Regression analysis is used to establish linear relationships between various baseball statistics and the runs scored by a player or by a team. Correlation analysis is used to find which baseball statistics are the best predictor of runs scored by a player or by a team.

Chapter 6 applies the concepts of descriptive statistics to one or more qualitative variables. Contingency tables and special graphs are used to find a relationship between two qualitative variables. The concept of when two variables are independent is introduced.

Chapter 7 introduces the concept of probability. Three types of probabilities are defined. They are classical probability, relative frequency probability, and subjective probability. The *Fundamental Counting Principle*, *permutations*, and *combinations* are defined. A contingency table for Aaron and Bonds is displayed, and simple and compound probabilities are computed. Independent and mutually exclusive events are defined. Conditional probability is introduced and used to compare qualitative variables. Side-by-side and stacked bar graphs are used to observe relationships between qualitative variables. Probability disks are constructed for Aaron and Bonds. Simulations are done using both physical and theoretical models.

Chapter 8 introduces the concepts involved in sports betting. The concept of odds is introduced. The relationship between odds and probability is explored. The expected gain and loss resulting from making a bet are explained. A discussion of how a casino profits from sports betting is presented. Examples demonstrating both casino betting and sports betting are presented. A discussion is made of how to approach sports betting.

Chapter 9 relates the standard descriptive measures used in any statistics course to the descriptive measures specific to baseball.

Chapter 10 uses the baseball descriptive statistics evaluated in the previous chapters to draw some conclusions on which player, Aaron or Bonds, was a better hitter.

Chapter 11 looks at discrete probability distributions. The idea of a theoretical mathematical model is introduced. The binomial distribution and the geometric distribution are defined. Baseball situations are modeled using both of these distributions. Probabilities using the geometric and binomial model are calculated.

Chapter 12 looks at continuous probability distributions. The most important continuous distribution is the normal distribution. Probabilities are calculated using the normal curve.

Chapter 13 looks at the concept of a sampling distribution. Sampling distributions for sample means and for sample proportions are studied. The role of a sampling distribution as a bridge to inferential statistics is explained.

Chapters 14 and 15 introduce the two major techniques used in statistical inference. They are *confidence intervals* and *hypothesis testing*.

Chapter 14 introduces the inferential technique of *confidence intervals*. A new continuous distribution, called the *t-distribution*, is introduced. This chapter differentiates between a sample baseball statistic and a population baseball parameter. Confidence intervals are used to estimate both the population batting average and population on-base percentage for Aaron and Bonds. The term *level of confidence* is introduced.

Chapter 15 introduces the inferential technique of *hypothesis testing*. Three methods of hypothesis testing are presented. They are classical hypothesis testing, the *p*-value approach, and the confidence interval approach. An eight-step method for classical hypothesis testing is presented. Hypothesis testing for one population mean and for one population proportion is illustrated. These techniques are used to explore such issues as whether either Aaron or Bonds was a *true* career .300 hitter. The term *statistical significance* is explained. The concept of *level of significance* is introduced.

The next two chapters are involved with two research studies in baseball. The information covered in the first 15 chapters is applied to these baseball research questions.

Chapter 16 studies different baseball batting streaks. Much of the chapter involves Joe DiMaggio's 56-game hitting streak. The 84-game on-base streak of Ted Williams and other lesser-known streaks are also discussed. The purpose of this chapter is to apply the concepts of probability and statistics, from the previous 15 chapters, to analyze the probability of these streaks happening. The probability of various players achieving each of these streaks is examined.

Chapter 17 looks at the fabulous baseball feat of batting .400 for a season. The last .400 hitter was Ted Williams, who accomplished this feat in 1941. All .400 hitters, since 1913, are presented. An attempt is made to present the baseball characteristics that seem to be necessary for a player to hit .400. The likelihood of this feat being duplicated in the future is analyzed.

Chapter 18 is the concluding chapter. In this chapter, the following questions are analyzed:

- What are the greatest hitting feats of all time?
- Which hitting feat will be the hardest to duplicate?
- Who are the top hitters in various baseball eras?
- Who are the top 10 hitters of all time?

In order to decide on the top 10 hitters of all time, a scoring system is developed that is based on the statistics covered in prior chapters in this book.

Chapter 18 concludes with a final set of chapter problems that involve the concepts covered in many of the prior chapters. The final problem asks the student to finalize their decision on whether their chosen player should be elected into the Hall of Fame.

BATTING PRACTICE: QUIZ ON SPECIAL NUMBERS IN BASEBALL

Connect these numbers to a ballplayer and his special baseball event:

Number	Event	Player
56:		
60:		
61:		
73:		
84:		
.406:		
714:		
755:		
762:		
2130:		
2632:		
4256:		
23:		
191:		
2295:		
7:		

Hint: The players come from this list: Henry Aaron, Barry Bonds, Ty Cobb, Joe DiMaggio, Bob Feller, Jimmie Foxx, Lou Gehrig, Rickey Henderson, Rogers Hornsby, Joe Jackson, Mickey Mantle, Roger Maris, Willie Mays, Mark McGwire, Stan Musial, Albert Pujols, Cal Ripken Jr., Pete Rose, Babe Ruth, Nolan Ryan, Tom Seaver, Duke Snider, Sammy Sosa, Ichiro Suzuki, Bill Terry, Ted Williams, and Hack Wilson.

Chapter 1

Basic Statistical Definitions

Baseball teams stage promotions to increase attendance. In 2006, the Los Angeles Dodgers ran a "free hot dogs" promotion. At Dodger Stadium, it's difficult to see the game from the right field stands, so fans are reluctant to buy tickets to sit there. Empty seats, though, mean empty coffers. To encourage people to fill the right field stands, the Dodgers offered free, unlimited hot dogs, nachos, peanuts, popcorn, and soda to any person buying a ticket in this section. But would the extra ticket sales cover the cost of the food and drinks they were giving away?

Promotions don't always lead to good results, as long-time Indians fans might remember:

> "Ten-Cent Beer Night" in Cleveland backfired after drunken and disorderly fans stumbled onto the field of play causing the Indians to forfeit the game to the Texas Rangers. With a five-all score in the ninth, Tribe fans poured onto the field and surrounded outfielder Jeff Burroughs while trying to take his hat and glove for souvenirs. After players from both sides rushed to his aid, the game was called in favor of the visitors.*

How do you transform a promotional *idea* into a promotional *event*? First, identify the group to target for this promotion. This group is called the *population*. Knowledge of the population is essential for advertising purposes. As there is no complete list for this population, this is an *infinite population*. Next, we choose the *variables* (characteristics of the population) that we wish to study. For the "hot dog promotion" there are two variables. One is *qualitative*: Will the promotion encourage people to buy a ticket in the right field stands (Y or N)? The second variable is *quantitative*:

* www.almanac-reference.com

If a person buys a ticket, how many (number) free hot dogs would that person eat?

Since the population is infinite, we can't ask everybody. Instead, we study a small part of the population, called a *sample*. The sample should be selected so that it is representative of the population. One method of selection is *simple random sampling* (every subject in the population has the same chance of being selected).

After the sample is selected, the people in the sample are given a questionnaire, which consists of two questions: Would you buy a ticket to sit in the right field stands because of this promotion? If you answered yes, how many hot dogs would you eat? The answers become the *data* to be studied. The sample data are summarized through the use of *descriptive measures*. Basically, the Dodgers want to know how many extra tickets they will sell, increasing their revenue, and at what cost (how much will the hot dogs cost them)?

It might be expensive. The National Hot Dog and Sausage Council polled concessionaires at Major League ballparks, who estimated that an average of 862702 hot dogs are eaten per ballpark, per year in the United States. Since a team plays half of its 162-game schedule at home, that's about 10651 dogs per game.

One application of a descriptive measure is to find the typical value (the middle) for the data. The *mean* is the descriptive measure used most often to find the typical value for *quantitative* data. The *proportion* is the descriptive measure most used to find the typical value for *qualitative* data. In our study, we are interested in the *sample proportion* of people who would buy tickets for the right field stands because of the promotion and the *sample mean* number of hot dogs they would eat. Using the techniques of *inferential statistics*, the *sample proportion* of people who answered yes can be used to estimate the *true proportion* of the population that would answer yes, and the *sample mean* number of free hot dogs a person would eat can be used to estimate the *true mean* number of free hot dogs a person in the population would eat. The sample mean and sample proportion will always be known. The population mean and population proportion are not known.

The *sample proportion* and *sample mean* are called statistics. Observe that there are two meanings attached to the word "statistics." Statistics can refer to the field of study or to descriptive measures of a sample. The *population proportion* and *population mean* are called *parameters*.

The "hot dog promotion" is an example of a *research study*. Some *confounding variables* (outside variables that can affect the results of the research study) are the win–loss record of each team, the difficulty in getting to the stadium, the day of the week, the opposing team, the weather, and conflicts due to other events scheduled on the same day as the promotion.

Based on the positive results of this *observational research study* (the researcher is a passive participant who plays the role of an observer in the study), the management of the Los Angeles Dodgers decided to proceed with the promotion as an event.

1.1. Variables and Data

A *variable* is a characteristic that can be associated with each subject or object in a group. The individual values produced from the variable are called the *data*. An uppercase X will represent a variable, and a lowercase x will represent a *value* of the variable. The individual values in the data are called data values, data points, or x-values. The x-values in the data can be the same or different. If the values of the variable are *numbers*, the variable is *quantitative*. If the values are *categories*, the variable is *qualitative*. Numbers that are used as a naming mechanism for categories are considered categories. For example, if a hit is assigned the number 1 and an out is assigned the number 0, the numbers 0 and 1 are categories. Zip codes, telephone numbers, and social security numbers are other examples of numbers used as categories. When numbers are used as categories, the normal operations of addition and multiplication have no meaning. The sum of two zip codes has no meaning.

1.1.1. Sources of Data

Three important sources of data are research studies, surveys, and external resources. A *research study* on a group of subjects or objects is interested in either certain observations of the subjects or responses of the subjects to a treatment variable. These observations or responses become the data. A *survey* is a study of various characteristics of a group of people performed by asking the people certain questions. These characteristics mentioned in the questions are called *survey variables*. The survey can be carried out by interviewing each person or by having the people complete a written questionnaire. The answers to the questions become the data. Data can also come from such *external sources* as research articles, books, or websites.

In the "hot dog promotion," the data come from a survey. The two characteristics of interest are whether the promotion would cause a person to buy a ticket for the right field stands (yes or no) and the number of free hot dogs a person would eat. The *yes or no* variable is a qualitative variable, and the number of hot dogs is a quantitative variable. The values of the *yes or no* variable would consist of only the words "yes" or "no." The values for the quantitative variable are such numbers as 0, 1, 2, 3, 4, and so on. (The record for hot dogs eaten in 10 minutes, set at Nathan's hot dog contest held each Fourth of July at Coney Island, stands at 68. If you could keep that rate up for a 3-hour ball game, you'd consume 1224 dogs.)

Other Examples of Quantitative Variables

Example 1.1. The number of home runs hit by a player for each year during his career.

Example 1.2. The batting average of a player for each year during his career.

Example 1.3. The age of each player on the 25-man roster of a team in a given year.

1.1.2. Discrete and Continuous Quantitative Variables

Quantitative variables can be further classified as either *discrete* or *continuous*. The data for a *discrete quantitative variable* consist of either a finite set of numbers or numbers that are only integers. The values of a *continuous quantitative variable* can be any real number between any two real numbers. There will be gaps between the values of a discrete variable. Since between any two real numbers there is another real number, continuous variables have no gaps. The easiest way to distinguish between these two types of variables is to ask the following question: Is the variable being counted or measured? If the answer is "counted," the variable is discrete; if the answer is "measured," the variable is continuous.

1.1.3. Discrete Qualitative Variables

Qualitative variables that produce a finite number of values (categories) are also considered *discrete*.

Examples of Discrete and Continuous Variables

- Since the number of home runs hit is counted, the variable in Example 1.1 is a discrete quantitative variable.
- For the "hot dog promotion," the number of hot dogs eaten by a patron would also be discrete.
- Since the variables in Examples 1.2 and 1.3 are measured, they are continuous quantitative variables. Even if the ages are rounded into integers, an age is still a measurement of time and consequently a continuous quantitative variable.
- In the "hot dog promotion," whether the promotion influences the purchase of a ticket (yes or no) is a qualitative discrete variable.
- To understand the difference between the *possible* values of a variable and the *actual* data values, assume that the survey consisted of only two questions:
    - Question 1: Would you buy a ticket to the right field stands for free hot dogs (yes or no)?

- Question 2: If you answered "yes" to question 1, how many free hot dogs would you eat?

Suppose that 100 people were surveyed, and 60 answered "yes" and 40 answered "no." Of the 60 people who answered "yes," 10 said they would eat one hot dog, 30 said they would eat two hot dogs, 12 said they would eat three hot dogs, and 8 said they would eat four hot dogs. Question 1 produced 100 data points. The two *possible* values for the variable in question 1 are "yes" and "no." The value "yes" appears 60 times and the value "no" appears 40 times. Question 2 produced 60 data points. The *possible* values of the variable in question 2 are any nonnegative integer. The four *actual* values for the variable in question 2 are the numbers 1, 2, 3, and 4. The value 1 appears 10 times, the value 2 appears 30 times, the value 3 appears 12 times, and the value 4 appears 8 times.

1.1.4. Scoring Difference between an At Bat and a Plate Appearance

The variables at bat and plate appearance for a player are used often in this book, so their meanings must be clarified.

In baseball scoring there is a difference between a plate appearance and an at bat. An *at bat* (AB) includes four outcomes from batting: (1) a hit (H); (2) safe on dropped third strike (SD); (3) safe on error (SE); and (4) any result that causes an out (including a fielder's choice but excluding a sacrifice fly [SF] or sacrifice hit [SH]). Our definition of a *plate appearance* (PA) includes an at bat (AB), a base on balls (BB), a sacrifice fly (SF), and a hit by pitch (HBP). Notice that a sacrifice hit or defensive interference is counted neither as a plate appearance nor as an at bat. In baseball scoring, our definition for a plate appearance means an official plate appearance. An official plate appearance is used as the denominator for such baseball statistics as on-base percentage. When establishing whether a player qualifies for a batting title, both a sacrifice hit and defensive interference are counted as plate appearances. However, when we refer to a plate appearance, it will be an official plate appearance. This is because most of the baseball statistics involving plate appearances use only official plate appearances in their calculations. The possible batting outcomes along with their symbols are shown in Figure 1.1.

Other Examples of Qualitative Variables

Example 1.4. The *experiment* is a PA by a player. The possible values of the variable PA are AB, SF, BB, and HBP. When there are more than two categories, the variable is called *multinomial*.

Example 1.5. The ranking of a player (1 being highest) based on the number of home runs compared with all other players for a given year. In 2008, the number 1 ranked player in this category was Albert Pujols of the St. Louis Cardinals.

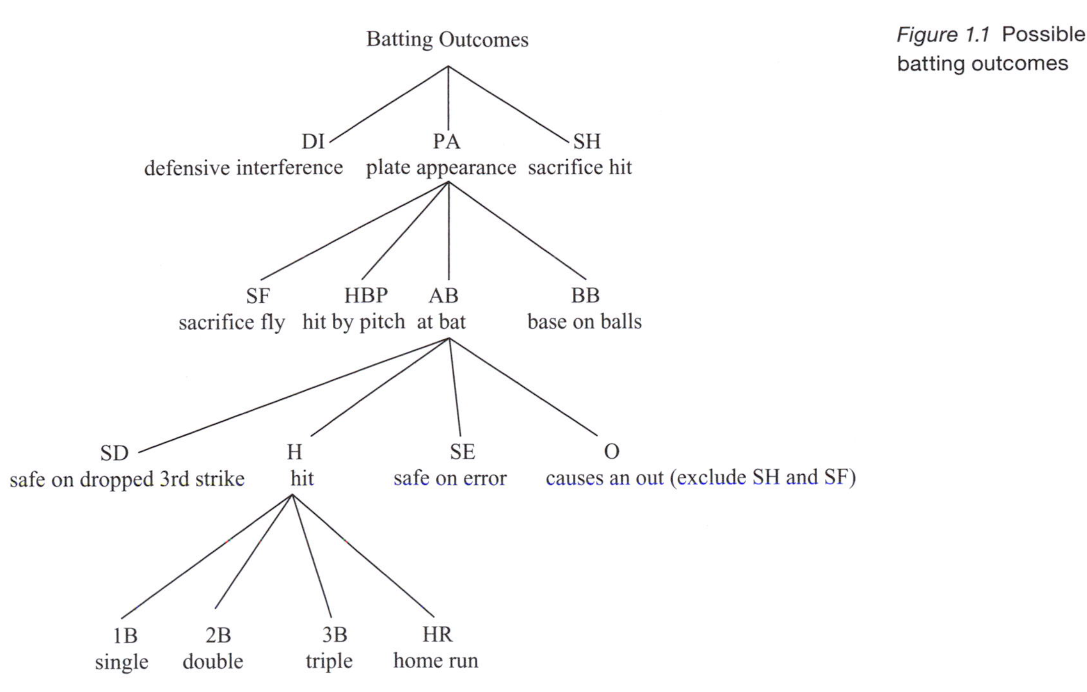

Figure 1.1 Possible batting outcomes

Example 1.6. The experiment is an at bat by a player. Group the at bats into two categories. One category is "H" for hit, and the other category is "Not H," which includes "O" for out, SE, and SD (see Fig. 1.1). When there are two categories, the *variable* is referred to as *dichotomous*.

Example 1.7. The experiment is one roll of a standard six-sided die. There are six possible values of the variable (1, 2, 3, 4, 5, and 6). The numbers are treated as categories.

Examples 1.4–1.7 are all discrete qualitative variables.

1.1.5. Random Variables

When the values of the variable are partially or entirely controlled by chance, the variable is called a *random variable*. Examples 1.4–1.7 are examples of random variables. When values of a variable are predetermined, the variable is said to be a *fixed variable*. If you study a player's batting performance in his rookie year, 5th year, 10th year, and 15th year, then you have a fixed variable—the years you choose to study. On the other hand, if we placed slips of paper containing all of a player's career years in a bowl and randomly selected four of them, the variable (years) would be a random variable.

Exercise 1.1. Identify three quantitative variables and three qualitative variables associated with baseball players, teams, or leagues.

1.2. Measurement Scales

A measurement scale provides a way to compare the difference between data points. A *tag* is associated with each data point. The tag can be the data point itself or a different value applied to the data point. By using the assigned tag, a comparison can be made between two data values. Ratio and interval scales have numeric tags, whereas ordinal and nominal scales have categorical tags.

For the data gathered from a variable, its level of measurement is significant. This is because the level of measurement can determine which statistical tests are meaningful. The four *measurement scales* from highest level to lowest level are *ratio*, *interval*, *ordinal*, and *nominal*.

A *nominal* measurement scale is a naming device and can only determine if two data values are in the same or different categories. An *ordinal* measurement scale, besides determining whether two data values are in the same or different categories, also provides an ordering or ranking system. A comparison can be made between two categories as to which is stronger or greater or to the right of the other category. (Hits can be classified as home runs or not home runs, which would have a nominal scale. But you can also rank a hit as a single, double, triple, or home run, which is an ordinal scale.)

In both the *ratio* and *interval* measurement scales, not only can the two data values be ranked, but the difference between two measurements is the same as their real-world difference as well.

Ratio and interval scales differ with regard to the presence or absence of a *true zero*. A true zero applied to a quantitative variable indicates the absence of the property. Ratio scales have a true zero; interval scales do not. The data values for the variable "weight" have a ratio scale since a zero weight means an absence of weight. The data values for Fahrenheit temperature have an interval scale since a zero Fahrenheit temperature does not mean an absence of temperature.

When data have a ratio scale, ratios of measurements have meaning in the real world. The heaviest player in Major League Baseball was Baltimore's Walter Young, who tipped the scales at 315 pounds. The lightest was Eddie Gadael, of the St. Louis Browns, who played only once and weighed 65 pounds. The ratio of Walter's weight to Eddie's is 315 to 65 or about 5 to 1.

Fahrenheit temperature data have an interval, not a ratio, measurement scale. A temperature of 100°F at Tropicana Field, home to the Tampa Bay Rays, is not twice as warm as 50°F at Fenway Park in Boston.

The data for the number of hot dogs eaten by patrons and the data mentioned in Examples 1.1, 1.2, and 1.3 all have a ratio measurement scale.

The data consisting of "yes" or "no" responses in the "hot dog promotion" have a nominal measurement scale.

The data in Examples 1.4, 1.6, and 1.7 have a nominal measurement scale, whereas the data in Example 1.5 have an ordinal measurement scale.

The next example illustrates the differences between the four measurement scales.

Example 1.8. The eight regular positional players for a fictional baseball team are classified, in the table below, by their position and their number of home runs for a given year. It is known that the home stadium for these players is a difficult place to hit home runs.

HR Total	40	45	31	44	47	30	20	42
Position	C	1B	2B	3B	SS	LF	CF	RF

The variable "home run totals" is a quantitative discrete variable, whose data values consist of 40, 45, 31, 44, 47, 30, 20, and 42. The variable "position" is a qualitative variable, whose data values consist of C, 1B, 2B, 3B, SS, LF, CF, and RF.

Case 1. Assuming that these recorded numbers are the actual home run totals, the tag assigned to each data point is the recorded number. The measurement scale is ratio because a zero value means an absence of hitting any home runs.

Case 2. The tag assigned to each data point is the recorded number itself. However, we later discover that the actual number of home runs was 10 less than the recorded number. The reason the recorded data were incremented by 10 is to adjust for the difficulty of hitting home runs in the stadium. The lowest possible recorded number is 10. Since a recorded value of 40 is actually 30 home runs and a recorded value of 20 is actually 10 home runs, a recorded 40 does not represent twice as many home runs as a recorded 20. Since there is no true zero, we cannot say that the recorded value of 40 is twice the recorded value of 20; the measurement scale is interval. Note that the real-world difference is preserved, that is, the difference between 40 and 20 is equal to the difference between 30 and 10.

Case 3. The tags assigned to the recorded numbers consisted of ranks (1 being the highest). After tagging, we have the following ranks for the data:

Ranks	5	2	6	3	1	7	8	4
Position	C	1B	2B	3B	SS	LF	CF	RF

The measurement scale is ordinal. Unlike an interval scale, the rank difference does not preserve the real-world difference. The difference between rank 1 and rank 2 is two home runs, whereas the difference between rank 2 and rank 3 is one home run.

Case 4. Each recorded number is tagged with an I, an O, or a C, depending on whether the player with that recorded number was an infielder, outfielder, or a catcher, respectively. After tagging the players, we have C, I, I, I, I, O, O, and O. Since the data are composed of categories without an ordering, the measurement scale is nominal.

Exercise 1.2. Four home runs were hit in a game at new Yankee Stadium. A radar gun was used to measure the distance of each home run. The recorded distances were 410, 370, 390, and 440 feet. (In the old Yankee Stadium, demolished in 2008, the distance to the center field wall was 461 feet, earning it the nickname "Death Valley.")

a. Assuming that the gun was accurate, the measurement scale assigned to the recorded distances would be ________.
b. Suppose that it was discovered that the gun was faulty and added an extra 20 feet to each measurement. The measurement scale assigned to the recorded distances would be ________.
c. Suppose that tags were used to rank the distances of the home runs. The measurement scale would be ________.
d. Suppose that each distance was tagged with either an "H" for a home run hit by a player on the home team or an "A" for a home run hit by a member on the away team. The measurement scale would be ________.

1.3. Research Studies: Experimental and Observational

A research study begins with a question about an entire group of subjects, called a *population*, and one or more variables associated with the subjects. Research studies are classified as either *experimental* or *observational.*

In *experimental* studies, the researchers are active participants. Prior to running the experiment, they choose their subjects by using a statistically valid sampling technique. They provide the necessary instructions and control the variables used in the study. An experimental study can have one or more treatment variables and one or more response variables.

In *observational* studies, the researchers play a passive role and are only interested in obtaining and observing the results of the study. An observational study can have one or more response variables.

Example 1.9 (Experimental Study). The research question involves comparing player A and player B with respect to their home run hitting ability. The experimenter notes that the two players are approximately the same age. He takes both players to the same Major League park for the experiment. The same pitching machine throws 50 pitches each to the players. The total number of home runs each player hits is recorded. Notice the active role played by the researcher. He chooses the park, the number of pitches, and the way the pitches are delivered. The treatment variable is the 50 pitches thrown to each player, and the response variable is the number of home runs hit by each player.

Example 1.10 (Observational Study). The research question is the same as in the above example. This time the two players are selected from different eras: one player (say, Joe Torre) was born in 1940, whereas the other player (say, Albert Pujols) was born in 1980. The only way to compare their home run hitting ability is to observe their recorded data on number of home runs. Notice the passive role played by the researcher.

Exercise 1.3. Is the "hot dog promotion" study experimental or observational? Give reasons for your answer. How about the home run hitting contest at the All-Star Game?

Baseball statistics is primarily involved with observational studies. That is, we use existing baseball records to investigate baseball questions.

1.4. Populations

In this book, the term *population* can have two meanings. A population can refer to the entire group of subjects or objects being studied or to the data obtained from a variable applied to the subjects or objects in the population. When population refers to the data, the data are called population data.

In a research study, a population is targeted. A *population* is the entire group about which we wish to draw conclusions.

Examples of Populations

Example 1.11. The population consists of all at bats for a baseball player for one year. The qualitative variable, the outcome of an at bat, is assigned two possible values: "H" or "Not H." Since this variable has only two values, it is dichotomous. The population data would consist of a certain number of data values; each value is either an "H" (hit) or a "Not H" (not a hit). The number of data values is called the population size and symbolized by an uppercase "N." In 2008, Chipper Jones, third baseman for the Atlanta Braves, had 439 at bats ($N = 439$). He had 160 hits and led the league with a batting average of .364. So he had 160 H and $439 - 160 = 279$ "Not H."

The next example will use the same population but with a different variable.

Example 1.12. The qualitative variable, the outcome of an at bat, is assigned five possible values: 1B (single), 2B (double), 3B (triple), HR (home run), and Not H (not a hit). This variable is multinomial and not dichotomous.

Example 1.13. The population consists of the years in the career of a player. The quantitative variable would be the number of home runs hit each year. If the player had a 20-year career, the *population data* would consist of the 20 numbers corresponding to his yearly home run totals. For this population, $N = 20$.

Exercise 1.4. Describe three other populations that could be studied in baseball.

1.5. Infinite and Finite Populations

If a list of all subjects of a population can be provided, the population is called *finite*. If no list exists or a list cannot be compiled, the population is called *infinite*.

The *populations* included in Examples 1.11–1.13 are all finite populations. Here we look at examples of infinite populations:

Example 1.14. Consider the age at which every Major League player appeared in his first game in the Major Leagues since 1876. The population consists of every person who has ever played in the Major Leagues. The quantitative variable is his age in his first game. A list would be difficult, if not impossible, to obtain. (Joe Nuxhall pitched at the age of 15 for the Cincinnati Reds in 1944. He is thought to be the youngest player ever to play in the Major Leagues.)

Example 1.15. Consider the total time necessary to complete a Major League game. The population is all Major League games played since 1876. The quantitative variable is the time it takes to complete the game. It would be impossible to obtain such a list since in the early years of baseball this statistic was not kept. (In the modern era, the longest game, a 7–6 victory by the White Sox over the Brewers in 1984, took 25 innings and over 8 hours and 6 minutes to settle.)

Example 1.16. Consider a player's *true* batting average for a given year. This population is theoretical and based on the player coming to bat over and over again, an infinite number of times. Since this population is theoretical and never can happen, no list could ever exist. Do not confuse a player's observed batting average for a year with his *true* batting average for that year. His observed batting average is based on all his actual at bats for the given year. His *true* batting average is theoretical and unknown. This idea of a *true* batting average will be addressed several times in this book.

Exercise 1.5. Describe the population in the "hot dog promotion" study. Is it finite or infinite? Give reasons for your answer.

1.6. Samples Taken from a Population

In most research studies, since the researcher is unable or unwilling to obtain the *population data,* a small portion of data is collected from the population. This small portion of data is called a *sample.* A *sample* is any proper (excludes at least one member of the *population*) subset selected from a *population*. The term *sample* can have two meanings. It can refer to the actual subjects in the study or to the data values for a variable applied to the subjects. If a population is finite with a large amount of data, the researcher can study either the entire population or a sample from the population. Since there is no list for infinite populations, samples must be used. For finite populations, expense and time constraints dictate whether to use a sample instead of the entire population. When a sample is used in place of the entire population, *sampling errors* will occur. A *sampling error* is the error that occurs because the entire population is not being considered. In the later chapters on *inferential statistics,* techniques will be established that enable us to draw conclusions about a population from a sample. For these methods to be successful, the sample selected must be representative of the population.

The same collection of subjects or objects can be considered a population or a sample from a population. In Example 1.11, the population consisted of the actual at bats of a player for a given year. In Example 1.16, the actual at bats of a player for a given year are considered a sample for the infinite population of a player coming to bat over and over again. Remember, a population is the entire group of objects about which a researcher wishes to draw conclusions.

Exercise 1.6. Give an example of a population in a research study on baseball that you feel would require the use of a sample.

1.7. Common Sampling Techniques and Biases

1.7.1. Introduction to Sampling and Biases

The most important characteristic of a good sample is that the sample is representative of the population. It should look like a mini version of the population. A sample should be free of any biases. In general, a *bias* is anything that systematically leads to either a favored result or a wrong result.

1.7.2. Sampling Bias

A *sampling bias* exists when some factor in the selection process leads systematically to either a favored population result or an error in estimating a population result. A sampling bias can occur for many reasons: some subjects of the population might have been excluded from the selection process, certain subjects of a population might have a greater chance of being selected, certain selected subjects of the sample might refuse to participate, or the information obtained from the selected subjects might be false.

The following is an example of a biased sample.

Example 1.17. In our "hot dog promotion," suppose that the sample only consisted of people who recently attended a Dodger's home game. Notice that those people who hadn't attended a recent home game had no chance of being included. Since part of the *targeted* population in the hot dog study was those people who had *not* been to a recent game and they had no chance of being included in the sample, this would create a sampling bias. This type of sampling is sometimes referred to as *convenience sampling*. (Not all *convenience sampling* is bad; in fact, there are times in which it must be used.)

To help eliminate sampling biases, some form of random sampling is needed.

1.7.3. Random Sampling

Random sampling is any technique where each individual has a known chance of being included in the sample.

One form of random sampling is *simple random sampling* (SRS). SRS is any technique that guarantees that each individual is chosen entirely by chance and each member of the population has an equal chance of being included in the sample. SRS provides not only that each individual has an equal chance of being selected but also that every possible sample of a given size has the same chance of being selected. A sample size will be denoted by the lowercase letter "n." One way of selecting a simple random sample of size n, where the population is known, is to put a slip for each member of the population into a bowl. Then n slips are selected from the bowl, after the slips are thoroughly mixed.

Other forms of random sampling often used include *stratified, cluster*, and *systematic sampling*. Later in this chapter, we will discuss some methods used to select a simple random sample.

Stratified sampling is a form of random sampling that is based on sampling from a population that has already been divided into homogenous subpopulations called strata. The idea behind this form of sampling is to build a sample that maintains the same proportions with respect to the strata as found in the population. We then use SRS to select the correct proportion from each of the strata. The combination of these proportions forms the stratified random sample.

Example 1.18 (Stratified Sampling). Suppose that we are studying the batting performance of Baltimore's Nick Markakis for 2008. The population for this study is his plate appearances for that year (with 595 ABs, 99 BBs, 0 SF, and 2 HBP, he had 696 PAs that year). We know that half of Nick's games were played at Camden Yards and the other half were played in equal numbers at 10 other stadiums. We thus have 11 strata for this population. We could form a stratified sample of size 100 by using SRS to pick a sample consisting of 50% of his plate appearances from his home stadium and 5% of his plate appearances from each of the other 10 stadiums. The stratified sample would consist of 50 plate appearances at his home stadium and 5 plate appearances at each of the other stadiums. In stratified sampling, all samples of a fixed size do not have an equal chance of being selected. There are many samples of size 100 that have no chance of being selected. For example, no sample can consist of 51 plate appearances from the home stadium. The reason for using a stratified sample is to take into account that some stadiums (such as the Rockies' Coors Field or the Reds' Cinergy Stadium) are more favorable to the batter. By maintaining the same proportions for the strata in the sample as in the population, the sample will be more representative of the population than if simple random sampling was used.

Cluster sampling can be done if we can divide the original population into subgroups, called clusters, each of which is representative of the population. We then use SRS to select one or more of these clusters as our random sample. The idea behind cluster sampling is that each cluster is a group of subjects that are in close proximity to each other. So instead of selecting individuals one at a time, a researcher can select one or more clusters.

Example 1.19 (Cluster Sampling). Given that the population of the study is all players with at least 300 at bats for the year 2006, we wish to study the characteristic of the ratio of strikeouts to at bats. We will use cluster sampling to select our random sample. The clusters are composed of the Major League teams. We use SRS to select a small number of teams. The players with at least 300 at bats on the selected teams are combined to form the random sample. In order to perform cluster sampling, the assumption is that each teams' players are representative of the population with respect to their ratio of strikeouts to at bats.

Systematic sampling is done by ordering your subjects and then selecting every nth one until the desired sample size is achieved.

Example 1.20 (Systematic Sampling). We will use the same population and characteristic as in Example 1.19. For this population, we assign a number starting with the number 1 to each member of the population. We will select every nth subject, where n is determined by dividing the population size by the desired sample size. Suppose that there were 200 players in the population and a sample of size 20 was needed. We would divide 200 by 20 to determine who the nth subject should be. In this case, we will select every 10th player. Now, to get the first player, we would use SRS to select a number between 1 and 10. Let us assume that the number selected is 6. Our systematic sample would consist of the following players: 6, 16, 26, 36, 46, 56, 66, 76, 86, 96, 106, 116, 126, 136, 146, 156, 166, 176, 186, and 196.

The goal of any statistically valid sampling method is to produce a sample that is representative of the population being sampled. All the above sampling techniques are considered statistically valid.

Even though random sampling may eliminate sampling *bias*, the possibility of sampling *errors* still exists. However, these sampling errors are due primarily to chance.

Since for an infinite population all the subjects or objects in a population are not known, random sampling theoretically cannot be done. However, random sampling can be applied to a very large sample from the population. If we believe that the sample is truly representative of the population, the techniques of inferential statistics can still be applied. In Example 1.16, we would use as a representative sample the player's observed at bats for that year. Remember, the population in Example 1.16 is theoretical and consists of an infinite number of at bats for that player.

Even though both biased and unbiased samples are subject to errors, the techniques of *inferential statistics* should only be used with either an unbiased random sample or a sample that is believed to be representative of the population. These inferential techniques will be explored in the later chapters.

We now look at other types of biases.

1.7.4. Other Types of Biases

A second type of bias is a *measurement bias.* A measurement process with a bias will systematically either overestimate or underestimate the true value of the property it measures. In baseball, an instrument, called a gun, is used to measure the distance of home runs at a stadium. Suppose that a gun was calibrated incorrectly and added 20 feet to each measurement. This causes a measurement bias, leading systematically to an overestimate of the true average distance of home runs at that stadium.

The design of many experiments is biased. A design is biased if it intentionally systematically leads to a desired result. Many of the new baseball stadiums are designed to produce more home runs. The bias is introduced by having smaller dimensions. Having smaller dimensions systematically increases the number of home runs.

The last bias mentioned above is an *estimator bias.* This type of bias will be discussed in the chapters that cover inferential statistics.

Exercise 1.7. In 2005, the governor of New Jersey signed an executive order that qualifies any student who reaches a state championship tournament for random steroid testing. If the population of this study is all high school students who play on a varsity high school team, would you consider this a biased sample? If so, what types of biases are there? What sampling method is being used?

Descriptive measures play an important role in the study of baseball records. Descriptive measures come in the form of counts and calculations involving the addition, subtraction, multiplication, and division of counts. With that in mind, the next section will provide a discussion of several descriptive measures used to describe quantitative and qualitative data.

1.8. Mean, Proportion, Ratio, Rate, Percentage, and Probability

We begin this discussion with data collected for a quantitative variable. Each data point for a quantitative variable is a number. The *mean* is calculated by summing the data values and then dividing that sum by the number of data values. Many people refer to the mean as the average. Actually, the mean is just one type of average. There are many types of averages that are not means. For instance, average speed is obtained by dividing the change in distance by the change in time. Another type of average is a weighted mean.

Each at bat during a player's season is either a hit (H) or not a hit (Not H). We assign the number 1 to an "H" and the number 0 to a "Not H," creating a data set that consists of ones and zeros. The mean for this data set is obtained by summing the ones and zeros and then dividing this sum by the total number of ones and zeros. This *mean* corresponds to a player's batting average for that season. Notice that the discrete quantitative variable, number of hits, is a count, which is divided by the number of at bats, giving us the continuous variable called a player's batting average.

For a qualitative data set, each data point is assigned to a particular category. The *proportion* of data points belonging to one category is calculated by dividing the number of data points in the given category by the total number of data points. The *probability* of a given data point belonging to a category is equal to the proportion of data points in the category. A *proportion* can be converted into a *percentage* by multiplying the proportion by 100. A *ratio* is formed by dividing a number representing one quantitative variable by a number representing a second quantitative variable.

A proportion must be a number between 0 and 1, but a ratio can be any nonnegative number. Since a proportion is equal to the quotient of the number of data points in one category by the total number of data points, every proportion is a ratio. A ratio is not necessarily a proportion. A *rate* is a ratio reduced to a new quotient having a designated number in the denominator.

Looking again at the concept of batting average, we can say that a batting average is the proportion of hits to at bats. The next two examples explore the different measures described above.

Example 1.21. The following two quantitative data sets consist of the number of hits and number of at bats in the past 10 games for a player.

The data for the number of hits in 10 games are (2, 0, 0, 0, 1, 3, 2, 1, 0, 0).

The data for the number of at bats in those 10 games are (4, 3, 5, 4, 4, 5, 3, 6, 3, 3).

The mean number of hits per games is equal to $(2+0+0+0+1+3+2+1+0+0)/10=9/10=.90$ (10 is the number of data points).

The ratio of the number of hits to the number of at bats is equal to $(2+0+0+0+1+3+2+1+0+0)/(4+3+5+4+4+5+3+6+3+3)=9/40=.225$.

The proportion of hits to at bats is (#hits/#at bats) = 9/40.

The ratio of the number of at bats to the number of hits is (#at bats/#hits) = 40/9 = 4.44.

This ratio is not a proportion.

Reducing the ratio 40/9 to 4.44/1 gives the rate of 4.44 at bats for each hit.

Reducing the ratio 9/10 to .90/1 gives the rate of .90 hits per game.

Reducing the ratio 9/10 to 4.5/5 gives the rate of 4.5 hits per five games.

The ratio of the number of at bats to the number of games is 40/10. The rate of at bats per game is 4 to 1.

Example 1.22. The following qualitative data set consists of the result of an at bat (H or Not H) for a player's past three games.

The data are (H, Not H, Not H, Not H, Not H, H, H, H, Not H, Not H).

The proportion of hits to at bats is 4/10. The probability of getting a hit in an at bat is 4/10. The percentage of hits to at bats is $(4/10)*100\%=40\%$. A 40% probability translates to a rate of 4 hits per 10 at bats. The ratio of hits to outs is 4/6 (we say that the ratio of hits to outs is 4 to 6). The ratio of at bats to hits is 10/4. The ratio of hits to outs and the ratio of at bats to hits are not proportions. The ratio of hits to at bats is a proportion.

Exercise 1.8. Below are some of the baseball statistics for the San Francisco Giants' Barry Bonds for the year 2001. Use these data to answer the questions that follow. All measurements should be rounded to three decimal places.

Yr	G	AB	R	H	1B	2B	3B	HR	RBI	BB	SF	HBP
2001	153	476	129	156	49	32	2	73	137	177	2	9

a. The total number of plate appearances is ________.
b. The yearly batting average is ________.
c. The proportion of hits to at bats expressed as a decimal is ________.
d. The rate of at bats per home run is ________.
e. The ratio of home runs to at bats is ________.
f. The percentage of at bats that result in a home run is ________.
g. The probability of a hit in an at bat expressed as a decimal is ________ and expressed as a fraction is ________.
h. The probability that an at bat results in a home run expressed as a fraction is ________.
i. The ratio of singles to hits expressed as a fraction is ________.
j. The number of runs per game is ________.

1.9. Parameters and Statistics

Data can be quantitative or qualitative. A *descriptive measure* is normally one number used to summarize some aspect of the data. A *parameter* is a descriptive measure of a *population*. A *statistic* is a descriptive measure of a *sample*.

Example 1.23. Batting averages, on-base percentages, and slugging percentages are examples of descriptive measures.

Exercise 1.9. In the hot dog promotion study, what are the two statistics and two parameters discussed?

1.10. The Role of Chance in Batting

A batting average is the *proportion* of the number of hits to the number of at bats. (George Herman "Babe" Ruth had 2518 hits in his 8398 at bats, for a lifetime batting average of .342. As a pitcher, his earned run average was a mere 2.28, of which he was rightly proud.) A batting average is one of many descriptive measures used to determine a baseball player's *ability* as a batter. His ability as a batter is determined by many variables. Some of these are personal variables, which include eyesight, hand-eye coordination, strength, speed, practice time, and mentoring. Other variables include external variables such as the dimensions of the home stadium, the area in foul territory of the home stadium, his place in the batting order, and the quality of pitching in the league. Since chance or luck also plays a role in the result of an at bat, an at bat is a random variable. If the ball was hit a few inches to the right or left, or the defensive setup of fielders was a little different, or the ability of the fielders was different, or the dimensions of the stadium were different, or the ability of the pitcher was different, the result of an at bat could certainly be different. To achieve a hit, the player first must make contact with the ball. His chance of success is increased by hitting the ball in such a way that it travels at a high speed and in a favorable direction. Hitting the ball hard in a direction where the fielders are not positioned (or, as Willie Keeler said, "Hit 'em where they ain't") is where the skill of the batter determines his *ability*.

In later chapters, we will see that different batting statistics have more or less skill associated with them. As the skill level related to a statistic increases, its reliability in assessing a player's batting ability increases.

Next, the concept of a *true batting average* is explained.

1.11. The Observed and True Batting Averages

A player's *true* batting average for a year or for a career is a combination of his ability and chance. For this reason, a player's *true* batting average is unknown and cannot be predicted ahead of time. The best we can do is to estimate his *true* batting average by using his *observed* batting average. In 2008, Derek Jeter's 596 at bats resulted in 179 hits. The observed batting average for that year would be 179/596 = .300. His 596 at bats that year can have two interpretations: either as a population for *all* observed at bats for that year, or as a sample to be used to estimate Jeter's *true* batting average for 2008. The population for his *true* batting average is an infinite population consisting of Derek coming to bat over and over again. Thus, his .300 batting average can be interpreted in two ways: as a *population proportion* for all observed at bats for that year or as a *sample proportion* for his *true* batting average for that year. In the first case, the observed batting average of .300 is a parameter, whereas in the second case, the observed value is a statistic. As the above example shows, population is a relative term.

Later in this book, statistical techniques will use the second case mentioned in the previous paragraph. A player's observed batting average will be used to estimate his *true* batting average. In this case, the observed batting average is a statistic; the *true* batting average is a parameter.

These statistical techniques rely on concepts from an area of mathematics called *probability*. In the next section, the concept of probability is introduced.

1.12. Classical Probability and Relative Frequency Probability

Given a process or an experiment that results in various observations, the likelihood of a particular observation occurring is called its *probability* of occurring.

Classical probability has its roots in gambling. Example 1.7, which involves the rolling of one die, is an example of a process involving classical probability. Since there is no skill involved in rolling a die, the outcome of the roll of a die occurs strictly by chance. In classical probability, the *true* probability that a particular outcome occurs can be predetermined before even undertaking the process. Assuming that the six sides of the die are assigned the numbers 1 through 6, the *true* probability of any number occurring in one roll of the die is 1/6. If we roll one die arbitrarily 1000 times, we would expect that each number would occur approximately $1000 * (1/6) = 167$ times.

The *Law of Large Numbers* states that when a probability process is repeated a large number of times, the proportion of times a particular outcome occurs will approach its *true* probability of occurring.

Relative frequency probability can involve processes where chance does not completely determine the outcome. For these processes, the *true* probability is unknown. Since the *true* probability is unknown, it cannot be predicted ahead of time. The process of a player's at bat involves relative frequency probability. His batting average is his probability of getting a hit. His *true* batting average for a given year or for his career is unknown and can only be estimated. The difference between the process of rolling a die and the process of an at bat is that in the former only chance determines the outcome, whereas in the latter the outcome is a combination of chance and skill. At the beginning of a year, a player has only a small number of at bats. Many players' observed batting averages are close to .500 or .100 for the first few weeks of the season. (After your first at bat, your batting average is either 1.000 or 0.000.) As the season progresses and the number of at bats increases, his observed batting average approaches his *true* batting average for that year. In a similar way, as his number of career at bats increases, his observed career batting average approaches his *true* career batting average. The concept of moving away from extreme values toward a true value is called *regressing toward the mean*. We say that as a season progresses, a player's observed batting average *regresses* to his *true* batting average.

Later in this book, when discussing how a player might have a batting average for a year of .400, we point out that the more at bats a player has, the harder it is for him to have a .400 batting average. The reason is that, if you believe that no player is a *true* .400 hitter, as his at bats increase his batting average regresses toward his *true* batting average, which is less than .400. In the modern era, only Red Sox great Ted Williams has hit over .400 (.406) for an entire season (1941). His lifetime batting average, however, was .344. (He was replaced by a pinch hitter only once, by Carroll Hardy.)

The difference between the processes of rolling a die and batting is that in rolling a die the *true* probability of an outcome occurring is known before the die is ever rolled, whereas in batting the *true* probability of an outcome occurring is unknown.

What is similar for these two processes is that the *Law of Large Numbers* applies to both. In either case, as the process is repeated a large number of times, the observed probabilities approach the *true* probabilities. During a baseball game the announcer may say, "It's no surprise that player X now has a batting average close to .300, considering his lifetime batting average on the back of his baseball card is close to .300."

Chapter 7 will deal in more depth with the concepts of probability.

1.13. Descriptive and Inferential Statistics

Descriptive statistics is best described by the acronym COPS, which stands for the *collection*, *organization*, *presentation*, and *summarization* of data. The data could be from a sample or population. Descriptive statistics will be studied in Chapters 1–10.

Inferential statistics is involved with using a sample result (a statistic) to estimate a population result (a parameter). Inferential statistics will be studied in the later chapters of this book.

1.14. Physical and Theoretical Models

Children love to build models of airplanes and cars, which are small-size representations of real-world objects. These are *physical models.*

Experiments where chance affects the outcomes are called *probability experiments.* A *simulation technique* uses a probability experiment to mimic a real-world situation. Simulation techniques can use either physical or theoretical models.

A *theoretical* model is not a *physical* model. It is usually based on a mathematical formula, which can only be used when certain assumptions are satisfied by the real-world process. When the real-world process satisfies these assumptions, the theoretical model can be used to predict results before the real-world process is executed. Later in this book, we will introduce several theoretical models. The normal, geometric, and binomial models are standard models used in any statistics course to evaluate probabilities. The regression model is used in any statistics course as a prediction tool. These models will be used to study baseball data.

As mentioned earlier, it is necessary to have techniques to produce simple random samples. One theoretical model that can be used as a simulation technique to produce a simple random sample is the mathematical formula called the *random number generator.* The random number generator produces integers in such a way that each digit of the integer has an equal probability of being any integer from 0 to 9.

Suppose that the population consists of $n = 1500$ subjects. We wish to select a random sample of a fixed size n. We will give two methods of selecting a simple random sample of size n. One method involves using the *physical* model of a disk (Box 1.1), and the other involves the *theoretical* mathematical formula called the random number generator.

Box 1.1 **Physical disk**

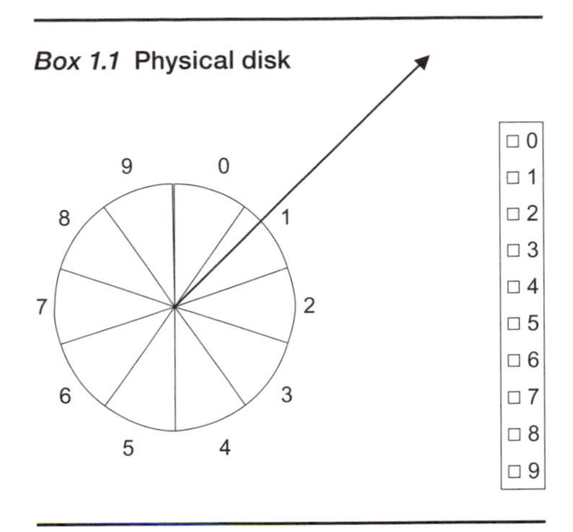

The physical disk consists of a pointer and the numbers 0–9; each number is assigned to a pie slice of the same area (online spinner available at http://illuminations.nctm.org/activitydetail.aspx?ID=79). To choose the random sample, each member of the population is assigned a four-digit number from 0001 to 1500. Four spins of the pointer will produce a four-digit number. If the number generated is already used or above 1500, disregard the result and spin again. Using this process, we can generate a simple random sample of any desired size n.

The other approach involves the theoretical model of the random number generator. The software package Microsoft Excel contains as a standard function the random number generator function, which is explained in Box 1.2.

Exercise 1.10. Use the RANDBETWEEN function in Excel to select samples of size 10, 20, and 30 games from the 56 games in Table 1.1, at the end of the chapter. Compute DiMaggio's batting average (BA) for each of these samples. Compare these BAs with DiMaggio's BA for the 56 games.

Box 1.2 **Procedure for generating a random sample using Excel**

The syntax for the random number generator function in Microsoft Excel Version 7 is as follows:

RANDBETWEEN(bottom, top), where bottom is the smallest number and top is the largest integer.
RANDBETWEEN(1, 1500) will generate a random number between 1 and 1500.

To obtain a simple random sample of size n, complete the following steps:

Step 1: First, execute RANDBETWEEN(1, 1500) in one cell.
Step 2: Right-click on the cell used in step 1.
Step 3: Then click on *copy.*
Step 4: *Highlight* n-cells and press *enter.*

The final result will be a random sample of size n.

Exercise 1.11. A study is performed for the 2010 Major League season. Our population is all players on all Opening Day rosters for all 30 Major League teams. The characteristic is the salary of each player on the 25-man Opening Day roster. We wish to pick a sample of 50 players by using cluster sampling. Use the RANDBETWEEN function to pick two teams. Should we expect this sample to be representative of all players in the Major Leagues? What could go wrong with this sample? (Search the Internet for these data.)

1.15. Our Hypothetical Hot Dog Promotion

Our research problem is to analyze whether offering free hot dogs will motivate people to buy seats in the right field stands, even though the views are poor from this location.

The *population* is all households within driving distance of the stadium. Since there is no list, this population is considered *infinite.* We need to select a *sample* that is representative of the population. To achieve this, we look at all households with a phone number whose exchange is in driving distance of the stadium. We will construct a sample of size $n = 1000$. The random number generator will select phone numbers in a random fashion. For each phone number selected, a person will call that household and inquire if someone in the household would be willing to buy a ticket in the right field stands for free hot dogs. Unfortunately, such a sample may have some *biases.* One bias is that households that use cell phones and have no landline can never be selected; another bias is that if no one is at home to answer the call, this household would be eliminated. Even with these biases, we still believe that the sample would be representative of the population and can be considered a *random sample.* After contacting 1000 households, we discover that 45 households have at least one person that would take advantage of the promotion. The *statistic* for this sample is the *sample proportion,* which is equal to 45 divided by 1000. The sample proportion is 45/1000. In Chapter 14, we learn how to use a technique from *inferential statistics* called *confidence intervals* to find an interval estimate for the *true* population proportion of households that would have at least one person willing to take advantage of the promotion.

Chapter Summary

A *variable* is a characteristic applied to a group of subjects or objects. The group can be a *population* (the entire group being studied) or a *sample* (part of the population). A variable can be *quantitative* or *qualitative.* Quantitative variables produce data that are numeric. Quantitative data can be further classified as *discrete or continuous.* Discrete quantitative data are integers, and continuous quantitative data can be any real number. Qualitative variables produce data in the form of *categories.*

Individual data values can be compared by using *measurement scales.* The four measurement scales are *nominal, ordinal, interval,* and *ratio.* Nominal and ordinal measurement scales are used with qualitative data. Interval and ratio measurement scales are used with quantitative data.

Statistics involves the analysis of *data.* The two areas of statistics are *descriptive statistics* and *inferential statistics.* Descriptive statistics involves the *collection, organization, presentation,* and *summarization* of data. Descriptive measures are used to summarize data. A descriptive measure summarizing sample data is called a *statistic*; a descriptive measure summarizing population data is called a *parameter.* Two of the most used descriptive measures are *mean* and *proportion.* The mean is used with quantitative data, and the proportion is used with qualitative data. *Inferential statistics* uses a statistic to estimate a parameter.

There are two basic types of statistical studies: *observational* and *experimental.* When conducting observational studies, the researcher plays a passive role and simply observes the data, drawing conclusions from these observations. When conducting experimental studies, the researcher plays an active role by designing and controlling the experiment that produces the data.

A hot dog promotion conducted by the Los Angeles Dodgers in 2006 was used as an example of a statistical study.

Since in most studies the populations are either too large or infinite, statisticians use subgroups called *samples* to obtain the necessary data for the study. Four basic methods to obtain samples include *simple random sampling, stratified sampling, cluster sampling,* and *systematic sampling.* These four methods attempt to produce a sample, free of any *biases,* that is *representative* of the population.

A *simulation technique* uses a probability experiment to mimic a real-world situation. Physical and theoretical models will be used throughout this book to simulate baseball experiments.

The techniques of statistics are used in many areas outside the world of baseball. Statistical methods are used in scientific research, in business research, and in our everyday life.

This book applies statistical methods to baseball research. However, many of the statistical techniques applied to baseball can also be applied to almost any other discipline.

CHAPTER PROBLEMS

Tables 1.1 and 1.2 are reproduced here from the website www.baseball-almanac.com.

Table 1.1 **Joe DiMaggio's hitting streak (in numerical order)**

Game	Date	Pitcher(s)	Team	Batting line						
				AB	R	H	2B	3B	HR	RBI
1	5/15/1941	Eddie Smith	Chicago	4	0	1	0	0	0	1
2	5/16/1941	Thornton Lee	Chicago	4	2	2	0	1	1	1
3	5/17/1941	Johnny Rigney	Chicago	3	1	1	0	0	0	0
4	5/18/1941	Bob Harris	St. Louis	3	3	2	1	0	0	1
		Johnny Niggeling				1				
5	5/19/1941	Denny Galehouse	St. Louis	3	0	1	1	0	0	0
6	5/20/1941	Elden Auker	St. Louis	5	1	1	0	0	0	1
7	5/21/1941	Schoolboy Rowe	Detroit	5	0	1	0	0	0	1
		Al Benton				1				
8	5/22/1941	Archie McKain	Detroit	4	0	1	0	0	0	1
9	5/23/1941	Dick Newsome	Boston	5	0	1	0	0	0	2
10	5/24/1941	Earl Johnson	Boston	4	2	1	0	0	0	2
11	5/25/1941	Lefty Grove (HOF)	Boston	4	0	1	0	0	0	0
12	5/27/1941	Ken Chase	Washington	5	3	1	0	0	1	3
		Red Anderson				2				
		Alex Carrasquel				1				
13	5/28/1941	Sid Hudson	Washington	4	1	1	0	1	0	0
14	5/29/1941	Steve Sundra	Washington	3	1	1	0	0	0	0
15	5/30/1941	Earl Johnson	Boston	2	1	1	0	0	0	0
16	5/30/1941	Mickey Harris	Boston	3	0	1	1	0	0	0
17	6/1/1941	Al Milnar	Cleveland	4	1	1	0	0	0	0
18	6/1/1941	Mel Harder	Cleveland	4	0	1	0	0	0	0
19	6/2/1941	Bob Feller (HOF)	Cleveland	4	2	2	1	0	0	0
20	6/3/1941	Dizzy Trout	Detroit	4	1	1	0	0	1	1
21	6/5/1941	Hal Newhouser (HOF)	Detroit	5	1	1	0	1	0	1
22	6/7/1941	Bob Muncrief	St. Louis	5	2	1	0	0	0	1
		Johnny Allen				1				
		George Caster				1				
23	6/8/1941	Elden Auker	St. Louis	4	3	2	0	0	2	4
24	6/8/1941	George Caster	St. Louis	4	1	1	1	0	1	3
		Jack Kramer				1				
25	6/10/1941	Johnny Rigney	Chicago	5	1	1	0	0	0	0
26	6/12/1941	Thornton Lee	Chicago	4	1	2	0	0	1	1
27	6/14/1941	Bob Feller (HOF)	Cleveland	2	0	1	1	0	0	1
28	6/15/1941	Jim Bagby	Cleveland	3	1	1	0	0	1	1
29	6/16/1941	Al Milnar	Cleveland	5	0	1	1	0	0	0
30	6/17/1941	Johnny Rigney	Chicago	4	1	1	0	0	0	0
31	6/18/1941	Thornton Lee	Chicago	3	0	1	0	0	0	0

Table 1.1 **Joe DiMaggio's hitting streak (in numerical order),** ***continued***

				Batting line						
Game	Date	Pitcher(s)	Team	AB	R	H	2B	3B	HR	RBI
32	6/19/1941	Eddie Smith	Chicago	3	2	1	0	0	1	2
		Buck Ross				2				
33	6/20/1941	Bobo Newsom	Detroit	5	3	2	1	0	0	1
		Archie McKain				2				
34	6/21/1941	Dizzy Trout	Detroit	4	0	1	0	0	0	1
35	6/22/1941	Hal Newhouser (HOF)	Detroit	5	1	1	1	0	1	2
		Bobo Newsom				1				
36	6/24/1941	Bob Muncrief	St. Louis	4	1	1	0	0	0	0
37	6/25/1941	Denny Galehouse	St. Louis	4	1	1	0	0	1	3
38	6/26/1941	Elden Auker	St. Louis	4	0	1	1	0	0	1
39	6/27/1941	Chubby Dean	Philadelphia	3	1	2	0	0	1	2
40	6/28/1941	Johnny Babich	Philadelphia	5	1	1	1	0	0	0
		Lum Harris				1				
41	6/29/1941	Dutch Leonard	Washington	4	1	1	1	0	0	0
42	6/29/1941	Red Anderson	Washington	5	1	1	0	0	0	1
43	7/1/1941	Mickey Harris	Boston	4	0	1	0	0	0	1
		Mike Ryba				1				
44	7/1/1941	Jack Wilson	Boston	3	1	1	0	0	0	1
45	7/2/1941	Dick Newsome	Boston	5	1	1	0	0	1	3
46	7/5/1941	Phil Marchildon	Philadelphia	4	2	1	0	0	1	2
47	7/6/1941	Johnny Babich	Philadelphia	5	2	2	1	0	0	2
		Bump Hadley				2				
48	7/6/1941	Jack Knott	Philadelphia	4	0	2	0	1	0	2
49	7/10/1941	Johnny Niggeling	St. Louis	2	0	1	0	0	0	0
50	7/11/1941	Bob Harris	St. Louis	5	1	3	0	0	1	2
		Jack Kramer				1				
51	7/12/1941	Elden Auker	St. Louis	5	1	1	1	0	0	1
		Bob Muncrief				1				
52	7/13/1941	Ted Lyons (HOF)	Chicago	4	2	2	0	0	0	0
		Jack Hallett				1				
53	7/13/1941	Thornton Lee	Chicago	4	0	1	0	0	0	0
54	7/14/1941	Johnny Rigney	Chicago	3	0	1	0	0	0	0
55	7/15/1941	Eddie Smith	Chicago	4	1	2	1	0	0	2
56	7/16/1941	Al Milnar	Cleveland	4	3	2	1	0	0	0
		Joe Krakauskas				1				
Totals				223	56	91	16	4	15	55

Table 1.2 **Barry Bonds's 73 home runs (in numerical order)**

#	Date	Pitcher	Team (AWAY)	Inn	Outs	Count	RBI	Length	Position
1	4/2/2001	Woody Williams	San Diego	5th	1	0–0	1	420	CF
2	4/12/2001	Adam Eaton	SAN DIEGO	4th	1	1–1	1	417	RCF
3	4/13/2001	Jamey Wright	MILWAUKEE	1st	1	0–1	2	440	RCF
4	4/14/2001	Jimmy Haynes	MILWAUKEE	5th	1	2–0	3	410	RF
5	4/15/2001	David Weathers	MILWAUKEE	8th	2	1–1	1	390	LCF
6	4/17/2001	Terry Adams	Los Angeles	8th	0	2–0	2	417	RF
7	4/18/2001	Chan Ho Park	Los Angeles	7th	2	0–0	1	420	RCF
8	4/20/2001	Jimmy Haynes	Milwaukee	4th	1	0–0	2	410	LCF
9	4/24/2001	Jim Brower	Cincinnati	3rd	1	3–2	2	380	RCF
10	4/26/2001	Scott Sullivan	Cincinnati	8th	2	3–2	2	430	CF
11	4/29/2001	Manny Aybar	Chicago	4th	2	2–1	1	370	RF
12	5/2/2001	Todd Ritchie	PITTSBURGH	5th	2	3–2	2	420	LCF
13	5/3/2001	Jimmy Anderson	PITTSBURGH	1st	1	0–1	2	420	LCF
14	5/4/2001	Bruce Chen	PHILADELPHIA	6th	1	0–1	2	360	RF
15	5/11/2001	Steve Trachsel	New York	4th	0	3–2	1	410	RCF
16	5/17/2001	Chuck Smith	FLORIDA	3rd	1	0–1	2	420	RCF
17	5/18/2001	Mike Remlinger	ATLANTA	8th	1	0–1	1	391	RF
18	5/19/2001	Odalis Perez	ATLANTA	3rd	0	3–2	1	416	RCF
19	5/19/2001	Jose Cabrera	ATLANTA	7th	0	2–0	1	440	RCF
20	5/19/2001	Jason Marquis	ATLANTA	8th	2	3–0	1	410	LCF
21	5/20/2001	John Burkett	ATLANTA	1st	2	2–2	1	415	CF
22	5/20/2001	Mike Remlinger	ATLANTA	7th	1	0–0	1	436	CF
23	5/21/2001	Curt Schilling	ARIZONA	4th	0	2–0	1	430	CF
24	5/22/2001	Russ Springer	ARIZONA	9th	0	0–0	2	410	LCF
25	5/24/2001	John Thomson	Colorado	3rd	2	2–2	1	400	RF
26	5/27/2001	Denny Neagle	Colorado	1st	1	3–1	2	390	RF
27	5/30/2001	Robert Ellis	Arizona	2nd	0	0–0	1	420	RF
28	5/30/2001	Robert Ellis	Arizona	6th	1	3–1	2	410	CF
29	6/1/2001	Shawn Chacon	COLORADO	3rd	1	2–0	2	420	RF
30	6/4/2001	Bobby Jones	San Diego	4th	0	2–1	1	410	CF
31	6/5/2001	Wascar Serrano	San Diego	3rd	1	1–1	2	410	LCF
32	6/7/2001	Brian Lawrence	San Diego	7th	2	1–1	2	450	CF
33	6/12/2001	Pat Rapp	Anaheim	1st	2	0–0	1	320	RF
34	6/14/2001	Lou Pote	Anaheim	6th	1	2–1	1	430	RCF
35	6/15/2001	Mark Mulder	Oakland	1st	2	3–1	1	380	LCF
36	6/15/2001	Mark Mulder	Oakland	6th	1	1–1	1	430	RCF
37	6/19/2001	Adam Eaton	SAN DIEGO	5th	1	2–1	1	375	RCF
38	6/20/2001	Rodney Myers	SAN DIEGO	8th	0	2–1	2	347	RF
39	6/23/2001	Darryl Kile	ST. LOUIS	1st	1	3–1	2	380	RF
40	7/12/2001	Paul Abbott	SEATTLE	1st	2	2–0	1	429	RCF
41	7/18/2001	Mike Hampton	Colorado	4th	0	2–1	1	320	RF
42	7/18/2001	Mike Hampton	Colorado	5th	1	0–1	2	360	LF
43	7/26/2001	Curt Schilling	ARIZONA	4th	0	0–0	1	375	RF

Table 1.2 **Barry Bonds's 73 home runs (in numerical order),** ***continued***

#	Date	Pitcher	Team (AWAY)	Inn	Outs	Count	RBI	Length	Position
44	7/26/2001	Curt Schilling	ARIZONA	5th	1	0–0	4	370	LCF
45	7/27/2001	Brian Anderson	ARIZONA	4th	0	1–1	1	440	RF
46	8/1/2001	Joe Beimel	Pittsburgh	1st	2	1–0	1	400	RCF
47	8/4/2001	Nelson Figueroa	Philadelphia	6th	1	2–2	2	405	RF
48	8/7/2001	Danny Graves	CINCINNATI	11th	0	1–1	1	430	RCF
49	8/9/2001	Scott Winchester	CINCINNATI	3rd	1	0–0	1	350	RF
50	8/11/2001	Joe Borowski	CHICAGO	2nd	2	2–2	3	396	CF
51	8/14/2001	Ricky Bones	Florida	6th	1	0–2	4	410	RF
52	8/16/2001	A.J. Burnett	Florida	4th	0	1–1	1	380	RF
53	8/16/2001	Vic Darensbourg	Florida	8th	0	1–2	3	430	RCF
54	8/18/2001	Jason Marquis	Atlanta	8th	1	2–2	1	415	RCF
55	8/23/2001	Graeme Lloyd	MONTREAL	9th	1	3–1	1	380	RCF
56	8/27/2001	Kevin Appier	NEW YORK	5th	0	1–1	1	375	RCF
57	8/31/2001	John Thomson	Colorado	8th	1	0–2	2	400	RF
58	9/3/2001	Jason Jennings	Colorado	4th	0	1–0	1	435	RCF
59	9/4/2001	Miguel Batista	Arizona	7th	2	0–0	1	420	RCF
60	9/6/2001	Albie Lopez	Arizona	2nd	2	2–2	1	420	RCF
61	9/9/2001	Scott Elarton	COLORADO	1st	1	1–1	1	488	RCF
62	9/9/2001	Scott Elarton	COLORADO	5th	2	2–2	1	361	RF
63	9/9/2001	Todd Belitz	COLORADO	11th	1	0–1	3	394	RCF
64	9/20/2001	Wade Miller	Houston	5th	2	1–0	2	410	CF
65	9/23/2001	Jason Middlebrook	SAN DIEGO	2nd	2	2–1	1	411	CF
66	9/23/2001	Jason Middlebrook	SAN DIEGO	4th	1	2–0	1	365	LF
67	9/24/2001	James Baldwin	LOS ANGELES	7th	2	1–1	1	360	RF
68	9/28/2001	Jason Middlebrook	San Diego	2nd	2	3–0	1	440	RCF
69	9/29/2001	Chuck McElroy	San Diego	6th	0	2–1	1	435	RF
70	10/4/2001	Wilfredo Rodriguez	HOUSTON	9th	0	1–1	1	480	RCF
71	10/5/2001	Chan Ho Park	Los Angeles	1st	2	1–0	1	440	RCF
72	10/5/2001	Chan Ho Park	Los Angeles	3rd	0	1–1	1	410	CF
73	10/7/2001	Dennis Springer	Los Angeles	1st	2	3–2	1	380	RF

Table 1.1 shows Joe DiMaggio's at bats for each game in his 56-game hitting streak. The population is the 56 games involved in Joe DiMaggio's hitting streak in 1941. The problems below all refer to this table.

1. Identify all the variables shown in this table.
2. For each variable determine if it is qualitative or quantitative.
3. For each variable determine its measurement scale.
4. For each quantitative variable determine if it is discrete or continuous.
5. Compute the proportion of his hits to at bats for this streak.
6. Compute his batting average during this streak.
7. Compute the ratio of at bats to home runs.

8. Compute the rate of at bats per home run.
9. Compute the percent of at bats that resulted in a home run.
10. What was the percentage of games in the streak in which Joe had exactly one hit?

Table 1.2 shows the games in which Barry Bonds hit his 73 home runs in 2001. The population is the 73 home runs for that season. The problems below all refer to this table.

11. Identify all the variables shown in this table.
12. For each variable determine if it is qualitative or quantitative.
13. For each variable determine its measurement scale.
14. For each quantitative variable determine if it is discrete or continuous.
15. Compute the proportion of his home runs hit with one out.
16. Compute the percentage of his home runs that traveled at least 400 feet.
17. Compute the mean number of RBIs produced by the home runs.
18. Compute the probability of a home run being hit in the first inning.
19. Compute the mean length of a home run.
20. What percentage of his home runs were hit against Atlanta?

The next set of questions deals with the population of all Major League players with at least 300 career plate appearances from 1999 through 2008.

21. Is this population finite or infinite?
22. List three quantitative variables and three qualitative variables you might be interested in studying.
23. For the quantitative variables identify which are discrete and which are continuous.
24. Find the measurement scale for the data collected for each of the variables in problem 23.
25. How would you go about selecting a simple random sample of 100 players?
26. How would you select a stratified sample of 100 players?
27. How would you select a cluster sample of 100 players?

The next questions deal with your own observational research study. The purpose of this study is to analyze the pitched balls in a game. Choose one complete baseball game for this study. The sample consists of each pitched ball in the game. Create a table consisting of the following column headings: column 1 is a player's name, column 2 is the pitch number for that player, column 3 is the result of the pitch, and column 4 is the result of the player's plate appearance. For the player's name just include last name, for the pitch number just include an integer (each player should begin with pitch number 1), for the result of a pitch use a letter (a—pitch is taken and is a ball [include hit by pitch]; b—pitch is taken and is a strike; c—pitch is swung at and missed; d—pitch is fouled off [include foul outs]; e—pitch is hit in fair territory), and for the result of the player's plate appearance also use a letter (a—single; b—double; c—triple; d—home run; e—walk; f—strikeout; g—sacrifice fly; h—sacrifice bunt; i—any other type of out; j—hit by pitch). Each of the four column headings is a variable for the sample.

28. Name the four variables in this study.
29. For each variable, classify it as qualitative or quantitative.
30. Find the measurement scale for each variable.
31. For each variable that is quantitative, classify the variable as discrete or continuous.
32. What sampling method is being used in this study?
33. Find the sample proportion of pitches that were taken as a strike.
34. Find the sample proportion of those plate appearances that resulted in a single.
35. What proportion of pitches were swung at and missed?
36. Find the sample mean number of pitches seen by a player in a plate appearance.
37. What proportion of players saw more than four pitches in a plate appearance?
38. What descriptive measures are used to summarize the results for this study?
39. Given that a player has two strikes, what is the mean number of pitches fouled off?

Chapter 2

Descriptive Statistics for One Quantitative Variable

In 1858, a game resembling baseball was played between the best ballplayers from New York and those from Brooklyn. The box score for that game is displayed in Table 2.1, as well as the inning-by-inning run production for the two teams.

Schwarz writes, "Brooklyn had won, 29 runs to 8, but the little chart told stories of Brooklyn's 29–8 victory and offered explanations as succinctly as his stock listings. Grum, Brooklyn's right fielder, clearly had led his team to victory, having scored 6 runs with just 1 'hands lost,' or out. Gelston, scored 3 runs with 2 hands lost, while several of his mates—De Bost, Bixby, and Davis—scored nary one run among them. It was all very clear; responsibility for the win or loss was identified, codified, and solidified, fired in the dispassionate kiln of numbers."* (A hit in which a player reached base but did not score a run was not displayed in the box score.)

The information from this game was collected and organized into three tables. There is a table for the performance of the Brooklyn players, a second for the New York players, and a third showing the inning-by-inning performance of the two teams. The tables not only describe what happened in the game but also summarize the results of the game.

The general form of a box score has been the table of choice in baseball and displayed in newspapers for over 150 years.

This leads to our definition for descriptive statistics. *Descriptive statistics* is involved with the *collection*, *organization*, *presentation*, and *summarization* of *data*.

We use the acronym COPS to remember this definition. In this chapter, you will learn how to *collect* baseball data by downloading a player's data

* Alan Schwarz, *The Numbers Game* (New York: St. Martin's Press, 2004), 2.

Table 2.1 **Box score for 1858 game between New York and Brooklyn**

Brooklyn			New York		
Names	H.L.	R's	Names	H.L.	R's
Masten, catcher	3	4	VanCott, pitcher	2	1
Pidgeon, pitcher	5	3	De Bost, catcher	3	0
Oliver, second base	3	3	Gelston, short	2	3
M. O'brien, third base	4	2	Bixby, first base	4	0
Pearce, short	2	4	Pinckney, second base	3	1
P. O'brien, left field	2	3	Davis, centre field	4	0
Grum, right field	1	6	Marsh, third base	3	1
Manolt, centre field	4	2	Tooker, left field	3	1
Price, first base	3	2	Hoyt, right field	3	1
Total		29	Total		8

	Runs made in each inning									
	1st	2nd	3rd	4th	5th	6th	7th	8th	9th	T
Brooklyn	6	0	5	6	2	3	4	2	1	29
New York	2	0	1	0	0	0	1	0	4	8

from a website into a Microsoft Excel spreadsheet, how to *organize* the data into tables, how to *present* the data in the form of several graphs, how to use important descriptive measures to *summarize* the data, and how to assign a *shape* to the data.

In this chapter, descriptive statistics will be applied to one set of quantitative data. In later chapters, descriptive statistics will be applied to more than one set of quantitative data and to one or more sets of qualitative data.

We begin the discussion of descriptive statistics by looking at the process of data collection.

2.1. Collection of Data

Data can be collected from many sources, including surveys, books, research papers, websites, and experiments. In Chapter 1, a survey was used to collect data for the "hot dog promotion." Most of the baseball data used in this book will come from websites, research articles, and books.

In downloading data from a baseball website into an Excel spreadsheet, one of the following two procedures will get the job done. Two procedures are needed because some websites store their data in text format with their columns separated by spaces, while other websites store the data in text with their columns separated by tabs.

Box 2.1 Procedure for downloading a player's data from www.baseball-reference.com

Step 1: Go to website www.baseball-reference.com.
Step 2: Highlight *players* (left pane of window) [double-click left mouse button].
Step 3: Find your *player* [double-click left mouse button].
Step 4: Highlight your *player's* statistics.
Step 5: Do the following:
  File [left-click]
  Save as: [left-click]
  Choose folder to *save in*: xxxxx
  File name: yyyyy [highlight then left-click]
  Save as type: text file
  Save: [left-click]
Step 6: Open Microsoft Word.
Step 7: Do the following:
  File [left-click]
  Open
  Go to *folder*: xxxxx
  Choose *file name*: yyyyy [highlight *open*, then double-click left mouse button]
  Text encoding: Windows (default)
  OK [left-click]
  Change view to Landscape [Do: Page Layout, Orientation, Landscape]
  Using Word, edit data, move left margin to end, and delete the columns not wanted
Step 8: Do the following:
  File [left-click]
  Save As: [left-click]
  Choose *folder*: xxxxx
  File name: yyyyynew
  Save as type: plain text
  Save [left-click]
  Windows Default
  OK [left-click]
Step 9: Open Excel.
Step 10: Do the following:
  File: open [then left-click]
  Go to *folder*: xxxxx
  Choose *file name*: yyyyynew
  File Type: all files
  Open: [left-click]
  Choose: fixed width
  Next
  Finish
  OK [left-click]
Step 11: Edit the spreadsheet by deleting columns that are not needed, adjusting the width of the columns, and performing other cosmetic changes.
Step 12: Do the following:
  File name: yyyyynewx
  Save As: Excel Workbook
  Save

The student is encouraged to use version 2003 of Excel or any later version. The procedures given below might have to be altered a little for later versions of Excel.

A procedure for downloading data from any website with columns separated by spaces is shown in Box 2.1.

Most websites store their data in text format with their columns separated by tabs (tab delimited). Two examples of such websites are www.baseball-almanac.com and www.mlb.com. These websites provide information on players, on the history of baseball, and on baseball records (including the top 100 players in many batting categories), as well as stats 101. Since these websites save the data in text, tab-delimited format, it can be down-

Box 2.2 **Procedure for downloading a player's data from www.baseball-almanac.com or www.mlb.com**

Step 1: Go to the website www.baseball-almanac.com.
Step 2: Left-click on *players* in top toolbar.
Step 3: Scroll down and left-click on *The Ballplayers*.
Step 4: Choose your player.
Step 5: Left-click on your player.
Step 6: Highlight player's statistics you wish to download.
Step 7: Do the following:
  Edit
  Copy
Step 8: Bring up Excel.
Step 9: Do the following:
  Edit
  Paste
Step 10: Edit the spreadsheet by deleting empty columns, adjusting the width of the columns, and making other cosmetic changes.
Step 11: Do the following:
  File Name: xxxxx
  Save as type: Excel Workbook
  Save

loaded directly into a Microsoft Excel spreadsheet. We outline this procedure in Box 2.2.

Of the two types of websites, downloading data from a website stored in tab-delimited form is clearly easier. Fortunately, most websites today use the tab-delimited format.

Table 2.2 shows Henry "Hank" Aaron's batting data for each year of his career. (These data were downloaded from the website www.baseball-almanac.com.)

The column headings of Year, G, AB, R, H, 2B, 3B, HR, RBI, BB, SO, SH, SF, HBP, AVG, OBP, and SLG are variables for the population consisting of all the Major League years for Henry Aaron. Throughout this book, either BA or AVG will be used as an abbreviation for batting average.

Exercise 2.1. Identify each column heading as qualitative or quantitative; for quantitative variables decide if they are discrete or continuous. Find the measurement scale for each of the variables.

2.2. Organization of the Data for One Quantitative Variable into a Table

When creating a table for data, the terms *frequency, proportion*, and *percentage* play a major role. These terms were introduced in Section 1.8. We now relate these terms to the outcomes of a quantitative variable.

If the data are *sample* data, the symbol n will be used for the number of objects in the *sample*. If the data are *population* data, the symbol N will be used for the number of objects in the *population*. The symbol N represents

Table 2.2 **Henry Aaron's yearly batting statistics**

Year	G	AB	R	H	2B	3B	HR
1954	122	468	58	131	27	6	13
1955	153	602	105	189	37	9	27
1956	153	609	106	200	34	14	26
1957	151	615	118	198	27	6	44
1958	153	601	109	196	34	4	30
1959	154	629	116	223	46	7	39
1960	153	590	102	172	20	11	40
1961	155	603	115	197	39	10	34
1962	156	592	127	191	28	6	45
1963	161	631	121	201	29	4	44
1964	145	570	103	187	30	2	24
1965	150	570	109	181	40	1	32
1966	158	603	117	168	23	1	44
1967	155	600	113	184	37	3	39
1968	160	606	84	174	33	4	29
1969	147	547	100	164	30	3	44
1970	150	516	103	154	26	1	38
1971	139	495	95	162	22	3	47
1972	129	449	75	119	10	0	34
1973	120	392	84	118	12	1	40
1974	112	340	47	91	16	0	20
1975	137	465	45	109	16	2	12
1976	85	271	22	62	8	0	10
Career	3298	12364	2174	3771	624	98	755

the *population size*, and the symbol n represents the *sample size*. A lowercase x represents an outcome of the variable X.

Definitions

- The *frequency at x* is the number of times x occurs in the data.
- The *frequency at x* is written F_x.
- The *proportion at x* is a quotient whose numerator is F_x and whose denominator is either n (sample size) or N (population size).
- The *proportion at x* is written Pr_x.
- The *percentage at x* is found by multiplying Pr_x by 100.
- The *percentage* at x is written P_x.

Pr_x can be expressed either as a decimal or as a fraction. The inequality $0 \le \mathrm{Pr}_x \le 1$ is always satisfied. In equation form,

$$\mathrm{Pr}_x = F_x/N \text{ or } F_x/n;\ P_x = (\mathrm{Pr}_x * 100)\%.$$

RBI	BB	SO	SH	SF	HBP	AVG	OBP	SLG
69	28	39	6	4	3	0.280	0.322	0.447
106	49	61	7	4	3	0.314	0.366	0.540
92	37	54	5	7	2	0.328	0.365	0.558
132	57	58	0	3	0	0.322	0.378	0.600
95	59	49	0	3	1	0.326	0.386	0.546
123	51	54	0	9	4	0.355	0.401	0.636
126	60	63	0	12	2	0.292	0.352	0.566
120	56	64	1	9	2	0.327	0.381	0.594
128	66	73	0	6	3	0.323	0.390	0.618
130	78	94	0	5	0	0.319	0.391	0.586
95	62	46	0	2	0	0.328	0.393	0.514
89	60	81	0	8	1	0.318	0.379	0.560
127	76	96	0	8	1	0.279	0.356	0.539
109	63	97	0	6	0	0.307	0.369	0.573
86	64	62	0	5	1	0.287	0.354	0.498
97	87	47	0	3	2	0.300	0.396	0.607
118	74	63	0	6	2	0.298	0.385	0.574
118	71	58	0	5	2	0.327	0.410	0.669
77	92	55	0	2	1	0.265	0.390	0.514
96	68	51	0	4	1	0.301	0.402	0.643
69	39	29	1	2	0	0.268	0.341	0.491
60	70	51	1	6	1	0.234	0.332	0.355
35	35	38	0	2	0	0.229	0.315	0.369
2297	1402	1383	21	121	32	0.305	0.374	0.555

Example 2.1. Hank Aaron held many records; perhaps the most well known was his 755 career home runs, which he held for 33 years until Barry Bonds surpassed it. But, from Table 2.2, let's look at Hank's yearly number of triples (3B). The data for this variable are

6, 9, 14, 6, 4, 7, 11, 10, 6, 4, 2, 1, 1, 3, 4, 3, 1, 3, 0, 1, 0, 2, 0.

For now, consider this *population data*. The population size is $N = 23$. Choose the outcome $x = 4$. Since the data point 4 occurs three times, $F_4 = 3$; $Pr_4 = (F_4/N) = (3/23) = .1304$; $P_4 = (.1304 * 100)\% = 13.04\%$.

Round-off rules: When expressing a proportion as a decimal, the decimal should be rounded to four decimal places. Percentages should be rounded to two decimal places. There can be exceptions to these round-off rules. For example, a batting average is usually expressed in the form *.xyz*.

Many different types of tables are used to organize data. A box score in baseball is specific to baseball and is used to both organize and summarize a team's performance in a game.

The tables in this chapter are general tables that can be used with any collection of data from any area of study. They are constructed for *quantitative data from one quantitative variable.* We begin our discussion with *frequency tables.*

2.2.1. Frequency Tables

A *frequency table* (often called a *frequency distribution*) has two columns. The first column consists of the different outcomes of the quantitative variable X sorted in ascending order from the smallest to the largest number. If the outcomes are only integers, *all* integers between the smallest and the largest integer must be included in the first column, even if the integer has a zero frequency. The second column consists of the *frequency at x*, F_x, for each *outcome x*. A *frequency table* for the yearly triples (3B) for Aaron is presented in Example 2.2.

Example 2.2. The population is the career years for Henry Aaron. The *variable* X = Yearly number of triples (3B) for Aaron. The *population data* consist of 23 numbers. The *frequencies*, F_X, are calculated using the data from Table 2.2. Table 2.3 shows the frequency table.

Table 2.3 Frequency table of yearly triples for Aaron

X	F_X
0	3
1	4
2	2
3	3
4	3
5	0
6	3
7	1
8	0
9	1
10	1
11	1
12	0
13	0
14	1
	$\Sigma F_X = 23$

The sum of the entries for any column is symbolized by using the summation notation Σ (the Greek capital letter *sigma*). This notation will be used throughout the book. The notation $\Sigma(\ldots)$ translates into summing the column headed by what is included within the parentheses $(\ldots)$. Usually, enclosed within $(\ldots)$ is a variable or arithmetic operation involving one or more variables. In Table 2.3, ΣF_X means to sum the frequency column, F_X. For this table, $\Sigma F_X = 23$ corresponds to the 23 years of Aaron's career.

New tables can be constructed by replacing the F_X column by a Pr_X column or by a P_X column. The two new tables are named, respectively, a *proportion table* or *relative frequency table* and a *percentage table.*

Since this chapter deals with quantitative data, *cumulative frequency tables, cumulative proportion tables*, and *cumulative percentage tables* can also be defined.

The *cumulative frequency at x* is equal to the sum of all *frequencies* for all values of the variable X less than or equal to x. It is symbolized by $CF_x = \Sigma F_X$ for $X \le x$.

From Table 2.3, $CF_4 = F_4 + F_3 + F_2 + F_1 + F_0 = 3 + 3 + 2 + 4 + 3 = 15$.

The *cumulative proportion* at x (sometimes referred to as the *cumulative relative frequency*) is equal to the cumulative frequency at x divided by the sum of the frequency column ($CF_x/\Sigma F_X$). It is written CPr_x.

From Table 2.3, we obtain

$$\mathrm{CPr}_4 = (CF_4/\Sigma F_X) = 15/23 = .6522.$$

The *cumulative percentage at x* is equal to ($\mathrm{CPr}_x * 100$)%, written CP_x.

From Table 2.4, we obtain

$$CP_4 = [(15/23) * 100]\% = 65.22\%.$$

Replacing the F_X column in a frequency table by the CF_X column yields the *cumulative frequency table*. Replacing the F_X column in a frequency table by the CPr_X column yields the *cumulative proportion table* or *cumulative relative frequency table*. Replacing the F_X column in a *frequency table* by the CP_X column yields the *cumulative percentage table*.

Exercise 2.2. Construct each of these new tables for X = Yearly number of home runs (HR) for Aaron.

CPr_x should be rounded to four decimal places, and CP_x should be rounded to two decimal places.

Instead of creating separate tables, we can combine F_X, Pr_X, P_X, CF_X, CPr_X, and CP_X into one table, called a *complete frequency table*. Table 2.4 is the *complete frequency table* for Aaron's yearly triples.

Table 2.4 **Complete frequency table for the yearly triples for Aaron**

X	F_X	Pr_X	P_X (%)	CF_X	CPr_X	CP_X (%)
0	3	.1304	13.04	3	.1304	13.04
1	4	.1739	17.39	7	.3043	30.43
2	2	.0870	08.70	9	.3913	39.13
3	3	.1304	13.04	12	.5217	52.17
4	3	.1304	13.04	15	.6522	65.22
5	0	.0000	00.00	15	.6522	65.22
6	3	.1304	13.04	18	.7826	78.26
7	1	.0435	04.35	19	.8261	82.61
8	0	.0000	00.00	19	.8261	82.61
9	1	.0435	04.35	20	.8696	86.96
10	1	.0435	04.35	21	.9130	91.30
11	1	.0435	04.35	22	.9565	95.65
12	0	.0000	00.00	22	.9565	95.65
13	0	.0000	00.00	22	.9565	95.65
14	1	.0435	04.35	23	1.000	100.00

Exercise 2.3. Construct the complete frequency table for X=Yearly number of home runs (HR) for Aaron.

Since the outcomes of X in Table 2.4 are integers, any integer between the smallest and largest outcome having zero frequency must be included. If there were gaps in the first column, the table can lead to a misleading graph. Observe in Table 2.4 that the x-values of 5, 8, 12, and 13 are listed even though they each have a frequency of zero.

When several different outcomes have only one occurrence, or the outcomes are spread out, or the outcomes are in the form of measurements, then listing individual outcomes in the first column may provide little insight into the meaning of the data. In these cases, a different approach is needed for the first column.

This approach replaces individual outcomes with *categories* called *class intervals.* A class interval includes all outcomes of the variable that lie within the interval. The process of constructing class intervals is called *grouping the data.* There are different approaches used to construct class intervals. The approach used in this book is shown in Box 2.3.

Example 2.3. Consider again the population consisting of the career years of Henry Aaron. Let X=Yearly number of runs scored. Using the data from the run (R) column in Table 2.2, we construct the class intervals following the eight steps described in Box 2.3.

Box 2.3 **Grouping data**

Step 1: Compute the *range* of the data by subtracting the smallest data point from the largest data point.

Step 2: Arbitrarily choose between 5 and 15 intervals.

Step 3: Divide the result of step 1 by the result of step 2.

Step 4: Choose one number, close to the number calculated in step 3, to be used as the width of each of the class intervals. This number should have the same number of decimal places as the numbers in the data set.

Step 5: Build the intervals from the top to bottom, starting with the smallest data number as the left endpoint of the first interval.

Step 6: Construct all left endpoints by adding the width from step 4 to each left endpoint. The process ends when the number for the next left endpoint exceeds the largest number in the data set.

Step 7: Subtract the appropriate number from the second left endpoint. Use the resulting number for the first right endpoint. If the data is in the form of integers, then subtract 1; if the data is in the form *y.x*, then subtract 0.1; if the data is in the form *y.xz*, subtract 0.01; and so on.

Step 8: Construct all right endpoints by adding the width to each right endpoint. This process ends at the last left endpoint.

As a final check of the suitability of the class intervals, every data point must lie in exactly one interval.

Step 1: The range = 127 − 22 = 105.

Step 2: Arbitrarily, we choose seven intervals.

Step 3: The result of dividing 105 by 7 is 15.

Step 4: Since the data consist of integers and the result in step 3 is an integer, no rounding is necessary and 15 will be used as the width for each of the class intervals.

Step 5: The first left endpoint is 22.

Step 6: Adding 15 to each left endpoint produces the left endpoints of 37, 52, 67, 82, 97, 112, ending at 127.

Step 7: The first right endpoint is calculated as 37 − 1 = 36.

Step 8: Adding 15 to each right endpoint produces the right endpoints of 51, 66, 81, 96, 111, 126, ending at 141.

The class intervals are [22, 36], [37, 51], [52, 66], [67, 81], [82, 96], [97, 111], [112, 126], [127, 141].

Example 2.4. Using the class intervals from Example 2.3, Table 2.5 represents a complete frequency table for X = Yearly number of runs.

When using class intervals, the *subscript X* is replaced by the *subscript I*. The letter I identifies the interval number from top to bottom. The first interval [22, 36] is $I=1$, the second interval [37, 51] is $I=2$, and so on.

The frequency, F_I, for each interval is calculated by tallying the number of outcomes that lie within the interval. We say that the interval I_k is less than the interval I_m when the right endpoint of I_k is less than the left endpoint of I_m. Although class intervals are categories, the ordering described above allows one to include the cumulative columns CF_I, CPr_I, and CP_I as part of a *complete frequency table*. CF_i is calculated by summing the F_I (ΣF_I) for all intervals $I \leq i$. As an example, $CF_3 = F_1 + F_2 + F_3 = 1 + 2 + 1 = 4$; $CPr_3 = CF_3/\Sigma F_X = 4/23 = .1739$; $CP_3 = 17.39\%$; $I=3$ refers to the third interval [52, 66].

Table 2.5 **Complete frequency table for the yearly number of runs scored by Aaron**

Class intervals	F_I	Pr_I	P_I (%)	CF_I	CPr_I	CP_I (%)
22–36	1	.0435	4.35	1	.0435	4.35
37–51	2	.0870	8.70	3	.1304	13.04
52–66	1	.0435	4.35	4	.1739	17.39
67–81	1	.0435	4.35	5	.2174	21.74
82–96	3	.1304	13.04	8	.3478	34.78
97–111	8	.3478	34.78	16	.6957	69.57
112–126	6	.2609	26.09	22	.9565	95.65
127–141	1	.0435	4.35	23	1.000	100.00

Notice that the process produced eight class intervals. This was one more than was specified, which can happen as a result of rounding. As long as the number of class intervals is in the acceptable range, this does not create a problem.

Exercise 2.4. Consider the population of career years for Henry Aaron. Let X=Yearly batting average (AVG). Construct the class intervals in two ways. First, use the AVG rounded to three decimal places as *.xyz*. Then, multiply each AVG by 1000 and use these numbers. Using the constructed class intervals, construct the complete frequency table.

A complete frequency table transforms the original data into a form that lends itself to summarizing, graphing, and answering "what if" questions.

Three conclusions can be drawn from Table 2.5:

1. By identifying the interval with the highest frequency (97–111), we can conclude that 35% of the time Aaron scored approximately 100 runs per year.
2. The cumulative percentage for the interval (82–96) is 35%, so we can conclude that approximately 65% of the time Aaron scored more than 96 runs per year.
3. Using the cumulative percentage for the interval (52–66), we can conclude that Aaron scored less than 67 runs for a year only 17.39% of the time.

The difference between Table 2.4 and Table 2.5 is the first column. In Table 2.4 the first column represents the actual values of the variable, whereas in Table 2.5 the first column is composed of categories of class intervals. Since class intervals are categories, the actual data values are lost.

In the next section, the information from the tables just constructed will be used to produce different types of graphs.

2.3. Presentation of Data for One Quantitative Variable Using Graphs

The graphs constructed in this section are *histograms*, *line graphs* or *polygons*, *cumulative polygons*, *stem-and-leaf graphs*, *time-series graphs*, and *box-and-whisker graphs*. They are constructed using Microsoft Excel (version 2003), but any of these graphs can be drawn with a pencil and ruler. Also, any later version of Excel can be used.

Histograms and *polygons* display the information from a *complete frequency table* as a graph in which the x-axis contains the *outcomes* of the first column (in the case of Table 2.4 the outcomes were integers; in the case of Table 2.5 the outcomes were class intervals) and the heights correspond to either *frequencies* or *percentages*.

2.3.1. Histograms

A *histogram* consists of rectangles that abut (have no gaps between them). A rectangle is specified by its base and height. The construction of the base of each rectangle depends on whether the *x*-axis consists of the actual outcomes or class intervals.

Case 1. If the *x*-axis consists of *numbers*, the center of the base is the number and the endpoints of the base extend halfway to the next number.

Case 2. If the *x*-axis consists of class intervals, the endpoints of the base extend halfway to the next interval.

For a *frequency histogram* the height of each rectangle is either F_X or F_I. For a *percentage histogram* the height of each rectangle is either P_X or P_I.

Figure 2.1 is a frequency histogram for Table 2.4. The *x*-axis displays the integers 0–14.

Figure 2.2 is a percentage histogram for Table 2.5. The *x*-axis displays the class intervals.

We can choose to use either frequencies or percentages for the heights of the rectangles. Both have their drawbacks. For one population, percentages alone can be misleading. Suppose that we are told that 50% of the ballplayers sampled approve a certain rule change. Later, we discover that the 50% came from a survey where 2 out of 4 approved the rule change. We would have been better informed if both the percentage and frequency were given. Suppose that we want to compare players in the American League with players in the National League regarding approval of this rule change. We are told that 10 players in the American League favor the change, whereas 20 players in the National League support the change. It seems that National League players are more inclined to favor the rule change. But given that only 20 American League players responded to the survey, whereas

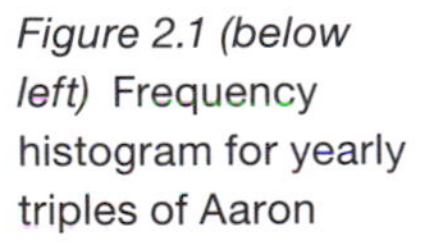

Figure 2.1 (below left) Frequency histogram for yearly triples of Aaron

Figure 2.2 (below right) Percentage histogram for yearly runs scored by Aaron

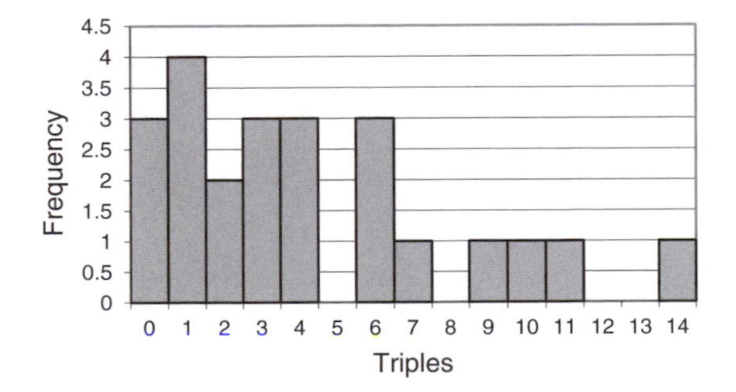

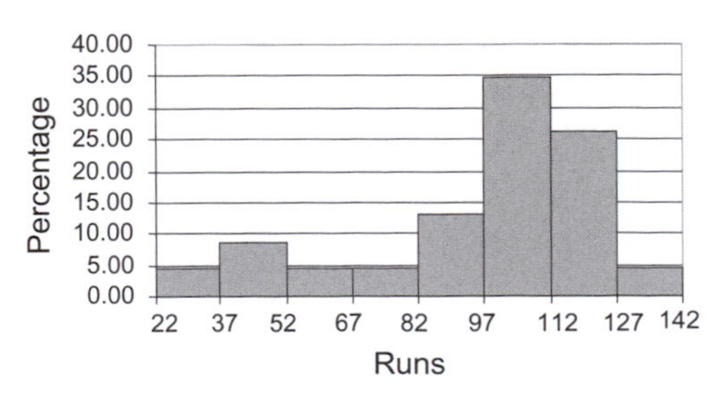

100 National League players responded, the frequencies by themselves would be misleading, for 50% of AL players like the rule change but only 20% of NL players welcomed it.

When in doubt, display both a frequency and a percentage histogram.

2.3.2. Polygons

A *polygon* or *line graph* consists of line segments joining ordered pairs (x, y). Here, the x-value is either the outcome itself or the midpoint of a class interval. The y-value corresponds to either the frequency or percentage of occurrences for that outcome or that class interval. The construction of a polygon begins by adding to the existing table either two outcomes or two class intervals both having frequency zero; one is added in front of the first outcome or first class interval and the other is added in back of the last outcome or last class interval. For Table 2.3, the number −1 would be added in front and the number 15 would be added at the end. For Table 2.4, the interval in front would be [7, 21] and the interval at the end would be [142, 156]. The reason for adding the two extra outcomes or class intervals is that the polygon should begin and end on the x-axis.

Figure 2.3 presents a *frequency* polygon for Table 2.4 using the actual outcomes on the x-axis.

Figure 2.4 presents a *percentage* polygon for Table 2.5 using the class intervals on the x-axis.

Figure 2.3 Frequency polygon for yearly triples of Aaron

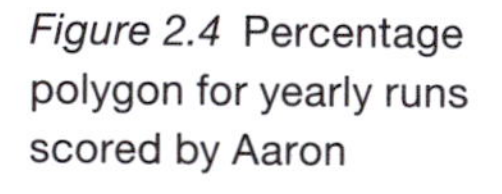
Figure 2.4 Percentage polygon for yearly runs scored by Aaron

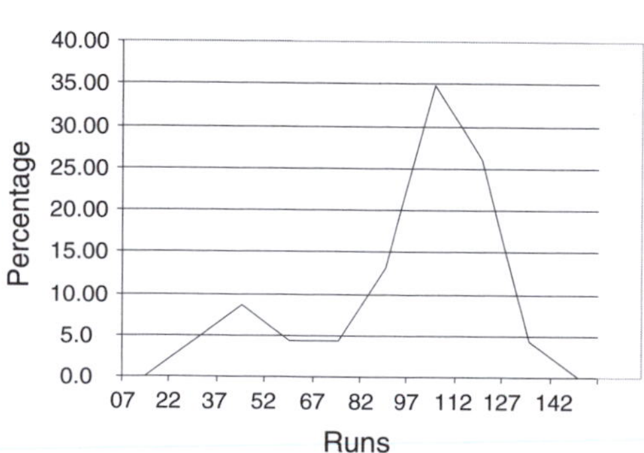

Exercise 2.5. Using the class intervals from Exercise 2.4, construct the percentage polygon and percentage histogram for the yearly AVGs for Aaron.

2.3.3. Cumulative Polygons

A *cumulative polygon* consists of line segments joining ordered pairs (x, y) where the x-value is either the outcome itself or the right endpoint of a class interval and the y-value corresponds to either the cumulative *frequency* or cumulative *percentage* of occurrences for either the outcome or the class interval. The construction of a cumulative polygon begins by adding to the existing table either one outcome or one class interval having frequency zero in front of the first outcome or class interval. For Table 2.5, the class interval in front would be [7, 21]. Figure 2.5 presents the cumulative percentage polygon.

Exercise 2.6. Using the class intervals from Exercise 2.4, construct the cumulative percentage polygon for the yearly AVGs for Aaron.

2.3.4. Stem-and-Leaf Graph

Before constructing a *stem-and-leaf graph*, the data must be sorted in ascending order from the smallest to largest number. After the numbers are sorted, each number is split at the same digit. The left side of the split becomes the *stem*, and the right side of the split becomes the *leaf*. The stems play the role of the class intervals in a histogram, and the number of leaves for a given stem represents the frequency in a histogram.

Example 2.5. The following sorted data consist of the 23 yearly batting averages (AVGs) for Aaron obtained from Table 2.2: .229, .234, .265, .268, .279, .280, .287, .292, .298, .300, .301, .307, .314, .318, .319, .322, .323, .326, .327, .327, .328, .328, .355.

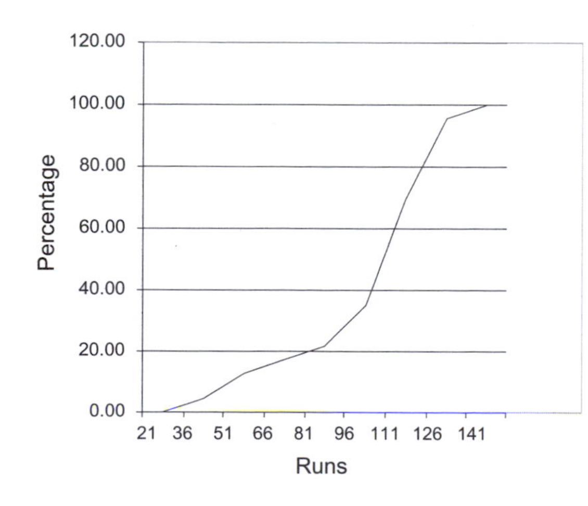

Figure 2.5 Cumulative percentage polygon for yearly runs scored by Aaron

Stem	Leaves
.22	9
.23	4
.24	
.25	
.26	5 8
.27	9
.28	0 7
.29	2 8
.30	0 1 7
.31	4 8 9
.32	2 3 6 7 7 8 8
.33	
.34	
.35	5

Figure 2.6 Stem-and-leaf graph for Aaron's yearly AVGs

In a *stem-and-leaf graph*, each number is split at the same digit. For the above *data*, each number *.xyz* can be split either as *.xy*|*z* or as *.x*|*yz*. In the first case the *stem* would be *.xy* and the *leaf* would be *z*, whereas in the second case the *stem* would be *.x* and the *leaf* would be *yz*. The splitting process is arbitrary and the decision is based on obtaining the best graph.

For the above data, if the split resulted in *.x* being the stem and *yz* being the leaf, there would only be two stems. Since this split results in too few stems, the best split would be *.xy*|*z*, which results in 14 stems. Figure 2.6 shows the stem-and-leaf graph for the yearly AVGs for Aaron from Table 2.2.

A *stem-and-leaf graph* can be thought of as a sideways histogram with the stems acting as class intervals. Consequently, the guideline of 5 to 15 can be used for the number of stems.

The next two examples provide stem-and-leaf graphs for the yearly number of triples for Aaron.

Example 2.6. Figure 2.7 presents a stem-and-leaf graph for the yearly triples for Aaron. Each number for yearly triples for Aaron is represented by two digits (1 is represented as 01). The two-digit numbers are split as *x*|*y*.

Figure 2.7 is a stem-and-leaf graph where each two-digit number "*xy*" was divided into *x*|*y*, with *x* being the stem and *y* being the leaf. The problem is too few stems. One way of resolving this problem is to further subdivide the stems.

Figure 2.8 shows that each existing stem from the previous graph is further subdivided into five equal parts.

Example 2.7. The difference between the two stem-and-leaf graphs is that in the first graph the two stems correspond to the two class intervals [0, 9] and [10, 19], whereas in the second graph the eight stems correspond to the eight class intervals [0, 1], [2, 3], [4, 5], [6, 7], [8, 9], [10, 11], [12, 13], and [14, 15]. The graph in Figure 2.8 is acceptable because the number of stems lies within the guidelines of 5 to 15.

Stem	Leaves
0	0001111122333344466679
1	014

Stem	Leaves
0	0001111
0	22333
0	444
0	6667
0	9
1	01
1	
1	4
1	
1	

Figure 2.7 (left) Stem-and-leaf graph for Aaron's yearly triples (two stems)

Figure 2.8 (right) Stem-and-leaf graph for Aaron's yearly triples (ten stems)

Exercise 2.7. Construct a stem-and-leaf graph for Aaron's yearly home runs, OBP, and SLG.

2.3.5. Time-Series Graphs

A *time-series graph* displays data that are observed over a given period of time. Using this graph, one can analyze the behavior of data over a period of time. A time-series graph consists of line segments joining ordered pairs (x, y), where the x-values are periods of time and the y-values are the outcomes of the variable being studied. The following are two examples of time-series graphs. Both examples use for the periods of time Aaron's yearly age. In 1954, Hank Aaron was 20 years old. The variable representing the periods of time is a fixed variable.

Example 2.8. This example presents two time-series graphs. For each time series the x-values represent the yearly age for Aaron. In Figure 2.9, the y-values are the number of Aaron's yearly triples. In Figure 2.10, the y-values represent the number of Aaron's yearly home runs. Interpret both of these time series.

Exercise 2.8. Using Aaron's age for the time periods, construct a time-series graph for his yearly runs scored.

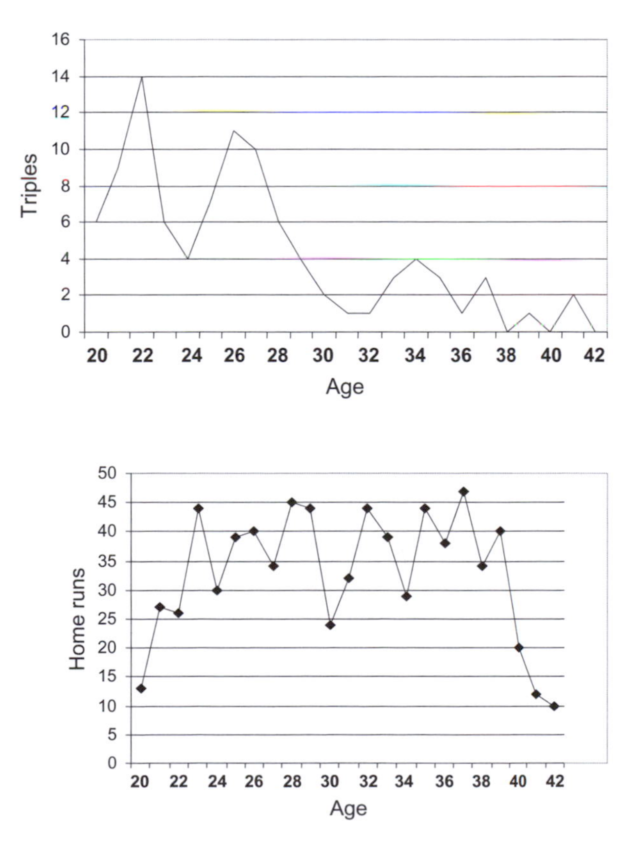

Figure 2.9 Time-series graph for yearly triples of Aaron

Figure 2.10 Time-series graph for yearly home runs of Aaron

2.3.6. *The Five-Number Summary*

Before undertaking the construction of a box-and-whisker graph, it is necessary to understand certain positional measures for our data.

For any set of quantitative data, there is a smallest number, *S*, a largest number, *L*, and the three quartiles, *Q1*, *Q2*, and *Q3*. Before computing these five measures, it is necessary to sort the data in ascending order from the smallest number to the largest number.

The three quartiles divide the data into four equal parts. (Many times it is not possible to divide the data into four equal parts. For example, if there are 11 data values, this cannot be done. In these cases, the quartiles will only be computational estimates.)

Q1, the *first quartile*, is a number that divides the data in such a way that 25% of the data are to the left of *Q1* and 75% of the data are to the right.

Q2, the *second quartile*, is a number that divides the data in such a way that 50% of the data are to the left of *Q2* and 50% of the data are to the right.

Q3, the *third quartile*, is a number that divides the data in such a way that 75% of the data are to the left of *Q3* and 25% of the data are to the right.

These five numbers *S*, *Q1*, *Q2*, *Q3*, and *L* are called the *five-number summary* for the data. The five-number summary is denoted by (*S*, *Q1*, *Q2*, *Q3*, *L*).

Procedure for Finding *Q1*, *Q2*, and *Q3*

Step 1: Sort the data from the smallest number to the largest number.
Step 2: Find *Q2*:

- If n is odd, $Q2 =$ the data value occupying the position $(n+1)/2$.
- If n is even, $Q2 =$ [the data value occupying position $(n/2)$ + the data value occupying position $(n/2+1)$]/2.

Step 3: Find *Q1* by repeating step 2 for $n =$ the number of data values to the left of *Q2*.
Step 4: Find *Q3* by repeating step 2 for $n =$ the number of data values to the right of *Q2*. (The first position to the right of *Q2* now becomes position 1.)

Example 2.9. Find the five-number summary for the yearly triples for Henry Aaron.

Step 1:

Values	0, 0, 0, 1, 1, **1**, 1, 2, 2, 3, 3, **3**, 4, 4, 4, 6, 6, **6**, 7, 9, 10, 11, 14
Position	1 2 3 4 5 **6** 7 8 9 10 11 **12** 13 14 15 16 17 **18** 19 20 21 22 23

Step 2: Since $n=23$ is an odd number, *Q2* is the occupant of position $(n+1)/2=(23+1)/2=12$; **Q2 = 3**.

Step 3: $n=11$ (the number of data values to the left of *Q2*); since 11 is odd, position $=(n+1)/2=6$. *Q1* is the occupant of position 6; **Q1 = 1**.

Step 4: $n=11$ (the number of data values to the right of *Q2*); since 11 is odd, position $=(n+1)/2=6$. *Q3* is the occupant of position 18; **Q3 = 6**. Observe that position 13 in the original data is the first position to the right of *Q2* and thus becomes position 1. Therefore, position 18 corresponds to the sixth position to the right of *Q2*.

The five-number summary is (0, 1, 3, 6, 14).

To illustrate finding the quartiles for an even number of data values, we use the data for Aaron's triples, excluding year 23 when he had zero triples.

Example 2.10. Find the five-number summary for the triple data for Aaron for years 1–22.

Step 1:

Values 0, 0, 1, 1, 1, **1**, 2, 2, 3, 3, **3**, **4**, 4, 4, 6, 6, **6**, 7, 9, 10, 11, 14

Position 1 2 3 4 5 **6** 7 8 9 10 **11** **12** 13 14 15 16 **17** 18 19 20 21 22

Step 2: Since $n=22$ is an even number, $Q2=$ [(the data value occupying position $n/2=11$) + (the data value occupying position $(n/2)+1=12$)]/2 $=[3+4]/2=3.5$; **Q2** occupies position 11.5 and is equal to **3.5**.

Step 3: $n=11$ (the number of data values to the left of *Q2*); since 11 is odd, position $=(n+1)/2=6$. *Q1* is the occupant of position 6; **Q1 = 1**.

Step 4: $n=11$ (the number of data values to the right of *Q2*); since 11 is odd, position $=(n+1)/2=6$. *Q3* is the occupant of position 17; **Q3 = 6**. Observe that position 12 in the original data is the first position to the right of *Q2* and thus becomes position 1. Therefore, position 17 corresponds to the sixth position to the right of *Q2*.

The five-number summary is (0, 1, 3.5, 6, 14).

2.3.7. Outliers

Outliers are extremely small or large data values that one may want to exclude from the data set because they are mistakes or are values that do not belong to the target population. *Outliers* are found by creating a closed interval, called the *outlier fence*, with left endpoint $Q1-3/2(Q3-Q1)$ and right endpoint $Q3+3/2(Q3-Q1)$. Data values that lie outside the outlier fence are called *outliers*. Outliers are only removed from the original data if we are sure that they are mistakes or do not belong to the target population. Arbitrarily removing data values could possibly bias the data.

What is meant by a data value not belonging to the population? An example of this is a player appearing in only 10 games for an entire year. Since his data represent the results of only 10 games, the number of home runs for that year would not represent a true year's performance. In many of our examples for Ted Williams, the years 1952 and 1953 are outliers. An example of a mistake is recording the number of triples as 41 instead of 14. If the mistake cannot be corrected, the value of 41 would be excluded from the original data.

Example 2.11. In Example 2.9, $Q1 = 1$ and $Q3 = 6$. The outlier fence would extend from $[1 - (3/2)(6 - 1)]$ to $[6 + (3/2)(6 - 1)] = [1 - 7.5, 6 + 7.5] = [-6.5, 13.5]$. Since the data point 14 lies outside the outlier fence, 14 would be the only outlier. However, there is no reason to exclude the data value 14 from our data set.

Outliers that either are mistakes or do not belong to the target population are removed from the data set before beginning the analysis. The revised data set will be used in the research study. The entire process must be repeated with the revised data set. However, those removed outliers should be mentioned in the write-up of the study.

The outliers that are not mistakes can provide valuable information about the data set.

Exercise 2.9. Find the five-number summary and all outliers (if any) for the yearly number of runs scored by Aaron.

Exercise 2.10. Using Table 2.2 for Aaron, compute the five-number summary and all outliers for his yearly OBP and SLG.

2.3.8. Box-and-Whisker Graph

The five-number summary (S, $Q1$, $Q2$, $Q3$, L) and the outliers are used in the construction of a box-and-whisker graph. The horizontal box-and-whisker graph consists of a box and two whiskers. The whiskers are horizontal line segments extending from the middle of the sides of the box to S and L. The interior box has its sides at $Q1$ and $Q3$. A vertical line segment is drawn inside the box at $Q2$. There is one exception to this entire configuration. The end of a whisker cannot be an outlier. Therefore, the ends of the whiskers consist of the smallest and largest numbers that lie within the outlier fence. Outliers are designated in the graph with the symbol *o*. The scale used for the x-axis is chosen to be appropriate for the values of the variable being studied.

Example 2.12. The box-and-whisker graph for the yearly triples for Aaron is displayed below.

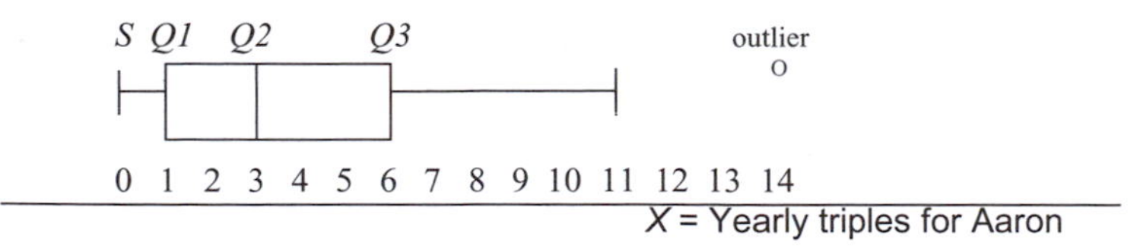

Exercise 2.11. Construct the box-and-whisker graph for the yearly runs scored by Aaron.

Exercise 2.12. Construct the box-and-whisker graphs for Aaron's yearly OBP and SLG.

2.4. Summarization of the Data for One Quantitative Variable

A *descriptive measure* is usually one number (in a rare instance, it is more than one number) used to summarize the data.

A descriptive measure of a sample is called a *statistic*, whereas a descriptive measure of a population is called a *parameter.* An easy way to remember these terms is that *p* is the first letter of the two words "population" and "parameter" and *s* is the first letter of the two words "sample" and "statistic." Therefore, the same *descriptive measure* can be referred to as a *statistic* or a *parameter.* In many cases, the computational formula for a given descriptive measure is the same whether it is a statistic or a parameter; however, the symbols usually are different. Greek letters are reserved for parameters. Since a descriptive measure is calculated using a computational formula, the number representing the descriptive measure may or may not be a member of the data set.

Descriptive measures should be rounded to at least one more decimal place than the number of decimal places for the numbers in the original data.

Unless otherwise stated, the collected data are assumed to be sample data. Statistics are used to summarize the collected sample data. In inferential statistics, statistics are used to estimate the corresponding *parameters* for the *population.*

In summarizing data, one is interested in measures that locate the *middle of the data*, describe the *spread* or *variability within the data*, and classify the *shape of the data*. The three key words here are *middle, spread,* and *shape.*

If Aaron's baseball manager (his first with the Braves was "Jolly Cholly" Grimm) were asked to describe his career yearly home run production, he might say that Hank averaged about 33 home runs each year and his home run production was consistent from year to year. The 33 represents the mean, and the term "consistent" means that the variability from year to year was small.

The three most used measures of the middle are the *mean*, the *median*, and the *mode*.

2.4.1. The Measures of the Middle

2.4.1.1. The Mean

The *mean* is the most used measure for the *middle of the data*. It is sometimes referred to as the average, but it is only one of several different types of averages used in statistics. It is called the balanced middle for the data. The computational formula for the mean is $\Sigma x/n$. The symbol $\bar{x}$ is used for the *sample mean*, and the symbol μ (the Greek letter *mu*) is used for the *population mean*.

Example 2.13. The yearly home run totals for Aaron from Table 2.2 are as follows: 13, 27, 26, 44, 30, 39, 40, 34, 45, 44, 24, 32, 44, 39, 29, 44, 38, 47, 34, 40, 20, 12, 10.

The uppercase X represents the variable. X = Number of yearly home runs for Aaron. The lowercase x represents the actual values of X. N = Number of years = 23.

Considered as *population data*, the *population mean* is $\mu = \Sigma x/N = 755/23 = 32.83$.

Considered as *sample data*, the *sample mean* is $\bar{x} = \Sigma x/n = 755/23 = 32.83$.

Properties of the Mean

1. It is unique. (The computational formula yields precisely one number.)
2. It can be used for any collection of quantitative data.
3. It is the balanced middle. Using algebra, it can be shown that $\Sigma(x - \bar{x})$ will always equal 0. If a number line is placed on a beam with 1 pound weights occupying each position of the numbers from the data set and a fulcrum is placed under the position of the number calculated as the mean, the beam would be balanced.
4. It can be affected by extreme values. (For a small set of numbers an extreme number will drastically shift the mean in the direction of the extreme value.)
5. The symbols $\bar{x}$ and μ are used, respectively, for the sample mean and the population mean (by definition).
6. The same computational formula is used for $\bar{x}$ and μ (by definition).

A batting average is the proportion of hits to at bats. An explanation of why a proportion is called an average is necessary. Suppose that a player has 10 at bats with six hits. His batting average is the proportion 6/10 = .600.

Since the data are qualitative (hit or out), one can assign the number 1 to a hit and the number 0 to an out. The data now consist of six ones and four zeros. The mean is $(1+1+1+1+1+1+0+0+0+0)/10=6/10$.

2.4.1.2. The Median

To find the median, first sort the data in ascending order from the smallest to the largest number. The median is the point that divides the data in half: 50% of the data fall below the median and 50% above it. It is called the *positional* middle of the data. The median is symbolized by M_d. The median is the second quartile, $Q2$.

Example 2.14. Find the median for Aaron's yearly home runs. Using the procedure for calculating $Q2$, we have the following steps:

Step 1: The data sorted in ascending order are

Values 10, 12, 13, 20, 24, 26, 27, 29, 30, 32, 34, **34**, 38, 39, 39, 40, 40, 44, 44, 44, 44, 45, 47

Position **12**

Step 2: Since $n=23$ is an odd number, $Q2$ is the occupant of position 12; $\mathbf{Q2=34}$. The median $M_d=Q2=34$.

Properties of the Median

1. It is unique. (The computational formula will yield only one number.)
2. It can always be calculated for quantitative data (by definition).
3. It is the positional middle. (Half the data lie on either side of the median.)
4. It is minimally affected by extreme values. (An extreme value does not change the number of data values on each side of the median.)
5. The symbol M_d is used for both the sample and population median (by definition).
6. It can be used with qualitative data having an ordinal measurement scale. (If the data consist of ranks, the middle rank is the median.)

2.4.1.3. The Mode

The *mode*, written M_o, is the data point that occurs most frequently. If each data point occurs the same number of times, there is no mode. Multiple modes exist when two or more data values tie for the most occurrences. The *mode* is referred to as the *frequency middle.*

Example 2.15. Using the data in Example 2.14, the mode is 44.

Properties of the Mode

1. It need not be unique (by definition).
2. It need not exist (by definition).

3. It is the frequency middle. (If the mode exists, it represents the outcome having the highest frequency.)
4. It is not affected by extreme values (by definition).
5. Only one symbol, M_o, is used (by definition).
6. It can be used for qualitative data with a nominal measurement scale. (Since the number of occurrences for each qualitative outcome can be counted, the outcome with the highest frequency will be the *modal* outcome.)
7. If the data are grouped into class intervals, we can talk about the *modal class interval*. (The modal class interval is the interval with the highest frequency. The logic for finding the modal interval is the same as we use for the individual data. The midpoint of the modal interval will be used as an estimate of the mode.)

Exercise 2.13. Using Example 2.9, show for Aaron's yearly triples that the mean = 98/23, the median = 3, and the mode = 1.

Exercise 2.14. Using the data from Table 2.2, find the mean, median, and mode for Aaron's yearly doubles (2B) and yearly OBP.

2.4.2. *Measures of Variability (Spread) within the Data*

Measures of variability, sometimes referred to as *measures of spread*, attempt to quantify how close data values are to each other or to some fixed point. If the variability of the data is small, more meaning can be attributed to such descriptive measures as the mean. If all data values have the same value, there is no variability within the data and the measure zero should be assigned. The measures of variability discussed in this section include the *range*, the *interquartile range*, the *variance*, the *standard deviation*, and the *coefficient of variation*.

2.4.2.1. Range

The *range* is the difference between the largest data value and the smallest data value. Range $= L - S$.

Properties of the Range

1. It is unique (by definition).
2. It can always be calculated for quantitative data (by definition).
3. It cannot be used for qualitative data (by definition).
4. The range supplies little information about the variability within the data. (Since the range is completely determined by only the two most extreme values, it does not utilize most of the data.)
5. No symbol is necessary; the word "range" is used (by definition).
6. It is used in the construction of class intervals (see Section 2.2).

2.4.2.2. Interquartile Range

The *interquartile range* equals (*Q3* – *Q1*); the first quartile is subtracted from the third quartile.

Properties of the Interquartile

1. It is unique (by definition).
2. It can always be calculated for quantitative data (by definition).
3. It cannot be used for qualitative data (by definition).
4. It is the range of the middle 50% of the data. The interval from *Q1* to *Q2* contains 25% of the data; the interval from *Q2* to *Q3* contains 25% of the data; the interval from *Q1* to *Q3* contains 50% of the data.
5. It is a much better measure of spread than the range. (The two values *Q3* and *Q1* are much closer to the middle of the data.)
6. The symbol *IQR* is used for both population data and sample data (by definition).
7. It is the preferred measure of spread when the median is used as the measure of the middle. (The smaller the interquartile range, the closer the middle 50% of the data are to the median.)

2.4.2.3. Variance and Standard Deviation

The *sample variance* $= [\sum(x-\bar{x})^2]/(n-1)$ (definition)
$= [\sum x^2 - (\sum x)^2/n]/(n-1)$ (computational formula).

The *population variance* $= [\sum(x-\mu)^2]/N$ (definition)
$= [\sum x^2 - (\sum x)^2/N]/N$ (computational formula).

The *sample standard deviation* is the square root of the sample variance.

The *population standard deviation* is the square root of the population variance.

Both the *variance* and the *standard deviation* measure the *spread of the data* around the *mean*.

The first formula of the pair is the definition, whereas the computational formula is easy to use with a calculator. The computational formula just requires two sums, and the mean need not be calculated. The verification that these two formulas are equivalent can be shown using algebra.

Remember, n represents the *sample size*, and N represents the *population size*; $\bar{x}$ is the *sample mean*, and μ is the *population mean*. Notice that the only difference between the formulas for the *sample variance* and *population variance* is in the denominator. The reason for dividing by $(n-1)$ instead of n is theoretical. In statistical inference, it makes the sample variance an unbiased estimator for the population variance. The term $(n-1)$ is called the *degrees of freedom* and is symbolized by df.

Properties of the Variance and Standard Deviation

1. Both are unique.
2. Both can always be calculated for quantitative data.
3. Both cannot be used for qualitative data.
4. Both measure how spread out the data are around the mean.
5. The symbol for the sample variance is s^2, and the symbol for the population variance is σ^2 (the Greek letter *sigma*). The symbol for the sample standard deviation is s, and the symbol for the population standard deviation is σ.
6. The population variance is another example of an average. It is the average of the numbers derived by squaring the distance of each data point from the mean. Since the numerator is divided by $(n-1)$, the sample variance is close to an average.
7. The variance and standard deviation are the preferred measures of spread when the mean is used for the measure of the middle.
8. To calculate the variance, it is easier to use the computational formula than the definition.
9. The smaller the variance or standard deviation, the closer more of the data are to the mean.

2.4.2.4. Coefficient of Variation

The coefficient of variation (CV) is equal to the standard deviation divided by the mean. This value is then multiplied by 100 to convert to a percent. The sample CV is equal to $(s/\bar{x}) * 100\%$; the population CV is equal to $(\sigma/\mu) * 100\%$. The CV is used to compare the spread within two populations or two samples. If the data consist of larger numbers, the standard deviation will naturally be larger than if the data consist of smaller numbers. Since data consisting of the larger numbers will also have a larger mean, dividing the standard deviation by the mean allows for a fair comparison of the spread within the two data sets.

Example 2.16. Using the data from Example 2.1 on Aaron's yearly triples, we compute the range, the interquartile range, the population variance, the population standard deviation, the sample variance, the sample standard deviation, the population coefficient of variation, and the sample coefficient of variation.

The range $= L - S = 14 - 1 = 13$.
The interquartile range $= Q3 - Q1 = 6 - 1 = 5$.

Table 2.6 provides the necessary columns needed for the calculation of the variance.

$\Sigma(x-\mu)^2 = 324.43$, $\Sigma x = 98$, $\Sigma x^2 = 742$.

Table 2.6 **Columns needed for the calculation of the variance**

X	Mean	(X – Mean)	(X – Mean)^2	X	X^2
0	4.26	−4.26	18.16	0	0
0	4.26	−4.26	18.16	0	0
0	4.26	−4.26	18.16	0	0
1	4.26	−3.26	10.63	1	1
1	4.26	−3.26	10.63	1	1
1	4.26	−3.26	10.63	1	1
1	4.26	−3.26	10.63	1	1
2	4.26	−2.26	5.11	2	4
2	4.26	−2.26	5.11	2	4
3	4.26	−1.26	1.59	3	9
3	4.26	−1.26	1.59	3	9
3	4.26	−1.26	1.59	3	9
4	4.26	−0.26	0.07	4	16
4	4.26	−0.26	0.07	4	16
4	4.26	−0.26	0.07	4	16
6	4.26	1.74	3.02	6	36
6	4.26	1.74	3.02	6	36
6	4.26	1.74	3.02	6	36
7	4.26	2.74	7.50	7	49
9	4.26	4.74	22.46	9	81
10	4.26	5.74	32.94	10	100
11	4.26	6.74	45.42	11	121
14	4.26	9.74	94.85	14	196
Sums			324.43	98	742

The population variance $= [\sum (x - \mu)^2]/N = 324.43/23 = 14.11$
$= [\sum x^2 - (\sum x)^2/N]/N)$
$= [742 - (98)^2/23]/23 = 14.11.$

The sample variance $= [\sum (x - \bar{x})^2]/(n - 1) = 324.43/22 = 14.75$
$= [\sum x^2 - (\sum x)^2/n]/(n - 1)$
$= [742 - (98)^2/23]/22 = 14.75.$

The population standard deviation $= \sqrt{14.11} = 3.76.$
The sample standard deviation $= \sqrt{14.75} = 3.84.$
The population CV $= (3.76/4.26) * 100\% = 88.26\%.$
The sample CV $= (3.84/4.26) * 100\% = 90.14\%.$

Exercise 2.15. Using Aaron's yearly home runs, compute the range, the interquartile range, the population variance, the population standard deviation, the sample variance, the sample standard deviation, the population coefficient of variation, and the sample coefficient of variation.

Exercise 2.16. Use the results of Example 2.16 and Exercise 2.15 to compare the population CV of Aaron's yearly triples with that of his yearly home runs. Which data set is more spread out around its mean?

In dealing with baseball data, many of the statistics are numbers expressed as decimals. For example, batting averages have the form of *.xyz*. By multiplying each batting average by 1000, we can look at the data as integers instead of decimals. There is an obvious advantage of performing calculations with integers. However, the mean and variance for the new data will change. Is there a formula that allows one to calculate the mean and standard deviation for the new data from the mean and standard deviation for the original data? The answer is yes.

2.4.3. Effect on the Mean and Standard Deviation of a Data Set

Write the mean and standard deviation for the original data as mean_o and std_o. After multiplying each data value by a fixed constant *c*, the mean for the new transformed data is $(c * \text{mean}_o)$, and the standard deviation for the new transformed data is $(|c| * \text{std}_o)$. If one adds a fixed constant *k* to each data value of an existing data set, the mean for the new data set is $(k + \text{mean}_o)$, and the standard deviation for the new data set is unchanged and equal to std_o.

Exercise 2.17. Using Aaron's yearly AVGs in Example 2.5, compute the population mean and population standard deviation. Multiply each data point by 1000 and compute the population mean and population standard deviation for the new data. Show that $(1000 * \text{mean}_o)$ is equal to the mean of the transformed data and $(1000 * \text{std}_o)$ is equal to the standard deviation for the transformed data.

Exercise 2.18. Add 10 points to each yearly AVG for Aaron. Compute the mean and standard deviation of the new data set. How do these statistics compare to the mean and standard deviation of the original data set?

2.5. The Shape of the Data for One Quantitative Variable

Polygons, constructed from frequency tables (distributions), provide a shape for the data. The most common shapes are *symmetrical*, *right skewed* (positively skewed), and *left skewed* (negatively skewed). These three shapes are studied in this section. For each shape, the theoretical properties are presented. Since these properties are theoretical, it may be the case that, even though all the requirements for a particular shape are not met, we may subjectively conclude that the data fit a certain shape.

For each of these three shapes, a sample polygon will be displayed.

Since polygons are made up of line segments, their shapes consist of ragged edges. Mathematicians replace these polygons with theoretical models consisting of smooth curves. The most important example of this is the family of normal curves.

Later in this chapter, a section will be dedicated to the properties of this important class of curves.

2.5.1. Symmetric Distributions

If folding in half a piece of paper with the polygon graph centered on it results in a mirror image, the shape of the polygon is said to be symmetric. In applications, if the polygon for a data set is close to symmetric, we will say that the data set is symmetric.

Properties for Symmetric Data Sets

- The mean is equal to the median.
- The five-number summary (S, $Q1$, $Q2$, $Q3$, L) theoretically has the following properties:

  $Q2 - Q1 = Q3 - Q2,$
  $Q1 - S = L - Q3.$

- There may be a unique mode, no mode, or more than one mode.

A *bimodal symmetric polygon*, displayed below, describes a symmetric data set with two modes.

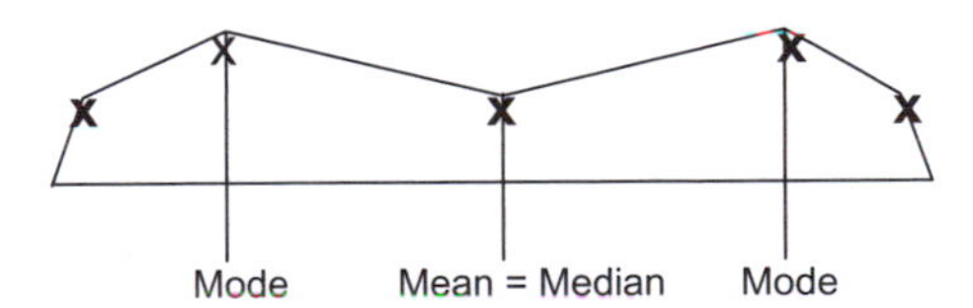

A *normal* or *bell-shaped symmetric polygon*, displayed below, describes a symmetric data set with a unique mode such that the mean = median = mode. Later in this chapter, an entire section will be devoted to a discussion of the properties of normal curves.

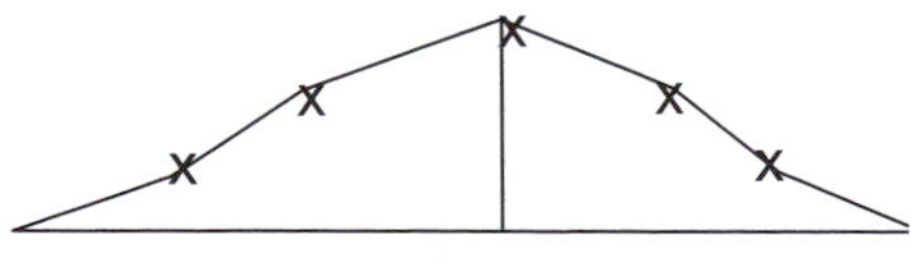

Can you give an example of a symmetric graph with no mode?

2.5.2. Right-Skewed Distributions

The polygon for this distribution rises quickly on the left side and falls slowly on the right side, producing a long tail to the right.

Properties for Right-Skewed Distributions

- The mean is greater than the median.
- The five-number summary (S, $Q1$, $Q2$, $Q3$, L) theoretically has the following properties:

$$Q3 - Q2 > Q2 - Q1,$$
$$L - Q3 > Q1 - S.$$

A distribution is right or positively skewed if most of the values tend to cluster at the lower end with a long tail of values to the right. The mean is pulled to the right of the median by the extreme values. A right-skewed distribution, along with its box-and-whisker graph, is shown below.

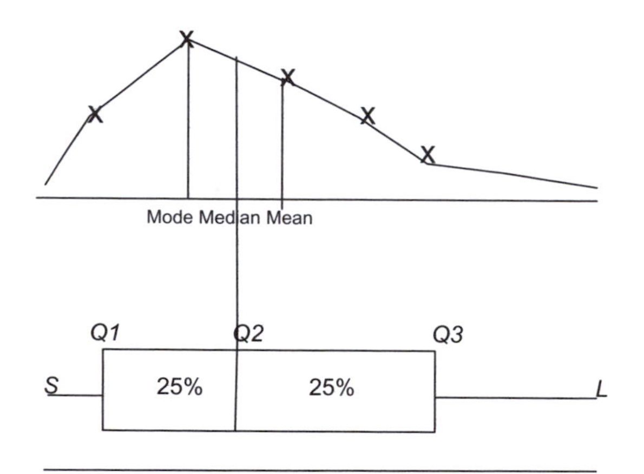

2.5.3. *Left-Skewed Distributions*

The polygon for this distribution has a long tail on the left rising slowly and then falling off sharply on the right side.

Properties of Left-Skewed Distribution

- The mean is less than the median.
- The five-number summary (S, $Q1$, $Q2$, $Q3$, L) theoretically has the following properties:

$$Q3 - Q2 < Q2 - Q1,$$
$$L - Q3 < Q1 - S.$$

A distribution is left or negatively skewed if most of the values tend to cluster at the upper end with a long tail of values to the left. The mean is pulled to the left of the median by the extreme values. A left-skewed distribution, along with its box-and-whisker graph, is shown on page 67.

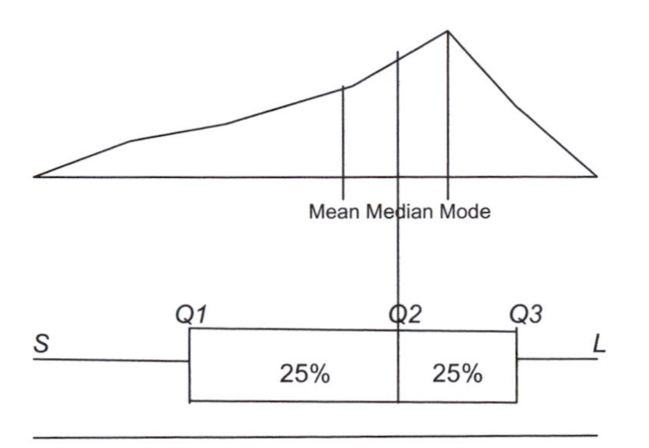

The definitions of symmetric, right-skewed, and left-skewed distributions are theoretical. Since the data are sample data ("real-world data"), we do not expect any of our data to perfectly fit all the theoretical conditions for a given shape. Therefore, we will just look for which shape (if any) closely fits our observed data.

2.6. Advantages and Disadvantages of the Different Graph Types

One of the decisions to be made in descriptive statistics is which graph or graphs to use to display the numeric data. We now explore the advantages and disadvantages of each graph type.

2.6.1. A Stem-and-Leaf Graph

Advantages

1. The individual data values are maintained.
2. Since all data values are displayed, the measures of the middle and spread can be calculated.
3. The shape of the data is displayed.

Disadvantages

1. It is cumbersome to use with a large amount of data.
2. It is harder to display multiple data sets.

2.6.2. A Polygon Graph

Advantages

1. Since the individual values are replaced by class intervals, these graphs can be used with any set of numeric data.
2. The shape of the data is displayed.
3. The measures of the middle and spread of the data can be estimated.

Disadvantages

1. The individual data values are lost.
2. The measures of the middle and spread can only be estimated.
3. It is harder to display multiple data sets.

2.6.3. A Box-and-Whisker Graph

Advantages

1. The five-number summary (S, $Q1$, $Q2$, $Q3$, L) is displayed.
2. This graph is as versatile as a "Swiss Army knife" in that it provides a measure for the middle of the data, a measure for the spread of the data, and a way to classify the shape of the data. The median is represented by the vertical line segment within the box. The interquartile range is the width of the box. The position of the median, $Q2$, in relation to the sides of the box, $Q1$ and $Q3$, allows an estimate for the shape of the data. If $Q2 - Q1 = Q3 - Q2$, a symmetry shape is indicated; if $Q2 - Q1 > Q3 - Q2$, a left-skewed shape is indicated; if $Q2 - Q1 < Q3 - Q2$, a right-skewed shape is indicated.
3. The important positions of $Q1$ and $Q3$ are displayed.
4. Outliers are displayed.
5. It is easy to display multiple data sets.

Disadvantages

1. The individual data values are lost.
2. The important descriptive measures of mean and variance are hard to estimate.

2.6.4. A Time-Series Graph

Advantages

1. This graph is the only graph that tracks the variable of interest with respect to time intervals.
2. The shape of the data with respect to time intervals is displayed.
3. The individual data values for the variable of interest are displayed as y-values on the graph.

Disadvantage

1. It can only be used when the x-values are intervals of time.

Exercise 2.19. Construct a stem-and-leaf graph, a box-and-whisker graph, and a time-series graph for Aaron's yearly doubles and yearly OBP.

Exercise 2.20. What shape would you assign to the samples of yearly doubles and yearly OBP for Aaron? What shape would you assign to the yearly runs scored for Aaron? What shape would you assign to the yearly triples for Aaron? In each case provide reasons.

2.7. The Properties of Normal Distributions

The most important symmetric distributions encountered in statistics are the family of normal distributions. The graph of any normal distribution is called a normal curve, sometimes called a "bell-shaped curve" or the "bell curve." The theoretical model for each member of the family of normal distributions is given by the following equation:

$$f(x;\mu,\sigma)=\frac{1}{\sigma\sqrt{2\pi}}e^{-\frac{(x-\mu)^2}{2\sigma^2}}.$$

The height for each x-value is $f(x;\ \mu,\ \sigma)$. The X-population of x-values is a continuous variable. The graph of $f(x;\ \mu,\ \sigma)$ is based on the mean (μ) and standard deviation (σ) for the underlying X-population. Since the mean and standard deviation are different for different normal populations, different normal populations have different normal curve graphs.

Properties for the Family of Normal Curves (Distributions)

1. The three measures of the middle for the X-population—the mean, the median, and the mode—are equal.
2. All normal curves have a bell shape with their inflection points (shape changes) occurring at $x=\mu+1\sigma$ and $x=\mu-1\sigma$.

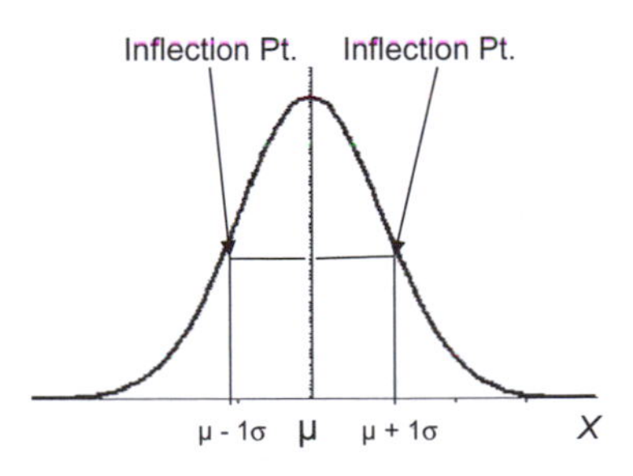

Any normal curve is symmetric around the mean of the X-population.

3. The "68-95-99.7 rule" applies to the data. Translated, 68% of the data lie in the interval $(\mu-1\sigma,\ \mu+1\sigma)$, 95% of the data lie in the interval $(\mu-2\sigma,\ \mu+2\sigma)$, and 99.7% of the data lie in the interval $(\mu-3\sigma,\ \mu+3\sigma)$.
4. For all normal distributions, the same percentage of data lie in the interval $(\mu-k\sigma,\ \mu+k\sigma)$, where $k>0$.
5. The peak (highest point) of any normal curve is located above the mean.

6. The larger the standard deviation is for the X-population, the flatter the normal curve.

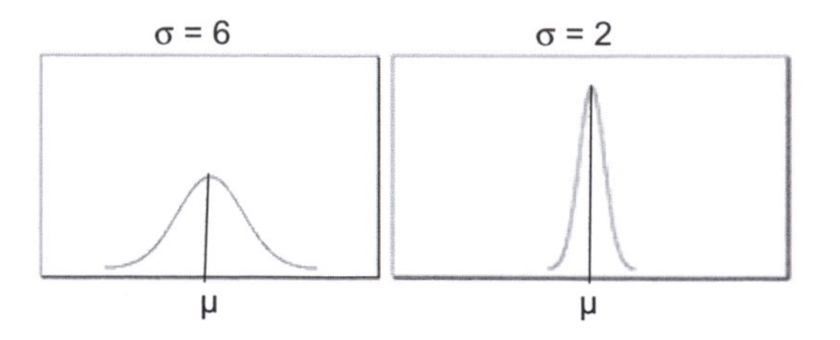

7. The total area of the region above the x-axis and below any normal curve is 1.
8. The cumulative polygon graph resembles an elongated S.

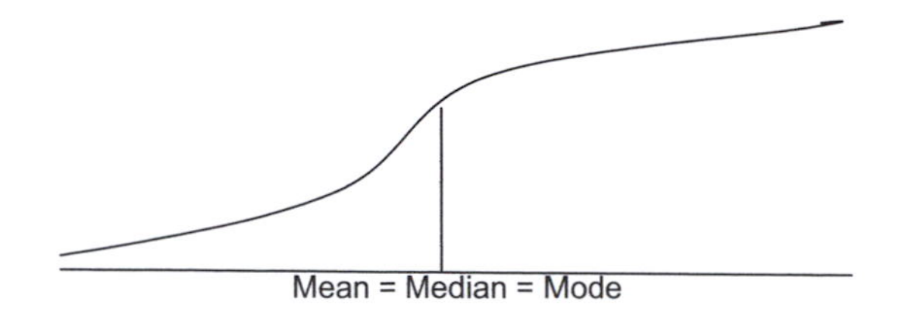

9. The ends of any normal curve flatten out and get closer to but never touch the x-axis. We say that the curve is asymptotic to the x-axis.
10. The range of the data set for a normal X-population is equal to approximately 6σ.
11. The special normal curve with $\mu=0$ and $\sigma=1$ is called the standard normal curve. For this normal curve the symbol for the variable is Z. All other normal curves use the symbol X for the X-variable.

2.7.1. Deciding When a Sample Data Set Comes from a Normal Population

Since the normal (bell) shape is a common shape for population data, it is not unreasonable to assume that sample data are collected from a normal population. Clearly, many populations are not normal. To continue with the assumption of normality for the population, the sample data should exhibit five properties. Since we are only dealing with a small sample from a population, these properties serve as guidelines only. For example, a theoretical normal curve has the property that the mean, median, and mode are all equal. However, one only needs to show for the sample that they are close to one another. Also, the mode for a small sample may not be relevant.

Five Properties of a Theoretical Normal Curve That Should Be Exhibited by the Sample

1. The mean and median for the sample should be close to each other.
2. The shape exhibited by the polygon, histogram, and stem-and-leaf graph for the sample data is close to a bell shape.
3. The interval ($\bar{x}-1s$, $\bar{x}+1s$) contains approximately 68% of the sample data. The interval ($\bar{x}-2s$, $\bar{x}+2s$) contains approximately 95% of the sample data. The interval ($\bar{x}-3s$, $\bar{x}+3s$) contains approximately 99.7% of the sample data. The range of the data is close to 6 times the standard deviation.
4. The box-and-whisker graph for the sample data resembles the figure below.

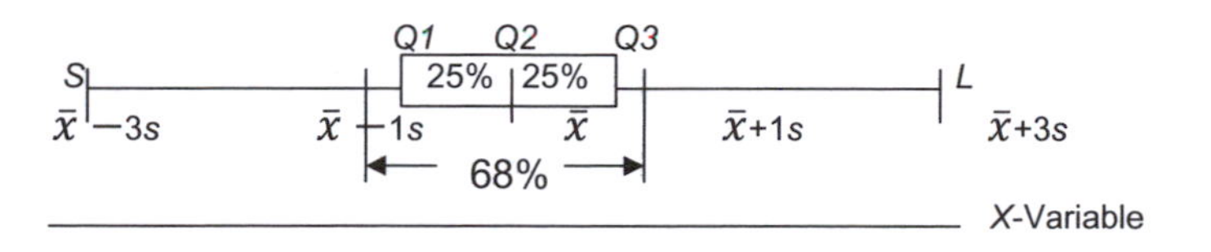

5. The shape of the cumulative polygon for the sample data is close to an elongated *S*.

Yearly batting averages for better hitters, those who have played several years in the majors, might well take a normal shape. The reason for this is age. Michael Schell states, "for the best hitters, hitting ability improves rapidly from age 20–22, climbs more slowly at ages 23–27, peaks at age 28, declines 3–4 points per year to age 32, and then remains relatively stable until age 35. The average declines like a rocket at age 38. A further 19-point drop occurs the next two years."* Aaron's yearly batting averages, treated as a sample, are examined for normality in Example 2.17 below. Even if a sample is chosen from a normal population, a small sample size may cause a sample to deviate from some of the normal properties. Therefore, for smaller sample sizes, a more liberal interpretation of these properties is necessary. In the next example, we demonstrate the first four properties above for the yearly AVGs of Aaron.

Example 2.17. Before looking at these properties, each yearly batting average is multiplied by 1000. A batting average of *.xyz* will become the integer *xyz*. Transforming the data into integers makes it easier to handle.

As discussed earlier, multiplying each data point by 1000 establishes a new mean equal to 1000 times the old mean and a new standard deviation equal

* Michael Schell, *Baseball's All-Time Best Hitters* (Princeton, NJ: Princeton University Press, 1999), 33.

Table 2.7 **Frequency distribution for Aaron's yearly batting averages (class intervals: width = 20)**

Class intervals	Midpoint	f(midpoint)	Frequency	Proportion	Proportion/20
209–228	219	0.0004	0	0.0000	0.0000
229–248	239	0.0018	2	0.0870	0.0043
249–268	259	0.0052	2	0.0870	0.0043
269–288	279	0.0100	3	0.1304	0.0065
289–308	299	0.0128	5	0.2174	0.0109
309–328	319	0.0108	10	0.4348	0.0217
329–348	339	0.0061	0	0.0000	0.0000
349–368	359	0.0023	1	0.0435	0.0022
369–388	379	0.0006	0	0.0000	0.0000
Sums			23	1.0000	0.0500

to 1000 times the old standard deviation. The converted yearly batting averages for Aaron are 229, 234, 265, 268, 279, 280, 287, 292, 298, 300, 301, 307, 314, 318, 319, 322, 323, 326, 327, 327, 328, 328, 355. The sample standard deviation $s = 31.16$. A frequency distribution is given in Table 2.7.

The First Four Properties of a Sample Taken from a Normal Population

1. The sample mean = 301.17, the sample median = 307, and the sample mode = 327, 328. Since the mean and median are relatively close to each other, symmetry is indicated.
2. Two polygons are shown in Figure 2.11. One is the *observed* polygon; the other is called the *theoretical* polygon. Both polygons are constructed using Table 2.7, which provides a frequency distribution for Aaron's yearly batting averages with class intervals having width equal to 20.

The observed polygon is constructed using for the x-values the midpoint of each class interval and for the y-values the adjusted proportion (adj. proportion) for each class interval. To calculate the adj. proportion, the sixth column in Table 2.7, multiply the proportion (the fifth column in Table 2.7) for each class interval by (1/20). Multiplying by (1/20) forces the sum of the areas of the rectangles to equal 1. An area equal to 1 is necessary to compare the observed polygon with a theoretical normal curve (Section 2.7 properties). The observed polygon appears left skewed.

The second polygon is constructed using the same x-values as in the observed polygon with the y-value for each x-value found by substituting the value x into the equation for $f(x; \mu, \sigma)$.The function $f(x; \mu, \sigma)$, given below, is the theoretical equation for a normal distribution with population

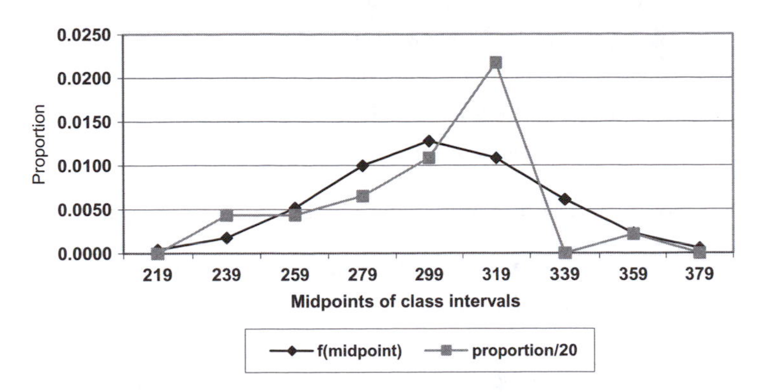

Figure 2.11 Observed and theoretical polygons for Aaron's yearly AVGs

mean equal to μ and population standard deviation equal to σ. Since the population is not known, the sample mean of 301 is used for μ and the sample standard deviation of 31.16 is used for σ. The form of the general normal distribution function is given by

$$f(x;\mu,\sigma)=\frac{1}{\sigma\sqrt{2\pi}}e^{-\frac{(x-\mu)^2}{2\sigma^2}}.$$

The polygon obtained by calculating $f(x; 301, 31.16)$ for the x-values (219, 239, 259, 279, 299, 319, 339, 359, 379) is called the polygon for the theoretical normal curve. These values are shown in the third column of Table 2.7. One can see the bell shape for the theoretical polygon.

Displaying both polygons in the same graph lets us see if the observed polygon is close in shape to the theoretical polygon.

3. Using $s=31.16$ and $\bar{x}=301$, the three intervals are as follows:

   - Interval 1 = $(\bar{x}-1s, \bar{x}+1s)=(301-31.16, 301+31.16)=(269.84, 332.16)$;
   - Interval 2 = $(\bar{x}-2s, \bar{x}+2s)=(301-62.32, 301+62.32)=(238.68, 363.32)$;
   - Interval 3 = $(\bar{x}-3s, \bar{x}+3s)=(301-93.48, 301+93.48)=(207.52, 394.48)$.

   Interval 1 contains 18 data values = 78.26% of the data. Interval 2 contains 21 data values = 91.30% of the data. Interval 3 contains 23 data values = 100% of the data.

4. The five-number summary for the sample is (229, 283.5, 307, 324.5, 355). The box-and-whisker graph for Aaron's yearly AVGs is shown below.

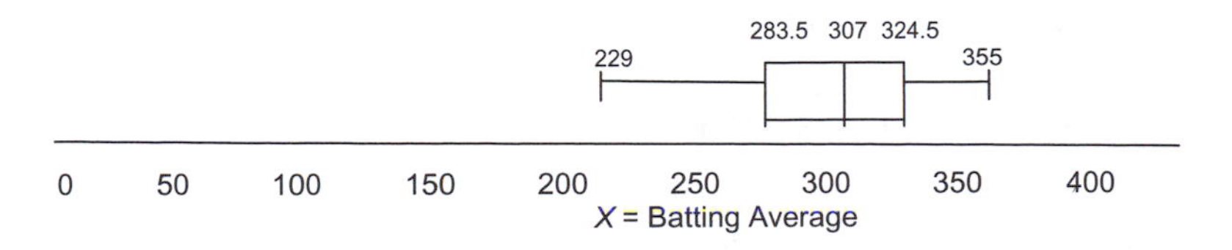

   The box-and-whisker graph indicates that the AVGs are slightly left skewed.

Looking at these four properties, it appears that the sample data are slightly left skewed. However, since the sample size (23) is small, one can argue that there is not enough evidence to conclude that the underlying population for this sample is not normal.

Exercise 2.21. For the example above verify the following facts:

a. Show that the sample standard deviation $s = 31.16$.
b. Show that the five-number summary is (229, 283.5, 307, 324.5, 355). Check for outliers.
c. Show that the computations in the third column of Table 2.6 are correct.
d. Construct the cumulative frequency table.
e. Construct the cumulative frequency polygon.
f. Does the shape of the cumulative frequency polygon resemble an elongated *S*?
g. Verify that the sum of the areas of the rectangles for the observed polygon is equal to 1.

Exercise 2.22. Can we assume that the samples of yearly doubles and yearly OBP for Aaron both come from normal populations? Use the five checks described in this section.

2.7.2. Why Is the Normal Distribution Important?

The normal distribution is important because it is the most common shape for population data. Many quantitative characteristics in the real world are normally distributed, such as the heights of American women or the annual rainfall at any specific location in the United States (the rainiest city home to a Major League baseball team is Miami).

If the data are normally distributed, knowing the mean and standard deviation of the data allows one to assign a percentile to any data value. (A percentile associated with a data value gives the percentage of data values less than the given data value.)

Knowing that a population is normally distributed allows one to calculate probabilities. If the yearly batting averages for a player are normally distributed, we could estimate the probability that for a given year he has a batting average of at least .300.

Many important inferential statistical techniques require that the underlying population is approximately normal.

2.7.3. Standard Units (z-Scores) for a Normal Population

We are often interested in comparing two values or scores between two subjects. For example, two students in two different classes take an exam. One student scores 88; the other scores 80. Which student did better? At first glance one might say that the student who scored 88 did better. What if we found out later that the student who scored 80 had the highest score in his class while the 88 was a low score in the other class? Would you still say that the 88 was better? The answer is no. The same situation can occur in baseball. For example, in 1961 Henry Aaron's yearly batting average at age 27 was .327, whereas in 1992 Barry Bonds's yearly batting average at age 27 was .311. Ignoring all chance factors and other variables such as their home stadiums, can one conclude, using only their respective batting averages, that Aaron at age 27 was a better hitter than Bonds at age 27? Just like the two test scores mentioned above, it is necessary to compare each batting average with its underlying population of batting averages.

A method of comparing each score with its population of all scores is described next. Each score is converted into a standard unit by subtracting the population mean from the score and then dividing the resulting number by the population standard deviation. The resulting number is called the standard unit or z-score. Symbolically the formula is

$$z\text{-score} = (x\text{-value} - \text{mean})/\text{standard deviation}. \quad (2.1)$$

The z-score for a data value tells you by how many standard deviations the data value is above or below the population mean. If the population is approximately normal, the z-score can be converted into a percentile. A *percentile P* is a number such that $0 < P < 100$, where P% of the data values are less than the z-score. (The method for doing this will be shown later in the book.)

For now, accept the fact that a larger z-score means a higher percentile. For example, a z-score of 2.00 corresponds to a 97.5 percentile, and a z-score of 0.75 corresponds to a 77 percentile. Remember, a 97.5 percentile means that the z-score, and thus the original score, was higher than 97.5% of the data values of the sample or population. The score corresponding to the larger standard unit will be the better score. A negative standard score indicates a score below the mean.

Additional information concerning the two classes that took a certain exam is given below.

Student Score	Class Mean	Class Standard Deviation	z-score
88	82	8	$(88 - 82)/08 = 0.75$
80	60	10	$(80 - 60)/10 = 2.0$

The scores of both classes are assumed normally distributed. The student who scored 88 actually scored .75 standard deviations above the class mean, whereas the student who scored 80 scored 2 standard deviations above the class mean. Consequently, the student with the score of 80 did better *relative to his class* than the student who scored 88. The student with the original score of 80 scored higher than 97.5% of the class, whereas the student with the original score of 88 scored higher than 77% of the class.

If we can assume that the two data sets are approximately normal, we can use z-scores to compare certain offensive batting categories between players.

This process can be applied to many baseball statistics. Assuming that the underlying population is normal, a player's baseball statistic is converted to a z-score. The z-score is then converted to a percentile. When comparing two players, the player with the higher percentile performed better relative to the players in his league for that year.

This process is used, in the next example, to compare Aaron's batting average for 1961 with Bonds's batting average for 1992.

Example 2.18. Using the batting averages for all National League players with at least 200 at bats for the years 1961 and 1992 from the website www.baseball-reference.com, we can complete Table 2.8. Again, the assumption is made that the shape of all National League (NL) batting averages (AVGs) for the years 1961 and 1992 is approximately normal.

Aaron's z-score of 1.82 translates into a 96.56 percentile, whereas Bonds's z-score of 1.58 translates into a 94.29 percentile. Aaron's batting average was higher than approximately 96% of the National League players in 1961. Bonds's batting average was higher than approximately 94% of the National League players in 1992. Because of the chance factors in batting and the closeness of their respective percentiles, one can conclude that relative to their populations there is either no difference or a slight edge to Aaron.

The disadvantage of this method is that the comparison is done through percentiles and not batting averages.

Since people are more at ease comparing actual batting averages, we can adjust this process to translate a batting average for a player from one year to what it would be in another year. Again, the following procedure depends on the assumption of normality for both populations. Using Formula (2.1), solving for the x-value in terms of the z-value gives the following formula:

Table 2.8 Calculation of z-scores for the batting averages for Aaron (1961) and Bonds (1992)

				Batting AVGs of NL players with 200+ ABs		
Year	Age	Player	AVG	Mean	Standard deviation	z-score
1961	27	Aaron	.327	.271	.0307	(.327 − .271)/.0307 = 1.82
1992	27	Bonds	.311	.262	.0311	(.311 − .262).0311 = 1.58

x-value = mean + [(z-score) * (standard deviation)]. (2.2)

We use Formula (2.2) to convert Aaron's batting average in 1961 to what it would be in 1992:

x-value = .262 + (1.82 * .0311) = .318602 (Aaron's AVG converted to 1992).

We use Formula (2.2) to convert the batting average of Bonds in 1992 to what it would be in 1961:

x-value = .271 + (1.58 * .0307) = .319506 (Bonds's AVG converted to 1961).

The .318692, rounded to .319, gives the batting average for Aaron in 1992 that would rank him in the same percentile as his .327 did in 1961. The .319506, rounded to .320, gives the batting average for Bonds in 1961 that would rank him in the same percentile as his .311 did in 1992. Aaron's converted .319 for 1992 is only 8 points higher than Bonds's .311 for that year. Bonds's converted .320 for 1961 is 7 points below Aaron's average of .327 in 1961. We observe a slight edge for Aaron over Bonds with respect to batting average.

The conversion of a player's batting average for a given year to what it would be in another year works because the same z-score gives the same percentile for any normal curve. This process takes a player's batting average for a given year and converts it into its z-score. This z-score is then converted into a batting average for a different year.

The next chapter will provide a different method for adjusting baseball statistics. This method will not require that the X-population is approximately normal. These adjustments will take into effect the year of the league and the home ballpark.

Exercise 2.23. Use the above process to compare the home run average (HRA) for Aaron in 1961 with the home run average for Bonds in 1992. The HRA is the number of home runs divided by the number of at bats. The data needed for the calculations in Table 2.9 were acquired from the website www.baseball-reference.com.

Table 2.9 **Calculation of z-scores for the home run averages for Aaron (1961) and Bonds (1992)**

				HRAs of NL players with 200+ Abs		
Year	Age	Player	HRA	Mean	Standard deviation	z-score
1961	27	Aaron	.056	.030	.0204	?
1992	27	Bonds	.072	.021	.0145	?

Use Formula (2.2) and the z-score for each player to convert each player's HRA for the given year to what it would be in the other year. Use these converted numbers to compare HRA between the two players.

2.8. Standard Units for Data Sets, Which Are Not Normally Distributed

The shape of the data set does not change the meaning of a z-score. A converted z-score, for an x-value in the data set, is the number of standard deviations the x-value is above or below the mean. If the underlying shape of the data set is close to normal, a percentile can be attached to the data value. What if the underlying data set is not close to normal? The best we can do is the following inequality provided by the Russian mathematician Pafnuty Chebyshev:

> Chebyshev's Inequality states that the proportion of observations falling within C standard deviations of the mean must be at least $[1 - (1/C^2)]$, for any positive number C greater than 1.

The "at least" part of the rule means that the proportion we calculate will be the minimum proportion falling in the prescribed interval. It is possible that a higher proportion will fall in the interval.

According to Chebyshev's Inequality for $C = 2$, at least 75% of the data values lie within 2 standard deviations of the mean. Knowing that the data set is normal gives the more powerful result of expecting 95% of the data values to fall within 2 standard deviations of the mean. The next exercise allows you to practice using Chebyshev's Inequality for various C-values.

Exercise 2.24. Applying Chebyshev's Inequality to the data set of Aaron's yearly batting averages for $C = 1.1$, $C = 2$, and $C = 2.2$, calculate the proportion of data values that fall within C standard deviations of the mean. Is the inequality satisfied in each case?

Chapter Summary

A sample research problem is presented (to analyze the data consisting of the yearly triples for Henry Aaron). A step-by-step procedure was presented to analyze the research problem.

The batting data for Aaron were *collected* from the website www.baseball-almanac.com and displayed in Table 2.2. The variable or characteristic to be studied is the number of triples for each year of Aaron's career. The step-by-step approach given in the chapter is reviewed. These steps can be used for any quantitative data set.

Step 1: Check for Outliers. Data point 14 was found to be an outlier, but there was no reason to remove this number from the data. Conclusion: All 23 data values should be used in the analysis.

Step 2: Find Summary Measures for Important Positions in the Data.

- The mean = 98/23 = 4.26.
- The median = $Q2$ = 3.
- The mode = 1.
- The first quartile $Q1$ = 1.
- The third quartile $Q3$ = 6.
- The largest value = 14.
- The smallest value = 0.

Step 3: Find Important Measures of Spread for the Data.

- The range = 14.
- The sample variance = 14.75.
- The sample standard deviation = 3.84.
- The interquartile range = 6 − 1 = 5.

Step 4: Organize Data into a Complete Frequency Table. The first column can be either the actual outcomes or class intervals. Since the numbers corresponding to yearly triples are integers close to each other, the actual outcomes can be used in the first column. Table 2.3 gives the complete frequency table for the yearly triples.

Step 5: Determine Shape of the Distribution. Figures 2.1 and 2.3 show that the shape of the frequency polygon and frequency histogram is right skewed. The stem-and-leaf graph in Figure 2.8 also shows that the shape is right skewed. Finally, the box-and-whisker graph (Fig. Q2.1) confirms that the shape is right skewed. The fact that the mean is greater than the median also supports the conclusion of right skewed.

Step 6: Possible Explanation for Yearly Triples Being Right Skewed in Shape. In the beginning years of Aaron's career, he was a very fast runner, which probably allowed him to extend doubles into triples. As he aged and bulked up, he slowed down and was less inclined to stretch a double into a triple. The conjecture is that age was the principal factor in determining his yearly triples.

Step 7: Analyze Triples as a Function of Age through a Time-Series Graph. Figure 2.9 shows the time-series graph for yearly triples. This graph clearly shows that as Aaron got older, his triple output diminished.

Step 8: Find Number to Use for Aaron's Typical Number of Yearly Triples. Theory says that since the shape of the distribution is right skewed, the median should be used instead of the mean as the measure of the middle. If this rule was followed, the number 3 would be used. However, owing to the time-series analysis, it might be better to divide the 23 years into two sets; the first set would consist of his first 10 years, while the second set would consist of his final 13 years. In the first 10 years his total was 77 triples, giving a mean of 7.7 triples per year. For the final 13 years of his career, his total was 21 triples, for a mean of 1.62 triples per year. An argument that can be given is that for the first 10 years his typical triple output was 7.7, whereas for the final 13 years his typical triple output was 1.6.

Step 9: Final Conclusions. Henry Aaron was a fast runner at the beginning of his career. His running speed diminished as he got older. This might account for his reduced number of triples later in his career. There may be other explanations, such as that the outfielders played deeper as he got older, or he did not try to extend doubles because he feared injuries to his legs. Another explanation could be the change in his home stadium from Milwaukee to Atlanta. His first 12 years were spent in Milwaukee. In Step 8, even though theory says that the median should be used as the measure of the middle, the time-series analysis suggests that we break up the 23 years into two sets. The number 3 just does not work. The conclusion of using the sample mean of 7.7 for the first 10 years and the sample mean of 1.6 for the remaining 13 years makes much more sense. There are times in statistics where one must go *outside the lines*.

This analysis, based on descriptive statistics, is often done by general managers before drafting a player. A player's high school, college, and Minor League data are collected. Many different batting statistics are evaluated. Graphs are constructed from these statistics. Different general managers emphasize different statistics.

Billy Beane became the assistant general manager of the Oakland Athletics in 1994 and the general manager in 1998. He succeeded Sandy Alderson, who began the practice of using statistical techniques to evaluate undervalued players. Beane learned his craft well. Having a limited budget, it was necessary for Beane to evaluate the less expensive players. Throughout his tenure as the GM of the Athletics, he crafted the Athletics into one of the most cost-effective teams in baseball.

As a result of his team's success, despite its low payroll, Beane was the subject of author Michael Lewis's 2003 best-selling book *Moneyball*.* This book provided the statistical and mathematical models Beane used to run

* Michael Lewis, *Moneyball* (New York: W. W. Norton, 2003).

his team in a cost-effective way. According to the book, these techniques allowed Beane to be successful despite his financial constraints. Lewis's book, which displayed Beane's methods, has influenced the way other general managers evaluate players.

CHAPTER PROBLEMS

Chapter 16 studies important streaks in baseball, such as Joe DiMaggio's 56-game hitting streak.

Questions 1–14 are based on the data from Table 2.10, acquired from the website www.baseball-almanac.com. The sample consists of the players who have hit safely in at least 30 consecutive games. Willie Keeler and Jimmy Rollins have two years listed in the year column. For Keeler use 1896 for his year, and for Rollins use 2005 for his year. The reason for the listing of two years is that their streaks extended over two years. For this example, a streak extending over two years will be allowed.

From this table, answer the following questions:

1. List all the variables (characteristics) for this sample.
2. For the variable representing number of games in the hitting streak, construct a complete frequency table. (Use class intervals.)

The variable representing number of games in the streak will be used for the remaining questions.

3. Graph a frequency histogram and polygon. (Use the class intervals in problem 2.)
4. Graph a percentage histogram and polygon. (Use the class intervals in problem 2.)
5. Graph a cumulative percentage polygon. (Use the class intervals in problem 2.)
6. Construct a stem-and-leaf graph.
7. Calculate the five-number summary.
8. Construct a time-series graph. (Time periods are needed.)
9. Construct the box-and-whisker graph.
10. Use the logic presented in this chapter to assign a shape to the data values for the variable.
11. Compute the mean, median, and mode for the values of the variable.
12. Compute the range, the interquartile range, the sample variance, the sample standard deviation, and the sample coefficient of variation for the values of the variable.
13. Using the five checks mentioned in this chapter, decide if the data values for the variable can be assumed to come from a normal population.

Table 2.10 **Players who hit safely in at least 30 consecutive games**

Rank	Year	Name	Team	League	Games
1	1941	Joe DiMaggio (AL Record)	New York	AL	56
2	1896/1897	Willie Keeler (NL Record)	Baltimore	NL	45
3	1978	Pete Rose	Cincinnati	NL	44
4	1894	Bill Dahlen	Chicago	NL	42
5	1922	George Sisler	St. Louis	AL	41
6	1911	Ty Cobb	Detroit	AL	40
7	1987	Paul Molitor	Milwaukee	AL	39
8	2005/2006	Jimmy Rollins	Philadelphia	NL	38
9	1945	Tommy Holmes	Boston	NL	37
10	1895	Fred Clarke	Louisville	NL	35
	1917	Ty Cobb	Detroit	AL	35
	2002	Luis Castillo	Florida	NL	35
	2006	Chase Utley	Philadelphia	NL	35
14	1925	George Sisler	St. Louis	AL	34
	1938	George McQuinn	St. Louis	AL	34
	1949	Dom DiMaggio	Boston	AL	34
	1987	Benito Santiago	San Diego	NL	34
18	1893	George Davis	New York	NL	33
	1907	Hal Chase	New York	AL	33
	1922	Rogers Hornsby	St. Louis	NL	33
	1933	Heinie Manush	Washington	AL	33
22	1899	Ed Delahanty	Philadelphia	NL	31
	1906	Nap Lajoie	Cleveland	AL	31
	1924	Sam Rice	Washington	AL	31
	1969	Willie Davis	Los Angeles	NL	31
	1970	Rico Carty	Atlanta	NL	31
	1980	Ken Landreaux	Minnesota	AL	31
	1999	Vladimir Guerrero	Montreal	NL	31
29	1876	Cal McVey	Chicago	NL	30
	1898	Elmer Smith	Cincinnati	NL	30
	1912	Tris Speaker	Boston	AL	30
	1934	Goose Goslin	Detroit	AL	30
	1950	Stan Musial	St. Louis	NL	30
	1976	Ron LeFlore	Detroit	AL	30
	1980	George Brett	Kansas City	AL	30
	1989	Jerome Walton	Chicago	NL	30
	1997	Sandy Alomar, Jr.	Cleveland	AL	30
	1997	Nomar Garciaparra	Boston	AL	30
	1998	Eric Davis	Baltimore	AL	30
	1999	Luis Gonzalez	Arizona	NL	30
	2003	Albert Pujols	St. Louis	NL	30
	2006	Willy Taveras	Houston	NL	30
	2007	Moises Alou	New York	NL	30

In Chapter 17, the topic discussed is, will we ever have another .400 hitter for a season? Since Ted Williams batted .406 in 1941, no player has batted .400 in the Major Leagues. For the next set of problems, we will look at the four players who came closest to batting .400 since 1941. They are Ted Williams (.388 in 1957), Rod Carew (.388 in 1977), George Brett (.390 in 1980), and Tony Gwynn (.394 in 1994). Notice that all four were left-handed hitters.

For each of these players, using the year their batting average was close to .400, the sample will be all the games they appeared in during that year. The variable will be their batting average at the conclusion of each game in which they appeared in. For each player, use the daily splits from the website www.retrosheet.org to answer the following questions:

14. Find the number of the last game in which they were still batting at least .400 at the end of the game.
15. For each player, assuming that his at bats stayed the same, how many more hits did they need to reach .400?
16. For each player, construct a time-series graph with the game number as the *X*-variable and the batting average at the end of the game as the *Y*-variable. The four time series should be on one graph.
17. Compare the four time-series graphs.
18. Find the number of the game in which the range of their four batting averages is smallest and largest.

Chapter 3

Descriptive Measures Used in Baseball

When people talk baseball, they use descriptive measures to evaluate a baseball player's current, past, and future performance. Descriptive measures are also used to compare two or more players in an attempt to conclude which player was a better hitter, which player was more valuable to his team, or which player should be inducted into Baseball's Hall of Fame.

This chapter provides definitions, symbols, and a brief timeline for the introduction of many baseball statistics. These baseball statistics are compared with the standard statistics of mean, proportion, ratio, and rate introduced in Section 1.8.

Before studying the new topics introduced in this chapter, the concepts introduced in Chapter 1 are reviewed. As mentioned in Chapter 1, a *variable* is a characteristic applied to each member of a group. A variable is either *quantitative* or *qualitative*. Quantitative variables are further classified as either *discrete* or *continuous*. The *measurement scale* for a variable provides a way of comparing the values of the variable. A *statistic* is a descriptive measure for a *sample*; a *parameter* is a descriptive measure for a *population*. A statistic can also be a variable. For example, a player's batting average is a statistic, since it describes a player's batting performance. However, if we are interested in studying the batting averages of each player on a team, a batting average is a variable for the players on the team. Since this book deals only with batting statistics, no pitching or fielding statistics will be studied.

3.1. A Timeline for Baseball Statistics

The early player statistics, in the 1850s, concentrated on counting the number of runs, outs, or hits. In the 1860s, ratios were created by dividing the counts by the number of games. This led to the number of runs per game and the number of hits per game. Hits per game were used as the criteria for crowning the batting champion until 1876. The inequity (better teams have more at bats in a game and so provide their players with more opportunities of getting more hits per game) was clear.

In 1876, the National League was created and a new statistic, the batting average, was introduced, where the number of hits was divided by the number of at bats. During that year, the batting statistics calculated for a player included number of games, number of runs, number of hits, number of at bats, number of runs per game, and batting average. That made a total of six statistics.

Other batting statistics introduced before 1900 for a player included the number of stolen bases, sacrifice hits, doubles, triples, home runs, strikeouts, base on balls, times hit by pitch, and times grounded into a double play.

In 1920, the runs batted in (RBI) statistic was adopted. Three years later the slugging average statistic was introduced. On-base percentage and isolated power were introduced in the 1950s. In the 1970s, the statistics of on-base plus slugging, runs created, and the batter's run average made their appearance.

These statistics are a partial list of the important baseball statistics used today.

3.2. Baseball Variables and Statistics

Statistics are descriptive measures for sample data. The two most used descriptive measures are *mean* and *proportion*. The *mean* describes the middle for *quantitative* data, whereas the *proportion* is used to find out what is typical for *qualitative* data. One major source of data used in this book comes from players' plate appearances. Such data can be qualitative or quantitative. A list of every outcome for every plate appearance for a player for an entire season is qualitative data, whereas the number of hits for a player in each game he played for an entire season would be quantitative data. Quantitative data come from quantitative variables. Qualitative data come from qualitative variables.

Since the outcomes of a variable are a set of subjects or objects, it is important to review sets and operations on sets.

3.2.1. Brief Introduction to Sets

The list of all possible outcomes of a variable is called a set. The members of the list are enclosed within braces. A pair of braces { } is read as "the set consisting of those outcomes listed between the two braces." The possible outcomes for the qualitative variable "hit" are single, double, triple, and home run. We express this relationship as H = {single, double, triple, home run}. We say that H is the set consisting of the outcomes single, double, triple, and home run. The symbol { } refers to the *empty set*. The empty set is the set consisting of no outcomes. An uppercase letter is usually used to name a set. For the set "hit" we can use "H" to name the set.

The members of a set are called elements or outcomes. In listing the elements of a set, two rules exist: the order in which the elements are listed is not important, and the same element is never repeated.

Many of the sets considered in this book come from samples. Some examples of sets studied in this book are a sample of at bats for a player, a sample of years for a league, a sample of teams, a sample of games, and so on.

3.2.2. Operations on Sets

The "+" symbol used between two sets represents the addition of two sets. The addition of two sets is a new set that consists of all outcomes that belong to either one of the two sets or both sets. Example 3.1 demonstrates the addition of two sets.

Example 3.1. Let

$A=\{1, 2, 3, 4\}$,
$B=\{1, 3, 5, 7, 9\}$,
$C=A+B=\{1, 2, 3, 4\}+\{1, 3, 5, 7, 9\}=\{1, 2, 3, 4, 5, 7, 9\}$.

Notice that elements 1 and 3 are only listed once in set *C*.

The "–" symbol represents the subtraction of one set from another set. The subtraction of a second set from a first set is a new set that consists of all outcomes that are in the first set but not in the second set. Example 3.2 demonstrates the subtraction of two sets.

Example 3.2. Let *A* and *B* be the sets from Example 3.1.

$C=A-B=\{1, 2, 3, 4\}-\{1, 3, 5, 7, 9\}=\{2, 4\}$.

Table 3.1 is arranged in three columns. The first column gives the name of the variable, the second column lists the set of possible outcomes of the variable, and the third column gives the symbol used for the qualitative variable and for the set of possible outcomes.

Example 3.3. We demonstrate the addition and subtraction of sets.

OB = H + {BB, HBP} = {1B, 2B, 3B, HR, BB, HBP}.
IPAB = AB – {SO} = O + H – {SO} = {FC, FO, GO, GDP, 1B, 2B, 3B, HR}.
AB = O + H + {SE} = {FC, FO, GO, GDP, SO, 1B, 2B, 3B, HR, SE}.
PA = AB + {BB, HBP, SF} = {FC, FO, GO, GDP, SO, 1B, 2B, 3B, HR, SE, BB, HBP, SF}.
Extra-base hit = H – {1B} = {2B, 3B, HR}.

Table 3.1 **Baseball's qualitative variables**

Variable	Outcomes	Symbol
Hit	{1B, 2B, 3B, HR}	H
	Single	1B
	Double	2B
	Triple	3B
	Home run	HR
Out	{FC, FO, GO, GIDP, SO}	O
	Fielder's choice	FC
	Fly out	FO
	Ground out	GO
	Ground into double play	GIDP or GDP
	Strikeout	SO or K
In-play out	O – {SO}	IPO
Extended out	O + {SF, SH}	EO
	Out	O
	Sacrifice fly	SF
	Sacrifice hit or bunt	SH
At bat	O + H + {SE}	AB
	Out	O
	Hit	H
	Safe on error	SE
On base	H + {BB, HBP}	OB
	Hit	H
	Base on balls	BB
	Hit by pitch	HBP
Plate appearance (official)	AB + {BB, HBP, SF}	PA
	Base on balls	BB
	Hit by pitch	HBP
	Sacrifice fly	SF
In-play at bat	AB – {SO}	IPAB
Extra-base hit	H – {1B}	XBH

Note: A home run is considered an in-play at bat.

Since catcher's interference, safe on a dropped third strike, and defensive interference are rare outcomes, they will not be considered in this book.

The difference between an out (O) and an extended out (EO) is its interpretation in scoring for the record books. Any member of O counts as a time at bat, whereas even though an SH and SF result in a player making an out, they do not count as an at bat in baseball scoring.

There is a difference between the intent of a batter in the outcomes SH and SF. A sacrifice hit (SH) is when a player intentionally makes an out for the purpose of advancing a runner; a sacrifice fly (SF) results in a runner scoring from third base on a fly out but the out is not intentional. For this reason, an SF is counted as a PA but not as an AB, whereas an SH will count as neither a PA nor an AB. However, for the purpose of qualifying for a batting title, a sacrifice hit is counted in the total plate appearances.

It should be noted that a safe on error (SE) is scored as an out (O). So many websites consider an at bat (AB) to be equal to H + O. We will follow this convention also.

The placing of the number symbol (#) in front of a symbol in column 3 of Table 3.1 represents the number of times the outcome occurs. For instance, #H is the number of hits, #1B is the number of singles, and so on. When #(outcome) is calculated for each subject or object of a group, #(outcome) becomes a discrete quantitative variable. Most websites just use the symbol without the # in front. For example, H is used for the number of hits. In this book, many times we will also omit the symbol #.

Table 3.2 **Baseball's discrete quantitative variables**

Discrete quantitative variables	Symbol
Number of games played	#G
Number of runs scored	#R
Number of home runs	#HR
Number of hits	#H
Number of singles	#1B
Number of doubles	#2B
Number of triples	#3B
Number of runs batted in	#RBI
Number of base on balls	#BB
Number of sacrifice hits	#SH
Number of sacrifice flies	#SF
Number of hit by pitch	#HBP
Number of strikeouts	#SO
Number of safe on error	#SE
Number of on base	#OB = #H + #BB + #HBP
Number of at bats	#AB = #O + #H + #SE
Number of plate appearances	#PA = #AB + #BB + #HBP + #SF
Number of outs	#O = #AB − #H − #SE
Number of in-play at bats	#IPAB = #AB − #SO
Number of extra-base hits	#XBH = #2B + #3B + #HR
Number of stolen bases	#SB
Number of caught stealing	#CS

Each quantitative variable in Table 3.2 can be associated with each subject or object in a group. A *discrete quantitative variable* is a characteristic whose values are *integers*.

Example 3.4. Let the set of objects be the games a team plays at home. The outcome we are examining is getting a hit (H). The discrete quantitative variable #H counts the number of hits for each home game. There would be 81 numbers generated by this variable.

Example 3.5. The quantitative variable #HR can be associated with each year of a player's career, each team in the AL for a given year, each year for the NL since 1876, and so on.

Table 3.3 gives a list of derived descriptive measures (statistics) used in baseball. These statistics are derived by performing the operations of addition, subtraction, multiplication, and division on the discrete variables in Table 3.2 or on other derived statistics in Table 3.3. These statistics are used to measure many aspects of batting performance for a player, a team, or a

Table 3.3 **Derived baseball statistics**

Derived statistics	Symbol	Definition
Batting average	BA or AVG	#H/#AB
Total bases	TB	#1B + 2(#2B) + 3(#3B) + 4(#HR)
On-base percentage	OBP or OBA	#OB/#PA
Slugging average	SLG	TB/#AB
On-base plus slugging	OPS	OBA + SLG
Runs created	RC	(#H + #BB) * (TB)/(#AB + #BB)
Batter's run average	BRA	(OBP) * (SLG)
Bases over PA	BOP	(TB + #BB + #HBP)/#PA
Home run average	HRA	#HR/#AB
Home run rate	HRR	#AB/#HR
Isolated power	ISO	(TB – #H)/#AB
Run production	RPR	#RBI + #R – #HR
Run production per game	RPR/G	RPR/#G
Run production per 27 outs	RPR/27	(25.5) * (RPR)/(#AB – #H)
Runs per game	R/G	#R/#G
Runs per 27 outs	R/27	(25.5) * (#R)/(#AB – #H)
RC per 27 Outs	RC/27	(25.5) * (RC)/(#AB – #H)
Strikeout average	SOA	#SO/#AB
Base-on-balls average	BBA	#BB/#PA
Strikeouts to base on balls	SO/BB	#SO/#BB
Strikeouts to home runs	SO/HR	#SO/#HR
In-play batting average	IPBA	#H/(#AB – #SO) or #H/#IPAB
In-play home run average	IPHR or IPHRA	#HR/#IPAB

league. Many of these derived baseball statistics are nothing more than the traditional statistics of mean, proportion, and ratio.

Statistics can also be variables. Obtaining a set of values by applying a particular statistic to each member of a group turns a statistic into a variable. The number of runs scored by a player (#R) for a season is a statistic since it is a descriptive measure of a player's contribution to a team's run production. However, if one were to look at the #R of a player for each year of his career, then #R would be a discrete quantitative variable. In Example 3.6 below, many of the discrete variables from Table 3.2 and the derived statistics from Table 3.3 are used to summarize a player *Z*'s batting outcomes for a four-game series.

Example 3.6. Player *Z*'s batting results for a four-game series:

Game 1	Game 2	Game 3	Game 4
1B	SE	HR	1B
FO	SO	3B	BB
1B	HBP	FO	1B
GO	2B	GDP	SO
BB	HR	SF	SH
#R = 1	#R = 2	#R = 1	#R = 0
#RBI = 0	#RBI = 3	#RBI = 2	#RBI = 0

In the above, the #R and #RBI are used as variables applied to each game.

Below #R and #RBI are used as statistics providing totals for the four-game series.

Descriptive measures for player *Z*'s batting results for the four-game series are as follows:

#PA = 19 (SH is not counted as a PA)
#SH = 1
#SF = 1
#AB = 15 (SF, SH, BB, HBP are not counted as at bats)
#1B = 4
#2B = 1
#3B = 1
#HR = 2
#SO = 2
#O = 7 (SE is considered an out)
#EO = 9 (SF and SH are added to the out total)
#H = 8
#R = 4
#RBI = 5

#BB = 2
#HBP = 1
#SE = 1
#IPAB = 13
R/G = 4/4 = 1
R/27 = 4 * 25.5/(15 − 8) = 14.57
BA or AVG = 8/15 = .533
TB = 1(#1B) + 2(#2B) + 3(#3B) + 4(#HR) = 1 * 4 + 2 * 1 + 3 * 1 + 4 * 2 = 17
#OB = 11
OBP = (8 + 2 + 1)/(15 + 2 + 1 + 1) = 11/19 = .580
RPR = 4 + 5 − 2 = 7
RPR/27 = 25.5 * RPR/(15 − 8) = 25.50
SLG = 17/15 = 1.133
BOP = (17 + 2 + 1)/19
OPS = .580 + 1.133 = 1.713
RC = (8 + 2) * 15/(15 + 2) = 8.82
RC/27 = (25.5) * (8.82)/7 = 32.13
HRA = 2/15 = .133
HRR = 1/HRA = 15/2 = 7.50
ISO = (17 − 8)/15 = 9/15 = .600
BRA = (.580) * (1.133) = .657
BBA = 2/19 = .105
SOA = 2/15 = .133
SO/BB = 2/2 = 1.000
SO/HR = 2/2 = 1.000
IPBA = 8/(15 − 2) = 8/13 = .615
IPHR = 2/(15 − 2) = 2/13 = .154

Exercise 3.1. For each derived statistic in Table 3.3, determine if it is a mean, a proportion, a ratio, or just an amount. Identify those derived statistics that are both a ratio and a proportion.

Exercise 3.2. Which of the derived statistics from Table 3.3, treated as variables, are discrete and which are continuous?

Exercise 3.3. Compute the derived statistics given in Table 3.3 for the batting data of Aaron for the year 1960 (see Table 2.2).

3.3. Evolution of Baseball Statistics

Many baseball statistics evolve from an experiment that results in an outcome, which lies in one of two possible nonoverlapping sets. The outcomes contained in one set are called successes, whereas the outcomes contained in the other set are called failures. If the experiment is repeated a certain number of times, we can count the number of successes. The proportion formed by dividing the number of successes by the number of times the experiment is repeated is the basis for many baseball statistics. If the number 1 is assigned to each success and the number 0 is assigned to each failure, then the proportion of successes corresponds to the mean number of successes. The next two examples illustrate the above concepts for the baseball statistics of AVG and OBP.

Example 3.7. The experiment leading to the calculation of an AVG is based on repeated ABs (AB = H + O). Each AB results in an outcome belonging to one of two possible sets. An outcome belonging to the set H is designated success; an outcome belonging to the set O (O includes {SE}) is designated failure. The number 1 is assigned to each outcome that is a success, and the number 0 is assigned to each outcome that is a failure. The duration of time can be a series, a season, a career, and so on. Using the results from Example 3.6, starting with the first at bat in game 1 and ending with the last at bat in game 4, we assign 0 and 1 as follows:

Game 1	Game 2	Game 3	Game 4
1B(1)	SE(0)	HR(1)	1B(1)
FO(0)	SO(0)	3B(1)	BB(-)
1B(1)	HBP(-)	FO(0)	1B(1)
GO(0)	2B(1)	GDP(0)	SO(0)
BB(-)	HR(1)	SF(-)	SH(-)

The computation of the AVG for the player as a *proportion* is AVG = #H/#AB = 8/15 = .533.

The computation of the AVG for this player as a *mean* is AVG = (1 + 0 + 1 + 0 + 0 + 0 + 1 + 1 + 1 + 1 + 0 + 0 + 1 + 1 + 0)/15 = 8/15.

The outcomes BB, HBP, SF, and SH are assigned a hyphen since they are not outcomes belonging to the set AB.

Example 3.8. The experiment leading to the calculation of an OBP is based on repeated PAs (PA = O + H + {BB, HBP, SF}). Each PA results in an outcome belonging to one of two possible sets. An outcome belonging to the set OB = H + {BB, HBP} is designated success; an outcome belonging to the set PA − OB = O + {SF} is designated failure. The number 1 is assigned to each outcome that is a success, and the number 0 is assigned to each outcome that is a failure. Using the results from Example 3.6, starting with the first at bat in game 1 and ending with the last at bat in game 4, we assign 0 and 1 as follows:

Game 1	Game 2	Game 3	Game 4
1B(1)	SE(0)	HR(1)	1B(1)
FO(0)	SO(0)	3B(1)	BB(1)
1B(1)	HBP(1)	FO(0)	1B(1)
GO(0)	2B(1)	GDP(0)	SO(0)
BB(1)	HR(1)	SF(0)	SH(-)

The computation of OBP for this player as a *proportion* is OBP = #OB/#PA = 11/19 = .579.

The computation of OBP for the player as a *mean* is OBP = (1 + 0 + 1 + 0 + 1 + 0 + 0 + 1 + 1 + 1 + 1 + 1 + 0 + 0 + 0 + 1 + 1 + 1 + 0)/19 = 11/19.

The outcome SH is assigned a hyphen because it does not belong to the set PA.

3.3.1. The Calculation for SLG

The experiment leading to the calculation of SLG is based on repeated at bats. Each AB results in an outcome that is treated as belonging to any one of five possible sets. The five possible sets are O, {1B}, {2B}, {3B}, {HR}. The computation of SLG is done by assigning a weight to each outcome and summing the product of the weight of each outcome times its frequency and then dividing this total by #AB. The weights are assigned according to the type of hit. A HR is given the highest weight of 4, a 3B is given a weight of 3, a 2B is given a weight of 2, and a 1B is given a weight of 1. A member of O is given the weight of 0. SLG is a special type of average called a *weighted mean.*

$$\begin{aligned} \text{SLG} &= \text{TB}/\#\text{AB}, \\ \text{SLG} &= \Sigma(\text{weight} * \text{frequency})/\#\text{AB} \\ &= \frac{1 * \#1B + 2 * \#2B + 3 * \#3B + 4 * \#HR}{\#AB}. \end{aligned}$$

The numerator of the SLG formula is called a *linear weight.* The linear weight part of the SLG formula gives the total bases (TB).

Example 3.9. The calculation for SLG for the player in Example 3.6 (weights are in parentheses).

Game 1	Game 2	Game 3	Game 4
1B(1)	SE(0)	HR(4)	1B(1)
FO(0)	SO(0)	3B(3)	BB(-)
1B(1)	HBP(-)	FO(0)	1B(1)
GO(0)	2B(2)	GDP(0)	SO(0)
BB(-)	HR(4)	SF(-)	SH(-)

SLG = (1 + 0 + 1 + 0 + 0 + 0 + 2 + 4 + 4 + 3 + 0 + 0 + 1 + 1 + 0)/15 = 17/15 = 1.133,

$$\text{SLG} = \frac{1*\#1B + 2*\#2B + 3*\#3B + 4*\#HR}{\#AB} = \frac{1*4 + 2*1 + 3*1 + 4*2}{15} = \frac{17}{15}.$$

The linear weight part of the formula is 17.

Both AVG and OBP are expressed as formulas in which the numerators are linear weights.

$$\text{AVG} = \frac{1*\#1B + 1*\#2B + 1*\#3B + 1*\#HR}{\#AB},$$

$$\text{OBP} = \frac{1*\#1B + 1*\#2B + 1*\#3B + 1*\#HR + 1*\#BB + 1*\#HBP}{\#PA}.$$

John Thorn and Peter Palmer discuss the concept of a generalized linear weight formula in their book *The Hidden Game of Baseball* (pp. 62–81).

Two of the several generalized linear weight formulas mentioned in this book are

$$a*\#1\text{B} + b*\#2\text{B} + c*\#3\text{B} + d*\#\text{HR} + m, \qquad \text{(LWTF1)}$$

$$e*\#1\text{B} + f*\#2\text{B} + g*\#3\text{B} + h*\#\text{HR} + i*\#\text{BB} + j*\text{HBP} + n. \qquad \text{(LWTF2)}$$

These two formulas will be called the "Linear Weight Formula 1" and "Linear Weight Formula 2."

The constants a, b, c, d, e, f, g, h, i, and j are the weights that are assigned to each of the outcomes. If $a=1$, $b=1$, $c=1$, $d=1$, $m=0$ in LWTF1, we obtain the linear weight used in the numerator of the formula for AVG; if $a=1$, $b=2$, $c=3$, $d=4$, $m=0$ in LWTF1, we obtain the linear weight used in the numerator of the formula for SLG; if $e=1$, $f=1$, $g=1$, $h=1$, $i=1$, $j=1$, $n=0$ in LWTF2, we get the linear weight used in the numerator of the formula for OBP.

The statistics AVG, SLG, and OBP are all ratios. The numerator of each of these is a linear weight. Linear weights are amounts and not ratios. Since all three of these statistics have the form of a linear weight divided by the number of points in the sample space (the number of opportunities satisfying the criteria in the denominator), they can all be considered averages. The difference is that AVG and OBP are means, whereas SLG is a special type of average called a weighted mean.

The advantage of a linear weight over a ratio is that a linear weight gives an amount.

One question investigated by baseball researchers is, which constant values substituted into the two formulas LWTF1 and LWTF2 will transform these formulas into the best predictors for the number of runs a player contributes to the total runs scored by his team? This question is studied by Thorn and Palmer in their book.

Thorn and Palmer present several specific linear weight formulas in their book. One example of such a linear weight formula is the formula introduced by George Lindsey in 1963. Lindsey's formula was

$$\text{Runs} = (.41) * 1B + (.82) * 2B + (1.06) * 3B + (1.42) * HR.$$

His formula was based on a combination of play-by-play data and basic probability theory.

Since the linear weight component of the batting average statistic assigns the weight of 1 to 1B, 2B, 3B, and HR, its value in predicting the runs a player contributes to his team is limited. Needless to say, the same value cannot be assigned to a single and to a home run. Lindsey, in his formula, assigns 1.42 as a value for HR and .41 as a value for 1B.

Since Lindsey's formula appeared in 1963, many other formulas have been developed. These formulas take into account such variables as stolen bases, grounded into double plays, and caught stealing.

The goal of any team in baseball is to win games. Clearly, an increase in runs scored by a team leads to an increase in wins for that team. Later, we will look at the ratio of the runs scored divided by the runs allowed as a predictor of team wins.

The most important role of a player as a batter is to produce runs for his team. General managers know that players with a high batting average, slugging percentage, or on-base percentage will produce more runs for their teams. The question that general managers debate is, which batting statistic is most important for run production? Later in this book, this question will be discussed and analyzed.

In the next section, we return to the topic of using baseball statistics for the purpose of comparing players, teams, and leagues from different eras.

Exercise 3.4. Using Lindsey's formula, for which year did Aaron contribute the most runs for his team? Use the data in Table 2.2.

3.4. Adjusted Baseball Statistics

In Section 2.7, we saw that a given x-value can be converted into a z-score (standard unit) by applying the formula

$$z\text{-score} = (x\text{-value} - \text{mean})/\text{standard deviation}.$$

Provided that the data are approximately normal, the z-score can then be associated with a percentile (the percent of data values that are less than the z-score). Because of the properties for normal curves mentioned in Section 2.7, the percent of data values less than the given x-value is equal to the percentile associated with its *z*-score. In Section 2.7, we saw that if we consider the data set to be all AVGs for players with a certain minimum number of at bats for a given year, we can assign a percentile to any player's AVG. This allowed us to compare batting averages between two players for two different years or different eras. The disadvantage of this method is that the underlying populations must be close to normal to assign percentiles. This same method can be used for any baseball statistic whose data are approximately normal.

We now look at other methods for adjusting baseball statistics that do not depend on the assumption of normality for the underlying population. These methods replace an existing baseball statistic with an adjusted baseball statistic of the same form. These adjusted baseball statistics can then be used to compare batting performance between two players. These adjustments do not depend on the underlying population being normal.

3.4.1. Special Terminology

In order to change the underlying sample of plate appearances, we can insert special words in front of the statistic. This is illustrated below for the batting average statistic. However, this special terminology can be used for any baseball statistic. Throughout this book we have used the two abbreviations AVG or BA for batting average. In what follows, we will use the BA abbreviation.

Some examples of specific criteria for at bats for the statistic BA are as follows:

- *xyzw* BA refers to the batting average for a player based on ABs for a specific year *xyzw*.
- *Homepark* BA refers to a player's BA based on his home ballpark ABs.
- *Awaypark* BA refers to a player's BA based on away ballpark ABs.
- *Career* BA refers to a player's BA based on a player's career ABs.
- *Team* BA is the BA for a sample consisting of all ABs for a team.
- *League* BA is the BA for a sample consisting of all ABs for a league.
- *Homepark team* BA refers to a BA based on all the home team's players' ABs at their home team's ballpark.
- *Awaypark team* BA refers to a BA based on all the home team's players' ABs at all away ballparks.

The insertion of the word *true* in front of a statistic will change the descriptive measure from a statistic (a descriptive measure of the sample) to a parameter (a descriptive measure for the population).

A *true xyzw* BA is a player's theoretical population batting average for the year *xyzw*. A *true career* BA is a player's theoretical population batting average for his career.

In evaluating a player's batting performance, there are situations where a player's baseball statistics will be adjusted by factors such as the year they played, ballpark effect, rule changes, and the number of games a team played for that year. The adjustments covered in this book are the *adjustment for number of games played*, the *adjustment for time (year played)*, and the *adjustment for ballpark effect.*

3.4.2. Adjustment Based on Number of Games Played by a Team

In 1961, American League teams shifted from a 154-game schedule to a 162-game schedule. The following year, National League teams made the same switch. The change was triggered by the addition of two new teams in the American League in 1961 and two new teams in the National League in 1962. Coincidentally, in the first year of the 162-game schedule, Roger Maris hit his 61 home runs in 1961. Many baseball people wanted an asterisk to be placed next to this number. Those people still considered Babe Ruth's 60 home runs as the home run record, because the Sultan of Swat had eight less games to achieve his feat. How would you adjust records for a change in number of games a team plays in a season?

3.4.3. Adjustment Based on the Time Period in Which the Player Played

In comparing the statistics for two players from different time periods, statistics should sometimes be adjusted for the different years in which they played. The pitching talent and changes to baseball rules could have a major effect on the batting performance of a player. The time adjustment will modify a batting statistic based on the batting performance of the entire league for that year. These time-adjusted statistics were first introduced by David Shoebotham in the 1970s. They will be symbolized by placing AdjT_ in front of the given statistic.

Calculation for AdjT_(statistic)

Step 1: Divide the given statistic for that player for that year by the same statistic calculated for all plate appearances in that league for that year.
Step 2: Multiply the ratio in step 1 by a *standard statistic*.

The *standard statistic* for a given statistic is defined as that statistic evaluated for all players over a specific interval of time. For the standard statistic, we will arbitrarily choose the years 1960 through 2007 as the interval of time. The reason for multiplying the ratio in step 1 by the standard statistic is to convert the ratio in step 1 into the same form as the given statistic. This adjustment technique can be used for almost any statistic. The ratio gives the relative value of the statistic for the given player with respect to the same statistic for all players in the league for a given year. Example 3.10 below illustrates the calculation for AdjT_BA.

Table 3.4, downloaded from the website www.baseball-reference.com (under the topic of leagues), represents the yearly batting results for the National League for the years 1960–2007. Using the cumulative totals at the bottom of the table for the years 1960–2007, we calculate the BA, OBA, and SLG as .258, .324, and .391, respectively. These statistics will be called the *standard BA*, the *standard OBA*, and the *standard SLG*, respectively.

The same adjustment technique can be used for other derived statistics. These adjusted statistics can then be used for the comparison of players.

Example 3.10. The Adj_T formulas for the three statistics BA, OBA, SLG are given next. This is followed by their actual calculations for the year 1961 for Henry Aaron.

AdjT_BA for *xyzw* = [(*xyzw* BA)/(league *xyzw* BA)] * (standard BA),
AdjT_OBA for *xyzw* = [(*xyzw* OBA)/(league *xyzw* OBA)] * (standard OBA),
AdjT_SLG for *xyzw* = [(*xyzw* SLG)/(league *xyzw* SLG)] * (standard SLG)
(*xyzw* represents the year used).

Below are the calculations of the AdjT_(statistic) for Aaron for 1961.
Since (*x*/*y*) * *z* = (*x* * *z*)/*y*, we will use the form *x* * *z*/*y* for our calculations below.

AdjT_BA for 1961 = (Aaron 1961 BA * standard BA)/league 1961 BA
= .327 * .258/.262 = .322,
AdjT_OBA for 1961 = (Aaron 1961 OBA * standard OBA)/league 1961 OBA
= .381 * .324/.327 = .378,
AdjT_SLG for 1961 = (Aaron 1961 SLG * standard SLG)/league 1961 SLG
= .594 * .391/.405 = .573.

The league 1961 BA, the league 1961 SLG, and the league 1961 OBP were taken from Table 3.4.

Using Table 2.2 and Example 3.10,

- Aaron's BA of .327 has an AdjT_BA for 1961 of .322;
- Aaron's OBA of .381 has an AdjT_OBA for 1961 of .378;
- Aaron's SLG of .594 has an AdjT_SLG for 1961 of .573.

Exercise 3.5. Compute for the year 1963 the AdjT_BA, AdjT_OBA, AdjT_SLG, and AdjT_HRA for Henry Aaron. (For the AdjT_HRA you must calculate the *standard* HRA by using the cumulative totals on the bottom of Table 3.4.)

Exercise 3.6. Compute for the year 1992 the AdjT_BA for Barry Bonds. Compare his results with Henry Aaron's results in Example 3.10. (The data can be obtained from Table 4.1.)

3.4.4. Adjustment Based on Ballpark Effect

A second adjustment for comparing two players allows for their home ballpark. A player plays half his games in his home ballpark, and not all ballparks are the same. They differ in their dimensions, the type of surface, the height of their walls, the amount of room in foul territory, open or domed, the altitude, and many other factors. Each of these factors can affect a player's batting statistics. Some ballparks are more favorable to hitters than others. The *ballpark effect* statistics will be symbolized by placing AdjP_ in front of the given statistic. The logic for adjusting for the ballpark effect is discussed in much more detail by Michael Schell in his book *Baseball's ALL-TIME BEST Hitters*, pp. 103–33.

In the calculations that follow, the *homepark team* refers to the home team of the player being considered. The *ballpark effect* adjustment for the BA of a player is outlined in the procedure below.

Clearly, a ballpark adjustment can be made for any baseball statistic. The procedure below can be used for any baseball statistic.

Calculation for AdjP_BA

Step 1: Calculate the homepark team BA. The BA is based on all ABs for the home team's players for all games played at the home team's ballpark.

Step 2: Calculate the awaypark team BA. The BA is based on all ABs for the home team's players for all games played at all the away ballparks.

Step 3: Calculate (homepark team BA + awaypark team BA)/2. *This number represents the BA for an average home team player.* Remember, the number of home games is the same as the number of away games.

Step 4: Calculate [homepark team BA + $(K-1)*$(awaypark team BA)]/K, where K is the number of teams in the league. This number represents what the BA would be for the average home team player, assuming that the home team plays the same number of games at every ballpark in the league, including his home ballpark. Of course, if each team played the same number of games at each ballpark, there would be no need in adjusting for the home ballpark.

Step 5: The Ballpark Factor (BF) is the result of step 3 divided by the result of step 4. If the BF is greater than 1, the home ballpark is considered favorable for that statistic.

Step 6: AdjP_BA = Player's BA/BF. If the BF is greater than 1, the AdjP_BA will be less than the player's actual BA. In other words, a favorable home ballpark for a statistic will lower the player's statistic. Since the home and away ballpark BAs will be different for different ballparks, there will be a different BF assigned to each ballpark.

Table 3.4 **Batting statistics for the National League for the years 1960–2007**

Year	R/G	R	G	AB	H	2B	3B
2007	4.71	12208	2594	89488	23796	4898	505
2006	4.76	12337	2590	88844	23501	4834	561
2005	4.45	11535	2594	88120	23058	4754	468
2004	4.64	12018	2590	88622	23271	4687	494
2003	4.61	11945	2590	88426	23126	4657	491
2002	4.45	11516	2588	87794	22753	4482	488
2001	4.70	12186	2592	88100	23027	4613	488
2000	5.00	12976	2593	88743	23594	4633	532
1999	5.00	12966	2591	89010	23880	4619	512
1998	4.60	11932	2596	88700	23213	4493	491
1997	4.60	10440	2268	77203	20300	3907	485
1996	4.68	10623	2268	77711	20398	3782	434
1995	4.63	9329	2014	69049	18184	3367	418
1994	4.62	7422	1606	55068	14695	2784	377
1993	4.49	10190	2270	77489	20427	3588	513
1992	3.88	7539	1944	65748	16538	2967	459
1991	4.10	7955	1940	65365	16363	2819	441
1990	4.20	8173	1944	65968	16917	2967	405
1989	3.94	7673	1946	65817	16215	2903	411
1988	3.88	7522	1938	65563	16277	2828	415
1987	4.52	8771	1942	66276	17275	3126	435
1986	4.18	8096	1938	65730	16643	2991	387
1985	4.07	7899	1942	65818	16596	2861	437
1984	4.06	7894	1942	65919	16842	2770	451
1983	4.10	7993	1948	65717	16781	2753	484
1982	4.09	7947	1944	66263	17085	2823	445
1981	3.91	5035	1288	43654	11141	1881	354
1980	4.03	7852	1946	66272	17186	2856	523
1979	4.22	8186	1942	66088	17229	2886	518
1978	3.99	7742	1942	65156	16556	2861	482
1977	4.40	8556	1944	66700	17465	3033	526
1976	3.98	7739	1944	65814	16778	2652	499
1975	4.13	8014	1942	66102	17002	2781	458
1974	4.15	8070	1944	66212	16907	2642	447
1973	4.15	8062	1942	66087	16817	2600	386
1972	3.91	7265	1860	63116	15683	2392	430
1971	3.91	7601	1944	65903	16590	2505	457
1970	4.52	8771	1942	66465	17151	2743	554
1969	4.05	7890	1946	65751	16461	2455	471
1968	3.43	5577	1626	54913	13351	1995	359
1967	3.84	6218	1620	55026	13698	2133	427
1966	4.09	6624	1618	55385	14202	2099	412
1965	4.03	6558	1626	55377	13794	2122	422
1964	4.01	6517	1624	55284	14032	2161	427
1963	3.81	6181	1622	54803	13434	1984	439
1962	4.48	7278	1624	55449	14453	2075	453
1961	4.52	5600	1238	42128	11029	1749	350
1960	4.24	5250	1238	42176	10745	1722	384
Total	4.31	413671	96074	3270412	842459	148233	21805

HR	BB	SO	BA	OBP	SLG	SH	SF	HBP
2705	8576	17449	0.266	0.334	0.423	1045	752	934
2840	8600	17426	0.265	0.334	0.427	1190	719	1030
2580	8396	16880	0.262	0.330	0.414	1151	669	980
2846	8736	17299	0.263	0.333	0.423	1190	704	944
2708	8666	16996	0.262	0.332	0.417	1093	660	967
2595	8921	17161	0.259	0.331	0.410	1134	689	895
2952	8567	17908	0.261	0.331	0.425	1074	739	969
3005	9735	17344	0.266	0.342	0.432	1062	809	898
2893	9602	17153	0.268	0.342	0.429	1098	751	799
2565	8710	17455	0.262	0.331	0.410	1167	691	824
2163	7704	15320	0.263	0.333	0.410	1030	643	773
2220	7501	15252	0.262	0.330	0.408	974	648	683
1917	6668	13309	0.263	0.331	0.408	947	531	624
1532	5193	10147	0.267	0.333	0.415	758	455	451
1956	7104	13358	0.264	0.327	0.399	1110	701	567
1262	5978	11342	0.252	0.315	0.368	983	545	395
1430	6254	11446	0.250	0.317	0.373	891	562	367
1521	6221	11164	0.256	0.321	0.383	876	564	352
1365	6251	11354	0.246	0.312	0.365	899	536	318
1279	5793	11032	0.248	0.310	0.363	938	589	368
1824	6577	11657	0.261	0.328	0.404	823	481	349
1523	6560	11648	0.253	0.322	0.380	869	519	312
1424	6373	10674	0.252	0.319	0.374	901	486	280
1278	6149	10929	0.255	0.319	0.369	809	570	249
1398	6424	10749	0.255	0.322	0.376	921	548	292
1299	5964	10300	0.258	0.319	0.373	978	580	305
719	4107	6332	0.255	0.319	0.364	688	379	185
1243	5969	9849	0.259	0.320	0.374	967	577	257
1427	6188	9920	0.261	0.325	0.385	949	549	332
1276	6279	9905	0.254	0.320	0.372	970	573	330
1631	6487	10488	0.262	0.328	0.396	847	522	330
1113	6263	9602	0.255	0.320	0.361	975	559	310
1233	6730	9793	0.257	0.327	0.369	1082	536	367
1280	6828	9971	0.255	0.326	0.367	983	552	360
1550	6453	10507	0.254	0.322	0.376	958	496	358
1359	5985	10544	0.248	0.315	0.365	865	481	358
1379	6059	10542	0.252	0.316	0.366	918	484	395
1683	6919	11417	0.258	0.329	0.392	819	501	393
1470	6397	11628	0.250	0.319	0.369	891	430	443
891	4275	9502	0.243	0.300	0.341	794	387	352
1102	4672	9468	0.249	0.310	0.363	729	390	354
1378	4404	9312	0.256	0.313	0.384	745	363	363
1318	4730	9649	0.249	0.311	0.374	709	366	404
1211	4394	9256	0.254	0.311	0.374	789	371	327
1215	4560	9545	0.245	0.306	0.364	732	381	372
1449	5265	9032	0.261	0.327	0.393	656	410	373
1196	3995	6622	0.262	0.327	0.405	572	313	254
1042	3937	6824	0.255	0.319	0.388	532	344	220
81245	312119	572460	0.258	0.324	0.391	44081	26105	23362

Example 3.11. We calculate the AdjP_BA for Aaron for the year 1961. The home and away batting averages for the Milwaukee Braves for the year 1961 were acquired from the website www.retrosheet.org.

Step 1: The *homepark team* BA for the Milwaukee Braves was .252.
Step 2: The *awaypark team* BA for the Milwaukee Braves was .264.
Step 3: (.252 + .264)/2 = .258.
Step 4: [.252 + (7 * .264)]/8] = .2625, for $K = 8$ teams.
Step 5: BF = .258/.2625 = .9829. Since .9829 < 1, the Milwaukee ballpark was unfavorable for hitters.
Step 6: AdjP_BA = Aaron's 1961 BA / BF = .327/.9829 = .333.

The reason that Aaron's adjusted BA for 1961 was higher was because the Milwaukee ballpark was unfavorable for hitters in 1961.

Aaron's BA for 1961 was .327. Applying the "adjustment for the year he played" lowered his BA from .327 to .322. Applying the adjustment for ballpark raised his BA from .327 to .333. We now apply both adjustments. We first adjust for the year and then adjust this result for the ballpark effect.

$$\text{AdjP\_(AdjT\_BA)} = \text{AdjP\_(AdjT\_.327)} = \text{AdjP\_.322} = .322/0.9829 = .3276 \text{ (rounded to .328).} \quad (3.1)$$

The question is, what happens if we reverse the two adjustments? Let's do it:

$$\text{AdjT\_(AdjP\_BA)} = \text{AdjT\_(.333)} = (.333 * .258)/.262 = .3279 \text{ (rounded to .328).} \quad (3.2)$$

Reversing the two adjustments has a minimal effect. The results in Formulas (3.1) and (3.2) both round to .328. Observe that even though these two numbers are very close to each other, the two numbers are not equal.

Formulas (3.1) and (3.2) demonstrate the operation of composition in algebra. It is known that composition is not commutative. So, in general, altering the order in which the adjustments are performed will yield different numbers.

Exercise 3.7. From the website www.retrosheet.org, we have, for the year 1971, the following splits for the Atlanta Braves:

Homepark team BA = .268; *homepark team* OBA = .323; *homepark team* SLG = .385.
Awaypark team BA = .246; *awaypark team* OBA = .300; *awaypark team* SLG = .355.

Using this information, calculate the following for Henry Aaron for 1971:

a. AdjP_BA
b. AdjP_OBA
c. AdjP_SLG
d. AdjP_(AdjT_OBA)
e. AdjT_(AdjP_OBA)
f. AdjP_(AdjT_SLG)
g. AdjT_(AdjP_SLG)
h. AdjP_(AdjT_BA)
i. AdjT_(AdjP_BA)

Exercise 3.8. Use Table 3.4 to calculate the *standard* IPBA, the *standard* SOA, and the *standard* IPHR. Then repeat Exercise 3.7 for these statistics for Aaron for 1971. You will need to obtain the homepark team and awaypark team IPBA, SOA, and IPHR from the website www.retrosheet.org.

Exercise 3.9. Repeat Exercise 3.7 for Barry Bonds for the year 1992. Again, you will need to obtain the homepark and awaypark team statistics from www.retro sheet.org. Compare Aaron's results for 1961 with the results for Bonds for 1992. Bonds played for the Pirates in 1992.

3.5. Explanation of Baseball Statistics

This section analyzes baseball statistics with respect to three questions:

1. What aspect of batting does the baseball statistic measure?
2. What are the weaknesses of the baseball statistic?
3. How does the baseball statistic relate to the traditional statistics of amount, mean, proportion, ratio, and rate?

For many years the highest batting distinction was to win the "Triple Crown," leading the league in batting average (BA), runs batted in (RBIs), and home runs (HRs). Another important trio is BA, OBP, and SLG. Table 3.5 lists all Triple Crown winners since 1900, with the following variables: year, league, team, player, BA, HR, RBI, OBP, and SLG.

Exercise 3.10. For each variable in Table 3.5 answer the following questions:

a. Are the variables qualitative or quantitative?
b. If the variable is quantitative, is it discrete or continuous?
c. What is the measurement scale for each variable?

Take the time to review Section 1.8, where the descriptive measures of mean, proportion, percentage, probability, and ratio were discussed.

Table 3.5 **Triple Crown winners since 1900**

Year	League	Team	Player	BA	HR	RBI	OBP	SLG
1901	AL	PHA	Nap Lajoie	.426	14	125	.463	.643
1909	AL	DET	Ty Cobb	.377	9	107	.431	.517
1922	NL	STL	Rogers Hornsby	.401	42	152	.459	.722
1925	NL	STL	Rogers Hornsby	.403	39	143	.489	.756
1933	AL	PHA	Jimmie Foxx	.356	48	163	.449	.703
1933	NL	PHI	Chuck Klein	.368	28	120	.422	.602
1934	AL	NYY	Lou Gehrig	.363	49	165	.465	.706
1937	NL	STL	Joe Medwick	.374	31	154	.414	.641
1942	AL	BOS	Ted Williams	.356	36	137	.499	.648
1947	AL	BOS	Ted Williams	.343	32	114	.499	.634
1956	AL	NYY	Mickey Mantle	.353	52	130	.464	.705
1966	AL	BAL	Frank Robinson	.316	49	122	.410	.637
1967	AL	BOS	Carl Yastrzemski	.326	44	121	.418	.622

An *experiment* is any process that yields *outcomes.* Some examples of experiments are rolling a die, flipping a coin, or a plate appearance in baseball. Experiments whose outcomes are influenced by chance are called *probability experiments.*

A *probability* measures the likelihood that a particular outcome or set of outcomes occurs during an experiment. A probability must always be a number between 0 and 1. The probability experiment needed for the calculation of a BA is repeated at bats, and the probability experiment needed for the calculation of an OBP is repeated plate appearances. Using "Pr(*E*)" as the notation for the probability of an event *E* occurring, we can express BA and OBP as the following probabilities:

BA or AVG = Pr(H) = #H/#AB,
OBP or OBA = Pr(OB) = #OB/#PA.

In the case of BA, the #H represents the number of successes and #AB represents the number of opportunities. For OBP, #OB represents the number of successes and #PA represents the number of opportunities.

As discussed in Section 3.3, BA and OBP have the following additional interpretations:

- BA is the proportion of #H to #AB.
- The numerator of the BA formula is a linear weight.
- OBP is the proportion of #OB to #PA.
- The numerator of the OBP formula is a linear weight.
- BA and OBP both can be interpreted as means.
- Multiplying by 100 transforms BA and OBP into percentages.

It is nice to remember that the three *Ps* (*proportion*, *probability*, and *percentage*) are interchangeable.

Slugging percentage, SLG = TB/#AB, does *not* fit the definition of probability because TB is a special calculation, not a count of the number of successes. SLG is a number between 0 and 4.

Other statistics that can be interpreted as a percentage, proportion, or a probability are HRA, SOA, BBA, IPBA, and IPHR (see Table 3.3).

3.5.1. BA or AVG

Batting average (BA) was introduced by H. A. Dobson in 1876. Before this, a batter was rated by the number of hits per game. Dobson was motivated by the fact that hits per game favored those players who hit at the beginning of the batting order and also was dictated by circumstances within the game. AVG represents the probability or likelihood of a player getting a hit in an at bat.

The batting average measures a player's ability as a hitter to make contact with the ball and direct it to where the fielders are not positioned. It is a proportion and also can be interpreted as a mean.

The critics of BA argue that the statistic does not provide a good measure of a batter's ability because it treats the different types of hits (1B, 2B, 3B, HR) as equally important. In addition, BA does not take into account the ability to reach base on a BB or HBP. So a great hitter who walks a lot, or who is often hit by a ball when the pitcher tries to back him off the plate, is not rewarded. (In modern times, Craig Biggio of the Houston Astros was hit the most often, 285 times.) Finally, the BA gives no indication of the value of the type of hit to the team's run production. A bases-empty single when you're ahead 10-0 is not the same as Kirk Gibson's game 1, World Series, bottom of the ninth, pinch-hit home run.

3.5.2. OBP or OBA

On-base percentage (OBP) or on-base average (OBA) was first introduced by Allan Roth and Branch Rickey in 1954 as part of a larger equation. It represents the probability of a player getting on base in a plate appearance. Originally, OBP did not include the SF. In 1984, the sacrifice fly was included in OBP.

OBP can be interpreted as a proportion or mean. It measures a player's ability to get on base in a plate appearance, and unlike the BA, it credits players for getting on base through a BB or HBP. However, it too does not distinguish types of hits and their respective values. In August 2009, Mets third baseman David Wright had 49 at bats, 15 hits, 6 walks, wasn't hit by a pitch, and didn't have a sacrifice fly. His OBP for the month was $(15+6+0)/(49+6+0+0)=0.382$.

Replacing H by TB in the OBP formula gives the formula for bases over plate appearances, abbreviated BOP. You will need this formula for the chapter problems.

3.5.3. SLG

In 1923, John Heydler added slugging average (SLG) to the National League baseball statistics. It is a weighted mean or ratio.

In the 2008 season, New York Yankees star Alex Rodriguez had 510 at bats, in which he hit 86 singles, 33 doubles, no triples, and 35 home runs. His SLG was $[86+(2*33)+(3*0)+(4*35)]/510$, so $\text{SLG}=0.573$.

One disadvantage is that it does not credit a player's ability to get on base via a BB or HBP. Another problem is the weighting factors. A home run is 4 times as valuable as a single, and a triple is 3 times as valuable as a single. A home run is twice as valuable as a double. These weights have been shown to be a poor predictor of team runs.

3.5.4. OPS

The statistic on-base plus slugging (OPS) was introduced in the mid-1970s by John Thorn and Pete Palmer.

On-base plus slugging measures a player's ability to get on base and advance runners. The formula is

$$\text{OPS}=\text{OBP}+\text{SLG}=\frac{\#H+\#BB+\#HBP}{\#AB+\#BB+\#HBP+\#SF}+\frac{TB}{\#AB}.$$

Adding OBP to SLG produces a number larger than both OBA and SLG. As SLG tends to be much larger than OBP, OPS emphasizes SLG. In 2008, then–Texas Ranger Milton Bradley had an OBP of .436 and an SLG of .563, leading to an OPS of .999, fourth best in the majors.

One disadvantage is that OPS ignores the number of runs a player contributes to his team: a player who played in 150 games can have the same OPS as a player who only played in one game. Can you provide an example of this happening?

3.5.5. RC

Runs created (RC) was developed by Bill James in the late 1970s. Recall that the formula is

$$\text{RC}=(\#\text{H}+\#\text{BB})*\text{TB}/(\#\text{AB}+\#\text{BB})=\frac{\#H+\#BB}{\#AB+\#BB}*\text{TB}.$$

Clearly, RC is a ratio.

Notice that the quotient part of the product, $(\#\text{H}+\#\text{BB})/(\#\text{AB}+\#\text{BB})$, is close to the OBP statistic, since it is only missing HBP in the numerator and HBP and SF in the denominator. Since these statistics are usually small in number compared with #H, #BB, and #AB, RC involves the multiplication of a good approximation for OBP with TB. By multiplying OBP by TB, Bill James rated a player by the product of his ability to get on base (OBP) with his ability to advance runners through TB. In later chapters, we will use OBP * TB as a good approximation for RC.

3.5.6. BRA

In the 1970s Dick Cramer and Pete Palmer multiplied on-base percentage and slugging percentage and named this statistic the batter's run average (BRA), even though this statistic was not an average. The formula is

$$\text{BRA} = \text{OBP} * \text{SLG} = \frac{\#H + \#BB + \#HBP}{\#AB + \#BB + \#HBP + \#SF} * \frac{TB}{\#AB}.$$

Since BRA is the product of two ratios, it is also a ratio.

The BRA formula looks a little like the RC formula but is different. The basic difference is that TB is divided by #AB. Normally the SLG is a number larger than the OBP. Although the SLG can be larger than 1, usually it is less than 1. The OBP is always less than 1. When two decimals between 0 and 1 are multiplied, the result will be less than the smaller of the two decimals. For the 2008 season, Rockies center fielder Willy Taveras had an OBP = .308 and SLG = .296, so his BRA was 0.308 * 0.296 = 0.091. Thus, the BRA takes into account both the OBP and SLG but emphasizes OBP.

3.5.7. R/G

By dividing the number of runs scored by the number of games, we obtain the rate of runs scored per game.

3.5.8. RC/27

The statistic RC/27 (runs created per 27 outs) estimates the runs created by a player for his team for each 27 outs the player makes. The formula is

$$\text{RC/27} = \frac{25.5 * RC}{\#AB - \#H}.$$

The reason 25.5 is used instead of 27 is that in a game on the average 25.5 outs occur because of a hitless at bat. Other outs occur because of caught stealing, thrown out by an outfielder, or a second out in a double play. Also, the home team may only have 24 outs in a game. Of course, an extra-inning game would require more than 27 outs.

With the assumption that the same player bats in all nine positions in the batting order, the RC/27 statistic is used to predict the number of runs a team would score for a game if the player was cloned and played all positions.

This provides a unique way to compare two players with regard to run production.

3.5.9. HRA

Home run average (HRA) is the proportion of a player's at bats that result in a home run. In 1971, Henry Aaron hit 47 home runs in 495 at bats. The calculation is

$$\text{HRA} = \frac{\#HR}{\#AB} = \frac{47}{495} = .095 = \frac{9.5}{100} = 9.5\%.$$

HRA is a proportion. It leads to these three interpretations for Aaron for his home run production for the year 1971: (1) he hit a home run at a rate of 9.5 times for every 100 at bats; (2) he hit a home run in 9.5% of his at bats; and (3) his probability of hitting a home run in an at bat was .095, or 95/1000.

3.5.10. HRR

Home run rate (HRR) is the ratio of the number of at bats to the number of home runs. The smaller this number, the better the player is as a home run hitter. In 1971, Aaron's ratio of at bats to home runs was 10.53. This says that we would expect Aaron to hit a home run in every 10.53 at bats.

$$\text{HRR} = \frac{\#AB}{\#HR} = 10.53.$$

3.5.11. ISO

Isolated power (ISO) was invented by Branch Rickey and Allan Roth in the 1950s. ISO is used to evaluate a player as a power hitter. ISO removed the flaw of including singles in the slugging average. We now breakdown the formula as follows:

$$\text{ISO} = \text{SLG} - \text{AVG} = \frac{(TB - \#H)}{\#AB} = \frac{(0 * \#1B + 1 * \#2B + 2 * \#3B + 3 * \#HR)}{\#AB}.$$

ISO is a ratio. The St. Louis Cardinals had an abundance of isolated power in 2008, when the ISO stat was headed by first baseman Albert Pujols (.296) and right fielder Ryan Ludwick (.292).

3.5.12. RPR

Run production (RPR) was introduced by Steve Mann in 1977. It stands for run productivity and attempts to measure a player's ability to produce runs for his team. The formula is

$$\text{RPR} = \#\text{R} + \#\text{RBI} - \#\text{HR}.$$

The critics of this statistic point out that if a player reaches base on an error, advances to third base on a double, and then scores on an out, his RPR would be incremented by 1. The person who hit the double, who was the main reason for the run scoring, would receive no credit in his RPR. The flaw with the RPR statistic is the belief that runs are produced only by those who score them or drive them home.

RPR is an amount.

3.5.13. SOA

Strikeout average (SOA) provides the probability of a player striking out in an at bat. If a player strikes out, his probability of getting a hit for that at bat is zero. If a player makes contact with the ball, there is at least a chance for a hit or a chance for advancing a runner.

SOA = #SO/#AB.

SOA is a proportion. Diamondback's third baseman Mark Reynolds led the league in strikeouts in 2008, with 204 to his credit in 539 at bats. His SOA = 204/539 = 0.379.

3.5.14. IPBA

In-play batting average (IPBA) gives a player's batting average based solely on batted balls put into play.

IPBA = #H/#IPAB (in our definition a home run is considered a ball put into play).

A player's BA can be expressed as follows:

$$\text{BA} = \frac{\#H}{\#AB} = \frac{\#IPAB}{\#AB} * \frac{\#H}{\#IPAB} = \left(1 - \frac{\#SO}{\#AB}\right) * \frac{\#H}{\#IPAB} = (1 - \text{SOA}) * \text{IPBA}.$$

Observe that #IPAB/#AB = (#AB – #SO)/#AB = (#AB/#AB) – (#SO/#AB) = 1 – (#SO/#AB).

In Chapter 17, the above formula will be used to analyze what it will take for a batter to hit .400 for a season.

The IPBA measures a player's batting ability to get a hit after the player hits the ball in fair territory. The chance element is always there. However, the speed, height, and direction of the ball after contact with the bat along with the batter's running speed are determined by the skill of the batter.

IPBA is a proportion.

3.5.15. IPHR

In-play home run average (IPHR) gives the proportion of home runs a player hits when considering only those at bats that resulted in the ball being hit into fair territory.

IPHR = #HR/#IPAB.

IPHR is a proportion.

3.5.16. BBA

Base-on-balls average (BBA) provides the probability of a player drawing a base on balls in a plate appearance. A base on balls allows a player to reach first base without making contact with the baseball. By reaching first base, there is always the possibility of him scoring a run. Also, if a player is already on first base, a base on balls will advance that runner to second base. With today's cutoff for a pitcher of approximately 100 pitches, a walk forces a pitcher to use more of his allotted pitches. Clearly, the BBA will help to determine where a player will bat in the batting order and also the ability of a player to produce runs for his team.

BBA = #BB/#PA.

BBA is a proportion. Jack Cust, who served as the designated hitter for the Oakland Athletics, walked 111 times in 2008, ending up with a BBA of .186, second only to Adam Dunn's .187.

3.5.17. LWTF1

The generalized formula is given below:

$$\text{LWTF1} = a * \#1\text{B} + b * \#2\text{B} + c * \#3\text{B} + d * \#\text{HR} + m.$$

3.5.18. LWTF2

The generalized formula is given below:

$$\text{LWTF2} = e * \#1\text{B} + f * \#2\text{B} + g * \#3\text{B} + h * \#\text{HR} + i * \#\text{BB} + j * \text{HBP} + n.$$

Both these generalized linear weight formulas produce amounts, not ratios.

Example 3.12. Find the AVG, OBP, SLG, RC, RC/27, and HRA for the career of Henry Aaron. The data from Table 2.2 are used.

$$\textit{Career}\ \text{AVG} = \frac{3771}{12364} = .305,$$

$$\textit{Career}\ \text{OBP} = \frac{3771 + 1402 + 32}{13918} = .374,$$

$$\textit{Career}\ \text{SLG} = \frac{2294 + 2 * 624 + 3 * 98 + 4 * 755}{12364} = .555,$$

$$Career\ RC = \frac{6856 * (3771 + 1402)}{12364 + 1402} = 2576.35$$ (runs created for his team for his career),

$$Career\ RC/27 = \frac{25.5 * 2576.35}{12364 - 3771} = 7.64$$ (the average runs scored in a game for a team, composed solely of Aaron batting in all nine positions in the batting order),

$$Career\ HRA = \frac{755}{12364} = .061$$ (for 6.1% of his career at bats, Aaron hit a home run).

Exercise 3.11. Calculate for Aaron his BOP, BBA, SOA, IPBA, IPHR, SO/BB, SO/HR, R/G, and R/27 for his career. Associate each of these baseball statistics with a traditional statistic (mean, proportion, ratio, rate, and amount). What batting skill does each of these statistics attempt to measure?

Chapter Summary

The term *statistics* has two meanings. *Statistics* can refer to the field of study involved with the analysis of data. A *statistic* is also defined as a descriptive measure used to summarize data for a sample. The two most used statistics are mean and proportion. The *mean* for a set of quantitative data gives the balanced middle of the data. The *proportion* for a set of qualitative data determines the percentage of the data that satisfy a certain characteristic.

Bill James introduced the term *Sabermetrics* in the 1970s. Sabermetrics is defined as the study of baseball data through the use of mathematics and statistics. The statistics used in Sabermetrics are simply means, proportions, rates, ratios, and amounts.

The first two statistics used in baseball in the 1850s were counting the number of runs and number of outs a player made during a game and for a season. The problem with amounts, when trying to evaluate a player's performance, is that these numbers are unduly influenced by a player's number of at bats or number of games played. To overcome these problems, ratios were introduced. Two of the first rates used to describe a player's performance were runs per game and hits per at bat (batting average).

At the end of the 1876 season, the six baseball statistics used to describe a player's yearly performance were number of games, number of runs, number of hits, number of at bats, runs per game, and batting average. The first four statistics are simply counts (amounts). The number of runs per game is a rate, and batting average can be interpreted as a mean, a proportion, or a probability. Other examples of proportions introduced in this chapter are OBP, HRA, SOA, BBA, IPBA, and IPHR.

Baseball batting statistics were developed for the purpose of evaluating a player's batting performance. Certain statistics are used to evaluate a player's ability to get on base; other statistics are used to measure a player's ability to hit with power; other statistics attempt to evaluate both of these abilities simultaneously; and other statistics attempt to measure a player's ability to produce runs for his team.

Adjusting our statistics is one way to control the effects of outside variables. Since these outside variables cannot be removed, techniques are introduced to adjust a player's batting statistics to take them into account. The three adjustments introduced in this chapter were as follows:

1. The time (year) in which the player played.
2. The home ballpark the player played in.
3. The number of scheduled games his team played in a season.

Adjusting for the variable of year allows one to compare players from different eras.

Since a player plays half of his games in his home ballpark and ballparks are not uniform, adjusting for the variable of home ballpark is necessary. A player is rewarded for hitting in a difficult home ballpark and penalized for hitting in a favorable home ballpark.

Other baseball books, which concentrate solely on baseball, look at the topic of adjusted statistics in much greater depth. One such book is *Baseball's ALL-TIME BEST Hitters*, by Michael Schell.

CHAPTER PROBLEMS

1. Using the data values from Tables 3.6 and 3.9, complete Tables 3.7, 3.8, 3.10, and 3.11.
2. Compare the HRA for Ruth (1927), Maris (1961), McGuire (1998), and Bonds (2001). What conclusions can you draw?
3. Compare the HRA for Ruth (1927), Maris (1961), McGuire (1998), and Bonds (2001) by calculating their AdjT_HRAs. Do any of your conclusions change? You can obtain the league data from the website www.baseball-reference.com.
4. For the ambitious student, I would like to pose an interesting research question. Considering the three statistics AVG, OBP, and SLG, produce a list of players who led their league in all three of these categories for the same season.

Table 3.6 **Joe DiMaggio's yearly batting statistics**

Year	Team	G	AB	R	H	2B	3B	HR	RBI	BB	SO	SH	SF	HBP
1936	Yankees	138	637	132	206	44	15	29	125	24	39	3	—	4
1937	Yankees	151	621	151	215	35	15	46	167	64	37	2	—	5
1938	Yankees	145	599	129	194	32	13	32	140	59	21	0	—	2
1939	Yankees	120	462	108	176	32	6	30	126	52	20	6	—	4
1940	Yankees	132	508	93	179	28	9	31	133	61	30	0	—	3
1941	Yankees	139	541	122	193	43	11	30	125	76	13	0	—	4
1942	Yankees	154	610	123	186	29	13	21	114	68	36	0	—	2
1946	Yankees	132	503	81	146	20	8	25	95	59	24	3	—	2
1947	Yankees	141	534	97	168	31	10	20	97	64	32	0	—	3
1948	Yankees	153	594	110	190	26	11	39	155	67	30	0	—	8
1949	Yankees	76	272	58	94	14	6	14	67	55	18	0	—	2
1950	Yankees	139	525	114	158	33	10	32	122	80	33	0	—	1
1951	Yankees	116	415	72	109	22	4	12	71	61	36	0	—	6
Career		1736	6821	1390	2214	389	131	361	1537	790	369	14	—	46

Table 3.7 **Exercise to fill in table for certain batting statistics for Joe DiMaggio**

Year	BA	OBP	SLG	OPS	RC	BRA	BOP	R/G	R/27	RC/27
Career										

Table 3.8 **Exercise to fill in table for certain batting statistics for Joe DiMaggio**

Year	HRA	HRR	RPR	SOA	BBA	SO/BB	IPBA	IPHRA	ISO	SO/HR
Career										

Table 3.9 **Ted Williams's yearly batting statistics**

Year	Team	G	AB	R	H	2B	3B	HR	RBI	BB	SO	SH	SF	HBP
1939	Red Sox	149	565	131	185	44	11	31	145	107	64	3	—	2
1940	Red Sox	144	561	134	193	43	14	23	113	96	54	1	—	3
1941	Red Sox	143	456	135	185	33	3	37	120	147	27	0	—	3
1942	Red Sox	150	522	141	186	34	5	36	137	145	51	0	—	4
1946	Red Sox	150	514	142	176	37	8	38	123	156	44	0	—	2
1947	Red Sox	156	528	125	181	40	9	32	114	162	47	1	—	2
1948	Red Sox	137	509	124	188	44	3	25	127	126	41	0	—	3
1949	Red Sox	155	566	150	194	39	3	43	159	162	48	0	—	2
1950	Red Sox	89	334	82	106	24	1	28	97	82	21	0	—	0
1951	Red Sox	148	531	109	169	28	4	30	126	144	45	0	—	0
1952	Red Sox	6	10	2	4	0	1	1	3	2	2	0	—	0
1953	Red Sox	37	91	17	37	6	0	13	34	19	10	0	—	0
1954	Red Sox	117	386	93	133	23	1	29	89	136	32	0	3	1
1955	Red Sox	98	320	77	114	21	3	28	83	91	24	0	4	2
1956	Red Sox	136	400	71	138	28	2	24	82	102	39	0	0	1
1957	Red Sox	132	420	96	163	28	1	38	87	119	43	0	2	5
1958	Red Sox	129	411	81	135	23	2	26	85	98	49	0	4	4
1959	Red Sox	103	272	32	69	15	0	10	43	52	27	0	5	2
1960	Red Sox	113	310	56	98	15	0	29	72	75	41	0	2	3
Career		2292	7706	1798	2654	525	71	521	1839	2021	709	5	20	39

Table 3.10 **Exercise to fill in table for certain batting statistics for Ted Williams**

Year	BA	OBP	SLG	OPS	RC	BRA	BOP	R/G	R/27	RC/27	SO/HR
Career											

Table 3.11 **Exercise to fill in table for certain batting statistics for Ted Williams**

Year	HRA	HRR	RPR	SOA	BBA	SO/BB	IPBA	IPHRA	ISO
Career									

Chapter 4

Comparing Two Quantitative Data Sets

This chapter uses the techniques of descriptive statistics, COPS, studied in Chapter 2, to compare two quantitative data sets. When Barry Bonds broke Henry Aaron's all-time home run record, there was heated debate over which player was better. So here, the two data sets chosen for comparison are the yearly batting data of Aaron and those of Bonds. The techniques introduced in this chapter can be applied to any situation involving multiple data sets, whether or not they deal with baseball. As we are looking only at batting statistics, the descriptive measures come from Chapter 3 and include, among others, BA, OBA, and SLG.

In 1977, WFAN, a radio station in New York, introduced all-talk sports radio broadcasting 24 hours a day, 7 days a week. What could keep people listening to a sports radio station all day? One answer is the love a fan (let's not forget that it is short for "fanatic") has for their home team and its players. Fans rage over the following:

- Who was the greatest hitter in the history of baseball?
- Will any player ever equal the 56-consecutive-game hitting streak of "Joltin'" Joe DiMaggio?
- Will a player ever repeat Ted Williams's 1941 BA and hit over .400 in a season?
- Which baseball record is the hardest to break?
- Which team was the best in the history of baseball? (The answer is easy: the New York Yankees.)
- Does a certain player belong in the Hall of Fame?
- If a player hits 500 home runs for his career, does he deserve an automatic ticket to the Hall of Fame? How about 3000 hits?
- In addition to these questions, another one, more statistical in nature, often crops up on WFAN: which batting statistic is most important in determining runs scored by a team?

One important concept necessary for understanding the difference between descriptive statistics and inferential statistics is recognizing the difference between a player's *batting performance* and his *batting ability.*

Let us look at the following two examples. Suppose that we toss a coin 100 times and the result is 53 heads and 47 tails. These 53 heads and 47 tails are called the *observed* results or the *performance* in tossing the coin 100 times. The *expected theoretical* results or the *ability* is 50 heads and 50 tails. If Mark Texeira gets 30 hits in 100 at bats, his 30 hits and 70 outs are his observed results or his batter's performance in 100 at bats. Of course, there is a big difference between 100 at bats and 100 tosses of a coin. The tossing of a coin involves no skill (I have the same ability in tossing a coin as Albert Pujols) and is completely determined by chance. A plate appearance involves a combination of skill and chance (of course, mostly skill). The *true* proportion of heads expected (the ability) can be determined before ever tossing the coin, whereas a player's *true* proportion of hits (the ability) can never be determined. However, in both cases, we can use the observed descriptive measure (performance) to estimate the theoretical descriptive measure (ability). Any descriptive measure of performance is based on a sample and is a *statistic.* Any descriptive measure of ability is based on a population and is a *parameter.*

A player's batting performance and batting ability are measured through many different descriptive measures. Some of these characteristics are his performance and ability to get a hit in an at bat (BA), to get on base in a plate appearance (OBA), to hit with power and advance runners (SLG), to produce runs for his team (RC or RPR), to put the ball in play (SOA), and to control the direction and power of a batted ball so that it becomes a hit (IPBA and IPHR).

When a baseball descriptive measure refers to *ability,* it is a *parameter* and is preceded by the word "true" or "population." The difference between a (batting average, home run average, and on-base percentage) and a *true* (batting average, home run average, and on-base percentage) is that the former are known numbers, affected by the combination of chance and skill, calculated from observed plate appearances, whereas the latter are unknown and can only be estimated from the observed results. This estimation is accomplished by using the techniques from *inferential statistics. Performance* is described through *descriptive statistics,* while *ability* is measured through *inferential statistics.* This chapter concentrates on performance, saving the analysis of ability for later chapters.

In the past, the most widely used and accepted measure of batting performance and ability was the batting average (AVG or BA). The observed, or sample, batting average of a player for a year, a sequence of years, or a career is used as one measurement of a player's batting performance.

In recent years, the BA has been replaced, or at least supplemented, by other statistics that are believed to be better predictors of a player's batting ability. Some of these statistics are OBA, OPS, and RC. In Chapter 3, all of these statistics were described, along with their corresponding strengths and weaknesses.

In evaluating a player for a draft or for a trade, a general manager tries to estimate a player's ability. Many players, when first exposed to Major League pitching, have horrible batting statistics. Their initial performance does not reflect their ability. For example, it took Mickey Mantle several at bats to get his first hit. Mantle felt so bad he called his father and told him he wanted to quit baseball. His father, using tough love, visited Mickey in New York, stopping him from quitting, and the rest is history. On the other hand, Kevin Maas, as a rookie in 1990, hit 21 home runs in 254 at bats. However, he lasted just 5 years in the majors, and for his career he hit just 65 home runs in 1248 at bats. Short-term performance does not necessarily measure ability. I thought Maas was the real deal and bought several packs of playing cards seeking his rookie card. Boy, was I wrong. Yes, I still have several of his worthless rookie cards.

Professional baseball statisticians view certain baseball statistics as more valuable than others when it comes to judging a player's ability. One theory is that those statistics that are more closely related to runs produced for their team are more valuable. After all, runs win games. Another theory is that those statistics that have less chance associated with them are more valuable. For example, strikeout average relies less on chance and therefore is a good indicator of a player's ability.

Traditionally, you judge the ability of a player as a hitter by calculating his yearly BA (batting average), OBA (on-base average), and SLG (slugging average). These three statistics will constitute the first group of statistics studied in this chapter.

A batter's primary job is to produce runs for his team, so this is a crucial element in evaluating a player's batting ability. The data obtained for RC/27 and RPR/27 can be used to gauge a player's run production. RC/27 and RPR/27 are the second group of statistics studied in this chapter.

In the article "A Batting Average: Does It Represent Ability or Chance?" Jim Albert conjectures that SOA (strikeout average), BBA (base-on-balls average), and IPHR (in-play home run average) are better predictors of a player's batting ability than BA. The premise is that chance influences SOA, BBA, and IPHR less than most other baseball statistics. Assuming that Albert's premise is correct, the yearly data for these three statistics, along with the statistic of SO/BB, can be used to evaluate a player's ability. These four statistics will constitute the third group studied in this chapter.

After providing a brief timeline for Aaron and Bonds, the rest of the chapter will compare these two players with respect to the following three groups of statistics:

1. BA, OBA, and SLG.
2. RC/27 and RPR/27.
3. SOA, BBA, IPHR, and SO/BB.

The comparison of batting performance between Aaron and Bonds will use the following eight presentations (referred to in this chapter as "the eight presentations") applied to the statistics in each of the three groups for each player:

1. A table showing the yearly batting data for each of the statistics in the group for each player.
2. A table showing the important descriptive measures for each of the statistics in the group for each player.
3. A table showing the five-number summary for each of the statistics in the group for each player.
4. A comparative box-and-whisker graph for each of the statistics in the group.
5. A table showing each of the yearly statistics in the group sorted in ascending order for each player.
6. A comparative stem-and-leaf graph for each of the statistics in the group.
7. A comparative time-series graph for each of the statistics in the group.
8. A comparative write-up for each group of statistics.

For group 1, the adjusted statistics will also be calculated. These were defined in Chapter 3. We limit our discussion to the adjustment for time (year), AdjT_. For this chapter, the standard BA is .257, the standard OBA is .324, and the standard SLG is .390.

In the problems at the end of this chapter, a fourth group of statistics will be introduced. Any collection of baseball statistics can be considered a group. The student will be encouraged to create their own group of statistics.

4.1. A Brief Timeline for Aaron and Bonds

Henry Louis Aaron was born on February 5, 1934, in Mobile, Alabama. His debut in the Major Leagues occurred on April 13, 1954, at the age of 20, as a member of the Milwaukee Braves (today known as the Atlanta Braves). Aaron batted and threw right-handed. He is 6' 0" tall and weighed 180 pounds. From 1954 until 1975 he played in the National League. His final two seasons, in 1975 and 1976, were spent playing for the Milwaukee Brewers in the American League. His final game was on October 3, 1976. He was inducted into the Hall of Fame in 1982.

Barry Lamar Bonds was born on July 24, 1964, in Riverside, California. His debut in the Major Leagues occurred on May 30, 1986, at the age of 21, with the Pittsburgh Pirates. Bonds batted and threw left-handed. He is 6' 1" tall and his weight ranged from 185 pounds (in his early years) to 228 pounds (in the later years of his career). He played for the Pirates from 1986 until he signed with the San Francisco Giants in 1993. His last season was in 2007 with the Giants.

The weight of the two players is based on an average for their careers. Clearly, weight does fluctuate. But it is fair to say that Bonds was physically bigger.

4.2. Data, Variables, and Baseball Statistics

The yearly baseball data for Barry Bonds presented in Table 4.1 were downloaded from www.baseball-almanac.com. The yearly baseball data for Henry Aaron were displayed in Table 2.2.

From Table 4.1, note that in 2005, Bonds had only 42 at bats. Normally, a year in which a player had so few at bats would be considered an outlier and excluded. However, since Bonds's derived statistics for that year were consistent with his other years, I have decided to include that year in the analysis. However, such counts as number of home runs and number of hits for that year should be excluded when considering yearly totals.

To compare Aaron with Bonds, we need data for the National League as a whole. Table 3.4 provided the statistics for all plate appearances for all players in the NL for each of the years from 1960 to 2007.

Table 4.2, downloaded from www.baseball-reference.com, provides the statistics for all plate appearances for the NL players for each of the years from 1920 to 1959.

4.3. Group 1: BA, OBA, and SLG

First, recall the definitions for these three statistics:

BA = #H/#AB,
OBA = (#H + #BB + #HBP)/#PA,
SLG = (1 * #1B + 2 * #2B + 3 * #3B + 4 * #HR)/#AB.

The columns of Table 4.3 represent the yearly data for BA, OBA, and SLG for Aaron and Bonds.

The columns of Table 4.4 represent the yearly data for AdjT_BA, AdjT_OBA, and AdjT_SLG for the two players.

Table 4.1 **Barry Bonds's yearly batting statistics**

Year	Team	G	AB	R	H	2B	3B	HR
1986	Pirates	113	413	72	92	26	3	16
1987	Pirates	150	551	99	144	34	9	25
1988	Pirates	144	538	97	152	30	5	24
1989	Pirates	159	580	96	144	34	6	19
1990	Pirates	151	519	104	156	32	3	33
1991	Pirates	153	510	95	149	28	5	25
1992	Pirates	140	473	109	147	36	5	34
1993	Giants	159	539	129	181	38	4	46
1994	Giants	112	391	89	122	18	1	37
1995	Giants	144	506	109	149	30	7	33
1996	Giants	158	517	122	159	27	3	42
1997	Giants	159	532	123	155	26	5	40
1998	Giants	156	552	120	167	44	7	37
1999	Giants	102	355	91	93	20	2	34
2000	Giants	143	480	129	147	28	4	49
2001	Giants	153	476	129	156	32	2	73
2002	Giants	143	403	117	149	31	2	46
2003	Giants	130	390	111	133	22	1	45
2004	Giants	147	373	129	135	27	3	45
2005	Giants	14	42	8	12	1	0	5
2006	Giants	130	367	74	99	23	0	26
2007	Giants	126	340	75	94	14	0	28
Career 22 Years		2986	9847	2227	2935	601	77	762

The next example provides sample calculations for the yearly BA, OBA, SLG, AdjT_BA, AdjT_OBA, and AdjT_SLG.

Example 4.1. To illustrate these calculations, year 4 is chosen for both players. This corresponds to 1957 for Aaron and 1989 for Bonds. The data are obtained from Tables 2.2, 3.4, 4.1, and 4.2.

Player	Year	BA	OBA	SLG
Aaron	1957	198/615	$\frac{(198+57+0)}{(615+57+0+3)}$	$\frac{(121+2*27+3*6+4*44)}{(615)}$
		.322	**.378**	**.600**
Bonds	1989	144/580	$\frac{(144+93+1)}{(580+93+1+4)}$	$\frac{(85+2*34+3*6+4*19)}{580}$
		.248	**.351**	**.426**

Below, the adjusted statistics are calculated for the years 1957 and 1989.

RBI	BB	IBB	SO	SH	SF	HBP	AVG	OBP	SLG
48	65	2	102	2	2	2	0.223	0.330	0.416
59	54	3	88	0	3	3	0.261	0.329	0.492
58	72	14	82	0	2	2	0.283	0.368	0.491
58	93	22	93	1	4	1	0.248	0.351	0.426
114	93	15	83	0	6	3	0.301	0.406	0.565
116	107	25	73	0	13	4	0.292	0.410	0.514
103	127	32	69	0	7	5	0.311	0.456	0.624
123	126	43	79	0	7	2	0.336	0.458	0.677
81	74	18	43	0	3	6	0.312	0.426	0.647
104	120	22	83	0	4	5	0.294	0.431	0.577
129	151	30	76	0	6	1	0.308	0.461	0.615
101	145	34	87	0	5	8	0.291	0.446	0.585
122	130	29	92	1	6	8	0.303	0.438	0.609
83	73	9	62	0	3	3	0.262	0.389	0.617
106	117	22	77	0	7	3	0.306	0.440	0.688
137	177	35	93	0	2	9	0.328	0.515	0.863
110	198	68	47	0	2	9	0.370	0.582	0.799
90	148	61	58	0	2	10	0.341	0.529	0.749
101	232	120	41	0	3	9	0.362	0.609	0.812
10	9	3	6	0	1	0	0.286	0.404	0.667
77	115	38	51	0	1	10	0.270	0.454	0.545
66	132	43	54	0	2	3	0.276	0.480	0.565
1996	2558	688	1539	4	91	106	0.298	0.444	0.607

For 1957:

Aaron AdjT_BA = (1957 BA / 1957 league BA) * (standard BA),
Aaron AdjT_OBA = (1957 OBA / 1957 league OBA) * (standard OBA),
Aaron AdjT_SLG = (1957 SLG / 1957 league SLG) * (standard SLG).

For 1989:

Bonds AdjT_BA = (1989 BA / 1989 league BA) * (standard BA),
Bonds AdjT_OBA = (1989 OBA / 1989 league OBA) * (standard OBA),
Bonds AdjT_SLG = (1989 SLG / 1989 league SLG) * (standard SLG).

Player	Year	AdjT_BA	AdjT_OBA	AdjT_SLG
Aaron	1957	(.322/.260) * .257	(.378/.322) * .324	(.600/.400) * .390
		.318	**.380**	**.585**
Bonds	1989	(.248/.246) * .257	(.351/.312) * .324	(.426/.365) * .390
		.259	**.365**	**.455**

Table 4.2 Statistics for all plate appearances for all NL players for the years 1920–1959

Year	R/G	R	G	AB	H	2B	3B	HR	BB	SO	BA	OBP	SLG
1959	4.40	5462	1240	42330	11015	1788	324	1159	3974	6525	0.260	0.325	0.400
1958	4.40	5419	1232	42143	11026	1769	365	1183	4065	6192	0.262	0.328	0.405
1957	4.38	5426	1238	42919	11162	1733	365	1178	3866	6150	0.260	0.322	0.400
1956	4.25	5275	1242	41849	10716	1659	372	1219	3982	5694	0.256	0.321	0.401
1955	4.53	5578	1232	41773	10808	1677	362	1263	4240	5419	0.259	0.328	0.407
1954	4.56	5624	1232	42027	11142	1816	403	1114	4414	5086	0.265	0.335	0.407
1953	4.75	5914	1244	42639	11342	1777	414	1197	4220	5307	0.266	0.335	0.411
1952	4.17	5158	1236	41878	10582	1672	338	907	4147	5240	0.253	0.323	0.374
1951	4.46	5552	1244	42704	11088	1746	367	1024	4362	4746	0.260	0.331	0.390
1950	4.66	5760	1236	42416	11085	1885	370	1100	4537	5007	0.261	0.336	0.401
1949	4.54	5650	1244	42711	11207	1865	370	935	4405	4587	0.262	0.334	0.389
1948	4.43	5487	1238	42256	11022	1840	384	845	4406	4729	0.261	0.333	0.383
1947	4.57	5666	1240	42434	11264	1860	392	886	4477	4529	0.265	0.338	0.390
1946	3.96	4916	1242	42094	10762	1752	382	562	4399	4474	0.256	0.329	0.355
1945	4.46	5512	1236	42823	11343	1823	336	577	4150	3864	0.265	0.333	0.364
1944	4.25	5295	1246	42918	11191	1882	395	575	3984	3941	0.261	0.326	0.363
1943	3.94	4898	1242	42491	10945	1739	388	432	4048	4059	0.258	0.324	0.347
1942	3.90	4784	1226	41769	10391	1680	323	538	4076	4189	0.249	0.318	0.343
1941	4.23	5266	1244	42729	11039	1892	359	597	4149	4414	0.258	0.326	0.361
1940	4.39	5421	1234	42986	11328	1934	416	688	3779	4329	0.264	0.326	0.376
1939	4.44	5472	1232	42285	11505	2032	418	649	3824	4195	0.272	0.335	0.386
1938	4.42	5388	1220	42513	11358	1913	450	611	3708	4093	0.267	0.329	0.376
1937	4.51	5567	1234	42660	11591	1922	458	624	3667	4553	0.272	0.332	0.382
1936	4.71	5837	1240	43891	12206	2071	431	606	3565	4203	0.278	0.335	0.386
1935	4.71	5806	1234	43438	12041	2053	462	662	3284	4072	0.277	0.331	0.391
1934	4.68	5695	1216	42982	11996	2108	433	656	3247	4150	0.279	0.333	0.394
1933	3.97	4908	1236	42559	11332	1854	422	460	2979	3528	0.266	0.317	0.362
1932	4.60	5680	1236	43763	12091	2293	502	651	3141	3857	0.276	0.328	0.396
1931	4.48	5537	1236	42941	11883	2188	532	493	3502	3862	0.277	0.334	0.387
1930	5.68	7025	1236	43693	13260	2386	625	892	3691	3848	0.303	0.360	0.448
1929	5.36	6609	1232	43030	12668	2253	569	754	3961	3465	0.294	0.357	0.426
1928	4.70	5769	1228	42336	11901	2021	518	610	3848	3410	0.281	0.344	0.397
1927	4.58	5651	1234	42344	11935	1888	540	483	3413	3491	0.282	0.339	0.386
1926	4.54	5612	1236	42009	11755	1948	589	439	3473	3359	0.280	0.338	0.386
1925	5.06	6195	1224	42859	12495	2120	614	636	3460	3373	0.292	0.348	0.414
1924	4.54	5581	1228	42445	12009	1881	622	499	3216	3408	0.283	0.337	0.392
1923	4.85	5987	1234	43216	12348	1912	588	538	3494	3408	0.286	0.343	0.395
1922	5.00	6194	1240	43050	12579	1911	662	530	3455	3380	0.292	0.348	0.404
1921	4.59	5632	1226	42376	12266	1839	670	460	2906	3380	0.289	0.338	0.397
1920	3.97	4893	1234	42197	11376	1604	644	261	3016	3632	0.270	0.322	0.357

Table 4.3 Yearly BA, OBP, and SLG for Aaron and Bonds

	BA		OBP		SLG	
Year	Aaron	Bonds	Aaron	Bonds	Aaron	Bonds
1	0.280	0.223	0.322	0.330	0.447	0.416
2	0.314	0.261	0.366	0.329	0.540	0.492
3	0.328	0.283	0.365	0.368	0.558	0.491
4	0.322	0.248	0.378	0.351	0.600	0.426
5	0.326	0.301	0.386	0.406	0.546	0.565
6	0.355	0.292	0.401	0.410	0.636	0.514
7	0.292	0.311	0.352	0.456	0.566	0.624
8	0.327	0.336	0.381	0.458	0.594	0.677
9	0.323	0.312	0.390	0.426	0.618	0.647
10	0.319	0.294	0.391	0.431	0.586	0.577
11	0.328	0.308	0.393	0.461	0.514	0.615
12	0.318	0.291	0.379	0.446	0.560	0.585
13	0.279	0.303	0.356	0.438	0.539	0.609
14	0.307	0.262	0.369	0.389	0.573	0.617
15	0.287	0.306	0.354	0.440	0.498	0.688
16	0.300	0.328	0.396	0.515	0.607	0.863
17	0.298	0.370	0.385	0.582	0.574	0.799
18	0.327	0.341	0.410	0.529	0.669	0.749
19	0.265	0.362	0.390	0.609	0.514	0.812
20	0.301	0.286	0.402	0.404	0.643	0.667
21	0.268	0.270	0.341	0.454	0.491	0.545
22	0.234	0.276	0.332	0.480	0.355	0.565
23	0.229		0.315		0.369	

Exercise 4.1. For the year 2004 (year #20) for Bonds, verify the results in Tables 4.3 and 4.4 for BA, OBA, SLG, AdjT_BA, AdjT_OBA, and AdjT_SLG.

Tables 4.5–4.7 and Figures 4.1–4.3 compare the statistics in group 1 for Aaron and Bonds. The stem-and-leaf graph provides information on the shape of the data. For BA, the numbers will be split after the second digit *xx*|*x*; for OBA, the numbers will be split after the first digit *x*|*xx* and then the first digit will be split again into two pieces; for SLG, the numbers will be split as for the OBA.

4.3.1. Comparison of BA, OBA, and SLG for Aaron and Bonds

We must make the decision whether to use the adjusted or the regular statistics. As the graphs for the time series in Figure 4.3 show little difference between the two sets of statistics, there is no reason to use the adjusted statistics. Therefore, the regular statistics of BA, OBA, and SLG will be used.

Table 4.4 **Yearly adjusted BA, OBP, and SLG for Aaron and Bonds**

	AdjT_BA		AdjT_OBP		AdjT_SLG	
Year	Aaron	Bonds	Aaron	Bonds	Aaron	Bonds
1	0.272	0.227	0.311	0.332	0.428	0.427
2	0.312	0.257	0.354	0.325	0.517	0.475
3	0.329	0.293	0.368	0.385	0.543	0.528
4	0.318	0.259	0.380	0.365	0.585	0.455
5	0.320	0.302	0.381	0.410	0.526	0.575
6	0.351	0.300	0.400	0.419	0.620	0.537
7	0.294	0.317	0.358	0.469	0.569	0.661
8	0.321	0.327	0.378	0.454	0.572	0.662
9	0.318	0.300	0.386	0.414	0.613	0.608
10	0.335	0.287	0.414	0.422	0.628	0.552
11	0.332	0.302	0.409	0.453	0.536	0.588
12	0.328	0.284	0.395	0.434	0.584	0.556
13	0.280	0.297	0.369	0.429	0.547	0.579
14	0.317	0.251	0.386	0.369	0.616	0.561
15	0.304	0.296	0.382	0.417	0.570	0.621
16	0.308	0.323	0.402	0.504	0.642	0.792
17	0.297	0.367	0.379	0.570	0.571	0.760
18	0.333	0.334	0.420	0.516	0.713	0.701
19	0.275	0.354	0.401	0.593	0.549	0.749
20	0.305	0.281	0.404	0.397	0.667	0.628
21	0.270	0.262	0.339	0.440	0.522	0.498
22	0.234	0.267	0.329	0.466	0.375	0.521
23	0.231		0.319		0.399	

4.3.1.1. Comparison for Batting Average

The BA reflects the ability to make contact with the ball and drive it into fair territory before striking out and the ability to direct the ball into an area where it does not lead to an out.

The descriptive measures for Aaron and Bonds with respect to their yearly batting averages (BAs) are similar (Table 4.5). The mean yearly BAs are .301 for Aaron and .299 for Bonds. This small difference can be attributed to chance. As their means are close, we can use their actual standard deviations to measure the spread for their yearly BAs. The standard deviations for the yearly BAs are .0312 (Aaron) and .0355 (Bonds). As the standard deviation shows how spread out the numbers (here the numbers are BAs) are about the mean, we conclude that both players had similar consistency in their yearly batting averages.

The same conclusions can be drawn by looking at their five-number summaries and their comparative box-and-whisker graph (Table 4.6 and

Table 4.5 **Descriptive measures for yearly BA, OBP, and SLG for Aaron and Bonds**

	BA		OBP		SLG	
Descriptive measures	Aaron	Bonds	Aaron	Bonds	Aaron	Bonds
Mean	0.301	0.299	0.372	0.442	0.548	0.615
Median	0.307	0.298	0.379	0.439	0.560	0.612
Mode	0.327,0.328	none	0.390	none	0.514	0.565
Q1	0.280	0.276	0.354	0.404	0.514	0.545
Q2	0.307	0.298	0.379	0.439	0.560	0.612
Q3	0.326	0.312	0.391	0.461	0.600	0.677
Largest value	0.355	0.370	0.410	0.609	0.669	0.863
Smallest value	0.229	0.223	0.315	0.329	0.355	0.416
Range	0.126	0.147	0.095	0.280	0.314	0.447
IQR	0.046	0.033	0.037	0.057	0.086	0.132
Sample standard deviation	0.031	0.001	0.026	0.073	0.079	0.118
Sample variance	0.001	0.035	0.001	0.005	0.006	0.014
Coefficient of variation	0.103	0.119	0.071	0.164	0.143	0.192
Outlier fence—lower value	0.211	0.222	0.299	0.319	0.385	0.347
Outlier fence—upper value	0.395	0.366	0.447	0.547	0.729	0.875
Outliers	none	0.370	none	0.582	0.355	none
Outliers				0.609	0.369	

Table 4.6 **Five-number summary of BA, OBP, and SLG for Aaron and Bonds**

	BA		OBP		SLG	
Summary	Aaron	Bonds	Aaron	Bonds	Aaron	Bonds
Smallest	.229	.223	.315	.329	.355	.416
Q1	.280	.276	.354	.404	.514	.545
Q2	.307	.298	.379	.439	.560	.612
Q3	.326	.312	.391	.461	.600	.677
Largest	.355	.370	.410	.606	.669	.863

Fig. 4.1). The median yearly BAs for Aaron and Bonds, respectively, are .307 and .298. The interquartile range for Aaron's yearly BAs is .046, whereas the interquartile range for Bonds's yearly BAs is .036.

The stem-and-leaf comparison graph (Fig. 4.2) reveals the interesting fact that Aaron had a batting average between .320 and .329 for years 3, 4, 5, 8, 9, 11, and 18. An estimate for the mode is .325, which is the midpoint of the longest stem at .32. On the other hand, Bonds had four years when his batting average was between .300 and .309, giving an estimated mode of .305.

Finally, the time-series comparison graph (Fig. 4.3) reveals another interesting result. After dividing their career years roughly in thirds, we

Table 4.7 **Sorted yearly BA, OBP, and SLG for Aaron and Bonds**

BA		OBP		SLG	
Aaron	Bonds	Aaron	Bonds	Aaron	Bonds
0.229		0.315		0.355	
0.234	0.223	0.322	0.329	0.369	0.416
0.265	0.248	0.332	0.330	0.447	0.426
0.268	0.261	0.341	0.351	0.491	0.491
0.279	0.262	0.352	0.368	0.498	0.492
0.280	0.270	0.354	0.389	0.514	0.514
0.287	0.279	0.356	0.404	0.514	0.545
0.292	0.283	0.365	0.406	0.539	0.557
0.298	0.286	0.366	0.410	0.540	0.565
0.300	0.291	0.369	0.426	0.546	0.577
0.301	0.292	0.378	0.431	0.558	0.585
0.307	0.294	0.379	0.438	0.560	0.609
0.314	0.301	0.381	0.440	0.566	0.615
0.318	0.303	0.385	0.446	0.573	0.617
0.319	0.306	0.386	0.454	0.574	0.624
0.322	0.308	0.390	0.456	0.586	0.647
0.323	0.311	0.390	0.458	0.594	0.667
0.326	0.312	0.391	0.461	0.600	0.677
0.327	0.328	0.393	0.497	0.607	0.688
0.327	0.336	0.396	0.515	0.618	0.749
0.328	0.341	0.401	0.529	0.636	0.799
0.328	0.362	0.402	0.582	0.643	0.812
0.355	0.370	0.410	0.609	0.669	0.863

have the following results: for the first third of their careers, Aaron had a higher BA; for the second third of their careers, Aaron's BA was higher but their averages were much closer; for the final third of their careers, Bonds's yearly batting average was much higher. In fact, Bonds's BA performance continually improved with age until age 40. Bonds batted .370 at the age of 37, and his best four consecutive years were between the ages of 36 and 39.

The time-series analysis of their respective batting average performances can lead one to speculate that, at the beginning of their respective careers, Aaron performed better. However, as they both got older, Bonds's performance, as measured by BA, kept improving, whereas Aaron's performance, as measured by BA, stayed pretty constant. It can be argued that from age 35 Bonds's performance, as measured by BA, was far better.

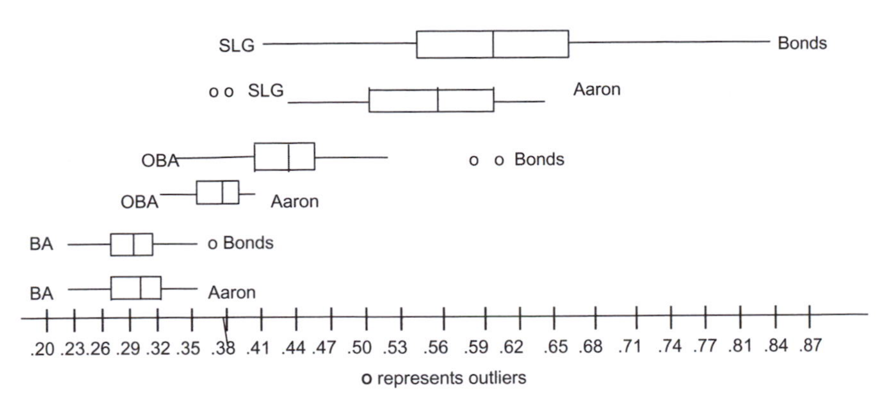

Figure 4.1 Comparative box-and-whisker graphs for group 1 statistics

Aaron Leaves	BA Stem	Bonds Leaves
9	.22	3
4	.23	
	.24	8
	.25	
8 5	.26	12
9	.27	09
70	.28	36
82	.29	124
710	.30	1368
984	.31	12
8877632	.32	8
	.33	6
	.34	1
5	.35	
	.36	2
	.37	0

Aaron Leaves	OBA Stem	Bonds Leaves
41 32 22 15	.3	29 30
96 93 91 90 90 86 85 81 79 78 69 66 65 56 54 52	.3	51 68 89
10 02 01	.4	04 06 10 26 31 38 40 46
	.4	54 56 58 61 97
	.5	15 29
	.5	82
	.6	09
	.6	

Aaron Leaves	SLG Stem	Bonds Leaves
69 55	.3	
47	.4	16 26
98 91	.4	91 92
46 40 39 14 14	.5	14 45
94 86 74 73 66 60 58	.5	57 65 77 85
43 36 18 07 00	.6	09 15 17 24 47
69	.6	67 77 88
	.7	49
	.7	99
	.8	12
	.8	63

Figure 4.2 Comparative stem-and-leaf graphs for group 1 statistics

Considering the sample of all their career y ears, it is difficult to decide which player's performance, based on yearly batting average, was superior.

4.3.1.2. Comparison for On-Base Average

On-base average (percentage) measures a player's ability to reach base by either a hit (H), a base on balls (BB), or being hit by a pitch (HBP). Unlike a batting average, the on-base average (OBA) takes into account the number

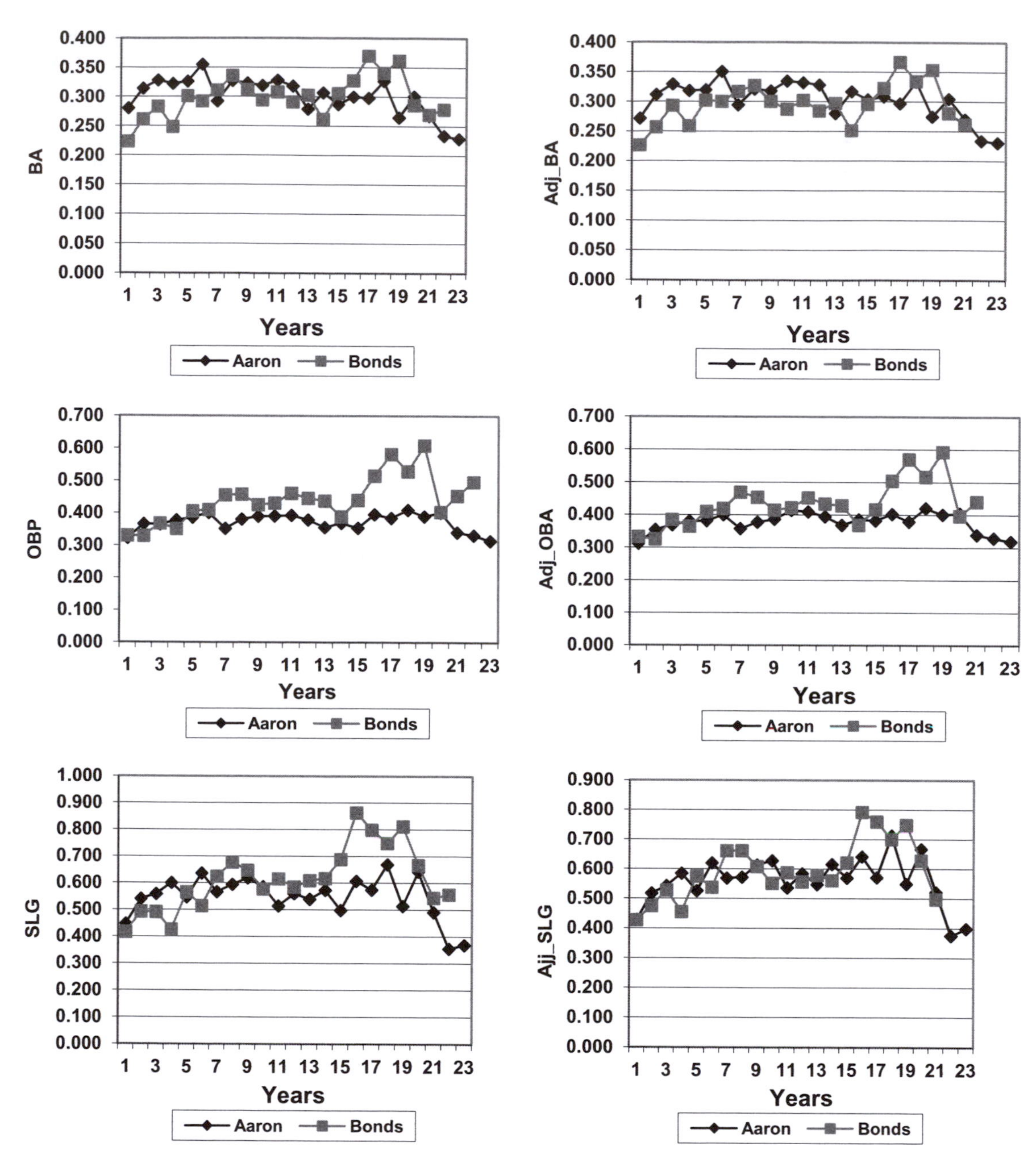

Figure 4.3 Comparative time-series graphs for group 1 statistics

of walks. Walks can be earned by skill (the batter knows the strike zone) or fear (the batter is intentionally walked to avoid a potential extra-base hit). In either case, it counts as a base on balls. Table 4.8 shows the total number of base on balls for each player and the number of intentional base on balls (IBB).

For our analysis, we won't worry about whether the base on balls are intentional or not. In other words, all walks are considered equal.

The analysis of OBA begins with the descriptive measures (Table 4.5). The mean and median for Aaron's yearly OBA were .372 and .379, respectively. The mean and median for Bonds's yearly OBA were .442 and .439, respectively. Clearly, the numbers are much higher for Bonds. The coefficient of variation for Bonds's yearly OBA is .1641. The coefficient of variation for Aaron's yearly OBA is .0705. The coefficient of variation shows that Bonds's yearly OBA had more variability than Aaron's yearly OBA.

The comparative box-and-whisker graph for OBA (Fig. 4.1) confirms these results: the OBA box for Bonds is shifted to the right of the OBA box for Aaron.

As with BA, we analyze the comparative times-series graph for OBA (Fig. 4.3) by dividing their careers into thirds. For the first third, their yearly OBAs are similar; for the second third, Bonds's yearly OBA performance was higher; and for the final third of their careers, Bonds's yearly OBA performance spiked far higher. This spike may be partially explained by the fact that Bonds's intentional base on balls (IBB) also spiked higher during that period.

The comparative stem-and-leaf graph (Fig. 4.2) for Aaron shows that the stem at .350 had 16 leaves, giving an approximate mode of .375. For Bonds, the stem at .400 had 8 leaves, giving an approximate mode of .425. The leaves for Bonds are shifted to the right of the leaves for Aaron. The comparative stem-and-leaf graph shows a clear superiority for Bonds with respect to yearly OBA. It also confirms the increased variability of Bonds's OBA.

From the above paragraphs, the yearly OBA for Bonds is superior to Aaron's yearly OBA: Bonds was far better at getting on base.

Table 4.8 **Yearly BB and IBB for Aaron and Bonds**

	Aaron		Bonds	
Year	BB	IBB	BB	IBB
1	28	—	65	2
2	49	5	54	3
3	37	6	72	14
4	57	15	93	22
5	59	16	93	15
6	51	17	107	25
7	60	13	127	32
8	56	20	126	43
9	66	14	74	18
10	78	18	120	22
11	62	9	151	30
12	60	10	145	34
13	76	15	130	29
14	63	19	73	9
15	64	23	117	22
16	87	19	177	35
17	74	15	198	68
18	71	21	148	61
19	92	15	232	120
20	68	13	9	3
21	39	6	115	38
22	70	3	132	43
23	35	1	—	—
Totals	1402	293	2558	688

Exercise 4.2. Calculate the following:

a. The mean, median, and mode for the yearly BB for Aaron and Bonds.
b. The mean, median, and mode for the yearly IBB for Aaron and Bonds.
c. For each year, the percent of BB that were IBB for Aaron and Bonds.
d. For their respective careers, the percent of BB that were IBB.

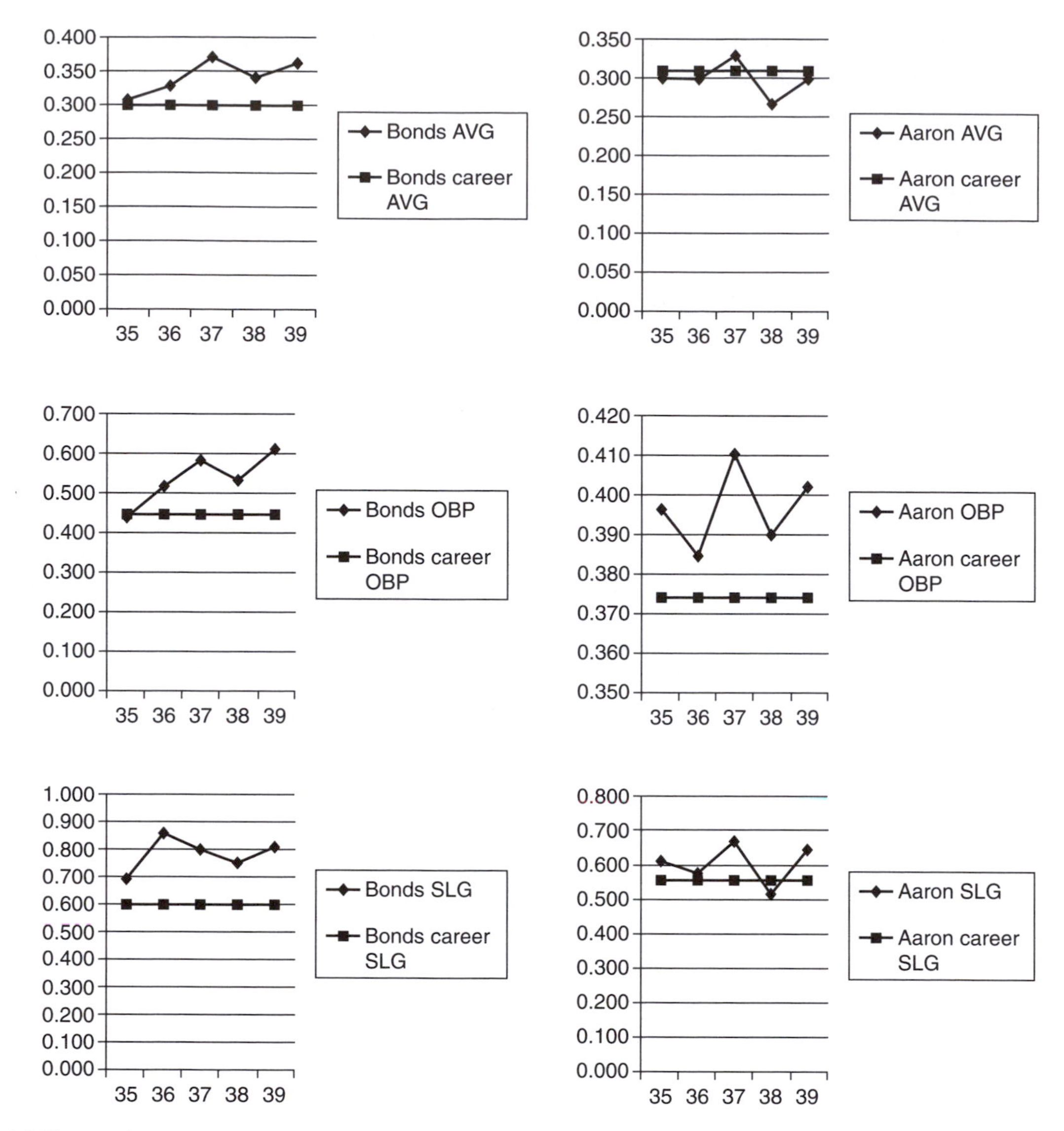

Figure 4.4 Time-series comparison (ages 35–39 to career results) for group 1 statistics for each player

Exercise 4.3. Construct the comparative box-and-whisker graph, the comparative stem-and-leaf graph, and the comparative time-series graph for OBA (use for the yearly number of base on balls the result of subtracting #IBB from the #BB). Compare these graphs with the graphs in Figures 4.1, 4.2, and 4.3.

4.3.1.3. Comparison for Slugging

Slugging follows a similar pattern to OBA. For that reason, the comparison is left as an exercise.

Exercise 4.4. Using the "the eight presentations" discussed at the beginning of the chapter, compare the slugging statistics of Aaron and Bonds.

4.3.2. Final Comparison for Group 1 Statistics

What I find most interesting in the comparison of BA, OBA, and SLG is how the three corresponding time-series graphs in Figure 4.3 follow the same pattern. The real difference appears in the last third of their careers, specifically the ages between 35 and 39. For those years, a pronounced upward spike occurred in BA, OBA, and SLG for Bonds, whereas Aaron's BA, OBA, and SLG show only a slight change.

Figure 4.4 provides time-series graphs to compare each player's career BA, OBA, and SLG with those for ages 35–39.

In Table 4.3, the years 16–20 correspond to Aaron's ages of 35–39. The years 15–19 correspond to Bonds's ages of 35–39.

Clearly, Bonds overachieved, compared with his career averages, for all three statistics. Aaron overachieved for OBA, slightly overachieved for SLG, and slightly underachieved for BA.

4.4. Group 2: RC/27 and RPR/27

Recall that runs created and runs produced measure the number of runs a player produces for his team and are defined by

RC = (#H + #BB) * (TB)/(#AB + #BB),
RPR = #R + #RBI − #HR.

The next two statistics measure the number of runs a player would produce for his team per game, assuming that he is cloned and bats in all nine positions in the batting order:

RC/27 = (25.5) * RC/(#AB − #H),
RPR/27 = (25.5) * (RPR)/(#AB − #H).

These statistics, described in Tables 4.9–4.12 and Figures 4.5–4.7, follow the same pattern as the statistics of group 1.

4.4.1. Comparison of RC/27 and RPR/27 for Aaron and Bonds

These statistics assume that the given player occupies all nine positions in the batting order. If so, then RC/27 uses Bill James's formula for RC to predict runs per game. The RPR/27 statistic uses the actual runs, runs batted in, and home runs produced by the player to predict his runs per game.

The time-series graphs show Bonds's superiority over Aaron for the last half of their careers. Again, the biggest spike occurred between the ages of 35 and 39. In fact, the outliers representing superior runs created occurred when Bonds was 36, 37, and 39 years of age.

Table 4.9 **Yearly RC/27 and RPR/27 for Aaron and Bonds**

	RC		RC/27		RPR		RPR/27	
Year	Aaron	Bonds	Aaron	Bonds	Aaron	Bonds	Aaron	Bonds
1	67.00	56.49	5.07	4.49	114	104	8.63	8.26
2	118.82	88.69	7.34	5.56	184	133	11.36	8.33
3	124.74	96.94	7.78	6.40	172	131	10.72	8.65
4	140.02	86.98	8.56	5.09	206	135	12.60	7.90
5	126.73	119.21	7.98	8.37	174	185	10.96	13.00
6	161.18	108.71	10.12	7.68	200	186	12.56	13.14
7	119.21	134.72	7.27	10.54	188	178	11.47	13.92
8	137.44	168.50	8.63	12.00	201	206	12.62	14.67
9	142.95	106.64	9.09	10.11	210	133	13.35	12.61
10	145.60	125.48	8.63	8.96	207	180	12.28	12.86
11	115.44	147.57	7.69	10.51	174	209	11.58	14.89
12	122.03	137.81	8.00	9.32	166	184	10.88	12.45
13	116.79	146.32	6.85	9.69	200	205	11.72	13.58
14	128.16	84.94	7.86	8.27	183	140	11.22	13.63
15	107.28	145.93	6.33	11.17	141	186	8.32	14.24
16	131.44	209.59	8.75	16.70	153	193	10.19	15.38
17	114.39	185.91	8.06	18.66	183	181	12.89	18.17
18	136.26	152.51	10.43	15.13	166	156	12.71	15.48
19	90.09	183.80	6.96	19.69	118	185	9.12	19.82
20	101.90	11.53	9.48	9.80	140	13	13.03	11.05
21	57.28	88.80	5.87	8.45	96	125	9.83	11.89
22	55.21	91.93	3.95	9.53	93	113	6.66	11.71
23	31.70		3.87		47		5.73	

The mean and median for RC/27 were, respectively, 7.59 and 7.86 for Aaron versus 10.29 and 9.72 for Bonds. For RPR/27, the respective mean and median were, respectively, 10.89 and 11.36 for Aaron versus 12.98 and 13.07 for Bonds.

4.4.2. Final Comparison for Group 2 Statistics

Taking all the above into account, Bonds's run production performance exceeded Aaron's. However, this superiority appears only in the last half of their careers.

Exercise 4.5. For the years for both Aaron and Bonds, compare RC/27 with RPR/27 by constructing a comparative stem-and-leaf graph. The headings should take the following form:

Aaron		Bonds	
RPR/27	RC/27	RPR/27	RC/27

Table 4.10 **Descriptive measures for yearly RC/27 and RPR/27 for Aaron and Bonds**

	RC/27		RPR/27	
Descriptive measures	Aaron	Bonds	Aaron	Bonds
Mean	7.59	10.29	10.89	12.98
Median	7.86	9.72	11.36	13.07
Mode	8.63	none	none	none
Q1	6.59	8.27	9.48	11.62
Q2	7.86	9.72	11.36	13.07
Q3	8.63	11.17	12.58	14.67
Largest value	10.43	19.69	13.35	19.82
Smallest value	3.87	4.49	5.73	7.90
Range	6.36	15.20	7.62	11.92
IQR	2.04	2.90	3.10	3.05
Sample standard deviation	1.71	4.06	2.04	3.02
Sample variance	2.93	16.50	4.16	9.10
Coefficient variation	0.22	0.39	0.19	0.23
Outlier fence—lower value	3.53	3.92	4.83	7.05
Outlier fence—upper value	11.69	15.52	17.23	19.25
Outliers	none	16.70	none	19.82
Outliers		18.66		
Outliers		19.69		

Table 4.11 **Five-number summary of RC/27 and RPR/27 for Aaron and Bonds**

	RC/27		RPR/27	
Summary	Aaron	Bonds	Aaron	Bonds
Smallest	3.87	4.49	5.73	7.90
Q1	6.59	8.27	9.48	11.62
Q2	7.86	9.72	11.36	13.07
Q3	8.63	11.17	12.58	14.67
Largest	10.43	19.82	13.35	19.82

4.5. Group 3: SOA, BBA, SO/BB, and IPHR (IPHRA)

This class of statistics has the following definitions:

SOA (strikeout average) = #SO/#AB,
BBA (base-on-balls average) = #BB/#PA,
SO/BB (ratio of strikeouts to base on balls) = #SO/#BB,
IPHR or IPHRA (in-play home run average) = #HR/#IPAB
[IPAB = (#AB − #SO)].

Table 4.12 **Sorted yearly RC, RC/27, RPR, and RPR/27 for Aaron and Bonds**

RC		RC/27		RPR		RPR/27	
Aaron	Bonds	Aaron	Bonds	Aaron	Bonds	Aaron	Bonds
31.70	11.53	3.87	4.49	47	13	5.73	7.90
55.21	56.49	3.95	5.09	93	72	6.66	8.26
57.28	60.42	5.07	5.56	96	104	8.32	8.33
67.00	84.94	5.87	6.40	114	125	8.63	8.65
90.09	86.98	6.33	7.68	118	131	9.12	11.05
101.90	88.69	6.85	8.27	140	133	9.83	11.62
107.28	88.80	6.96	8.37	141	133	10.19	11.89
114.39	96.94	7.27	8.45	153	135	10.72	12.45
115.44	106.64	7.34	8.96	166	140	10.88	12.61
116.79	108.71	7.69	9.32	166	156	10.96	12.86
118.82	119.21	7.78	9.69	172	178	11.22	13.00
119.21	125.48	7.86	9.75	174	180	11.36	13.14
122.03	134.72	7.98	9.80	174	181	11.47	13.58
124.74	137.81	8.00	10.11	183	184	11.58	13.63
126.73	145.93	8.06	10.51	183	185	11.72	13.92
128.16	146.32	8.56	10.54	184	185	12.28	14.24
131.44	147.57	8.63	11.17	188	186	12.56	14.67
136.26	152.51	8.63	12.00	200	186	12.60	14.89
137.44	168.50	8.75	15.13	200	193	12.62	15.38
140.02	183.80	9.09	16.70	201	205	12.71	15.48
142.95	185.91	9.48	18.66	206	206	12.89	18.17
145.60	209.59	10.12	19.69	207	209	13.03	19.82
161.18		10.43		210		13.35	

Tables 4.13–4.16 and Figures 4.8–4.11 provide "the eight presentations" for the statistics in group 3.

Exercise 4.6. Give a comparative write-up for Aaron and Bonds with respect to SOA, BBA, SO/BB, and IPHR.

Exercise 4.7. Give a final statement for the group 3 statistics.

Chapter Summary

This chapter used the tools of descriptive statistics to compare two quantitative data sets. Measures of the middle, measures of spread, stem-and-leaf graphs, box-and-whisker graphs, and time-series graphs were all used.

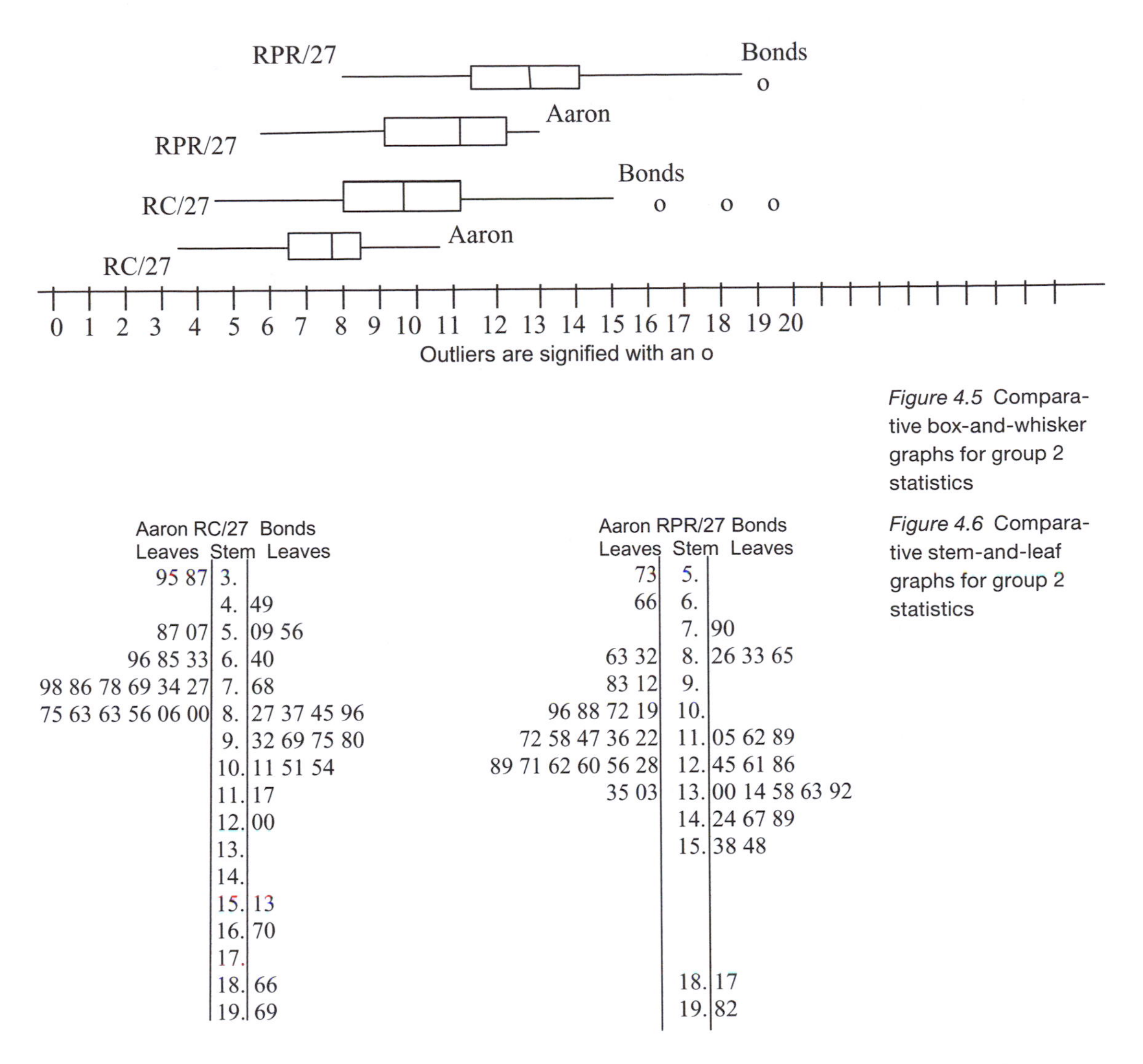

Figure 4.5 Comparative box-and-whisker graphs for group 2 statistics

Figure 4.6 Comparative stem-and-leaf graphs for group 2 statistics

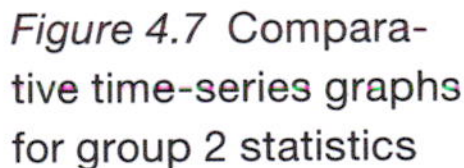

Figure 4.7 Comparative time-series graphs for group 2 statistics

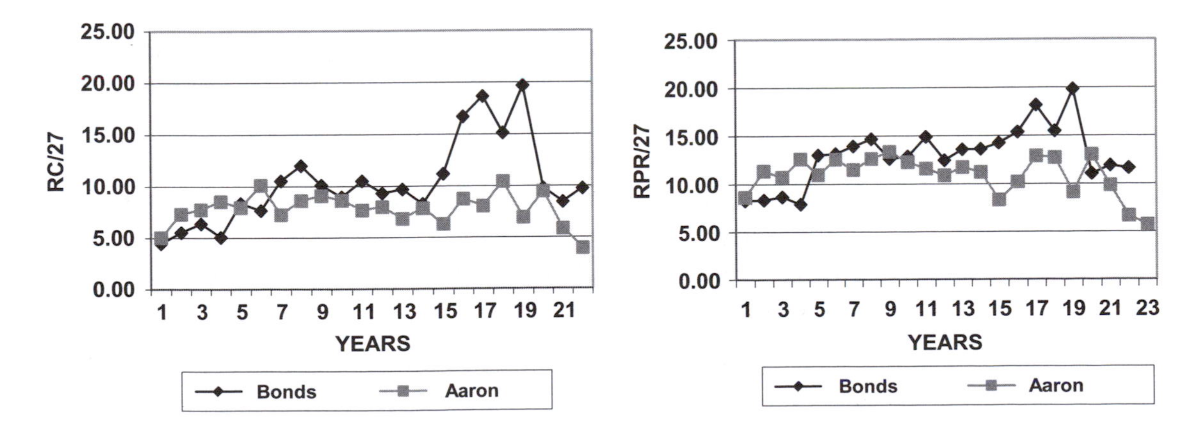

Table 4.13 **Yearly SOA, BBA, SO/BB, and IPHR for Aaron and Bonds**

	Aaron					Bonds				
Year	IPAB	SOA	BBA	IPHR	SO/BB	IPAB	SOA	BBA	IPHR	SO/BB
1	429	0.083	0.056	0.030	1.393	311	0.247	0.135	0.051	1.569
2	541	0.101	0.074	0.050	1.245	463	0.160	0.088	0.054	1.630
3	555	0.089	0.056	0.047	1.459	456	0.152	0.117	0.053	1.139
4	557	0.094	0.084	0.079	1.018	487	0.160	0.137	0.039	1.000
5	552	0.082	0.089	0.054	0.831	436	0.160	0.150	0.076	0.892
6	575	0.086	0.074	0.068	1.059	437	0.143	0.169	0.057	0.682
7	527	0.107	0.090	0.076	1.050	404	0.146	0.208	0.084	0.543
8	539	0.106	0.084	0.063	1.143	460	0.147	0.187	0.100	0.627
9	519	0.123	0.099	0.087	1.106	348	0.110	0.156	0.106	0.581
10	537	0.149	0.109	0.082	1.205	423	0.164	0.189	0.078	0.692
11	524	0.081	0.098	0.046	0.742	441	0.147	0.224	0.095	0.503
12	489	0.142	0.094	0.065	1.350	445	0.164	0.210	0.090	0.600
13	507	0.159	0.110	0.087	1.263	460	0.167	0.187	0.080	0.708
14	503	0.162	0.094	0.078	1.540	293	0.175	0.168	0.116	0.849
15	544	0.102	0.095	0.053	0.969	403	0.160	0.193	0.122	0.658
16	500	0.086	0.136	0.088	0.540	383	0.195	0.267	0.191	0.525
17	453	0.122	0.124	0.084	0.851	356	0.117	0.324	0.129	0.237
18	437	0.117	0.124	0.108	0.817	332	0.149	0.269	0.136	0.392
19	394	0.122	0.169	0.086	0.598	332	0.110	0.376	0.136	0.177
20	341	0.130	0.146	0.117	0.750	36	0.143	0.173	0.139	0.667
21	311	0.085	0.102	0.064	0.744	316	0.139	0.233	0.082	0.443
22	414	0.110	0.129	0.029	0.729	181	0.174	0.297	0.094	0.404
23	233	0.140	0.114	0.043	1.086					

Specifically, we compared the yearly batting performance of Aaron and Bonds with respect to three groups of batting statistics. The baseball statistics, represented in each group, measure a specific type of batting performance:

- Group 1 consisted of BA, OBA, and SLG. This group represented the traditional statistics used to judge a player's overall batting ability.
- Group 2 consisted of RC/27 and RPR/27. These two statistics measure a player's ability to produce runs for his team.
- Group 3 consisted of SOA, BBA, SO/BB, and IPHR. These four statistics are considered by some baseball experts to have the least amount of chance associated with them and thus provide a better measure of a player's batting ability.

The comparison of the statistics, within each of these groups, is done using "the eight presentations."

Table 4.14 **Descriptive measures for yearly SOA, BBA, IPHR, and SO/BB for Aaron and Bonds**

	SOA		BBA		IPHR		SO/BB	
Descriptive measures	Aaron	Bonds	Aaron	Bonds	Aaron	Bonds	Aaron	Bonds
Mean	0.112	0.156	0.102	0.203	0.069	0.096	1.021	0.705
Median	0.107	0.159	0.098	0.188	0.068	0.092	1.050	0.643
Mode	0.086	0.160	many	0.187	0.870	0.136	none	none
Q1	0.086	0.146	0.084	0.168	0.049	0.078	0.747	0.525
Q2	0.107	0.159	0.098	0.188	0.068	0.092	1.050	0.643
Q3	0.127	0.164	0.119	0.233	0.085	0.849	1.225	0.849
Largest value	0.162	0.247	0.169	0.376	0.117	0.191	1.540	1.630
Smallest value	0.081	0.110	0.056	0.088	0.029	0.039	0.540	0.177
Range	0.081	0.137	0.113	0.288	0.088	0.152	1.000	1.453
IQR	0.041	0.018	0.035	0.065	0.036	0.771	0.478	0.324
Sample standard deviation	0.026	0.029	0.027	0.070	0.023	0.036	0.279	0.036
Sample variance	0.001	0.001	0.001	0.005	0.001	0.001	0.078	0.001
Coefficient variation	0.228	0.186	0.268	0.344	0.331	0.381	0.273	0.052
Outlier fence–lower value	0.025	0.119	0.032	0.071	−0.005	−1.079	0.030	0.039
Outlier fence–upper value	0.189	0.191	0.172	0.331	0.139	2.006	1.942	1.335
Outliers		0.110		0.376				1.569
Outliers		0.110						1.630
Outliers		0.117						

Table 4.15 **Five-number summary of SOA, BBA, IPHR, and SO/BB for Aaron and Bonds**

	SOA		BBA		IPHR		SO/BB	
Summary	Aaron	Bonds	Aaron	Bonds	Aaron	Bonds	Aaron	Bonds
Smallest	.081	.110	.056	.088	.029	.039	.540	.177
Q1	.086	.146	.084	.168	.049	.078	.747	.525
Q2	.107	.159	.098	.188	.068	.092	1.050	.643
Q3	.127	.164	.119	.233	.085	.122	1.225	.849
Largest	.162	.247	.169	.376	.117	.191	1.540	1.630

The yearly *statistics* of BA, OBA, SLG, RC/27, RPR/27, SOA, BBA, SO/BB, and IPHR represent different aspects of a player's *batting performance* for a given year. In later chapters, the statistics of BA, OBA, SOA, BBA, and IPHR will be used to find a confidence interval estimate for the *true* (BA, OBA, SOA, BBA, and IPHR) for that year. The *true* (BA, OBA, SOA, BBA, and IPHR) represent different aspects of a player's *batting ability*. These descriptive measures are called *parameters*.

Table 4.16 **Sorted yearly SOA, BBA, IPHR, and SO/BB for Aaron and Bonds**

IPAB		SOA		SO/BB		BBA		IPHR	
Aaron	Bonds	Aaron	Bonds	Aaron	Bonds	Aaron	Bonds	Aaron	Bonds
311	36	0.081	0.110	0.540	0.177	0.056	0.088	0.029	0.039
341	181	0.082	0.110	0.598	0.237	0.056	0.117	0.030	0.051
394	293	0.083	0.117	0.729	0.392	0.074	0.135	0.043	0.053
414	311	0.085	0.139	0.742	0.404	0.074	0.137	0.046	0.054
429	316	0.086	0.143	0.744	0.443	0.084	0.150	0.047	0.057
437	332	0.086	0.143	0.750	0.503	0.084	0.156	0.050	0.076
453	332	0.089	0.146	0.817	0.525	0.089	0.168	0.053	0.078
489	348	0.094	0.147	0.831	0.543	0.090	0.169	0.054	0.080
500	356	0.101	0.147	0.851	0.581	0.094	0.173	0.063	0.082
503	383	0.102	0.149	0.969	0.600	0.094	0.187	0.064	0.084
507	403	0.106	0.152	1.018	0.627	0.095	0.187	0.065	0.090
519	404	0.107	0.160	1.050	0.658	0.098	0.189	0.068	0.094
524	423	0.110	0.160	1.059	0.667	0.099	0.193	0.076	0.095
527	436	0.117	0.160	1.086	0.682	0.102	0.208	0.078	0.100
537	437	0.122	0.160	1.106	0.692	0.109	0.210	0.079	0.106
539	441	0.122	0.164	1.143	0.708	0.110	0.224	0.082	0.116
541	445	0.123	0.164	1.205	0.849	0.114	0.233	0.084	0.122
544	456	0.130	0.167	1.245	0.892	0.124	0.267	0.086	0.129
552	460	0.140	0.174	1.263	1.000	0.124	0.269	0.087	0.136
555	460	0.142	0.175	1.350	1.139	0.129	0.297	0.087	0.136
557	463	0.149	0.195	1.393	1.569	0.136	0.324	0.088	0.139
575	487	0.159	0.247	1.459	1.630	0.146	0.376	0.108	0.191
233		0.162		1.540		0.169		0.117	

While this chapter was devoted to the comparison of Aaron's yearly batting performance with Bonds's yearly batting performance, the same techniques can be used for *any* two quantitative data sets. Of course, the underlying subject matter will dictate which statistics to use.

In the comparison between Aaron and Bonds, the role played by a time-series graph was critical. The box-and-whisker graph, the stem-and-leaf graph, and the descriptive measures looked at the set of yearly batting data for Aaron and for Bonds as two samples without regard to the order of the years. By comparing the yearly data for all nine statistics, we concluded that Bonds's batting performance exceeded Aaron's batting performance in all but two statistics. In BA, there difference was minimal. With respect to SOA, Aaron performed better. For all the other statistics (OBA, SLG, RPR/27, RC/27, BBA, IPHR, and SO/BB), Bonds had a decisive advantage. The reason for this advantage became much clearer from the time-series graphs.

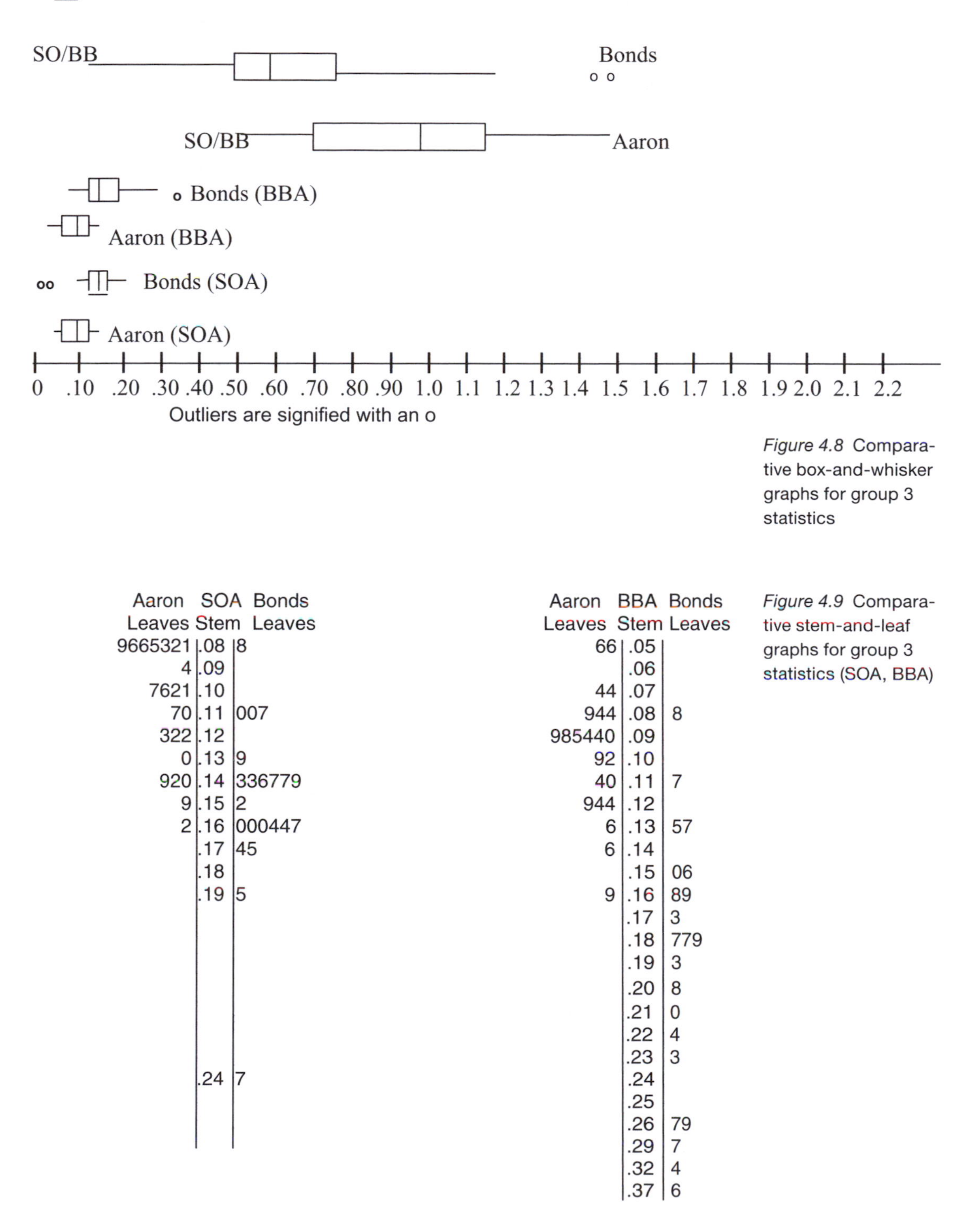

Figure 4.8 Comparative box-and-whisker graphs for group 3 statistics

Aaron Leaves	SOA Stem	Bonds Leaves
9665321	.08	8
4	.09	
7621	.10	
70	.11	007
322	.12	
0	.13	9
920	.14	336779
9	.15	2
2	.16	000447
	.17	45
	.18	
	.19	5
	.24	7

Aaron Leaves	BBA Stem	Bonds Leaves
66	.05	
	.06	
44	.07	
944	.08	8
985440	.09	
92	.10	
40	.11	7
944	.12	
6	.13	57
6	.14	
	.15	06
9	.16	89
	.17	3
	.18	779
	.19	3
	.20	8
	.21	0
	.22	4
	.23	3
	.24	
	.25	
	.26	79
	.29	7
	.32	4
	.37	6

Figure 4.9 Comparative stem-and-leaf graphs for group 3 statistics (SOA, BBA)

Figure 4.10 Comparative stem-and-leaf graphs for group 3 statistics (IPHR, SO/BB)

Aaron IPHR Bonds

Leaves	Stem	Leaves
9	.02	
0	.03	9
763	.04	
430	.05	347
8543	.06	
986	.07	68
877642	.08	024
	.09	045
8	.10	06
7	.11	6
	.12	29
	.13	669
	.14	
	.15	
	.19	1

Aaron SO/BB Bonds

Leaves	Stem	Leaves
	.1	77
	.2	37
	.3	92
	.4	04 43
40 98	.5	03 25 43 81
	.6	00 27 58 67 82 92
50 44 42 29	.7	08
51 31 17	.8	49 92
69	.9	
86 59 50 18	1.0	
43 06	1.1	39
63 45 05	1.2	
93 50	1.3	
59	1.4	
40	1.5	69
	1.6	30

Figure 4.11 Comparative time-series graphs for group 3 statistics

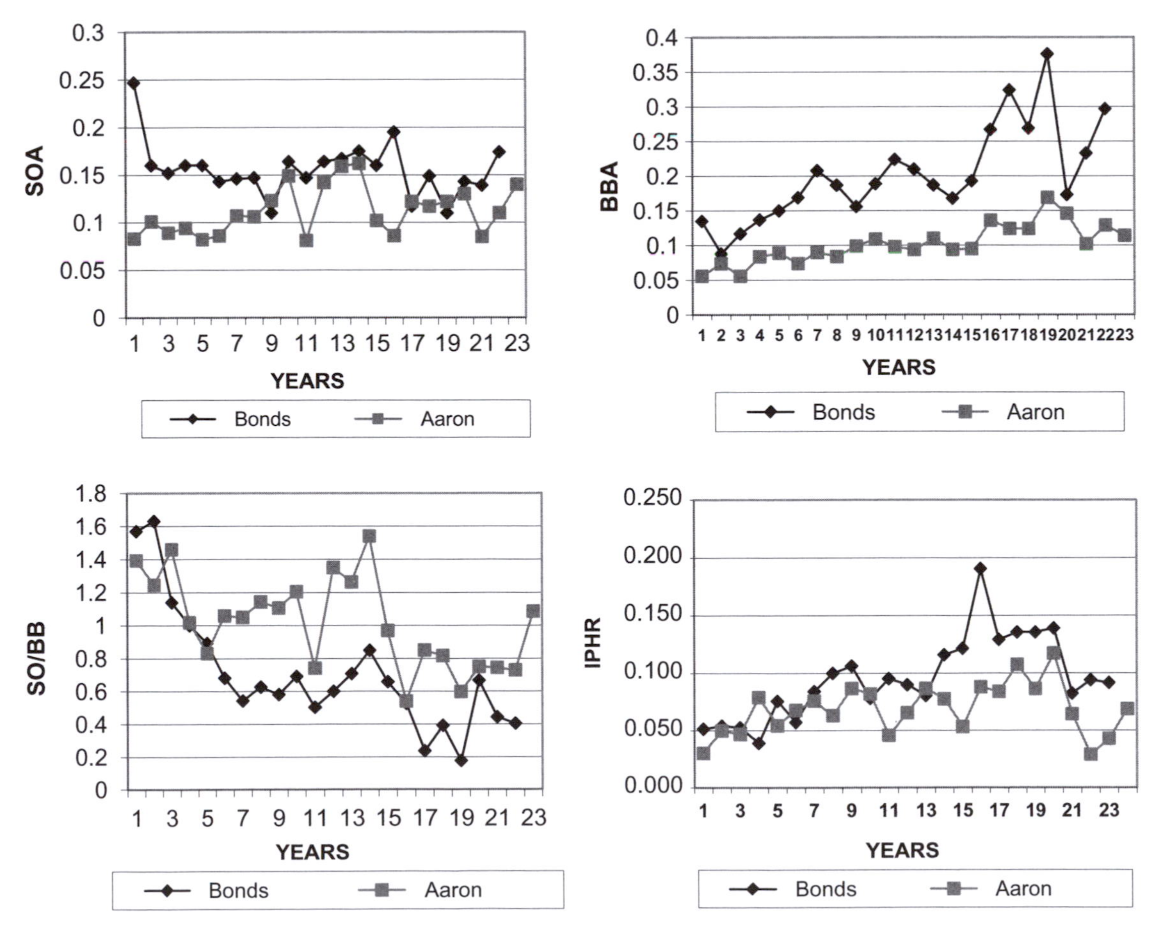

Time-series graphs pair the yearly data for Aaron and Bonds by their year number. That is, the result of year 1 for each player is paired; the results of year 2 are then paired, and so on. Looking at these data through the pairing indicates that the real difference between the two data sets occurred in the last third of their respective careers. For that period, Bonds's performance, as measured through these statistics, spiked sharply in a positive fashion. This spike occurred between the ages of 35 and 39. Bonds did the unthinkable in baseball history. While most players show a decrease in performance near the end of their careers, Bonds had his most productive batting performance in all areas in the latter years of his career. As mentioned earlier, he performed so well that his outliers, in the right direction, occurred during those years.

CHAPTER PROBLEMS

1. Create a new group 4, which consists of statistics used to measure a player's batting performance as a power hitter. The statistics in group 4 are isolated power (ISO), total power quotient (TPQ), and home run average (HRA). ISO and HRA were defined in Chapter 3; TPQ will be defined below. Their definitions are

   $$\text{ISO} = \text{SLG} - \text{AVG} = (2 * \#2\text{B} + 3 * \#3\text{B} + 4 * \#\text{HR})/\#\text{AB},$$
   $$\text{HRA} = \#\text{HR}/\#\text{AB},$$
   $$\text{TPQ} = (\#\text{HR} + \#\text{RBI} + \text{TB})/\#\text{AB}.$$

   Compare Aaron and Bonds as power hitters using "the eight presentations" for the statistics in group 4.
2. Using Tables 3.6 and 3.9, compare the batting performance of Ted Williams with that of Joe DiMaggio by using "the eight presentations" on any one of the four groups of statistics mentioned in this chapter.
3. Find a player whose batting performance for a consecutive 5-year period in their middle to late thirties is close to Bonds's. Does such a player exist? If you can't find one, use my candidate, Ted Williams. Compare this player or Ted Williams with Bonds for the group 1 statistics, using "the eight presentations."
4. Many baseball people believe that being a member of the 500-home-run club is a player's ticket to the Hall of Fame. Frank Thomas, Jim Thome, and Gary Sheffield are all members of this club. Using a basket of statistics consisting of BA, RC/27, OPS, and SOA, compare these three players by using the box-and-whisker graphs and the time-series graphs for each of these statistics. Which player or players do you think should be admitted to the Hall of Fame?

5. What group of batting statistics would you create to judge whether a player deserves admittance into the Hall of Fame? Explain what each statistic in your basket measures about a player's batting performance.

As mentioned in the introduction, you are now ready to start the process of comparing the batting performances of your two chosen players. If you have not already selected your two players, please refer back to the introduction.

6. Complete the following tasks:
   a. Download from the website www.baseball-almanac.com the yearly baseball statistics for your two players.
   b. Using Excel, construct two tables similar to Table 4.1.
   c. Compare the batting performance of the two players you have chosen by using "the eight presentations" on the group of statistics you chose in problem 5.

Chapter 5

Linear Regression and Correlation Analysis for Two Quantitative Variables

In Chapter 4 comparisons were made between two quantitative data sets. In particular, the two quantitative data sets were the yearly batting data (#AB, #H, #HR, #SO, and so on) of Henry Aaron and Barry Bonds. Their yearly batting performance was compared by using such descriptive measures (statistics) as BA, OBP, SLG, SOA, OPS, and RC/27. These statistics look at different aspects of a player's batting performance. Graphs such as the box-and-whisker, stem-and-leaf, and time-series were used to visually display their yearly statistics side by side.

In this chapter we look at one data set of objects and two quantitative variables associated with the objects. For example, the set of objects could be the career years for a player and the two quantitative variables could be BA and R (runs scored). One of the quantitative variables is called the *X*-variable, and the other is called the *Y*-variable. The resulting data set becomes the set of ordered pairs, (x, y), where the number x comes from the *X*-variable and the number y comes from the *Y*-variable. The *X*-variable is the *independent* or *explanatory* variable, and the *Y*-variable is the *dependent* or *response* variable. The question is, can we find a relationship between the *X*-variable and the *Y*-variable? The major issue throughout this chapter is to find which baseball statistics best predict the number of runs scored by a player or by a team.

Let's look at two examples, one from inside and one from outside of baseball. In the non-baseball example, the objects are a certain sample of adult males. For each male in the sample, the *X*-variable is his height and the *Y*-variable is his weight. For each male we form the ordered pair (x, y). We expect taller men to weigh more. For our baseball example, the objects are the years of Henry Aaron's career. The *X*-variable is his yearly statistic BA (batting average), and the *Y*-variable is his yearly statistic R (runs scored by player). The data set consists of the ordered pairs (x, y). We would expect that a higher batting average would produce more runs.

In both examples, we wish to find the best linear equation $(y = a * x + b)$ for predicting a y-value from an x-value.

The Goals of This Chapter

- To learn how to construct a *scatter-plot graph* for the ordered pairs.
- To learn how to find the best linear equation for predicting a y-value from an x-value (*linear regression analysis*).
- To learn how to evaluate the strength and direction of the best linear equation (*correlation analysis*).
- To understand the difference between a strong correlation, a weak correlation, and no correlation between the two variables.
- To understand the meaning of a positive and a negative correlation between the two variables.
- To learn how and when to use the linear regression equation as a predicting tool.
- To understand how regression analysis can be misused.

We begin with a game based on making a prediction. Place a fixed number of slips of paper, each with a number written on it, into a bowl. The player knows, in advance, what numbers are on the slips, but each slip is folded so that the number on it does not show. The player chooses a slip from the bowl. The player has to guess the number on the slip. The player wins if the number is the "best possible guess."

We define the best possible guess as the guess that minimizes the sum of numbers calculated by squaring the difference of the number on each slip in the bowl from the guess. For any guess, this sum is called the "error of the guess." The error of the guess is written $\Sigma(y - \text{guess})^2$, where y takes all the possible numbers on the slips. The best possible guess is the guess that has the smallest "error of the guess." The next example illustrates this game.

Example 5.1. Suppose that a bowl contains four slips of paper. Each slip is folded with one of the numbers 3, 5, 7, and 9 written inside. One of the slips is randomly selected from the bowl. The player knows only that the numbers 3, 5, 7, and 9 are on the slips. Which number is the best possible guess? We now look at a few examples of guesses.

Guess #1 = 3

Error of the guess $= (3-3)^2 + (5-3)^2 + (7-3)^2 + (9-3)^2 = 56$.

Guess #2 = 7

Error of the guess $= (3-7)^2 + (5-7)^2 + (7-7)^2 + (9-7)^2 = 24$.

Guess #3 = 6

Error of the guess $= (3-6)^2 + (5-6)^2 + (7-6)^2 + (9-6)^2 = 20$.

Guess #4 = 5

Error of the guess $= (3-5)^2 + (5-5)^2 + (7-5)^2 + (9-5)^2 = 24$.

Guess #5 = 9

Error of the guess $= (3-9)^2 + (5-9)^2 + (7-9)^2 + (9-9)^2 = 56$.

Of the five guesses, the guess of 6 produced the smallest sum. Because we defined the best guess as the one that minimizes this sum, 6 is a better guess than 3, 5, 7, or 9. As the object of the game is to find the guess that will give the smallest error, is there a better guess than 6? Some people might think that the guess of 6 is illegal because it isn't one of the possible numbers. However, the rules of the game did not say that your guess must be one of the numbers on the slips of paper! Do you think there is a better guess than 6? The answer is no, but the proof is beyond the scope of this book. In the next exercise, you are asked to play the game.

Exercise 5.1. Given five slips of papers containing the numbers 2, 8, 9, 4, and 17, find the best possible guess. Make at least 10 guesses. Make one of your guesses the number 8. Are any of your guesses better than 8? What do you think the number 6 in Example 5.1 has in common with the number 8 in this exercise?

From Chapter 2, we observed that the expression $\Sigma(y-$"mean of the y-values"$)^2$ is the numerator of the variance ratio. Do you see the connection between the "mean of the y-values" in $\Sigma(y-$"mean of the y-values"$)^2$ and the "best guess" in $\Sigma(y-$"best guess"$)^2$? The best guess is the mean of the numbers on the slips.

Suppose that we observe that the numbers 3, 5, 7, and 9 are related to the numbers 1, 2, 3, and 4 by the equation $y = 2x + 1$, where the x-values are 1, 2, 3, and 4 and the y-values are 3, 5, 7, and 9. How can we use this extra information to improve our guess? Suppose that the four slips had the x-values visible on the outside of the slip. Using this additional information (knowing both the x-value on the slip and the relationship given by the equation $y = 2x + 1$), we could guess the number on the other side of the slip. In fact, the guess would be perfect.

Summarizing, if we have no other knowledge, our best guess would be the mean. However, additional information, in the form of an equation relating the two variables, can guide us to a better guess.

5.1. Definition of a Scatter-Plot Graph

Suppose that we have a sample of objects and two quantitative variables, X and Y, associated with the objects. For each object, the ordered pair (x, y) is formed. The x-value comes from the X-variable, and the y-value comes from the Y-variable.

A scatter-plot graph is formed from the ordered pairs, (x, y), associated with the objects. After sketching a scatter plot, we analyze it to determine whether there is a relationship between the two variables. A linear relationship between two variables is the most common. A linear relation means that the scatter-plot graph resembles a straight line. Even though a linear pattern between two variables is often found, sometimes the pattern is not linear and can be described better by an equation for a curve. There are also times when no relationship exists. This chapter will only be interested in patterns that are linear.

5.2. Regression Analysis

Regression analysis is used to find an equation to describe the relationship between the X-variable and the Y-variable. The most common relationship between two variables is a *linear relationship*. Regression analysis used to find a *linear equation* between one X-variable and a Y-variable is called *simple linear regression*. The word "simple" indicates that only one X-variable is used. When more than one X-variable is used, we speak of *multiple regression analysis*.

5.3. Correlation Analysis

Correlation analysis measures the strength and direction of the linear equation. The strength of the linear relationship is determined by how close the regression line is to the points in the scatter plot. The direction shows a *positive relationship* when an increase in the numbers for the X-variable results in an increase in the numbers for the Y-variable. The direction shows a *negative relationship* when an increase in the numbers for the X-variable results in a decrease in the numbers for the Y-variable.

5.4. Procedure for Finding a Linear Relationship

Step 1: Draw the scatter-plot graph.

Step 2: If the scatter plot resembles a linear pattern, then proceed to step 3. If the scatter plot resembles a curve, then the methods of this chapter cannot be used. If the scatter plot shows no specific pattern, we also stop here.

Step 3: Find the equation of the regression line.

Step 4: Use correlation analysis to determine the strength and direction of the linear relationship.

Step 5: If a strong linear relationship exists, we can use the regression equation as a tool for predicting an unknown y-value for a known x-value.

We now return to baseball. One of the most important ways to evaluate a player's batting performance is through the number of runs the player produces for his team. One method for evaluating the run production of a player assumes that the player occupies each position in the batting order. The statistic R/27, defined in Chapter 3 as (25.5 * #R)/(#AB − #H), gives the number of runs per game for a team composed solely of the player being analyzed. The variable R/27 will be the *Y*-variable or dependent variable. This is the variable we wish to predict. Suppose that we had 23 slips of paper, corresponding to the 23 years of Aaron's career, with each slip containing the number R/27 for one of his years. If we selected a slip at random and had no other information, our best guess of the slip would be the mean of the 23 numbers. To improve our guess, we will look for a linear relationship between the *X*-variable and the *Y*-variable. The corresponding linear equation allows us to use a known *x*-value to predict the *y*-value. This method enables us to improve the accuracy of our guess.

But what is the best independent variable to explain Aaron's R/27 statistic? In this chapter, the *X*-variables BA, OBA, SLG, BRA, RC/27, OPS, and ISO (see Chap. 3) will be used to predict the *Y*-variable R/27. The two variables *X* and *Y* are associated with the career years for a given player. For each of the above *X*-variables and the *Y*-variable, R/27, we will find the best linear equation $y = m * x + b$. At this point, we need to develop the mathematics necessary to find the equation for the "best line."

5.5. Finding the Equation for the Best Line

The equation for any line can be expressed in the slope-intercept form $y = m * x + b$, where m is the slope of the line and b is the y-intercept. In Figure 5.1, (1) is a line with a positive slope, (2) is a line with a negative slope, (3) is a line with slope 0, and (4) is a line with slope undefined.

In the next example, we review the process of finding the equation of a line in slope-intercept form given two ordered pairs.

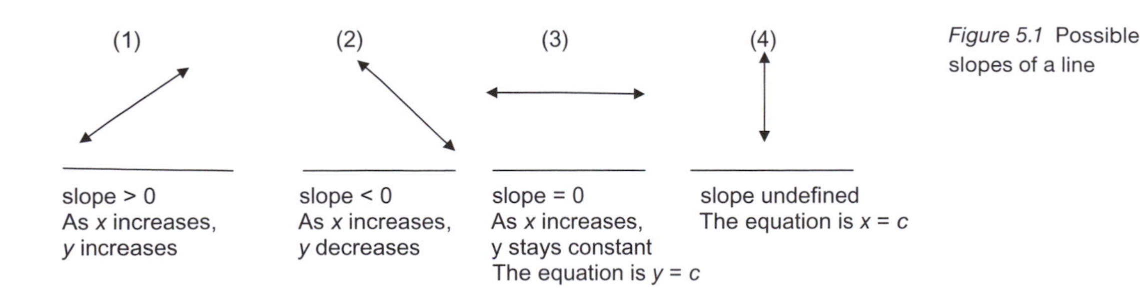

Figure 5.1 Possible slopes of a line

Example 5.2. Find the equation of the line, in slope-intercept form, joining the two ordered pairs (1, –4) and (–5, 2).

Solution:

Step 1: Find the slope of the line. The slope is equal to the change in the *y*-values divided by the change in the *x*-values. We use the letter "*m*" to represent the slope.

$m=(y_2-y_1)/(x_2-x_1)=(2--4)/(-5-1)=(6/-6)=-1;$
$(x_1, y_1)=(1, -4)$ and $(x_2, y_2)=(-5, 2)$.

We now have the equation $y=-1*x+b$.

Step 2: Find the *y*-intercept "*b*." To find *b*, we can use either one of the two original ordered pairs. Without a loss of generality, we use the ordered pair (1, –4) in our calculation. Substituting 1 for *x* and –4 for *y* in the above equation, we obtain

$-4=-1*1+b,$
$b=-4--1=-3.$

The final equation is $y=-1*x+-3=-x-3$.

As a final check, we see if the ordered pair that was *not* used to find *b* satisfies the equation. Substituting for *x* the value –5 and for *y* the value 2, we get

$2=-(-5)-3.$

Exercise 5.2. Find the equation, in slope-intercept form, of the line joining (–3, –6) to (3, 4). Sketch the graph for the line.

We now return to the original problem of finding the equation for the best line. We write the equation for the best line in the slope-intercept form:

$y'=m*x+b.$

For each *x*-value, the *y*-value is called the *observed* value and the y'-value is called the *expected* value. The y'-value is calculated by substituting an *x*-value into the above equation.

The best line is the line that minimizes $\Sigma(y-y')^2$. From the area of mathematics called calculus, the values *m* and *b* defined in the equation for the best line are calculated as follows:

$b=[(\Sigma y)(\Sigma x^2)-(\Sigma x)(\Sigma xy)]/[n(\Sigma x^2)-(\Sigma x)^2],$
$m=[n(\Sigma xy)-(\Sigma x)(\Sigma y)]/[n(\Sigma x^2)-(\Sigma x)^2].$

The best-line equation is $y'=m*x+b$.

Example 5.3. Find the equation for the line whose graph comes closest to the points (1, 3), (2, 4), (4, 10), and (5, 12). This line is referred to as the best line.

Table 5.1 Calculation of coefficients (m and b) for the best line

x	y	xy	x^2	y^2
1	3	3	1	9
2	4	8	4	16
4	10	40	16	100
5	12	60	25	144
$\Sigma x = 12$	$\Sigma y = 29$	$\Sigma xy = 111$	$\Sigma x^2 = 46$	$\Sigma y^2 = 269$

Note: n = the number of ordered pairs ($n = 4$); $b = (29 * 46 - 12 * 111)/(4 * 46 - 12^2) = 462/40 = 0.05$; $m = (4 * 111 - 12 * 29)/(4 * 46 - 12^2) = 2.4$; $y' = 2.4x + 0.05$ is the best line.

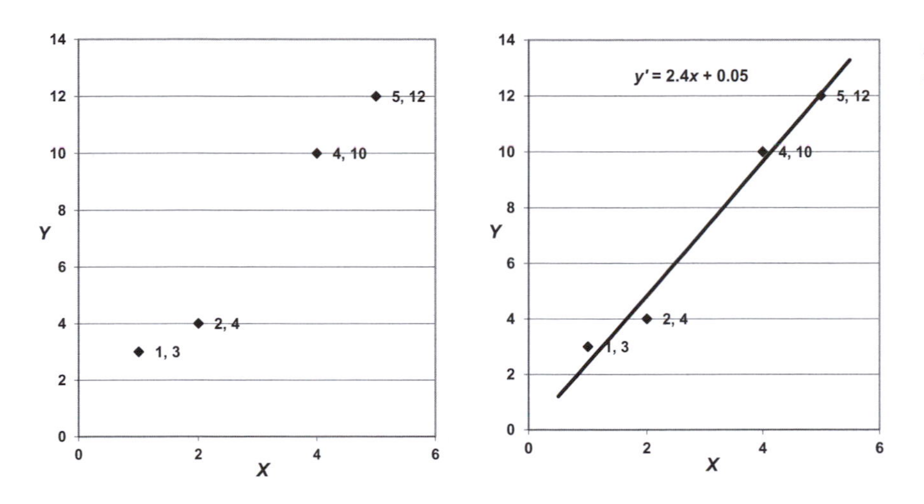

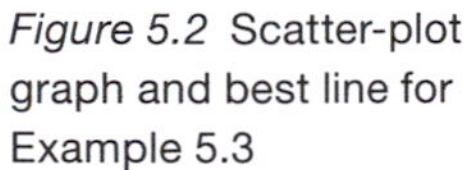
Figure 5.2 Scatter-plot graph and best line for Example 5.3

The calculation of b and m for the best line to fit the points (1, 3), (2, 4), (4, 10), and (5, 12) is shown in Table 5.1.

Figure 5.2 presents the scatter-plot graph and the best line for the ordered pairs in Example 5.3.

Exercise 5.3. Now that we know that the equation for the best line for the ordered pairs (x, y) in Table 5.1 is $y' = 2.4x + 0.05$, find the y'-values for each of the x-values 1, 2, 4, and 5. Add a new column to Table 5.1 with y' as its heading. Put these y'-values under the y'-column.

Exercise 5.4. Add a column with the heading $(y - y')^2$ to Table 5.1. Fill in this column and then compute $\Sigma(y - y')^2$ for the best line in Table 5.1.

5.6. Scatter-Plot Graphs

The time-series graphs encountered in Chapter 4 are examples of one type of scatter-plot graph. In the time-series graphs for Aaron and Bonds, the X-variable corresponded to the number of the year. The Y-variable came from various baseball statistics such as BA and OBP. To plot an ordered pair

(x, y), the x-value is located on the x-axis and the y-value is located on the y-axis. The intersection of the vertical line through the x-value and the horizontal line through the y-value gives one point on the graph. This point represents the ordered pair, (x, y). The graph representing all such points is the scatter-plot graph.

Exercise 5.5. Find the equation for the best line for the ordered pairs (–3, 4), (–5, 7), (5, –8), and (4, –2). Sketch the scatter-plot graph and the best line together on one graph.

5.7. Graphing the Regression Line

For any set of ordered pairs, the above formulas can always find the best line. Be warned: A best line can *always* be found, whether the graph for the given ordered pairs (x, y) looks linear or not. So, be careful not to use the best line as a model for a graph that clearly is *not* linear. This best line is called the *sample regression line*, since the ordered pairs are considered a sample.

The sample regression line, $y' = m * x + b$, is the line that makes the sum $\Sigma(y - y')^2$ as small as possible.

The equation for the regression line can be graphed by following these steps:

1. Substitute any two numbers, between the smallest and largest x-values of the ordered pairs, into the equation to find their corresponding y'-values.
2. Plot the two resulting ordered pairs on your graph.
3. Using a ruler, connect the two ordered pairs.

For the regression line calculated in Table 5.1, $y' = 2.4x + 0.05$, we chose arbitrarily $x = 2$ and $x = 3$. Substituting these x-values into the equation, the y'-values of $y' = 4.85$ and $y' = 8.38$ were obtained. The line drawn through the points (2, 4.85) and (3, 8.38) is the graph of the regression line.

A regression line is used to estimate an unknown y-value from a known x-value. But since a regression line can be found for *any* set of ordered pairs, how do we measure when a regression line fits the sample data well enough to be used as a tool for prediction?

Correlation analysis was created to find descriptive measures to determine how well the regression line fits the observed ordered pairs.

Exercise 5.6. Find the equation for the regression line given the following sample points: (2, 5), (–1, –4), (0, –1), (–2, –8), (1, 2). Sketch the graph of the regression line and the scatter plot on the same graph.

5.8. Statistics Used for Correlation Analysis

The two statistics used to measure the strength of the regression line are called the *correlation coefficient* and the *coefficient of determination*. The correlation coefficient is symbolized by "r" and the coefficient of determination by "r^2." These two statistics are defined later in this section.

In Table 5.2, $\Sigma(y-y')^2$ is computed for the ordered pairs in Table 5.1.

In Table 5.3, we compute two other important sums, $\Sigma(y-\bar{y})^2$ and $\Sigma(y'-\bar{y})^2$. The sample mean is $\bar{y}=29/4=7.25$.

Using the results of Tables 5.2 and 5.3, we see that the following equation is satisfied:

$$\Sigma(y-\bar{y})^2=\Sigma(y'-\bar{y})^2+\Sigma(y-y')^2. \qquad \text{(a)}$$

Verify that Formula (a) is satisfied for the regression line $y'=2.4x+0.05$.

The sum $\Sigma(y-\bar{y})^2$ is the numerator of the expression for the sample variance, which measures how spread out the y-values are around $\bar{y}$. In a similar way, $\Sigma(y'-\bar{y})^2$ measures how spread out the y'-values are around $\bar{y}$. The final sum, $\Sigma(y-y')^2$, is called the *residual error*. It turns out that Formula (a) is true for any set of ordered pairs.

The closer the sum $\Sigma(y'-\bar{y})^2$ is to the sum $\Sigma(y-\bar{y})^2$, the more the variation of the y-values is explained by the regression line.

We now define the *coefficient of determination* r^2 (r-squared) as follows:

$$r^2=\Sigma(y'-\bar{y})^2/\Sigma(y-\bar{y})^2.$$

Table 5.2 **$\Sigma(y-y')^2$ computed for the ordered pairs in Table 5.1**

x	y	$y'=2.4x+0.05$	$y-y'$	$(y-y')^2$
1	3	2.45=2.4*1+0.05	0.55	.3025
2	4	4.85=2.4*2+0.05	−0.85	.7225
4	10	9.65=2.4*4+0.05	0.35	.1225
5	12	12.05=2.4*5+0.05	−0.05	.0025
				$\Sigma(y-y')^2=1.15$

Table 5.3 **Computation of the two sums $\Sigma(y-\bar{y})^2$ and $\Sigma(y'-\bar{y})^2$**

y	y'	$\bar{y}$	$(y-\bar{y})$	$(y-\bar{y})^2$	$(y'-\bar{y})$	$(y'-\bar{y})^2$
1	2.45	7.25	−4.25	18.0625	−4.80	23.04
2	4.85	7.25	−3.25	10.5625	−2.40	5.76
4	9.65	7.25	2.75	7.5625	2.40	5.76
5	12.05	7.25	4.75	22.5625	4.80	23.04
				$\Sigma(y-\bar{y})^2=58.75$		$\Sigma(y'-\bar{y})^2=57.60$

For Example 5.3, $r^2 = 57.60/58.75 = 0.98$.

Since each of the sums in Formula (a) is nonnegative, $0 \le r^2 \le 1$. Can you explain this? The product of $100 * r^2$ is the percent of variation of the y-values explained by the regression line. Therefore, for Example 5.3, we can say that 98% of the variation of the y-values is explained by the regression line.

If $r^2 = 1$, we have $\Sigma(y' - \bar{y})^2 = \Sigma(y - \bar{y})^2$ and the residual error is zero. This says that $y = y'$ for each y-value. As a result, each y-value lies on the regression line. This tells us that the variation of the y-values is completely determined by the regression line. In real life, this result probably will never occur.

If $r^2 = 0$, then $\Sigma(y' - \bar{y})^2 = 0$, which says that each y' is equal to $\bar{y}$. The regression line would be the horizontal line $y' = \bar{y}$. A horizontal regression line contributes *nothing* to the explanation for the variation of the y-values.

If r^2 is close to 1, then the regression line can be used to explain the variation of the y-values. This allows us to use the regression line to predict an unknown y-value for a given x-value.

We define the sample correlation coefficient r as follows:

$$r = +\sqrt{r^2} \text{ if } m > 0; r = -\sqrt{r^2} \text{ if } m < 0,$$

where m is the slope of the regression line.

The sample correlation coefficient, besides determining the strength of the linear relationship, also provides a direction for the linear relationship.

If $r > 0$, we say that there is a *positive* correlation (as the x-values increase the y-values increase) between the X-variable and the Y-variable; if $r < 0$, we say that there is a *negative* correlation (as the x-values increase the y-values decrease) between the X-variable and the Y-variable; if $r = 0$, we say that there is *no* correlation between the variables.

Since $m > 0$ in Example 5.3, the correlation coefficient $r = +\sqrt{.98} = 0.99$. This shows a strong positive correlation between the X-variable and the Y-variable.

A short-cut *computational formula* to calculate r directly is

$$r = [n(\Sigma xy) - (\Sigma x)(\Sigma y)] / \sqrt{[n(\Sigma x^2) - (\Sigma x)^2][n(\Sigma y^2) - (\Sigma y)^2]}.$$

One advantage of the short-cut formula is that it automatically provides the right sign for r.

In general, we can use the value of the sample correlation coefficient to categorize the strength of the linear relationship as follows (m is the slope of the regression line):

Strong positive correlation	$.70 \le r \le 1.00$, $.49 \le r^2 \le 1.00$, and $m > 0$
Moderate positive correlation	$.40 \le r < .70$, $.16 \le r^2 < .49$, and $m > 0$
Weak positive correlation	$.20 < r < .40$, $.04 < r^2 < .16$, and $m > 0$
No or poor relationship	$-.20 \le r \le .20$, $0 \le r^2 \le .04$, and any m

Weak negative correlation	$-.40 < r < -.20$, $.04 < r^2 < .16$, and $m < 0$
Moderate negative correlation	$-.70 < r \leq -.40$, $.16 \leq r^2 < .49$, and $m < 0$
Strong negative correlation	$-1.00 \leq r \leq -.70$, $.49 \leq r^2 \leq 1.00$, and $m < 0$

Below are examples of various graphs of regression lines with their correlation coefficients *r*.

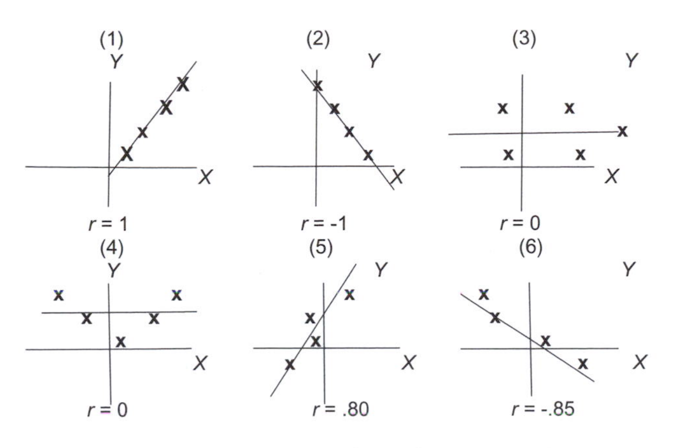

Graph 1: $r=1$; each observed *y*-value lies on the regression line whose slope is positive.
Graph 2: $r=-1$; each observed *y*-value lies on the regression line whose slope is negative.
Graph 3: $r=0$; horizontal regression line (no relationship).
Graph 4: $r=0$; horizontal regression line (no linear relationship but there is a quadratic relationship).
Graph 5: $r=.80$; strong positive linear relationship.
Graph 6: $r=-.85$; strong negative linear relationship.

If your scatter plot resembles graphs 3 or 4, then linear regression should not be used. Other forms of regression analysis do exist but are beyond the scope of this book.

Exercise 5.7. For the sample data in Exercise 5.6, complete the following:

a. Create a table similar to Table 5.2.
b. Create a table similar to Table 5.3.
c. Show that Formula (a) above is satisfied.
d. Compute r^2.
e. Compute *r* (two ways).
f. How would you classify the strength of the linear relationship?

For the rest of this chapter we apply regression analysis to baseball data. We begin by seeing how well various baseball statistics predict the statistic of R/27 for both Aaron and Bonds.

Figure 5.3 For Aaron, scatter-plot graphs, graphs of the regression lines, equations of regression lines, and *r*-squared statistics for each *X*-variable with *Y*=R/27

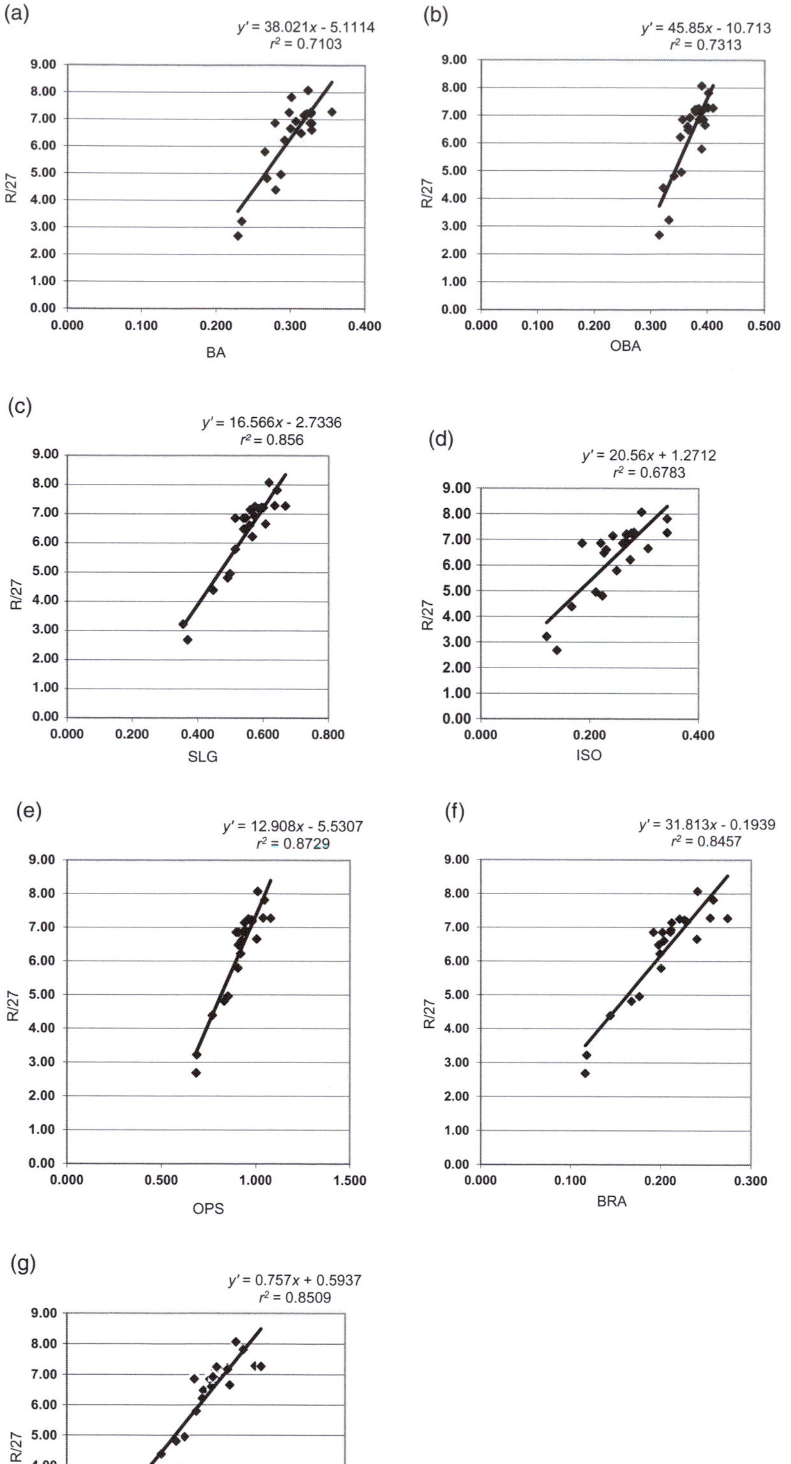

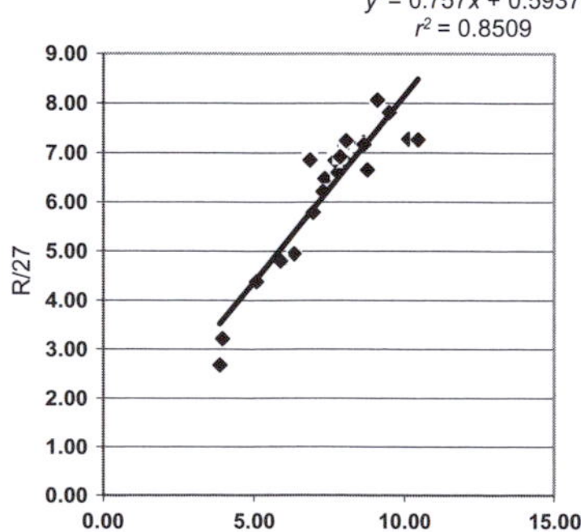

5.9. Best Statistics for Predicting R/27 for Aaron

R/27 represents the number of runs scored by a player per 27 outs, assuming that that player occupies each position in the batting order. The question is, which batting statistics can be used reliably to predict R/27 for that player? We investigate this using simple linear regression. We are attempting to predict R/27, the dependent *Y*-variable, using the independent *X*-variables BA, OBA, SLG, ISO, OPS, BRA, and RC/27.

Instead of finding the equation of the regression line and value of r and r^2 by using a calculator, we use the software package Excel.

We are now ready to compare the strength of the linear relationship between each of the *X*-variables BA, OBA, SLG, ISO, OPS, BRA, and RC/27 and the *Y*-variable R/27 for Hank Aaron. Figure 5.3 shows the corresponding scatter-plot graphs, the graphs of the sample regression lines, the equations for the regression lines, and the corresponding r-squared statistics for each *X*-variable with Y=R/27. The data come from Table 5.4.

Table 5.4 **Yearly values for the *X*-variables and the *Y*-variable for Henry Aaron**

	X-variables							*Y*
Year	BA	OBA	SLG	OPS	ISO	RC/27	BRA	R/27
1954	0.280	0.322	0.447	0.769	0.167	5.07	0.144	4.39
1955	0.314	0.366	0.540	0.906	0.226	7.34	0.198	6.48
1956	0.328	0.365	0.558	0.923	0.230	7.78	0.204	6.61
1957	0.322	0.378	0.600	0.978	0.278	8.56	0.227	7.22
1958	0.326	0.386	0.546	0.932	0.220	7.98	0.211	6.86
1959	0.355	0.401	0.636	1.037	0.281	10.12	0.255	7.29
1960	0.292	0.352	0.566	0.918	0.275	7.27	0.199	6.22
1961	0.327	0.381	0.594	0.975	0.267	8.63	0.226	7.22
1962	0.323	0.390	0.618	1.008	0.296	9.09	0.241	8.08
1963	0.319	0.391	0.586	0.977	0.268	8.63	0.229	7.18
1964	0.328	0.393	0.514	0.907	0.186	7.69	0.202	6.86
1965	0.318	0.379	0.560	0.939	0.242	8.00	0.212	7.15
1966	0.279	0.356	0.539	0.895	0.260	6.85	0.192	6.86
1967	0.307	0.369	0.573	0.942	0.267	7.86	0.211	6.93
1968	0.287	0.354	0.498	0.852	0.211	6.33	0.176	4.96
1969	0.300	0.396	0.607	1.003	0.307	6.75	0.240	6.66
1970	0.298	0.385	0.574	0.959	0.275	8.06	0.221	7.26
1971	0.327	0.410	0.669	1.079	0.341	10.43	0.274	7.27
1972	0.265	0.390	0.514	0.904	0.249	6.96	0.200	5.80
1973	0.301	0.402	0.643	1.045	0.342	9.48	0.258	7.82
1974	0.268	0.341	0.491	0.832	0.224	5.87	0.167	4.81
1975	0.234	0.332	0.355	0.687	0.120	3.95	0.118	3.22
1976	0.229	0.315	0.369	0.684	0.140	3.87	0.116	2.68

Table 5.5 Ranking from strongest to weakest correlations for Henry Aaron

Variables	Rank	r^2	r
OPS	1	.8729	.93
SLG	2	.8560	.93
RC/27	3	.8509	.92
BRA	4	.8457	.92
OBA	5	.7313	.86
BA	6	.7103	.84
ISO	7	.6783	.82

Exercise 5.8. For the year 1967, verify the values in the eight columns in Table 5.4. The data in Table 2.2 will be helpful.

Table 5.5 provides a sorted list of statistics for Aaron from the strongest to the weakest correlation.

Table 5.5 establishes that *each* of the X-variables would be an excellent predictor of R/27. Remember, R/27 represents the number of runs per game produced by a team, composed of Aaron batting in each position in the batting order. The sample mean for R/27 is 6.34.

Returning to the game at the beginning of the chapter, if we selected at random from a bowl one slip from the 23 slips containing the 23 values of R/27, with no other information available, the best guess for the number on the slip is the sample mean 6.34 for the values of R/27. Suppose that the chosen slip showed an OPS value of .978 for the year 1957. By substituting the x-value of 0.978 into the equation for the regression line in Figure 5.3e, we obtain the following y'-value:

$$y' = 12.908x - 5.5307 = (12.908)(.978) - 5.5307 = 7.09.$$

For OPS, the r-squared value is .8729, so that approximately 87% of the variation of the y-values is explained by the regression line. Therefore, the y'-value of 7.09 should be very close to the R/27 number on the slip. In fact, for the year 1957, the R/27 value is 7.22. Clearly, 7.09 is a better estimate for 7.22 than 6.34. The same type of analysis could be done with any of the X-variables above. The closer the r-squared value is to 1, the better the X-variable will predict R/27.

Since there is a very strong positive correlation between X=OPS and Y=R/27 for Aaron, we can use the OPS as a strong estimator for R/27. Suppose that we asked the question, what OPS would Aaron have to achieve in a season to give him an R/27 of 8.00? To solve this, we let $y' = 8.00$ and solve the following equation for x=OPS:

$$8.00 = 12.908 * \text{OPS} - 5.531,$$
$$\text{OPS} = (8.00 + 5.531)/12.908 = 1.048.$$

Exercise 5.9. For the year 1967, use SLG to predict R/27 for Aaron. For the year 1967, use BA to predict R/27. Which of these two predictions is better? Why? Are both these predictions better than using the sample mean of 6.34 for R/27? Using the regression equation, what BA and SLG would Aaron need to have R/27 = 8.00?

The top four r-squared values for Aaron correspond to the X-variables OPS, SLG, BRA, and RC/27. These four variables are each involved with SLG. Recapping from Chapter 3,

- OPS = OBA + SLG (the sum emphasizes SLG more than OBA);
- SLG = TB/#AB;
- BRA = OBA * SLG (the product emphasizes OBA more than SLG);
- RC/27 = 25.5 * [(#H + #BB) * TB/(#AB + #BB)]/(#AB − #H)
  = 25.5 * (OBA * TB)/(#AB − #H)
  = 25.5 * (OBA * SLG * #AB)/(#AB − #H).

5.10. Best Statistics for Predicting R/27 for Bonds

We now compare the corresponding strength of the linear relationship between each of the *X*-variables BA, OBA, SLG, ISO, OPS, BRA, and RC/27 and the *Y*-variable R/27 for Bonds. Figure 5.4 shows the corresponding scatter-plot graphs, the graphs of the sample regression lines, the equations of the regression lines, and the *r*-squared statistics for each *X*-variable with *Y* = R/27. The data come from Table 5.6.

Exercise 5.10. For the year 2005, verify the values in the eight columns in Table 5.6.

Exercise 5.11. For the year 2005, use BRA and BA to predict R/27. Calculate the sample mean for R/27 for the 22 years of Bonds's career. Which of the three estimates comes closest to the actual R/27 for Bonds for the year 2005? Why?

Table 5.7 provides a sorted list of variables for Bonds from the strongest correlation to the weakest correlation.

Similar to Aaron, the top four *r*-squared values correspond to RC/27, BRA, OPS, and OBA. For Bonds, the RC/27 variable had the highest *r*-squared value and the OPS variable was the third highest. For Aaron, OPS had the highest value and RC/27 was the third highest value. This flip-flop may have been caused partially by the fact that Aaron's *r*-squared OBA value of .7313 was smaller than Bonds's *r*-squared OBA value of .8009 and Aaron's *r*-squared SLG value of .8560 was greater than Bonds's *r*-squared SLG value of .8043.

As in the case of Aaron, each of the seven *X*-variables for Bonds had a strong positive correlation with *Y* = R/27.

5.11. Comparison of *r*-squared between Aaron and Bonds

We make the following observations:

- The correlation coefficient, *r*, for every statistic for both Aaron and Bonds was greater than .80. Thus, all statistics had a strong positive correlation with *Y* = R/27.
- The highest four values of *r* come from the same four statistics for Aaron and Bonds. These are OPS, BRA, RC/27, and SLG.

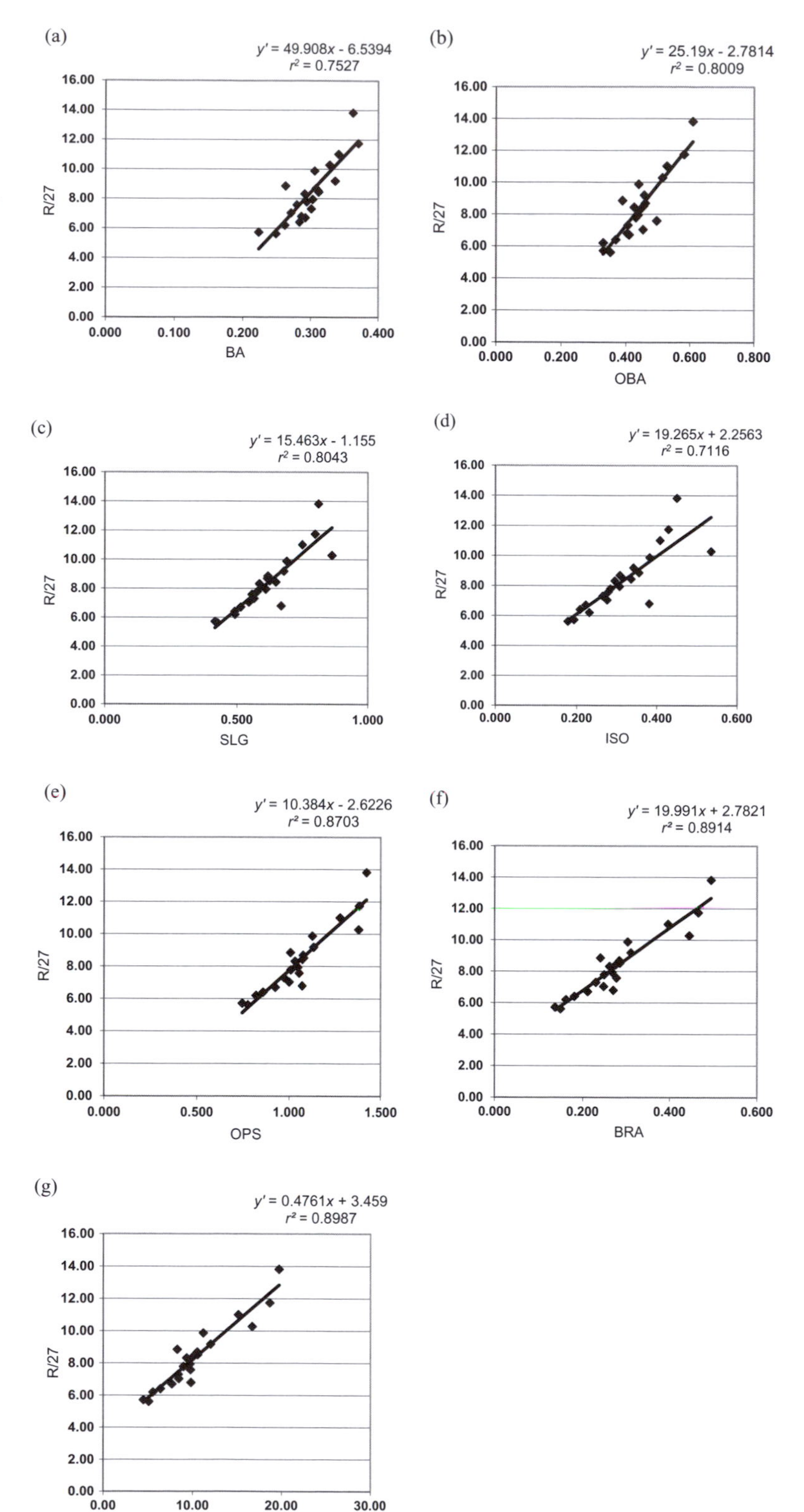

Figure 5.4 For Bonds, scatter-plot graphs, graphs of the regression lines, equations of regression lines, and *r*-squared statistics for each *X*-variable with $Y = R/27$

Table 5.6 **Yearly values of the *X*-variables and the *Y*-variable for Barry Bonds**

	X-variables							*Y*
Year	BA	OBA	SLG	OPS	ISO	RC/27	BRA	R/27
1986	0.223	0.330	0.416	0.746	0.194	4.49	0.137	5.72
1987	0.261	0.329	0.492	0.821	0.230	5.56	0.162	6.20
1988	0.283	0.368	0.491	0.859	0.208	6.40	0.181	6.41
1989	0.248	0.351	0.426	0.777	0.178	5.09	0.150	5.61
1990	0.301	0.406	0.565	0.971	0.264	8.37	0.229	7.31
1991	0.292	0.410	0.514	0.924	0.222	7.68	0.211	6.71
1992	0.311	0.456	0.624	1.080	0.313	10.54	0.285	8.53
1993	0.336	0.458	0.677	1.135	0.341	12.00	0.310	9.19
1994	0.312	0.426	0.647	1.073	0.335	10.11	0.276	8.44
1995	0.294	0.431	0.577	1.008	0.283	8.96	0.249	7.79
1996	0.308	0.461	0.615	1.076	0.308	10.51	0.284	8.69
1997	0.291	0.446	0.585	1.031	0.293	9.32	0.261	8.32
1998	0.303	0.438	0.609	1.047	0.306	9.69	0.267	7.95
1999	0.262	0.389	0.617	1.006	0.355	8.27	0.240	8.86
2000	0.306	0.440	0.688	1.128	0.381	11.17	0.303	9.88
2001	0.328	0.515	0.863	1.378	0.536	16.70	0.444	10.28
2002	0.370	0.582	0.799	1.381	0.429	18.66	0.465	11.75
2003	0.341	0.529	0.749	1.278	0.408	15.13	0.396	11.01
2004	0.362	0.609	0.812	1.421	0.450	19.69	0.495	13.82
2005	0.286	0.404	0.667	1.071	0.381	9.80	0.269	6.80
2006	0.270	0.454	0.545	0.999	0.275	8.45	0.247	7.04
2007	0.279	0.497	0.557	1.054	0.279	9.75	0.277	7.59

- The lowest three values of r came from the same three statistics. These were OBA, BA, and ISO.
- OPS had the highest r-value for Aaron, while RC/27 had the highest r-value for Bonds.
- RC/27 had the third-highest r-value for Aaron, while OPS had the third-highest r-value for Bonds.
- SLG had the second-highest r-value for Aaron, but BRA had the second-highest r-value for Bonds.
- BRA had the fourth-highest r-value for Aaron, while SLG had the fourth-highest r-value for Bonds.

We now form the ordered pairs (x, y) where the x-value and y-value are the corresponding r-squared numbers for each variable. Table 5.8 shows each variable and its corresponding ordered pair.

The ranks of 1 and 3 flip-flop the two statistics OPS and RC/27, and the ranks 2 and 4 flip-flop the statistics

Table 5.7 **Ranking from strongest to weakest correlations for Barry Bonds**

Variables	Rank	r^2	r
RC/27	1	.8987	.95
BRA	2	.8914	.94
OPS	3	.8703	.93
SLG	4	.8043	.90
OBA	5	.8009	.89
BA	6	.7527	.87
ISO	7	.7113	.84

Table 5.8 Comparison of *r*-squared values between Aaron and Bonds

	Aaron		Bonds		
Statistic	r^2	Rank	r^2	Rank	Ordered pairs
OPS	.8729	1	.8703	3	(.8729, .8703)
SLG	.8560	2	.8043	4	(.8560, .8043)
RC/27	.8509	3	.8987	1	(.8509. .8987)
BRA	.8457	4	.8914	2	(.8457, .8914)
OBA	.7313	5	.8009	5	(.7313, .8009)
BA	.7103	6	.7527	6	(.7103, .7527)
ISO	.6779	7	.7116	7	(.6779, .7116)

SLG and BRA. The statistics OBA, BA, and ISO rank 5, 6, and 7 in the same order for both players.

Figure 5.5 shows the scatter plot, where the ordered pairs (x, y) represent the corresponding r-squared values for the statistics of Aaron and Bonds, along with the regression line for these ordered pairs. The r-squared value is 0.7530, and the correlation coefficient is $r=0.8677$. This shows a strong positive correlation between the X-variable consisting of the r-squared numbers of Aaron and the Y-variable consisting of the corresponding r-squared numbers of Bonds.

5.12. Warnings concerning Linear Regression

The first piece of advice is simple: sketch the scatter-plot graph before using the regression equation.

If the scatter-plot graph does *not* show a linear relationship, do *not* use linear regression.

Figure 5.5 Regression line for *r*-squared numbers for Aaron versus *r*-squared numbers for Bonds

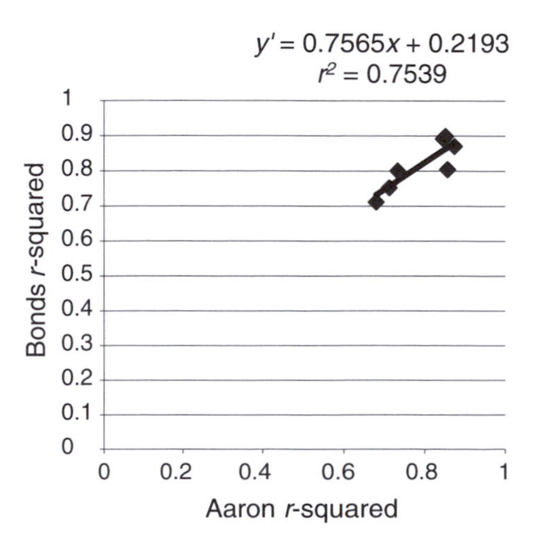

5.12.1. Be Careful of Outliers

In Chapter 2, the concept of outliers for one quantitative variable was introduced. Outliers are extreme values that can affect many descriptive measures such as the mean and standard deviation. Outliers that are not subjects of the targeted population or are mistakes should be removed from the data set. Retained outliers should be identified in your analysis and may supply important information about the data.

In simple linear regression, the data points are in the form of ordered pairs (x, y); the x-value comes from one variable, and the y-value comes from a second variable. An outlier can be an extreme x-value, an

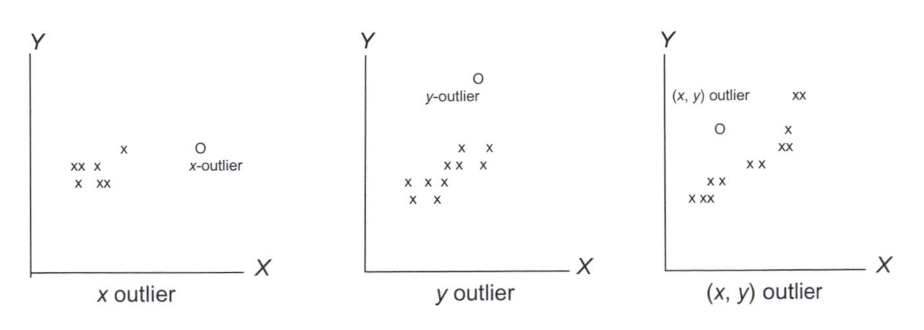

Figure 5.6 Outliers for regression line

extreme y-value, or an extreme ordered pair (x, y). Figure 5.6 shows the three types of outliers. One way of handling outliers is to find the regression line and correlation coefficient r with and without the outliers. A decision has to be made on whether to include or exclude the outliers or present both results.

In baseball, a common outlier occurs in either the first year or the last year in a player's career. This is because his number of plate appearances can be limited for those two seasons. It is not uncommon for a player to have less than 100 at bats in his first season (Derek Jeter, for example, had 48). Also, a player will have limited at bats in any season where he sustains a major injury. Many baseball people say that a player should have a minimum of 150 at bats for his yearly statistics for that season to count. However, the plate appearances for all years are part of his career statistics.

5.12.2. Extrapolation Should Not Be Used

A regression line estimate for the sample data is true only for the range of x-values observed in the sample. To illustrate this, assume that a player for the first 5 years of his career hit 20, 30, 35, 38, and 45 home runs. We would have the ordered pairs (1, 20), (2, 30), (3, 35), (4, 38), and (5, 45). Figure 5.7 shows the equation for the regression line and the r^2-value for these ordered pairs. Clearly, there is a strong positive correlation. However, if we used the regression line as an estimator for year 10 and year 15, we would be estimating the player hitting, respectively, 74 and 103 home runs.

5.12.3. Do Not Assume a Cause-and-Effect Relationship

What conclusions can be made about a strong linear relationship between a variable X and a variable Y? These are the possibilities:

- The variable X may cause Y.
- The variable Y may cause X.
- A third variable Z may cause the strong linear relationship between X and Y.
- The relationship is just a coincidence.

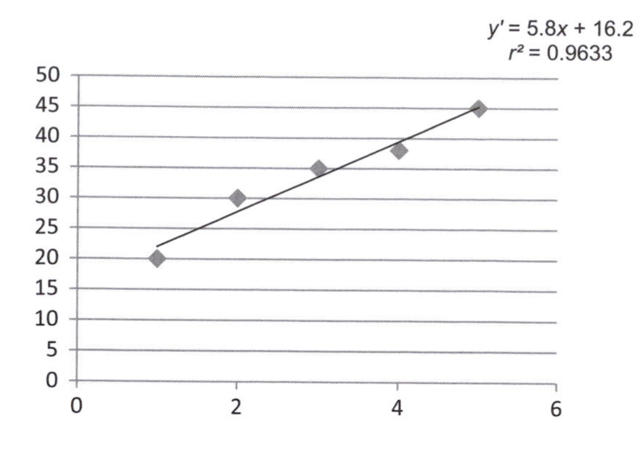

Figure 5.7 Regression line and *r*-squared value for first five years of home run totals

Both OPS and BRA had a strong linear relationship with R/27. Since both these statistics involve SLG, it may be that the SLG is really causing the strong linear relationship. Here SLG is a third variable that may have caused the strong linear relationship between both OPS and BRA and R/27.

5.13. The Pythagorean Theorem of Baseball

Bill James's formula, $W/L = (RS/RA)^2$, describes how the ratio of a team's wins (W) to losses (L) equals the square of the ratio of a team's runs scored (RS) to a team's runs allowed (RA). Later, James revised his formula by replacing the exponent 2 with 1.82.

James's formula means that if we know the runs scored and runs allowed by a certain team, we can predict its wins and losses.

First, we use algebra to convert the *ratio* of a team's wins to losses to a team's *percentage* of wins.

$$W/L = (RS/RA)^2 = (RS)^2/(RA)^2.$$

We divide the left-hand side by $(1 + W/L)$ and do the same for the right-hand side. But as $W/L = (RS)^2/(RA)^2$, we divide the right-hand side not by $(1 + W/L)$ but by $[1 + (RS)^2/(RA)^2]$. We get

$$(W/L)/[1 + (W/L)] = [(RS)^2/(RA)^2]/[1 + (RS)^2/(RA)^2].$$

We now find a common denominator for $[1 + (W/L)]$ and for $[1 + (RS)^2/(RA)^2]$:

$$(W/L)/[(L + W)/L] = [(RS)^2/(RA)^2]/\{[(RA)^2 + (RS)^2]/(RA)^2\}.$$

This simplifies to

$$W/(L + W) = (RS)^2/[(RS)^2 + (RA)^2].$$

Multiplying by 100 gives

$$100 * W/(L+W) = 100 * (RS)^2/[(RS)^2 + (RA)^2].$$

Since (L+W) is the total number of games, the left-hand side of the equation is the winning percentage of a team. Thus,

$$\text{Winning percentage} = 100 * W/(L+W) = 100 * (RS)^2/[(RS)^2 + (RA)^2].$$

Example 5.4. In 2008, the White Sox scored 811 runs and allowed 729 runs. We use James's formula to estimate Chicago's ratio of wins to losses and the team's percentage of wins.

The predicted ratio of wins to losses is $W/L = (811)^2/(729)^2 = 1.237618$.

The predicted percentage of wins is $100 * W/(W+L) = 100 * (811)^2/[(811)^2 + (729)^2] = 55.31\%$.

The White Sox ended the season with an 89–74 record, so their actual win-to-loss ratio was $89/74 = 1.2027$ and their actual winning percentage was 54.60%.

James's formula provides an opportunity to use a standard trick in regression analysis. If the equation is not linear, it may be possible to convert it into a linear form by use of a new set of variables. In what follows, by using algebra along with the properties of logarithms, we can convert the power-law equation $W/L = (RS/RA)^m$ with $X = (RS/RA)$ and $Y = W/L$ into the linear equation $\log(W/L) = m * \log[(RS)/(RA)] + 0$ with $X = \log[(RS)/(RA)]$ and $Y = \log(W/L)$. The "m" found for the linear equation is the desired "m" in the exponential equation.

We now wish to examine how accurate the coefficients $m = 2$ and $m = 1.82$ are in predicting the win-loss results for all the Major League teams for the 2008 season. In what follows, $\log x$ refers to $\log_{10} x$ (log to the base 10 of x). A property of logarithms is $\log x^m = m * \log x$. This property is used in the next calculations.

From James's formula,

$$W/L = (RS/RA)^m.$$

Take the logs of both sides to obtain

$$\log(W/L) = \log[(RS)/(RA)]^m.$$

Now use $\log x^m = m * \log x$ to rewrite the right-hand side:

$$\log(W/L) = m * \log[(RS)/(RA)] + 0.$$

Set $y = \log(W/L)$ and $x = \log[(RS)/(RA)]$ to obtain the linear equation $y = m * x + 0$.

The derivation above converts a power-law equation into a linear equation. James's suggestions for m were first $m = 2$ and later $m = 1.82$. We are now in position to use linear regression to find the best m for all the Major League teams for the 2008 season.

For the 2008 season, Table 5.9 displays for each Major League team their number of wins, their number of losses, their win-loss ratio, the log of their win-loss ratio, their number of runs scored, their number of runs allowed, and the log of the ratio of their number of runs scored to their number of runs allowed.

Figure 5.8 shows the graph for the regression line for the variables X and Y. Also shown is the equation of the regression line. The equation is

Table 5.9 Statistics for the Major League teams for the 2008 season

Team	W	L	W/L	Log(W/L)	RS	RA	log(RS/RA)
Tampa Bay	97	65	1.4923	0.1739	774	671	0.0620
Boston	95	67	1.4179	0.1516	845	694	0.0855
New York	89	73	1.2192	0.0861	789	727	0.0355
Toronto	86	76	1.1316	0.0537	714	610	0.0684
Baltimore	68	93	0.7312	−0.1360	782	869	−0.0458
Chicago	89	74	1.2027	0.0802	811	729	0.0463
Minnesota	88	75	1.1733	0.0694	829	745	0.0464
Cleveland	81	81	1.0000	0.0000	805	761	0.0244
Kansas City	75	87	0.8621	−0.0645	691	781	−0.0532
Detroit	74	88	0.8409	−0.0753	821	857	−0.0186
Los Angeles	100	62	1.6129	0.2076	765	697	0.0404
Texas	79	83	0.9518	−0.0215	901	967	−0.0307
Oakland	75	86	0.8721	−0.0594	646	690	−0.0286
Seattle	61	101	0.6040	−0.2190	671	811	−0.0823
Philadelphia	92	70	1.3143	0.1187	799	680	0.0700
New York	89	73	1.2192	0.0861	799	715	0.0482
Florida	84	77	1.0909	0.0378	770	767	0.0017
Atlanta	72	90	0.8000	−0.0969	753	778	−0.0142
Washington	59	102	0.5784	−0.2377	641	825	−0.1096
Chicago	97	64	1.5156	0.1806	855	671	0.1052
Milwaukee	90	72	1.2500	0.0969	750	689	0.0368
Houston	86	75	1.1467	0.0594	712	743	−0.0185
St. Louis	86	76	1.1316	0.0537	779	725	0.0312
Cincinnati	74	88	0.8409	−0.0753	704	800	−0.0555
Pittsburgh	67	95	0.7053	−0.1516	735	884	−0.0802
Los Angeles	84	78	1.0769	0.0322	700	648	0.0335
Arizona	82	80	1.0250	0.0107	720	706	0.0085
Colorado	74	88	0.8409	−0.0753	747	822	−0.0416
San Francisco	72	90	0.8000	−0.0969	640	759	−0.0741
San Diego	63	99	0.6364	−0.1963	637	764	−0.0790

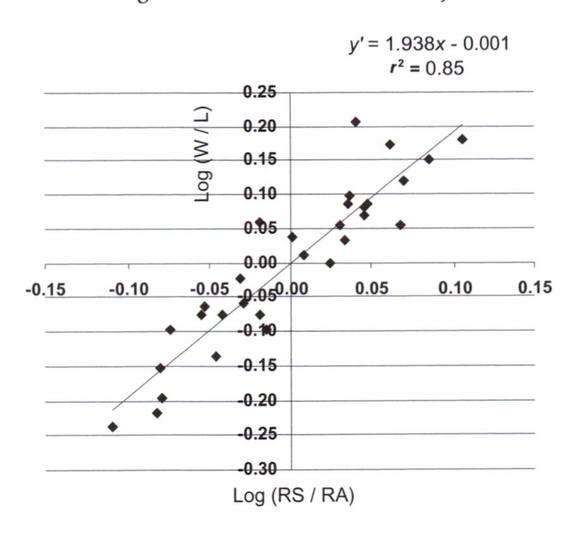

Figure 5.8 Graph of regression line for Example 5.4 (James's formula applied to Major League teams in 2008)

Table 5.10 **Actual and expected win/loss records of the AL East teams for the 2008 season**

Team	W	L	Win% (Obs.)	Standings (Obs.)	Win% (Exp.)	Standings (Exp.)
Tampa Bay	97	65	59.88	1	56.87	3
Boston	95	67	58.64	2	59.42	1
New York	89	73	54.94	3	53.96	4
Toronto	86	76	53.09	4	57.57	2
Baltimore	68	93	42.24	5	44.91	5

$y = 1.938 * x - 0.001$, with $y = \log$ (W/L) and $x = \log$ [(RS)/(RA)]. The coefficient, $m = 1.938$, is the slope of the regression line. Observe that $m = 1.938$ is between 1.82 and 2.00.

Since $r^2 = .85$, we have a correlation coefficient of $r = .92$. This indicates a strong linear relationship between $X = \log$ [(RS)/(RA)] and $Y = \log$ (W/L). Using the coefficient $m = 1.938$, we predict the winning percentage for the AL East teams and the NL East teams. The prediction is based on James's formula, which uses each team's runs scored and runs allowed. Table 5.10 displays in the first five columns the actual results for the 2008 season; columns 6 and 7 display the results based on James's equation $\text{W/L} = (\text{RS/RA})^{1.938}$. Table 5.11 does the same thing for the NL East teams.

In the two tables, "Obs." stands for the observed (actual) result, and "Exp." stands for the expected value based on James's formula.

We now show the calculations that produce columns 4 and 6 for both Tampa Bay and Boston in Table 5.10.

Table 5.11 **Actual and expected win/loss records of the NL East teams for the 2008 season**

Team	W	L	Win% (Obs.)	Standings (Obs.)	Win% (Exp.)	Standings (Exp.)
Philadelphia	92	70	56.79	1	57.75	1
New York	89	73	54.94	2	55.36	2
Florida	84	77	52.17	3	50.19	3
Atlanta	72	90	44.44	4	48.42	4
Washington	59	102	36.65	5	38.01	5

Tampa Bay:

Win% (Obs.) = 100 * [97/(97 + 65)] = 59.88%,
$(RS/RA)^{1.938} = (774/671)^{1.938} = 1.3188$ (expected ratio of wins to losses predicted by James's formula),
Win% (Exp.) = 100 * [1.3188/(1 + 1.3188)] = 56.87%.

Boston:

Win% (Obs.) = 100 * [95/(95 + 67)] = 58.64%.
$(RS/RA)^{1.938} = (845/694)^{1.938} = 1.4645$.
Win% (Exp.) = 100 * [1.4645/(1 + 1.4645)] = 59.42%.

Even though each observed winning percentage is close to each predicted winning percentage, the standings would change if James's formula was used to calculate the winning percentages of the teams in the AL East: the Boston Red Sox would have won the AL East!

Table 5.11 for the NL East shows that using James's formula with $m = 1.938$ would not change the standings.

Exercise 5.12. Verify all the results for the NL East in Table 5.11.

Exercise 5.13. Construct a table similar to Tables 5.10 and 5.11 for the NL West. Use the data in Table 5.9.

Exercise 5.14. Produce a table similar to Table 5.9 and a figure similar to Figure 5.8 for the 2007 Major League season. How does the constant *m* for 2007 compare with $m = 1.938$ used for 2008? Then produce a table similar to Table 5.10. One site where the necessary data can be obtained is www.retrosheet.org.

Exercise 5.15. Perform a new regression analysis for the Major League teams of 2007. Let the *X*-variable = RS/RA and the *Y*-variable = the winning percentage for each team. Find the regression equation and the correlation coefficient. Compare these results with the results using Bill James's Pythagorean theorem.

One of the tasks in the Chapter Problems section provides another application of linear regression analysis. This application uses various baseball statistics to predict the number of runs a team scores.

5.14. The Pythagorean Theorem of Baseball Applied to a Player

How can we apply the Pythagorean theorem of baseball to compare players? Remember, the important statistic RC/27 represents the average number of runs per game created by a player, assuming that he occupies each position in the batting order.

We now compare the highest RC/27 years for Aaron and Bonds using the Pythagorean theorem. Aaron's highest RC/27 was 10.43 in 1971. Bonds's highest RC/27 was 19.69 in 2004. These two numbers represent the average runs scored for a team composed solely of that player for the given year. In order to find the average runs allowed per team for the National League's years of 1971 and 2004, we use the league data from the website www.retrosheet.org. The average runs per game allowed by a National League team in 1971 were 3.91, and the average runs per game allowed by a National League team in 2004 were 4.65.

We apply the Pythagorean theorem to the above data. For a team composed of nine Henry Aarons, assuming average pitching and defense for the year 1971, we get the following winning percentage:

$$\begin{aligned}\text{Winning Percentage} &= 100 * (RS)^2/[(RS)^2 + (RA)^2] \\ &= 100 * (10.43)^2/[(10.43)^2 + (3.91)^2] \\ &= 87.7\%.\end{aligned}$$

This winning percentage times 162 games gives Aaron's team a total of 142 wins and 20 losses for the year 1971.

For a team consisting of only Barry Bonds's clones, assuming average pitching and defense for the year 2004, we get the following winning percentage:

$$\begin{aligned}\text{Winning Percentage} &= 100 * (RS)^2/[(RS)^2 + (RA)^2] \\ &= 100 * (19.69)^2/[(19.69)^2 + (4.65)^2] \\ &= 94.7\%.\end{aligned}$$

This winning percentage times 162 games gives Bonds's team a total of 153 wins and 9 losses for the year 2004.

Exercise 5.16. Find the highest yearly RC/27 for Babe Ruth, Ted Williams, and Joe DiMaggio. Then use the Pythagorean theorem to compare the winning percentage for each of their teams made up of that player occupying each position in the batting order.

Chapter Summary

The chapter began by asking the question, what is the "best guess" of the number on a folded slip of paper, drawn at random from a bowl consisting of many folded slips with known numbers on them? Given that the numbers on the slips are known, the *best guess* was defined to be the number that minimizes $\Sigma(y-\text{guess})^2$. It turns out that the best guess is the mean of the numbers on the slips. We can consider our "best guess" as a prediction.

How can we improve our prediction? We can do this by finding a quantitative variable that has a linear relationship to the variable we are trying to predict. A random sample of objects is selected, and the values for the two variables are observed. Since we want to use the new variable as our predictor, it will be designated as the X-variable. The variable we wish to predict will be designated as the Y-variable. A scatter-plot graph of the ordered pairs is created. If we observe a linear relationship, we proceed to find the equation of the line that comes closest to the points in the scatter plot. This line is called the *sample regression line.* Since a regression line can be found for any set of points, it was necessary to evaluate how good the regression line is.

Correlation analysis measures the strength and direction of the linear relationship. The two statistics r (the sample correlation coefficient) and r-squared (the sample coefficient of determination) are used for this purpose. We called the y-value the observed value for the x-value and the y'-value, the number calculated by substituting the x-value into the equation for the regression line, the expected value. The following equation is always satisfied:

$$\Sigma(y-\bar{y})^2 = \Sigma(y'-\bar{y})^2 + \Sigma(y-y')^2.$$

Using the sums in the above equation, r-squared was defined as follows:

$$r^2 = \Sigma(y'-\bar{y})^2/\Sigma(y-\bar{y})^2.$$

If $r^2=0$, the regression line would be a horizontal line passing through the sample mean. Since y' is constant for all x-values, the X-variable has no effect on the Y-variable. In this case, the sample mean would be our best guess.

If $r^2=1$, each y'-value would equal each y-value, which implies that each of the observed y-values lies on the regression line. In this case, choosing the y'-value for that x-value would be a perfect guess for our number. The ordered pairs (1, 3), (2, 5), (3, 7), and (4, 9), given in the introduction to this chapter, have $r^2=1$.

If the slope of the regression line is positive, the sample correlation coefficient r is defined as the positive square root of r-squared; if the slope of the regression line is negative, r is defined as the negative square root of r-squared.

A regression line is considered a strong predictor if either $r \geq .70$ or $r \leq -.70$. A regression line is a moderate predictor if either $r \geq .40$ or $r \leq -.40$.

For the remainder of the chapter, this prediction model was applied to baseball. In particular, we looked at how well the seven baseball statistics BA, OBA, OPS, BRA, SLG, RC/27, and ISO predicted R/27 for Aaron and Bonds. The objects were the years Aaron and Bonds played in the Major Leagues. For each year, the Y-variable was R/27 and the X-variable came from one of the seven baseball statistics. The results were similar for both players. Even though all the statistics had a strong positive correlation with R/27, the top four for both players were SLG, BRA, OPS, and RC/27 and the bottom three for both men were BA, OBA, and ISO. Since SLG was involved in some way with the top four statistics, it seemed to have a special importance.

It would be interesting to apply this prediction model to other great hitters such as Joe DiMaggio, Ted Williams, Alex Rodriguez, and Albert Pujols. Would we get similar results? You will get a chance to answer this question in the Chapter Problems section.

The chapter closed with warnings about misusing regression analysis.

CHAPTER PROBLEMS

It would be helpful to use Microsoft Excel for these problems.

1. Complete Tables 5.12 and 5.13 for the career years of Joe DiMaggio and Ted Williams. You can use Tables 3.6 and 3.9 at the end of Chapter 3 to help complete the tables.
2. For the career years of Joe DiMaggio use each of the independent variables X = BA, OBA, SLG, OPS, ISO, BRA, and RC/27 with the dependent variable Y = R/27 to complete the following:
   a. Plot each of the scatter-plot graphs.
   b. Find an equation for each of the regression lines.
   c. Graph the corresponding regression lines.
   d. Compute r and r^2 for each variable X.
   e. Make a table in descending order for the correlation coefficients r.
   f. Use the regression equation to predict R/27 for BA = .350.
   g. Using the regression equation, calculate what OPS would be necessary to have R/27 = 7.50.
3. For the career years of Ted Williams (excluding 1952 and 1953) use each of the independent variables X = BA, OBA, SLG, OPS, ISO, BRA, and RC/27 with the dependent variable Y = R/27 to complete the following:
   a. Plot each of the scatter-plot graphs.
   b. Find an equation for each of the regression lines.

Table 5.12 **Yearly batting statistics for Joe DiMaggio**

	X-variables							*Y*
Year	BA	OBA	SLG	OPS	ISO	BRA	RC/27	R/27
1936								
1937								
1938								
1939								
1940								
1941								
1942								
1946								
1947								
1948								
1949								
1950								
1951								

Table 5.13 **Yearly batting statistics for Ted Williams**

	X-variables							*Y*
Year	BA	OBA	SLG	OPS	ISO	BRA	RC/27	R/27
1939								
1940								
1941								
1942								
1946								
1947								
1948								
1949								
1950								
1951								
1952								
1953								
1954								
1955								
1956								
1957								
1958								
1959								
1960								

c. Graph the corresponding regression lines.
d. Compute r and r^2 for each variable X.
e. Make a table in descending order for the correlation coefficients r.
f. Use the regression equation to predict R/27 for BA = .350.
g. Using the regression equation, calculate what OPS would be necessary to have R/27 = 7.50.

4. Perform a comparison between DiMaggio and Williams with respect to their r-squared values in problems 2 and 3 above. How do these results compare with the results for Aaron and Bonds? Of the four players, who can be paired as most similar?
5. Repeat problems 2 and 3 for your chosen two players (Hall of Famer and potential Hall of Famer you chose in Chap. 4). What do the results say about predicting the two players' yearly R/27?
6. Table 5.14 provides the statistics for each of the Major League teams for the year 2007, downloaded from the website www.mlb.com. Instead of determining a relationship between baseball statistics and R/27 for a player, we want to analyze which statistics are related best to runs scored per game (R/G) for the teams in 2007.
   a. Fill in Table 5.15.
   b. For the Major League teams, use each of the independent variables X = BA, OBA, SLG, OPS, ISO, BRA, and RC/G with the dependent variable Y = R/G to complete the following:
      i. Plot the scatter-plot graphs.
      ii. Find an equation for each of the regression lines.
      iii. Graph the corresponding regression lines.
      iv. Compute r and r^2 for each variable X.
      v. Make a table in descending order for the correlation coefficients r.
      vi. Using the regression equation, a team with BA = .300 would be predicted to have an R/G of ________.
      vii. Using the regression equation, in order for a team to produce 8.50 R/G the team would need an RC/27 = ________.
7. Download the team data for the year 2008 from the website www.mlb.com. Repeat problem 6 for this data. In particular, discuss the Tampa Bay Rays with respect to the two years 2007 and 2008.
8. For the career years of Alex Rodriguez (exclude 1994 and 1995), use each of the independent variables X = BA, OBA, SLG, OPS, ISO, BRA, and RC/27 with the dependent variable Y = R/27 to complete the following:
   a. Plot the scatter-plot graphs.
   b. Find an equation for each of the regression lines.
   c. Graph the corresponding regression lines.
   d. Compute r and r^2 for each variable X.

Table 5.14 **Batting statistics for the Major League teams for the 2007 season**

Team	G	AB	R	H	2B	3B	HR	RBI	BB	SO	OBP	SLG	AVG
New York Yankees	162	5717	968	1656	326	32	201	929	637	991	0.366	0.463	0.290
Detroit Tigers	162	5757	887	1652	352	50	177	857	474	1054	0.345	0.458	0.287
Seattle Mariners	162	5684	794	1629	284	22	153	754	389	861	0.337	0.425	0.287
Los Angeles Angels	162	5554	822	1578	324	23	123	776	507	883	0.345	0.417	0.284
Colorado Rockies	163	5691	860	1591	313	36	171	823	622	1152	0.354	0.437	0.280
Boston Red Sox	162	5589	867	1561	352	35	166	829	689	1042	0.362	0.444	0.279
Atlanta Braves	162	5689	810	1562	328	27	176	781	534	1149	0.339	0.435	0.275
Los Angeles Dodgers	162	5614	735	1544	276	35	129	706	511	864	0.337	0.406	0.275
New York Mets	162	5605	804	1543	294	27	177	761	549	981	0.342	0.432	0.275
Philadelphia Phillies	162	5688	892	1558	326	41	213	850	641	1205	0.354	0.458	0.274
St. Louis Cardinals	162	5529	725	1513	279	13	141	690	506	909	0.337	0.405	0.274
Baltimore Orioles	162	5631	756	1529	306	30	142	718	500	939	0.333	0.412	0.272
Chicago Cubs	162	5643	752	1530	340	28	151	711	500	1054	0.333	0.422	0.271
Cleveland Indians	162	5604	811	1504	305	27	178	784	590	1202	0.343	0.428	0.268
Tampa Bay Devil Rays	162	5593	782	1500	291	36	187	750	545	1324	0.336	0.433	0.268
Cincinnati Reds	162	5607	783	1496	293	23	204	747	536	1113	0.335	0.436	0.267
Florida Marlins	162	5627	790	1504	340	38	201	749	521	1332	0.336	0.448	0.267
Minnesota Twins	162	5522	718	1460	273	36	118	671	512	839	0.330	0.391	0.264
Pittsburgh Pirates	162	5569	724	1463	322	31	148	694	463	1135	0.325	0.411	0.263
Texas Rangers	162	5555	816	1460	298	36	179	768	503	1224	0.328	0.426	0.263
Milwaukee Brewers	162	5554	801	1455	310	37	231	774	501	1137	0.329	0.456	0.262
Kansas City Royals	162	5534	706	1447	300	46	102	660	428	1069	0.322	0.388	0.261
Houston Astros	162	5605	723	1457	293	30	167	700	547	1043	0.330	0.412	0.260
Toronto Blue Jays	162	5536	753	1434	344	24	165	719	533	1044	0.327	0.419	0.259
Oakland Athletics	162	5577	741	1430	295	16	171	711	664	1119	0.338	0.407	0.256
Washington Nationals	162	5520	673	1415	309	31	123	646	524	1128	0.325	0.390	0.256
San Francisco Giants	162	5538	683	1407	267	37	131	641	532	907	0.322	0.387	0.254
San Diego Padres	163	5612	741	1408	322	31	171	704	557	1229	0.322	0.411	0.251
Arizona Diamondbacks	162	5398	712	1350	286	40	171	687	532	1111	0.321	0.413	0.250
Chicago White Sox	162	5441	693	1341	249	20	190	667	532	1149	0.318	0.404	0.246

e. Make a table in descending order for the correlation coefficients r.

Use the data in Table 5.16, downloaded from the website www.mlb.com.

9. Download the baseball data for Ryne Sandberg for the years 1981–1997 from the website www.mlb.com. Let $X = \text{BA}$ and $Y = \text{R}/27$.
   a. Using the years from 1981 to 1997, sketch the scatter-plot graph. Then, find the equation for the regression line, find the correlation coefficient r, and sketch the graph for the regression line.
   b. Repeat part (a) for the years 1982–1997.
   c. Discuss why the year 1981 is an outlier and how the results in (a) are affected by the outlier.

Table 5.15 **Batting statistics for the Major League teams for the 2008 season**

	X-variables							*Y*
Team	BA	OBA	SLG	OPS	ISO	BRA	RC/G	R/G
Rockies								
Braves								
Dodgers								
Mets								
Phillies								
Cardinals								
Cubs								
Reds								
Marlins								
Pirates								
Brewers								
Astros								
Nationals								
Giants								
Padres								
Diamondbacks								
Yankees								
Tigers								
Mariners								
Angels								
Red Sox								
Orioles								
Indians								
Rays								
Twins								
Rangers								
Royals								
Blue Jays								
Athletics								
White Sox								

Table 5.16 **Yearly batting statistics for Alex Rodriguez**

Year	Team	G	AB	R	H	2B	3B	HR	BB	SO	SH	SF	HBP	AVG	OBP	SLG
1994	Mariners	17	54	4	11	0	0	0	3	20	1	1	0	0.204	0.241	0.204
1995	Mariners	48	142	15	33	6	2	5	6	42	1	0	0	0.232	0.264	0.408
1996	Mariners	146	601	141	215	54	1	36	59	104	6	7	4	0.358	0.414	0.631
1997	Mariners	141	587	100	176	40	3	23	41	99	4	1	5	0.300	0.350	0.496
1998	Mariners	161	686	123	213	35	5	42	45	121	3	4	10	0.310	0.360	0.560
1999	Mariners	129	502	110	143	25	0	42	56	109	1	8	5	0.285	0.357	0.586
2000	Mariners	148	554	134	175	34	2	41	100	121	0	11	7	0.316	0.420	0.606
2001	Rangers	162	632	133	201	34	1	52	75	131	0	9	16	0.318	0.399	0.622
2002	Rangers	162	624	125	187	27	2	57	87	122	0	4	10	0.300	0.392	0.623
2003	Rangers	161	607	124	181	30	6	47	87	126	0	6	15	0.298	0.396	0.600
2004	Yankees	155	601	112	172	24	2	36	80	131	0	7	10	0.286	0.375	0.512
2005	Yankees	162	605	124	194	29	1	48	91	139	0	3	16	0.321	0.421	0.610
2006	Yankees	154	572	113	166	26	1	35	90	139	0	4	8	0.290	0.392	0.523
2007	Yankees	158	583	143	183	31	0	54	95	120	0	9	21	0.314	0.422	0.645
Career		1904	7350	1501	2250	395	26	518	915	1524	16	74	127	0.306	0.389	0.578

d. If you wanted to predict Y by using X, would you use the regression line in (a) or the regression line in (b)? Explain.

For the next problems, we look at the equation LWTF3 = (.46) * 1B + (.80) * 2B + (1.02) * 3B + (1.40) * HR + (.33) * BB + .33 * (HBP) + (.30) * SB − (.60) * CS − (.25) * (AB − H), where SB is the number of stolen bases and CS is the number of caught stealing. LWTF3 was introduced by Thorn and Palmer to predict runs scored. We define the X-variable to be LWTF3/27 = (25.5 * LWTF3)/(AB − H) and the Y-variable to be R/27.

10. Complete the following:
   a. For Aaron, Bonds, Williams, DiMaggio, and Rodriguez perform a regression analysis where X = LWTF3/27 and Y = R/27.
   b. Compare the correlation coefficient results for these players in part (a) with the correlation coefficient results for X = RC/27 and Y = R/27. Which is the best predictor for R/27 for each of the players?
11. Repeat problem 10 for your two chosen players.
12. Using X = LWTF3/G and Y = R/G, perform a regression analysis on the teams in the Major Leagues in Table 5.14. Then compare your results with those using X = RC/G and Y = R/G. Which is the best predictor for R/27?

Chapter 6

Descriptive Statistics Applied to Qualitative Variables

In Chapter 2 we studied descriptive statistics applied to one quantitative variable and looked at different quantitative baseball statistics for Henry Aaron and Barry Bonds. In Chapter 4 we used descriptive statistics to compare the yearly data for baseball quantitative variables between Aaron and Bonds. In this chapter we switch our attention from quantitative variables to qualitative variables.

Qualitative variables were introduced in Chapter 1. In this chapter we explore "COPS," the collection, organization, presentation, and summarization, applied to qualitative data. The graphs used to present the data are constructed using Microsoft Excel.

A variable is a characteristic that assigns values to the subjects or objects in a sample or population. The associated values are the values of the variable. In Chapter 2 we studied quantitative variables. Quantitative variables have outcomes or values that are numbers. Some examples of quantitative variables for Aaron and Bonds are their yearly batting averages, their yearly number of home runs, their yearly number of runs scored, and their yearly on-base averages.

Qualitative variables have outcomes or values that are categories. In this chapter we study qualitative variables for baseball, such as the outcome of a plate appearance, the type of hit, or the result of an at bat. We consider as possible outcomes of a plate appearance (PA) only the following: 1B, 2B, 3B, HR, BB, HBP, SF, O (out). The possible outcomes of an at bat (AB) are 1B, 2B, 3B, HR, and O (out).

For the rest of the book, the outcome of safe on error (SE) will be included in the category O. The justification for this is that SE is treated the same as an out in the scorebook.

Qualitative variables occur outside of baseball, as the next example shows.

Example 6.1. Given a sample of people, the following are qualitative variables:

a. Marital status
b. Gender
c. Zip code for home address
d. Age category (under 20, 21–40, over 40)
e. Social security number

Even though they are numbers, zip codes and social security numbers are qualitative: these numbers are used as identifiers, and the operations of addition and multiplication on them have no meaning. If the numbers represent categories, then the numbers are considered categories.

Example 6.2. The following are values for each of the variables in Example 6.1:

a. Married
b. Male
c. 05618
d. 21–40
e. 234-34-5678

Some variables from baseball are given next.

Example 6.3. Given a sample of plate appearances for a player, the following are qualitative variables:

a. Result of the plate appearance (1B, 2B, 3B, HR, BB, HBP, SF, O)
b. Result of an at bat (1B, 2B, 3B, HR, O)
c. SO (Y or N)
d. Type of hit (1B, 2B, 3B, HR)

Example 6.4. Given a sample of Major League players, the following are qualitative variables:

a. Batting side (L, R, or SW)
b. Throwing side (L or R)
c. Current team
d. Played college ball (Y or N)

6.1. Organizing Data for One Qualitative Variable into Tables

This section deals with one qualitative variable. We defined a *complete frequency table* in Chapter 2 and applied it to quantitative variables. We now apply it to qualitative variables. The same notations can be used for both. The terms F_x, Pr_x, and P_x will have the same meaning: F_x is the frequency at

x; Pr_x is the proportion at x; and P_x is the percentage at x. The difference is that values x are categories.

Table 2.4 was our first example of a complete frequency table for a *qualitative* variable. When *quantitative* data are grouped into class intervals, the class intervals are categories. These categories are values for a *qualitative* variable. Since there is a natural ordering for class intervals, we could compute the cumulative results CF_x, CPr_x, and CP_x. Normally, a qualitative variable has no ordering. If there is no ordering, the cumulative columns would make no sense and therefore are omitted from the table. Example 6.5 gives a complete frequency table for the outcomes for the career plate appearances of Aaron. A complete frequency table for a player can also be constructed for his plate appearances for a series, a season, or a group of seasons.

Table 6.1 **Complete frequency table for the career plate appearances of Henry Aaron**

X	F_X	Pr_X	P_X
1B	2294	0.1648	16.48
2B	624	0.0448	4.48
3B	98	0.0070	0.70
HR	755	0.0542	5.42
BB	1402	0.1007	10.07
HBP	32	0.0023	0.23
SF	121	0.0087	0.87
O	8593	0.6174	61.74
Totals	13919	1.0000	100.00

Example 6.5. Table 6.1 provides a complete frequency table for Aaron's career PAs. The sample is the career plate appearances (PAs) for Aaron. The *X*-variable represents the possible outcomes of a plate appearance. (If needed, you can review the definition of a plate appearance by consulting Chap. 2.)

Exercise 6.1. Construct a complete frequency table for Aaron's plate appearances for the year 1971 (see Table 2.2).

Exercise 6.2. Construct a complete frequency table for Bonds's plate appearances for the year 2001 (see Table 4.1).

Exercise 6.3. Construct a complete frequency table for Ted Williams's plate appearances for the year 1941 (see Table 3.9).

Exercise 6.4. Construct a complete frequency table for Joe DiMaggio's plate appearances for the year 1941 (see Table 3.6).

6.2. Presentation of Qualitative Data for One Variable

The graphs constructed in this section for the qualitative data for one qualitative variable are a bar graph and a pie graph. Figure 6.1 is a percentage bar graph for the career PAs of Aaron, and Figure 6.2 is a percentage pie graph for the career PAs of Aaron. Both graphs were constructed using Microsoft Excel. These graphs could have been constructed by hand.

For the bar graph, all the categories for a PA lie below the x-axis. The rectangles, which are the bars, have their bases over their categories on the x-axis. The bases must have the same width, be disjoint, and be the same

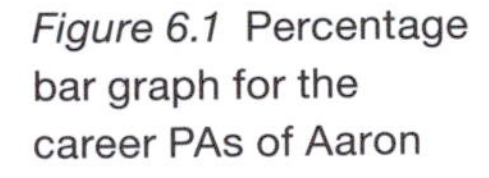
Figure 6.1 Percentage bar graph for the career PAs of Aaron

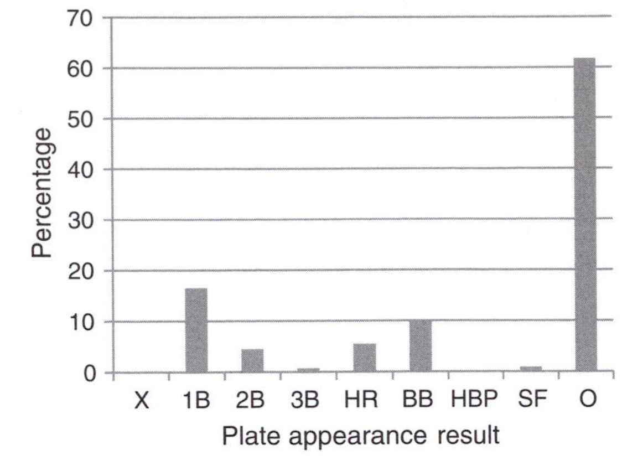

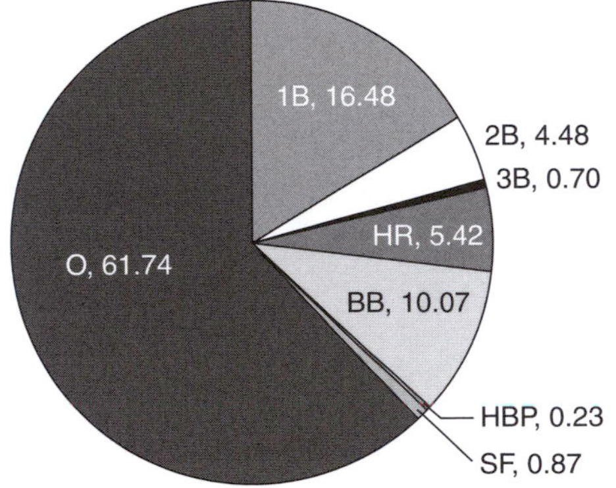

Figure 6.2 Percentage pie graph for the career PAs of Aaron

distance apart. The height of each bar is either the frequency or percentage of occurrences for that category.

The pie graph divides a circle into slices called sectors; the area of each slice corresponds to the percentage of occurrences for that category.

The pie graph, representing the career plate appearances for a player, will be used later in the book as a physical model to simulate a player's career batting performance.

Exercise 6.5. Construct a percentage bar graph and pie graph for the plate appearances in Exercise 6.1.

Exercise 6.6. Construct a percentage bar graph and pie graph for the plate appearances in Exercise 6.2.

Exercise 6.7. Construct a percentage bar graph and pie graph for the plate appearances in Exercise 6.3.

Exercise 6.8. Construct a percentage bar graph and pie graph for the plate appearances in Exercise 6.4.

6.3. Summarizing Qualitative Data for One Variable

The typical category for a qualitative variable is the category with the greatest frequency. This is called the *modal* category. The modal category for Example 6.5 is O (out).

Exercise 6.9. Find the modal category for the categories in Exercises 6.1, 6.2, 6.3, and 6.4.

6.4. Organizing Two Qualitative Variables into a Contingency Table

When two qualitative variables for one set of objects are organized into a table, it is called a *contingency* table.

Using the sample of all career plate appearances for Aaron and Bonds (Bonds's plate appearances are through the first 81 games of 2007), the two qualitative variables are the choice of player (Aaron or Bonds) and the outcome of a plate appearance (1B, 2B, 3B, HR, BB, HBP, SF, O). The complete contingency table for these two qualitative variables is presented in Table 6.2. The contingency table itself would consist of only the first three columns, X, F_X (Aaron), F_X (Bonds). The addition of $\Pr_X$ and P_X for each player is what we call the *complete* contingency table.

Exercise 6.10. Construct a complete contingency table for the yearly plate appearances for Aaron (1971) and Bonds (2001).

Exercise 6.11. Construct a complete contingency table for the yearly plate appearances for DiMaggio (1941) and Williams (1941).

6.5. Presentations for the Contingency Table for Two Qualitative Variables

Three types of graphs will be used to present the data for a complete contingency table for two qualitative variables. These graphs are called the *side-by-side* bar graph, the *stacked* bar graph, and the *100% stacked* bar graph. For each of these graphs, we can use either frequencies or percentages for the

Table 6.2 **Contingency table for the career PAs of Aaron and Bonds (through first 81 games of 2007)**

	F_X		$\Pr_X$		P_X (%)	
X	Aaron	Bonds	Aaron	Bonds	Aaron	Bonds
1B	2294	1477	0.1648	0.1187	16.48	11.87
2B	624	597	0.0448	0.0480	4.48	4.80
3B	98	77	0.0070	0.0062	0.70	0.62
HR	755	751	0.0542	0.0604	5.42	6.04
BB	1402	2520	0.1007	0.2026	10.07	20.26
HBP	32	105	0.0023	0.0084	0.23	0.84
SF	121	90	0.0087	0.0072	0.87	0.72
O	8593	6824	0.6174	0.5485	61.74	54.85
Totals	13919	12441	1.0000	1.0000	100.00	100.00

values on the y-axis. The variable whose categories are below the x-axis will be called the *first* variable. We refer to the other variable as the *second* variable. The y-axis could also be used for the first variable. As pointed out earlier in this book, both frequencies and percentages should be considered when comparing two data sets. These three types of graphs are options available in Excel.

These three graphs for Table 6.2 are presented in Figures 6.3, 6.4, and 6.5. The construction of these three types of graphs by hand requires one to select an axis for the first variable. For a vertical bar graph, the x-axis is used; for a horizontal bar graph, the y-axis is used. The categories for the first variable are then placed below that axis. For Figures 6.3, 6.4, and 6.5, we choose the x-axis for the first variable PA. We now describe the rest of the construction for each of the graphs.

6.5.1. Side-by-Side Percentage Bar Graph

Above each category for the first variable, the variable used for the x-axis, we construct abutting rectangles, one for each of the categories of the second variable. Each rectangle is of height P_x, which is the product of 100 times the ratio formed by dividing the frequency of the intersection of that category of the second variable and the category from the first variable by the sum of all the frequencies for all the intersections of that category of the second variable with each of the categories of the first variable. An example makes this clearer.

In Figure 6.3, the first variable for the x-axis is PA. The second variable is the choice of player (Aaron or Bonds). We look at the calculation of the two abutting rectangles, one for Aaron and one for Bonds, over the category of 1B for the first variable PA. The intersection of the category 1B and the category Aaron has a frequency of 2294, while the intersection of the category 1B and the category Bonds has a frequency of 1477. Since the sum of all the frequencies for the intersection of Aaron with each PA category is 13919 and the sum of all the frequencies for the intersection of Bonds with each PA category is 12441, the height of the rectangle for the category Aaron is $100 * (2294/13919) = 16.48\%$. The height of the rectangle for Bonds is $100 * (1477/12441) = 11.87\%$. The graph is completed when the same procedure is done for each category of PA. Observe from Table 6.2 that P_{1B} for Aaron equals 16.48% and P_{1B} for Bonds equals 11.87%.

6.5.2. Stacked Percentage Bar Graph

The only difference between this graph and the side-by-side bar graph is that instead of placing the rectangles side by side, we place the rectangles on top of each other (we are stacking the rectangles). Look at Figure 6.4 to

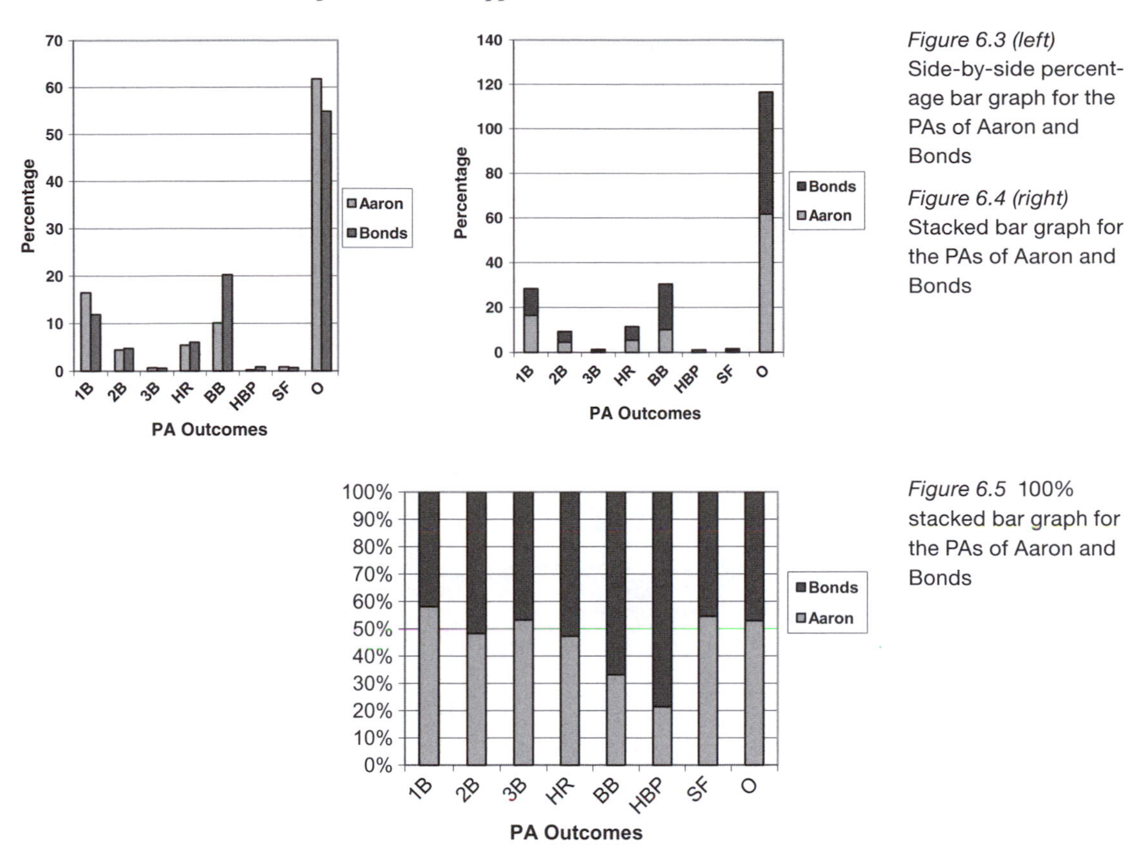

Figure 6.3 (left) Side-by-side percentage bar graph for the PAs of Aaron and Bonds

Figure 6.4 (right) Stacked bar graph for the PAs of Aaron and Bonds

Figure 6.5 100% stacked bar graph for the PAs of Aaron and Bonds

see this stacking process. Observe that the combined height of a stacked bar can exceed 100%.

6.5.3. 100% Stacked Percentage Bar Graph

Above each category for the first variable, the variable used for the x-axis, we stack rectangles, one for each of the categories of the second variable. The height of a rectangle for each category of the second variable is equal to the product of 100 times the ratio formed by dividing the percentage, calculated for that second category in the side-by-side percentage graph, by the sum of all the percentages for all the categories of the second variable, constructed in the side-by-side graph.

In Figure 6.5, the first variable for the x-axis is PA. The second variable is the choice of player (Aaron or Bonds). We look at the calculation of the heights for the two stacked rectangles over the category of 1B for the first variable, PA. The percentage calculated for the category Aaron, from the side-by-side graph, is 16.48, and the percentage calculated for the category Bonds, from the side-by-side graph, is 11.87. Adding these two percentages gives a total of 28.35. We divide 16.48 by 28.35 and multiply by 100 to get 58.13% and divide 11.87 by 28.35 and multiply by 100 to get 41.87%. In Figure 6.5, Aaron's rectangle has a height of 58.13, while Bonds's rectangle stacked on top has a height of 41.87. The graph is completed when the same procedure is done for each category of the first variable, PA.

Of the three types of graph, the 100% stacked bar graph gives the clearest picture of the relationships between the two variables. For example, we see that Aaron had a greater percentage of singles in his plate appearances, whereas Bonds had a greater percentage of base on balls in his plate appearances. Using Figure 6.5, what other major differences (if any) can you observe?

Exercise 6.12. Construct the three types of graphs for the contingency table in Exercise 6.10. What do these graphs tell us about the plate appearances of Aaron (1971) and Bonds (2001)?

Exercise 6.13. Construct the three types of graphs for the contingency table in Exercise 6.11. What do these graphs tell us about the plate appearances of DiMaggio (1941) and Williams (1941)?

Exercise 6.14. Using the variable "choice of player (Aaron or Bonds)" as the first variable for the *x*-axis, construct the three types of graphs for Table 6.2.

Exercise 6.15. Using only those plate appearances for Aaron and Bonds that fit one of the categories 1B, 2B, 3B, HR, BB, SO, or IPO, construct a contingency table in the form of Table 6.2. Then construct the three types of graphs. Observe that the categories in this table differ from those in Table 6.2 because HBP and SF are excluded. Also, the category O is divided into SO (strikeout) and IPO (in-play out).

Exercise 6.16. Complete Table 6.2 by including the plate appearances for Barry Bonds through the entire 2007 season. Table 4.1 contains the necessary data. Repeat Exercise 6.14 for this updated table. How similar are the graphs for the two tables?

6.6. Finding a Relationship between Two Qualitative Variables

In Chapter 5 we looked for a relationship between two quantitative variables. The graph used to help determine a functional form for such a relationship was the scatter-plot graph. For qualitative variables, a scatter-plot graph cannot be constructed. The side-by-side percentage bar graph, the stacked percentage bar graph, and the 100% stacked percentage bar graph are used instead. Any one or all three can be inspected to see the possibility of a relationship between the two variables. If there is no relationship between the

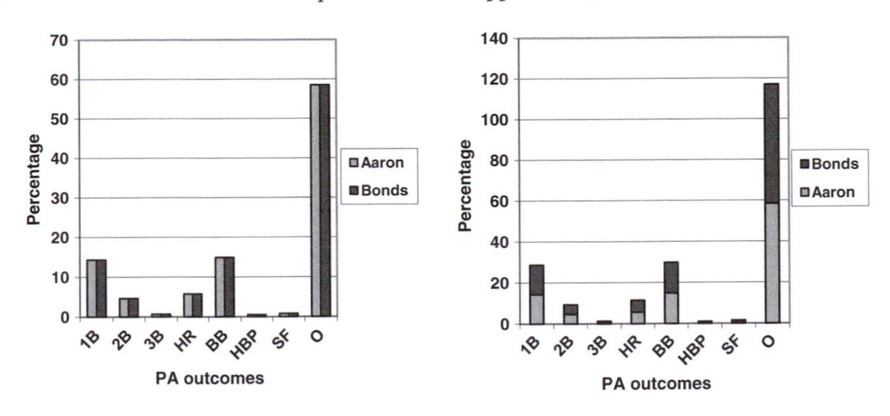

Figure 6.6 (left) Side-by-side percentage bar graph for the PAs of Aaron and Bonds (no relationship between player and PA)

Figure 6.7 (right) Stacked percentage bar graph for the PAs of Aaron and Bonds (no relationship between player and PA)

two qualitative variables, we say that the two variables are *independent*. Theoretically, if the choice of player is perfectly independent of the PA outcome, we would expect the three graphs to appear as in Figures 6.6, 6.7, and 6.8. Notice that, in all three of the graphs, the percentages over each category for PA are equal.

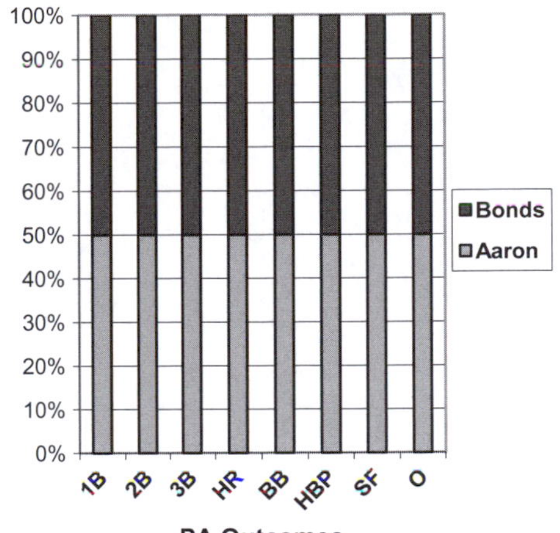

Figure 6.8 100% stacked bar graph for the PAs of Aaron and Bonds (no relationship between player and PA)

That is, under the assumption of perfect independence, the three graphs (Figs. 6.6, 6.7, and 6.8) show that for each category of PA the corresponding rectangles for Aaron and Bonds have the same height. Since we are dealing with a sample of plate appearances, the elements of chance and skill are both present. For this reason, even if *choice of player* and the *outcome of a plate appearance* were independent variables, we would not expect perfect independence. In determining whether the two variables are independent, we ask the following question: Are the differences between the graphs in Figures 6.3, 6.4, and 6.5 and their corresponding graphs in Figures 6.6, 6.7, and 6.8 enough to be attributed to more than just chance? The answer "yes" supports the idea of relationship between the qualitative variables *choice of player* and *outcome of a plate appearance* for our sample. An answer "no" indicates that there is no relationship between the two variables for our sample (the two variables are independent). Since we are dealing with a sample of plate appearances, the best we can do is to use the graphs to make a subjective opinion.

The purpose of our analysis is to compare the type of hitter Aaron was with the type of hitter Bonds was. Looking at Table 6.2 and Figure 6.3, the categories with the biggest difference between Aaron and Bonds are 1B, BB, and O. If we think that skill, more than just chance, affects the results, we might conclude that for their careers, Aaron was a better singles hitter and Bonds was much better at drawing walks. For 20% of his plate appearances, Bonds reached base on a walk, whereas Aaron drew a walk only 10% of the

Table 6.3 Contingency table for type of hit for NL players for the years 1965, 1975, 1985, 1995, and 2005

	F_Y					P_Y				
Year	1B	2B	3B	HR	Total H	1B	2B	3B	HR	Total %
1965	9932	2122	422	1318	13794	72.00	15.38	3.06	9.55	100.00
1975	12530	2781	458	1233	17002	73.70	16.36	2.69	7.25	100.00
1985	11874	2861	437	1424	16596	71.55	17.24	2.63	8.58	100.00
1995	12482	3367	418	1917	18184	68.64	18.52	2.30	10.54	100.00
2005	15256	4754	468	2580	23058	66.16	20.62	2.03	11.19	100.00

time. Looking at Table 4.8, we see that 21% of Aaron's walks were intentional, compared with 27% of Bonds's walks. However, no matter what the reason for the higher percentage of walks for Bonds, this helps to account for his smaller percentage of outs. Looking at the percentage of home runs, Bonds has a slight advantage. For the next example, we look at two variables, each of which has more than two categories.

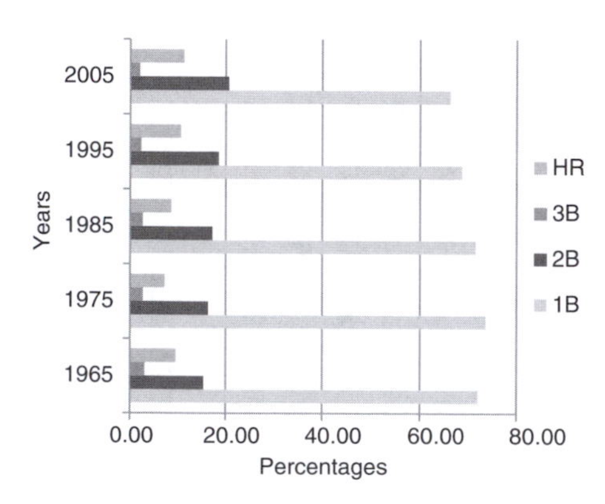

Figure 6.9 Side-by-side percentage bar graph for the years and type of hit for the NL.

Example 6.6. Is there a relationship between certain *years* (1965, 1975, 1985, 1995, 2005) and the *type of hit* (1B, 2B, 3B, HR) in the National League?

In this example, our sample is the set of all plate appearances resulting in a hit (H) for all National League players for the years 1965, 1975, 1985, 1995, and 2005. The two qualitative variables are the *years* (1965, 1975, 1985, 1995, and 2005) and the *type of hit* (1B, 2B, 3B, HR). The variable "years" has five categories: 1965, 1975, 1985, 1995, and 2005. The variable "type of hit" has four categories: 1B, 2B, 3B, and HR. The contingency table for these two variables is shown in Table 6.3. Figures 6.9 and 6.10 are, respectively, the side-by-side percentage bar graph and the stacked percentage bar graph. These are presented as horizontal bar graphs. The variable "years" is placed along the *y*-axis. The percentages P_Y are calculated using the row totals instead of the column totals.

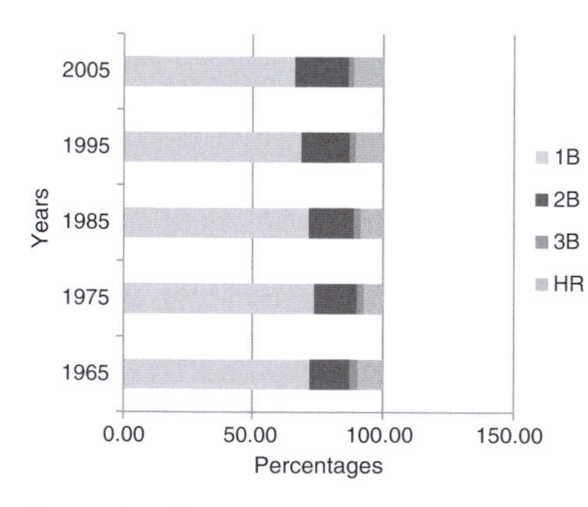

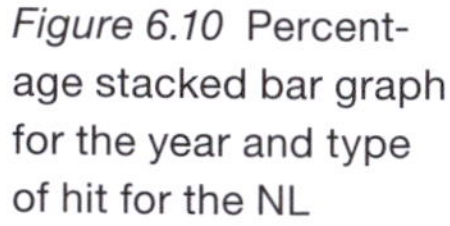

Figure 6.10 Percentage stacked bar graph for the year and type of hit for the NL

Inspecting Table 6.3, the percentage of hits that were home runs or doubles was highest in 1995 and 2005, while the percentage of hits that were singles was highest in 1965, 1975, and 1985. Since the chosen years were the midpoints of five decades, one might conclude that, beginning in 1990, players in the NL concentrated on hitting for power.

Exercise 6.17. This exercise looks for a relationship between the same five years and type of out. We divide the type of out into the two categories of SO (strikeout) and IPO (in-play out). We expect that the relationship is that the percent of strikeouts increased for the last two decades. Using the five years 1965, 1975, 1985, 1995, and 2005 and the categories SO and IPO, construct a table similar to Table 6.3 for the NL. After completing the table, construct the side-by-side percentage bar graph and the stacked percentage bar graph. Describe any relationship you observe between these years and type of out. The data can be retrieved from the website www.retrosheet.org. (Once you reach the site, click on Games, followed by Regular season, followed by the year, followed by NL Splits.)

Exercise 6.18. What effect did the designated hitting rule have on hitting? We would expect that the ratio of hits to at bats would favor the American League. Using the four years 1975, 1985, 1995, and 2005 and the categories O and H, construct a table similar to Table 6.3 for the NL and one for the AL. For each table, construct the side-by-side percentage bar graph and the stacked percentage bar graph. Describe any relationship you observe between these years and the two leagues with respect to the outcome of an at bat.

Exercise 6.19. Using the graphs from Exercise 6.12, can we conclude that there is a relationship between choice of player and type of hit? If so, describe it.

Exercise 6.20. Using the graphs from Exercise 6.13, can we conclude that there is a relationship between choice of player and type of hit? If so, describe it.

Chapter Summary

In this chapter we studied descriptive statistics (COPS) applied to one or two qualitative variables. The outcomes of a qualitative variable are called categories. The complete frequency table was used to organize the outcomes for *one* qualitative variable. The table used to organize the outcomes for *two* qualitative variables is called a contingency table.

The graphs used to present *one* qualitative variable are the bar graph and the pie graph. For *two* qualitative variables, the side-by-side percentage bar graph, the stacked percentage bar graph, and the 100% stacked percentage bar graph are used.

The study of two qualitative variables, applied to a sample or a population, is used to determine whether there is a relationship between the two variables. If no relationship exists, the two variables are said to be independent.

When dealing with two qualitative variables associated with a sample instead of the population, in deciding whether there is or is not a relationship between two variables, the chance factor must be considered. We saw how these graphs would appear if there was perfect independence between the two variables. Since we are dealing with a sample and not the population, relying on the graphs alone will enable us only to make a subjective opinion on the independence of the variables. The greater the 100% stacked percentage bar graph based on the observed outcomes of the sample is different from the 100% stacked percentage bar graph based on the theoretical independence of the two variables, the more likely there is a relationship between the two variables.

Research studies are interested in making a decision about the independence of the two variables not for the sample, but for the population. The continuous distribution used for this study is called the chi-square distribution. For those students interested in this inferential technique, we include the chi-square test for the independence of two qualitative variables in Appendix B.

CHAPTER PROBLEMS

1. Consider one qualitative variable to be the special years 1925, 1938, 1957, 1981, and 2001. The other qualitative variable is the type of hit (1B, 2B, 3B, HR). The sample will be the set of all plate appearances resulting in a hit for these years for the National League players. Obtain the data from the website www.retrosheet.org.
   a. Construct the complete contingency table.
   b. Construct the side-by-side percentage bar graph and the 100% stacked percentage bar graph.
   c. Is there a relationship between these years and the type of hit?
   d. If you think there is a relationship, describe it.
2. Consider the sample of all career plate appearances (excluding HBP and SF) for Joe DiMaggio and Ted Williams. The two qualitative variables are the choice of player (DiMaggio or Williams) and the outcome of a plate appearance (1B, 2B, 3B, HR, BB, SO, IPO). Remember that IPO is in-play out.
   a. Construct the complete contingency table.
   b. Construct the side-by-side percentage bar graph and the 100% stacked percentage bar graph.
   c. Is there a relationship between the choice of player and the outcome of a plate appearance?
   d. If you think there is a relationship, describe it.
3. Many baseball people believe that left-handed batters have higher batting averages than right-handed batters. Two possible reasons are that they are closer to first base and there are more right-handed pitchers. An

actual study looked at 5-year intervals beginning in 1950. In each interval, a batting average was calculated for all left-handed batters and a batting average was calculated for all right-handed batters. The results are given in Table 6.4.

The two qualitative variables are the yearly intervals and which side of the plate a batter hits from. The values are the average batting averages for each interval. Remember, a batting average multiplied by 100 represents the percentage of at bats that resulted in a hit.

a. Construct the side-by-side percentage bar graph and the 100% stacked percentage bar graph.
b. Is there a relationship between the yearly intervals and which side of the plate a batter hits from?
c. If you think there is a relationship, describe it.

Table 6.4 **Batting average for left-handed and right-handed batters in 5-year intervals**

Years	Left-handed AVG	Right-handed AVG
1950–1954	.270	.256
1955–1959	.267	.253
1960–1964	.260	.250
1965–1969	.250	.241
1970–1974	.260	.248
1975–1979	.269	.256
1980–1984	.267	.257
1985–1989	.263	.253
1990–1994	.268	.257
1995–1999	.276	.264
2000–2004	.270	.262
2005–2009	.267	.263

4. Consider the sample of all career plate appearances for your chosen two players. The two qualitative variables are the choice of player (Hall of Famer or potential Hall of Famer) and the outcome of a plate appearance (1B, 2B, 3B, HR, (BB + HBP), SO, IPO).
   a. Construct the complete contingency table.
   b. Construct the side-by-side percentage bar graph and the 100% stacked percentage bar graph.
   c. Is there a relationship between the choice of player and the outcome of a plate appearance?
   d. If you think there is a relationship, describe it.

Chapter 7

Probability

We live in an uncertain world. Every day decisions are made, based on certain known facts. Why are there times when we make the wrong decision? The answer usually lies with things over which we have no control. For example, we decide to leave our house at 8:00 a.m. to arrive at work at 9:00 a.m. We know that the distance is 30 miles to work and we can drive legally at a speed of 60 mph. Our calculations show that we have more than enough time to get to work on time. But we sometimes arrive late because we have no control over many chance occurrences. Your car could get a flat tire, an accident along the route can cause a long delay, or a torrential downpour slows up traffic. We would like to measure the *likelihood* that we arrive on time, for we cannot predict with 100% certainty that we will be on time. If we were asked to assign a number to the likelihood that we arrive on time, how would we do this? If you've ever had a job where you have to clock in, your employer recorded when you arrived late. Suppose that in the previous 500 workdays you were late 10 times. It would be reasonable to assign the number 0.98 = (490/500) to the likelihood of arriving on time. We can convert this number to a percent by multiplying the number 0.98 by 100. Considering our past performance, we might claim that the likelihood of you arriving on time is 98%.

Probability measures the likelihood that a particular outcome will occur. In a person's everyday life, decisions are made on the basis of the likelihood of a certain successful outcome occurring. Should we buy a certain stock? Should we change jobs? What university should we attend? Should we settle in a certain city? Should we marry a certain person? These are just a few of the many decisions we have to make in our lifetime. Can we ever be 100% certain that our decision is the right one? The answer to this question is usually no. The best we can do in making a decision is to consider many factors and then choose the decision that we feel is most likely to be the right one.

Decision making is used not only in our everyday life but in the world of business, in science, and in almost every vocation. Decision making is used in conjunction with probability. Before a decision is made, we attempt to assign a probability that the decision we make is the correct one. Only the future will show whether we were correct.

In baseball, some decisions a manager makes are as follows: Should I send in a pinch hitter? If so, which player? Should the batter bunt or hit away? Should I platoon two players? Should I yank my starting pitcher? And, in the National League, should I do a double switch? A manager is guided in making these and many other decisions by probability.

It's the bottom of the ninth, and the manager has to decide which player to use as a pinch hitter. Suppose that there are only two available bench players. Against the pitcher on the mound, player A has had two hits in 14 at bats and player B has had six hits in 12 at bats. For this season, batter A's AVG is .280 in 450 at bats and batter B's AVG is .282 in 500 at bats.

Exercise 7.1. If you were the manager, which player would you choose and why? Would your decision change if the pitcher was right-handed, batter A hits left-handed, and batter B hits right-handed? What other variables might influence your decision?

Inferential statistics is used to make a decision about a population using a sample result obtained from that population. Probability is used to measure the error involved in making a certain decision. In particular, the *p-value*, often called the chance error, is the probability of obtaining a certain sample result, strictly by chance, that seems to contradict the assumptions about the population that are believed *true*. When the p-value is small, the sample result can be used to make the decision that the assumptions about our population are really *false*. Using Table 6.3, we looked at two qualitative variables, certain years and type of hit. Under the assumption that these two qualitative variables are independent, we will be able to assign a p-value to the sample result. If the sample result has a p-value less than 5%, we can conclude that the assumption of independence of the two variables is false. The techniques used in inferential statistics replace a subjective decision by a decision based on a valid statistical test.

Probability models, based on certain assumptions, can be used to predict results. In medicine, scores on screening tests are used to assign a probability to a patient having a particular type of cancer. The patient can then decide whether to undergo surgery. In baseball, probability models can be used to predict the number of wins for a team, the batting average for a player or for a team, and which batting statistic is the best predictor of runs scored by a team.

A probability experiment that mimics a real baseball situation is called a *baseball simulation*. Simulations will be used in this book to mimic plate appearances for players.

7.1. Basic Definitions

Suppose that we wish to observe an experiment that may have many different possible outcomes. If we were to watch someone perform the experiment, we could state with complete certainty what outcome occurred. If, however, we were asked to *predict* which outcome would occur before the experiment was performed, we would be faced with a situation that involves uncertainty. Probability is nothing more than a measure of the uncertainty associated with any such prediction. For our purposes, probability is a measure of the likelihood that a particular outcome occurs.

A *probability experiment* is any process whose results are influenced by chance. The result of the experiment is called an *outcome*. Theoretically, an experiment can be repeated as many times as we want. Each time the experiment is executed, it will be called a *trial*. Each trial of the experiment results in one outcome.

The *sample space* for an experiment is the set of all possible outcomes of the experiment. We use an uppercase S as the naming symbol for the sample space. We express S by using a pair of braces, $S=\{\text{outcomes}\}$. A pair of braces { } is read as "the set consisting of" what lies within the braces. $S=\{1, 3, 5\}$ is read as "the set consisting of the outcomes of 1, 3, and 5."

$S=\{\ \}$ is the set consisting of no items; it is called the empty set or null set. (Please review the discussion of sets in Chap. 3.)

An *event* is a set of one or more outcomes from the sample space. We symbolize an event by an uppercase letter (E, F, G, etc.). A *simple* event consists of only one of the possible outcomes of the experiment. A *compound* event is one that consists of a combination of events. Compound events are discussed later in this chapter.

Example 7.1. Here are some examples of experiments with their corresponding outcomes and sample spaces.

Experiment	Outcomes	Sample Space
Toss a coin once	H, T	$S=\{$H, T$\}$
Roll a die once	1, 2, 3, 4, 5, 6	$S=\{1, 2, 3, 4, 5, 6\}$
Tossing two coins	HH, TT, HT, TH	$S=\{$HH, TT, HT, TH$\}$
One plate appearance	O, 1B, 2B, 3B, HR, BB, HBP, SF	$S=\{$O, 1B, 2B, 3B, HR, BB, HBP, SF$\}$
Tossing a coin and rolling a die	H1, H2, H3, H4, H5, H6, T1, T2, T3, T4, T5, T6	$S=\{$H1, H2, H3, H4, H5, H6, T1, T2, T3, T4, T5, T6$\}$

Some *events* for the experiment of tossing two coins are

E = getting exactly one head = {HT, TH},
F = getting exactly two heads = {HH} (a simple event).

Some *events* for the experiment of one plate appearance are

E = getting on base = {1B, 2B, 3B, HR, BB, HBP},
F = getting a base on balls = {BB} (a simple event).

7.2. Definition of Probability

Probability is a numerical measure of the likelihood of the occurrence of a particular event. The probability of an event *E* occurring is symbolized by Pr(*E*). Recall that the symbol "Pr" was used in previous chapters to stand for a proportion. Pr_x was interpreted as the proportion of occurrences of the outcome *x*. (Yes, a proportion is a probability.)

Two properties of any probability experiment are as follows:

1. For any event *E*, the probability that event *E* occurs is always a number between 0 and 1.
2. The sum of the probabilities of all the simple events of any experiment is equal to 1.

The above two properties hold for all the interpretations of probability discussed in the next section.

7.3. The Three Interpretations of Probability

In this section we look at three interpretations of probability. These are *classical*, *relative frequency*, and *subjective* probability.

7.3.1. Classical Probability

Classical probability evolved from gambling. Many games of chance have the property that all the outcomes of the game are equally likely. As a consequence, we say that any outcome occurs solely by chance. Rolling a die, tossing a coin, and selecting a card from an ordinary deck are all experiments in which all the outcomes are equally likely. This property of equally likely outcomes of an experiment distinguishes classical probability from other forms of probability. Because of the property of equally likely outcomes, mathematicians can provide formulas to calculate the theoretical probability for a particular event occurring. In classical probability, probabilities can be computed without ever performing the experiment. The definition for the probability of an event *E* occurring as a result of a probability experiment with equally likely outcomes in classical probability is given by

$$\Pr(E) = \frac{\text{number of elements in event E}}{\text{number of elements in the sample space S}}.$$

The *probability* of event E occurring is a *proportion* since E is a subset of S.

Examples 7.2 and 7.3 below contain examples from classical probability.

Example 7.2. Experiment: Rolling a die one time.

Sample space: $S = \{1, 2, 3, 4, 5, 6\}$.
Event E is the result of getting a 2; $E = \{2\}$.
Event F is the result of getting an even number; $F = \{2, 4, 6\}$.
$\Pr(E) = 1/6$; $\Pr(F) = 3/6$.
Simple events are {1}, {2}, {3}, {4}, {5}, and {6}.
$\Pr(1) + \Pr(2) + \Pr(3) + \Pr(4) + \Pr(5) + \Pr(6) = 1$.

Example 7.3. Experiment: Tossing two coins.

Sample space: $S = \{HH, HT, TH, TT\}$.
E is the result of getting exactly one head; $E = \{HT, TH\}$.
F is the result of getting two tails; $F = \{TT\}$.
$\Pr(E) = 2/4$; $\Pr(F) = 1/4$.
Simple events are {HH}, {HT}, {TH}, and {TT}.
$\Pr(HH) + \Pr(HT) + \Pr(TH) + \Pr(TT) = 1$.

Exercise 7.2. The experiment is tossing one coin. List the sample space if this experiment is repeated four times. (Hint: One possible outcome is HTHH.)

Exercise 7.3. Using the sample space from Example 7.3, find the probability that each of the events below occurs.

- Let $E = \{$those outcomes that resulted in exactly two heads$\}$.
- Let $F = \{$those outcomes that resulted in more than two heads$\}$.
- Let $G = \{$those outcomes that resulted in at least one head$\}$.

7.3.2. Relative Frequency Probability

Relative frequency probability involves probabilities associated with the empirical results of experiments that are repeated a certain number of times. The outcomes from each trial of the experiment may or may not be equally likely. These experiments can be repeated as often as we want. The *relative frequency probability* of an event E occurring from an experiment, repeated n times, is defined as follows:

$$\Pr(E) = \frac{\text{frequency at } E}{n} = F_E/n.$$

The frequency that event E occurs is denoted by F_E.

The *relative frequency probability* of an event E is equal to the *proportion* of times event E occurs after a fixed number of repetitions of an experiment.

Since an experiment can be performed an infinite number of times, a relative frequency probability for an experiment performed a fixed number of times is a *sample* proportion or *sample* probability. The *true* proportion or *true* probability of an event E occurring is based on the experiment being performed an infinite number of times.

If the outcomes of an experiment are equally likely, as in the tossing of a coin or rolling of a die, the population (*true*) probability can be calculated using mathematical formulas. However, there are gambling games in which the skill of the player has a great effect on the outcome of the game. Such games include poker and blackjack. These games involve a combination of skill and chance.

A baseball experiment, such as a plate appearance for a particular player, does not result in equally likely outcomes. The result of a plate appearance depends on the combination of the skill of the batter and chance. For experiments that are determined by a combination of skill and chance, it is impossible to have a formula to calculate the population (*true*) proportion. The *true* probability of a particular player getting a hit in an at bat is unknown.

The proportion of hits to at bats for a player for a year or a career is his *sample* proportion. The unknown *true* proportion of hits to at bats is based on an infinite number of at bats and is called the *population* proportion. The *sample* proportion of hits to at bats is his *sample* batting average; the *population* proportion of hits to at bats is his *true* batting average. The *true* batting average for a player is his *true* probability of getting a hit in an at bat. In the next example, we apply relative frequency probability to the baseball statistic OBA (OBP).

Example 7.4. The experiment was a plate appearance of Henry Aaron in his Major League career. The experiment was repeated 13919 times.

The sample space is the set consisting of the 13919 career plate appearance outcomes. *Note that the same outcome can occur many times in his plate appearances; however, a 1B in a second plate appearance is treated as a different outcome than a 1B in a fifth plate appearance.*

The event E is equal to the set of those career plate appearances that resulted in Aaron getting on base. For a plate appearance to result in getting on base, it must be either a H, BB, or HBP (see Tables 2.2, 3.1, and 3.2).

$$\Pr(E) = \frac{\text{frequency at } E}{n} = F_E/n,$$

$$\Pr(E) = \frac{\#OB}{\#PA} = \frac{\#H + \#BB + \#HBP}{\#1B + \#2B + \#3B + \#HR + \#O + \#BB + \#HBP + \#SF}$$

$$= \frac{3771 + 1402 + 32}{2294 + 624 + 98 + 755 + 8593 + 1402 + 32 + 121} = \frac{5205}{13919}$$

$$= .374.$$

The probability evaluated in Example 7.4 is the baseball statistic called on-base average or percentage and is denoted by OBA or OBP. This is considered a *sample* career OBA for Aaron. Aaron's *true* career OBA is the theoretical probability of Aaron getting on base in an infinite number of plate appearances.

The definition for relative frequency probability has already appeared in this book in Chapters 2 and 6. Both these chapters provided a complete frequency table with a column called $\Pr_X$. $\Pr_X$ was defined as $F_X/\Sigma F_X$, where ΣF_X is equal to the total number of elements in the sample being studied. In our definition of relative frequency probability, ΣF_X represents the total number of times the experiment is repeated.

A principle called the *Law of Large Numbers* is important in understanding relative frequency probability. This principle states that *for an experiment repeated many times, the relative frequency probability that an event occurs will approach its true probability.*

The Law of Large Numbers, applied to OBA, says that the more plate appearances a player has, the closer his sample OBA is to his *true* OBA.

The Law of Large Numbers also applies to classical probability. If the experiment consists of the tossing of a coin and the event E is getting a head, the more times we toss the coin, the closer the relative frequency probability of getting a head will be to the *true* probability of ½. Since the tossing of a coin results in one of two equally likely outcomes, the *true* probability of E occurring is known to be ½. (In the Second World War, South African statistician and prisoner of war John Kerrich killed time by tossing a coin 10000 times. He got 5067 heads, for a probability of 0.5067, close to the true probability of 0.5.)

Exercise 7.4. Using the sample space in Example 7.4, find the Pr(E) and Pr(F), where E is equal to those career plate appearances for Aaron that resulted in either a 1B or HR and F is equal to those career plate appearances for Aaron that resulted in either a 1B, 2B, 3B, or HR. Is the Pr(F) equal to Aaron's career batting average? Explain. If you answered no, what sample space would you use so that Pr(F) is his career batting average?

7.3.3. Subjective Probability

Subjective probability is the probability that a person, using their experience, expertise, or knowledge of past history, assigns to a particular event occurring.

This type of probability is used for experiments whose process cannot be repeated. Weather forecasting is an example of subjective probability. A weather forecaster will conclude that the probability of rain today is 70%. A baseball scout uses his expertise, his previous experience of evaluating similar players, and his judgment to assign a probability that a high school player has a chance of being a Major Leaguer. Assigning a probability to a team winning a game is another example of subjective probability.

Referring back to Exercise 7.1, the manager could use either relative frequency probability or subjective probability to choose his pinch hitter. If he chose player B, the manager may have noted that player B had a higher probability of getting a hit against that pitcher than player A. We would say that "the manager followed the book." If he chose player A, we would say that "he went against the book" and based his decision on his subjective probability derived from his "gut feeling." Every team keeps a book in the dugout that consists of all these relative frequency probabilities.

Of the three interpretations of probabilities, the one we will use the most is the relative frequency probability. Such statistics as BA (batting average), OBA (on-base average), HRA (home run average), BBA (base-on-balls average), IPBA (in-play batting average), and IPHR (in-play home run average) are all examples of relative frequency probabilities. Each relative frequency probability is a *sample* proportion. In Chapters 14 and 15 we will examine how to use these *sample* proportions to estimate a player's *true* (BA, OBA, HRA, BBA, IPHR, and IPBA).

To evaluate a probability, which is a ratio of two numbers, it is necessary to be able to count the number of subjects in a sample space and the number of subjects that belong to a given subset of the sample space.

7.4. Counting Principles

The three basic methods of counting the number of outcomes are as follows:

1. A direct listing
2. A tree diagram
3. A formula

7.4.1. A Direct Listing

As the name implies, this method refers to listing all possible outcomes of an experiment. This was done in Examples 7.2 and 7.3. In Example 7.4, we used the total number of career plate appearances of Aaron provided by a website. The website had access to a listing of each plate appearance for Aaron.

Example 7.5. The 2008 World Series between the Phillies and the Rays ended in five games, with the Phillies emerging as champions. Suppose that we don't have a list of which team won each game. Use P for the Phillies winning and R for the Rays winning. The possible results were PPPRP, PPRPP, PRPPP, and RPPPP. PPPPR could not be a result because the series would have ended after four games. The listing shows that there were four ways for the Phillies to win the series in five games. (The actual result was PRPPP.)

Exercise 7.5. The New York Yankees returned to winning ways in the 1996 World Series, which they won in six games against the Atlanta Braves. List all possible ways this could have happened. (Hint: The actual way was BBYYYY.)

7.4.2. Tree Diagrams

A tree diagram uses lines (called branches) and points (called nodes) to show all possible outcomes of each stage of a multistage experiment. At each stage a node represents the start of that stage, and we draw one branch for each possible outcome of that stage of the experiment. Each outcome ends with a node. The ending node of each stage becomes the starting point of the next stage. The next two examples provide tree diagrams.

Example 7.6. Suppose that Oakland's center fielder Rajai Davis had three at bats in a game against the Angels and each at bat resulted in either a hit (H) or an out (O). List all possible ways this can happen. A *listing* for all possible ways would be HOO, OOH, OHO, HHO, OHH, HOH, OOO, HHH. Notice that there are eight possible ways this can happen. Figure 7.1 presents a *tree diagram* listing all possible outcomes.

Example 7.7. Figure 7.2 provides a tree diagram for Example 7.5.

Exercise 7.6. Provide a tree diagram for Exercise 7.5.

Figure 7.1 (below left) Tree diagram for three at bats for Rajai Davis

Figure 7.2 (below right) Tree diagram for Phillies winning the 2008 World Series in five games

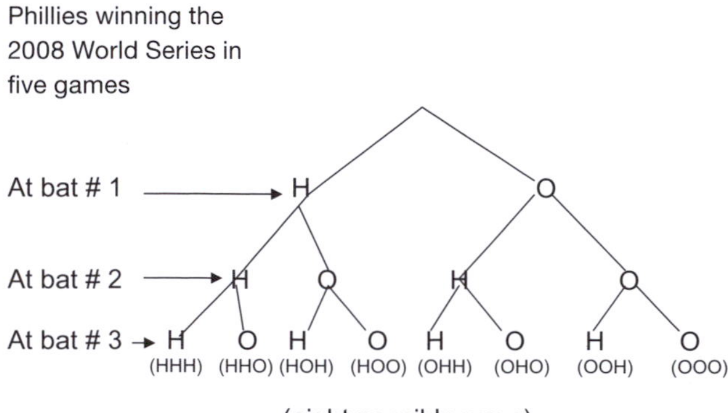

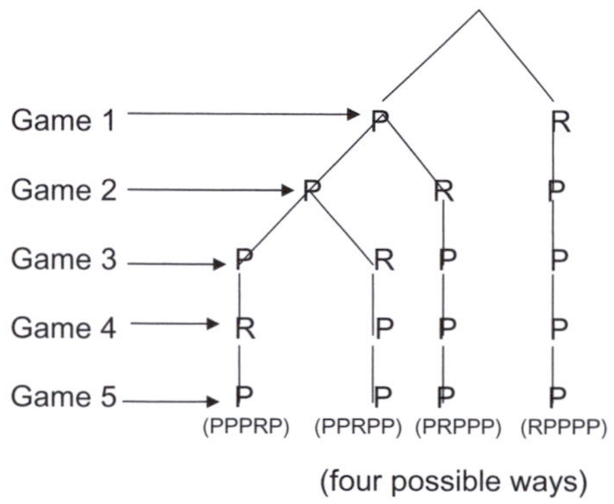

Game 1
Game 2
Game 3
Game 4
Game 5
(PPPRP) (PPRPP) (PRPPP) (RPPPP)
(four possible ways)

7.4.3. Formulas

Formulas can be used to obtain the number of outcomes without actually listing or showing them. There are three basic formulas.

7.4.3.1. Fundamental Law of Counting (FLC) (Formula 1)

If an experiment consists of two stages in which the first stage can occur in p ways and, after it has occurred, the second stage can occur in q ways, then both stages can occur together in $p * q$ ways. This can be extended to an experiment with *any* number of stages.

Formula 1. If stage 1 can occur in p_1 ways, stage 2 can occur in p_2 ways, . . . , and stage n can occur in p_n ways, then all n stages can occur together in $p_1 * p_2 * \ldots * p_n$ ways.

Example 7.8. In Example 7.6, the three at bats represent a three-stage experiment where each stage can occur in two ways (H or O). This gives us eight ($2 * 2 * 2$) ways in which all three stages can occur. Notice that applying the formula provides the number of outcomes without listing them.

Example 7.9. A cafeteria at the ballpark offers two choices for soup, four choices for the main course, three beverages, and five different desserts. How many different meals consisting of soup, a main course, a beverage, and a dessert are possible?

There are four stages, resulting in the number of meals $= 2 * 4 * 3 * 5 = 120$.

Exercise 7.7. A plate appearance has eight possible outcomes (1B, 2B, 3B, HR, O, BB, HBP, SF). A player makes four plate appearances in a game. How many different possible outcomes for the four plate appearances are there?

Exercise 7.8. A player had four at bats in a game. All we know is that each at bat resulted in either a hit (H) or an out (O). How many different possible outcomes are there for the four at bats?

Exercise 7.9. Draw a tree diagram to show all the possible outcomes in Exercise 7.8.

7.4.3.2. Permutations (Formula 2)

The symbol "!" (read as "factorial") is used to represent *factorials*. The number $n!$ (n factorial) is what you get if you multiply all the positive integers, up to and including n, together.

By definition, we have $0! = 1$.

For example, $1! = 1$; $2! = 2 * 1 = 2$; $3! = 3 * 2 * 1 = 6$; $4! = 4 * 3 * 2 * 1 = 24$; and so on.

In general, $n! = n * (n-1) * (n-2) * (n-3) \ldots * 4 * 3 * 2 * 1$.

If we have a set of n distinct items (no item repeats) and we wish to select only r of them (without replacement), each *arrangement* of the r items in a specific order is called a *permutation*.

The symbol ${}_nP_r$ denotes the number of permutations of n distinct items taken r at a time.

Formula 2. ${}_nP_r = n!/(n-r)!$; note that ${}_nP_n = n!$

Example 7.10. A manager has decided which nine players will start the game. How many different batting orders can he post?

Solution: The manager is arranging the nine players, nine at a time. So the answer is

$${}_9P_9 = 9! = 9*8*7*6*5*4*3*2*1 = 362880.$$

Example 7.11. A manager has five outfielders who can play each outfield position. How many different outfield alignments can he have to start a game?

Solution: The manager is selecting three outfields ($r=3$) from five players ($n=5$). He can do this ${}_5P_3 = 5!/(5-3)! = 5!/2! = 120/2 = 60$ ways.

This is a permutation problem, since the same three outfielders changing positions constitutes a different outfield alignment.

Exercise 7.10. On April 13, 2009, Orlando Hudson of the L.A. Dodgers hit for the cycle (1B, 2B, 3B, HR) in his first four at bats against the Giants. However, we do not know the order of the four hits. How many different permutations are there for the four hits? (O-Dog, as he's known, was the first Dodger since Gil Hodges in 1962 to hit for the cycle. His order of hits was 1B, HR, 2B, 3B.)

Exercise 7.11. Four days later, Jason Kubel of the Twins hit for the cycle and made one out in five at bats in a game. How many ways can this happen? (Jason actually went 2B, 1B, 3B, SO, HR against the Angels. His home run was a grand slam.)

Exercise 7.12. Draw a tree diagram and list all possible outcomes for the permutations in Exercise 7.10.

Exercise 7.13. Draw a tree diagram and list all possible outcomes for the permutations in Exercise 7.11.

7.4.3.3. Combinations (Formula 3)

Combinations differ from permutations because the order of selection is not important.

Suppose that we have a set of n distinct items and we wish to select only r of them without regard to the order in which they are chosen. Each set of r items selected is called a *combination*. The notation used to denote the combinations of n objects taken r at a time is ${}_nC_r$ (read as "the number of combinations of n objects taken r at a time").

Formula 3. ${}_nC_r = n!/[r! * (n-r)!]$; note that ${}_nC_n = 1$.

To see the difference between permutation and combination problems, we change Example 7.11 in the following way; suppose that we were only interested in which three outfielders started the game (assuming that they all can play each outfield position well). In that case, the position they occupied would not affect the result. The same three outfielders occupying different positions would not give a new result. Therefore, this would be a combination problem and we would have ${}_5C_3 = 5!/[3! * (5-3)!] = 10$ different combinations of three outfielders starting the game.

Example 7.12. A manager knows that he will need to use two pitchers for an exhibition game. He has five pitchers available. How many different ways can he choose the two pitchers for the game?

Solution:

$${}_5C_3 = 5!/[2! * (5-2)!] = 120/(2 * 6) = 120/12 = 10.$$

Since the order of selection is *not* important, this is *not* a permutation problem.

Exercise 7.14. Draw a tree diagram and list all possible outcomes for the combinations in Example 7.12.

7.4.3.4. The Relationship between Combinations and Permutations

Can you give a formula that provides a relationship between ${}_nP_r$ and ${}_nC_r$?

Exercise 7.15. Calculate the following: ${}_6P_4$, ${}_6C_4$, ${}_7P_3$, ${}_7C_3$, ${}_8P_3$, ${}_8C_3$. Give a formula that provides a relationship between ${}_nP_r$ and ${}_nC_r$. (Hint: Cross out those arrangements belonging to ${}_nP_r$ that would not belong to ${}_nC_r$.)

7.4.3.5. Permutations of Sets with Repeated Items

Suppose that we have two different objects; one object is repeated m times, the second object is repeated w times. Let $n = m + w$. Then the formula for the number of permutations of the n objects is

$${}_nP_{m,w} = n!/(m! * w!).$$

If $n = m + w + z$, where one object is repeated m times, a second object is repeated w times, and the third object is repeated z times, then the formula for the number of permutations of the n items is

$${}_nP_{m,w,z} = n!/(m! * w! * z!).$$

Clearly, this definition can be extended to more than three repeated elements. If either *m*, *w*, or *z* is 1, the formula will also work.

Example 7.13. A baseball player has five at bats resulting in two outs and three hits. In how many ways could this have happened?

Solution: The out is repeated twice, the hit is repeated three times. So the five objects can occur in ${}_nP_{m,w}$ ways. Here $m=2$, $w=3$, $n=m+w=5$. So we seek ${}_5P_{2,3}$. We can list the possible outcomes, which are as follows:

HHHOO, OOHHH, HOOHH, HHOHO, OHHHO, HHOOH, HOHOH, HOHHO, OHHOH, OHOHH.

By definition,

${}_5P_{3,2}=5!/(3!*2!)=10$ ways.

Even though this is a permutation problem, ${}_5P_{3,2}$ is equal to ${}_5C_2$ and ${}_5C_3$. Therefore, when an arrangement of *n* objects consists of *m* objects of one outcome and *w* objects of a second outcome ($n=m+w$), ${}_nP_{m,w}$ is equal to ${}_nC_m={}_nC_w$. This fact will be used in Chapter 11 when we discuss the Binomial Model for probability.

Exercise 7.16. A player has seven at bats in a game, resulting in two home runs, three doubles, and two outs. How many different permutations are there for the seven at bats? How would the result change if an eighth at bat resulted in a triple?

7.5. Counting Formulas and Sampling

After designing a statistical study, the study begins with a selection of a sample from the population. Two questions have to be answered about the actual sampling process:

1. Do we sample with or without replacement? That is, after a member of the population is selected for the sample, do we replace the member selected before the next selection?
2. Is the order in which the elements are selected important? That is, if two samples have the same members but the members are selected in a different order, do we consider the two samples as different?

If a population is infinite (no complete list exists for the population), the sampling will be assumed without replacement and the order of selection is not important.

For a finite population (all subjects of the population are known), we can assume that each member of the population appears on a slip of paper and all the slips are put into a bowl. If we sample *with replacement*, after a slip is selected, it is returned to the bowl before the next slip is selected. If we sample *without replacement*, after a slip is selected from the bowl, it is not returned to the bowl before the next slip is selected.

If two selected samples have exactly the same elements, but in a different order, and the order of selection is important, they would be considered two different samples. On the other hand, if the order was not important, the two samples would be considered as the same sample.

In our work in baseball statistics, the populations can be finite or infinite. For example, if we were to consider the actual plate appearances of a player as the population, the population would be finite. On the other hand, if we considered a player's population to be the theoretical population, consisting of an infinite number of plate appearances, the population would be infinite. In this case, his actual plate appearances would be a sample.

For this book, we assume that the sampling technique used is *simple random sampling*.

7.5.1. Simple Random Sampling

Simple Random Sampling (SRS) is any technique that guarantees that each member of the population has the same probability of being selected in the sample.

For infinite populations, no complete list exists, so if we can assume that a sample is representative of the population, we can treat it as a simple random sample. For our baseball samples, consisting of a player's yearly at bats or career at bats, we assume that these samples are representative of the theoretical population, consisting of an infinite number of at bats.

We now look at three ways of performing SRS.

Simple Random Sampling (with Replacement and the Order of Selection Is Important)

Example 7.14. Suppose that a bowl consists of these five results of a plate appearance: H, O, BB, SF, and HBP. Three are chosen *with replacement* and the *order of selection is important*. How many different samples can be selected?

Solution: For this example we would use the *Fundamental Law of Counting* (FLC). This is a three-stage experiment: stage 1 selects the first item, stage 2 selects the second item, and stage 3 selects the third item. Stage 1 can result in any one of five choices, stage 2 can result in any one of five choices, and stage 3 can result in any one of five choices. By the FLC, the number of possible samples is $5*5*5=125$ (Formula 1).

Sampling with replacement, when the order of selection is important, is an FLC problem.

Simple Random Sampling (without Replacement and the Order of Selection Is Important)

Example 7.15. Same problem as in Example 7.14, except that the sampling is done without replacement.

Solution: Stage 1 can result in any one of five choices, stage 2 can result in any one of four choices, and stage 3 can result in any one of three choices. The number of possible samples is $5*4*3=60$. Observe that ${}_5P_3=60$ (Formula 2).

Sampling without replacement, when the order of selection is important, is a permutation problem.

Simple Random Sampling (without Replacement and the Order of Selection Is Not Important)

Example 7.16. Same problem as in Example 7.14, except that the order is not important.

Solution: Sampling without replacement, when the order of selection is not important, is a combination problem.

${}_5C_3={}_5P_3/3!=60/6=10$ possible samples (Formula 3).

Exercise 7.17. We now look at one type of SRS not covered in the prior examples. Using the data in Example 7.14, find the number of different samples of size 3 that can be selected, assuming that the selection is with replacement and the order of selection is not important. Provide a list of all possible samples.

There are situations when the items being considered are not distinct. We now look at such a situation.

7.6. Relative Frequency Probabilities for Simple Events

We now apply relative frequency probability to contingency tables.

Each career PA for Aaron and Bonds will be represented by a slip of paper with either an A for Aaron or a B for Bonds on one side and the outcome of their plate appearance (1B, 2B, 3B, HR, BB, SF, HBP, O) on the other side. The plate appearances for Bonds will be through the first 81 games of 2007. The slips of paper for Aaron will be placed in bowl A; the slips of paper for Bonds will be placed in Bowl B. A third bowl S will contain the contents of both bowls. Bowl A contains the sample of Aaron's career plate appearances, bowl B contains the sample of Bonds's career plate appearances, and bowl S contains the sample of the combined career plate appearances for both Aaron and Bonds. The two variables (characteristics) associated with each item in bowl S are *choice of player* and *plate appearance outcome*. In Table 7.1, the

Table 7.1 Contingency table for the career PAs of Henry Aaron and Barry Bonds (through first 81 games of 2007)

Player	1B	2B	3B	HR	BB	SF	HBP	Out	PA Total
Aaron	2294	624	98	755	1402	121	32	8593	13919
Bonds	1477	597	77	751	2520	90	105	6824	12441
Totals	3771	1221	175	1506	3922	211	137	15417	26360

items in bowl S are organized into a *contingency table* for the two variables *choice of player* and *plate appearance outcome*.

This table presents the same sample and variables as in Table 6.2. How has the form of the table changed?

The *simple* events correspond to the choice of player, Aaron or Bonds, and the PA outcome, 1B, 2B, 3B, HR, BB, SF, HBP, or O. A cell is formed as the combination of one of the players and one of the outcomes of a PA. The number in the cell represents the frequency or the number of slips with both results. As an example, there are 2294 slips with Aaron on one side and 1B on the other side. The two players appear as the row headings, and the PA outcomes are the column headings. The number of slips in S with a particular outcome on one side corresponds to its row or column total. The row or column total will be designated by placing the number symbol "#" in front of the heading. The number of slips in set S, the total combined plate appearances, is 26360. The set S is called the sample space. The number of elements in S is the sample size and is symbolized by $\#S$. If X is any of the simple events in Table 7.1, $\Pr(X) = \#X/\#S$. As an example, $\Pr(1B) = \#1B/\#S$.

The probability that a simple event occurs is obtained by dividing its row or column total by the sample size. The next example computes the probabilities for all the simple events in Table 7.1.

Example 7.17. Calculation of probabilities for simple events.

Probabilities for Simple Events for Plate Appearance Results

$\Pr(1B) = 3771/26360 = .1430$, $\Pr(BB) = 3922/26360 = .1488$,
$\Pr(2B) = 1221/26360 = .0463$, $\Pr(SF) = 211/26360 = .0080$,
$\Pr(3B) = 175/26360 = .0068$, $\Pr(HBP) = 137/26360 = .0052$,
$\Pr(HR) = 1506/26360 = .0571$, $\Pr(O) = 15417/26360 = .5849$.

Probabilities for Simple Events for Player Choice

$\Pr(A) = 13919/26360 = .5280$, $\Pr(B) = 12441/26360 = .4720$.

If one slip is selected at random from the bowl S, the *probability* of a *simple event* is the likelihood that the slip contains the event.

Exercise 7.18. Construct a contingency table similar to Table 7.1 using the following two variables: choice of player (Joe DiMaggio [JD] or Ted Williams [TW]) and PA outcome. Since from 1936 to 1953 the SF rule did not exist, treat the SFs for Williams from 1954 to 1960 as outs. Therefore, the variable PA will only have seven categories. Use Tables 3.6 and 3.9 to obtain the necessary data. Compute the probabilities for each of the simple events.

Exercise 7.19. Repeat Exercise 7.18, except use the following PA outcomes: 1B, 2B or 3B, HR, BB, SO, and IPO (in-play out). Remember, IPO = O − {SO}.

7.7. Introduction to Compound Events

A *compound* event is a set created using operations to combine one or more events. If X and Y are events, we can create the compound events (X or Y), (X and Y), and (not X). In this section we define the addition of two events (X or Y), the multiplication of two events (X and Y), and the complement of an event (not X).

7.7.1. The Addition of Two Events (Union of Two Sets)

The addition of two events is accomplished by creating a new event (set) from the two existing events (sets) by selecting all elements that appear in either or both of the events. The three symbols used for the operation of the addition of two events are "+," "∪," and the word "or." The symbol "or" will be used most. The next example demonstrates the addition of two sets.

Example 7.18. If $A = \{1, 3, 4, 6, 7\}$ and $B = \{2, 3, 4, 8\}$, then the set (A or B) = {1, 3, 4, 6, 7, 2, 8}. If $C = \{1, 2, 5\}$ and $D = \{3, 6, 8, 9\}$, then (C or D) = {1, 2, 5, 3, 6, 8, 9}.

From the above we can calculate the following: $\#A = 5$, $\#B = 4$, $\#(A$ or $B) = 7$, $\#C = 3$, $\#D = 4$, and $\#(C$ or $D) = 7$. The $\#(A$ or $B)$ does not equal $(\#A + \#B)$ $(7 \neq 4 + 5)$, whereas $(\#C + \#D) = \#(C$ or $D)$.

We say that events C and D are *mutually exclusive* (ME). Events A and B are not ME. There is no number that appears in both C and D. The numbers 3 and 4 appear in A and in B.

> Two events X and Y are mutually exclusive if the same element does not appear in both events. Alternately, two events X and Y are mutually exclusive if $(\#X + \#Y) = \#(X$ or $Y)$.
>
> Note, $\#(X$ or $Y) = \#(Y$ or $X)$ will always be true.

7.7.2. The Multiplication of Two Events (Intersection of Two Sets)

The multiplication of two events is accomplished by creating a new event from the two existing events by selecting all items that are common to both events. The three symbols used for the operation of multiplication are "$*$," "$\cap$," and the word "and." Using the sets A, B, C, and D from Example 7.18, we have (A and B) = {3, 4}, (C and D) = { }, #(A and B) = 2, and #(C and D) = 0.

Mutually exclusive events X and Y have the property that #(X and Y) = 0.

Note that #(X and Y) = #(Y and X) is always true. We now provide a formula for calculating the number of elements in an event created as the sum of two events.

7.7.3. The General Addition Formula for Two Events

This formula handles the addition of two events whether the events are mutually exclusive or not:

#(X or Y) = #X + #Y − #(X and Y).

The subtraction is necessary so that an element belonging to both events X and Y won't be counted twice. The next example illustrates this formula.

Example 7.19. Using the events A, B, C, and D from Example 7.18, we have

7 = #(A or B) = #A + #B − #(A and B) = 5 + 4 − 2 = 7.
7 = #(C or D) = #C + #D − #(C and D) = 3 + 4 − 0.

7.7.4. The Complement of an Event

The complement (outside) of an event is formed by selecting the elements that are in the sample space but do not belong to the event. The complement of an event can be thought of as subtracting out from the sample space those elements in the given event that also belong to the sample space. We use the word "not" in front of the event to represent the complement. If the sample space S = {1, 2, 5, 8, 9} and E = {1, 2, 5}, then (not E) = {8, 9}.

#(not E) = #S − #E = 5 − 3 = 2.

7.8. Relative Frequency Probabilities and Probability Formulas for Compound Events

The simple events that will be used in this section come from Table 7.1 and are 1B, 2B, 3B, HR, BB, SF, HBP, O, A, and B. We use the letters "A" and "B" for Aaron and Bonds, respectively. The sample space, the set of all career plate appearances for Aaron and Bonds, will be symbolized by "S." Observe that the event A corresponds to the contents of bowl A, which contains all career plate appearances of Aaron; the event B corresponds to the contents of bowl B, which contains all career plate appearances of Bonds; and the set *S* corresponds to the contents of bowl S, which contains all plate appearances of both Aaron and Bonds. Each column or row total represents the total number of slips in Bowl S, which have the column heading outcome or row heading outcome on one side.

Suppose that one slip is randomly selected from bowl S. On one side of the slip is either A or B; the other side of the slip is one of the eight possible results of a plate appearance. The probability of a simple event *X* is the likelihood that the slip contains *X* on one of its sides. The next example reviews the calculations for the probabilities for simple events.

Example 7.20. One slip is randomly selected from bowl S. Here are the probabilities for some simple events:

a. Pr(HR) = #HR/#*S* = 1506/26360 (1506 slips in bowl S have HR on them),
b. Pr(A) = #A/#*S* = 13919/26360 (13919 slips in bowl S have Aaron on them),
c. Pr(BB) = #BB/#*S* = 3922/26360,
d. Pr(B) = #B/#*S* = 12441/26360.

We now provide a definition for the probability for each of the compound events. These definitions are based on counting the number of objects in the compound event. Following these definitions, a formula based on probabilities is provided for each of the compound events.

The Probability for the Compound Event (X and Y)

The probability for the compound event (*X* and *Y*) is the number of objects in the compound event (*X* and *Y*) divided by the number of objects in the sample space *S*. Symbolically,

Pr(*X* and *Y*) = #(*X* and *Y*)/#*S*.

We now compute probabilities for a compound event formed from "and."

The formula for Pr(*X* and *Y*), based on probabilities, is provided in Section 7.9. The next example illustrates the definition.

Example 7.21. The probabilities below show that the Pr(*X* and *Y*) may not equal the Pr(*X*) * Pr(*Y*).

a. Pr(HR and A) = [#(HR and A)/#*S*] = 755/26360 = .0286. A total of 755 slips in bowl S have HR on one side and Aaron on the other side.
b. Pr(1B and HR) = #(1B and HR)/#*S* = 0/26360 = 0. There are no slips in bowl S with 1B on one side and HR on the other side.
c. Pr(HR and A) = 755/26360.
d. Pr(HR) = 1506/26360; Pr(A) = 13919/26360.
e. Pr(HR) * Pr(A) = (1506/26360) * (13919/26360) = .0301.

Converting fractions to decimals, we get Pr(HR and A) = .0286 and Pr(HR) * Pr(A) = .0301. This shows for the events HR and A, Pr(HR and A) ≠ Pr(HR) * Pr (A).

We shall see in Section 7.9 that two events X and Y are independent if Pr(X and Y) = Pr(X) * Pr(Y).

The Probability for the Compound Event (X or Y)

The probability for the compound event (X or Y) is defined as the number of objects in the compound event (X or Y) divided by the number of objects in the sample space S. Symbolically,

Pr(X or Y) = #(X or Y)/#S = [#X + #Y − #(X and Y)]/#S.

By dividing each term in the *Addition Formula for Two Events* by the #S, we derive the following formula:

Addition Formula for the Probability of Two Events

Pr(X or Y) = [#X + #Y − #(X and Y)]/#S
= #X/#S + #Y/#S − #(X and Y)/#S
= Pr(X) + Pr(Y) − Pr(X and Y).

Example 7.22. This example calculates Pr(X or Y) first by using the definition for the compound event (X or Y), followed by the Addition Formula for Two Events.

a. Pr(HR or A) = #(HR or A)/#S
= [#HR + #A − #(HR and A)]/#S
= [1506 + 13919 − 755]/26360
= 14670/26360.

b. Pr(HR or A) = Pr(HR) + Pr(A) − Pr(HR and A)
= 1506/26360 + 13919/26360 − 755/26360
= (1506 + 13919 − 755)/26360
= 14670/26360.

If two events X and Y are mutually exclusive, then Pr(X and Y) = 0. The *Addition Formula for Mutually Exclusive Events* becomes

Pr(X or Y) = Pr(X) + Pr(Y).

c. Pr(HR or 1B) = #(HR or 1B)/#S
= [#HR + #1B − #(HR and 1B)]/#S
= [1506 + 3771 − 0]/26360
= 5277/26360.

d. Pr(HR or 1B) = Pr(HR) + Pr(1B)
= 1506/26360 + 3771/26360
= 5277/26360.

The events HR and A are not ME because there are slips with HR on one side and A on the other side. The events HR and 1B *are* ME. There are no slips with HR on one side and 1B on the other side.

The Probability for the Complement of an Event (not X)

The probability for the complement of an event X is defined as the ratio of the number of objects in the sample space S that are outside of the event X to the number of objects in the sample space S. Symbolically,

$$\Pr(\text{not } X) = \#(\text{not } X)/\#S$$
$$= (\#S - \#X)/\#S.$$

Since $(\#S - \#X)/\#S = 1 - (\#X/\#S) = 1 - \Pr(X)$, we have the *Probability Formula for the Complement of an Event*:

$$\Pr(\text{not } X) = 1 - \Pr(X).$$

The reason the complement of an event is considered a compound probability is that it involves the sample space S and the event X.

The next example provides two ways to calculate the "not" probability.

Example 7.23. This example calculates Pr(not X) first by using the definition for the compound event (not X), followed by a probability formula for Pr(not X).

a. $\Pr(\text{not HR}) = (\#S - \#\text{HR})/\#S = (26360 - 1506)/26360 = 24854/26360$,
b. $\Pr(\text{not HR}) = 1 - \Pr(\text{HR}) = 1 - (1506/26360) = 24854/26360$.

Exercise 7.20. Compute the following probabilities using the contingency table you constructed in Exercise 7.18: Pr(JD and BB), Pr(TW and BB), Pr(JD or BB), Pr(TW or BB), Pr(JD or TW), Pr(JD and HR), Pr(TW and HR), Pr(JD or HR), Pr(TW or HR), Pr(not HR), Pr(not JD), Pr(HR or 1B). Which pairs of events are mutually exclusive?

7.9. Conditional Probability and Independence

A slip is randomly selected from bowl S, and we are able to see the simple event X on one side of the slip. What is the probability that the other side contains the simple event Y? This is an example of a *conditional probability* problem. What makes this a conditional probability is that one of the events has already occurred.

Suppose in our example that we see HR on one side of the slip. What is the probability that A is on the other side? We separate out all the slips with

HR on one side and count them. This number corresponds to #HR. Next, we count the number of these slips with A on the other side. This number corresponds to #(HR and A). By dividing #(HR and A) by #HR, we obtain the conditional probability of event A appearing on one side of a slip given that we know in advance that the other side of the slip has HR. This conditional probability is written Pr(A|HR).

For any two events X and Y, given that the event X has already occurred, the probability that the event Y occurs is symbolized by $\Pr(Y|X)$.

The conditional probability of Y given X is defined as

$$\Pr(Y|X) = \#(X \text{ and } Y)/\#X.$$

If we divide both the numerator and denominator of the fraction by $\#S$, we do not change the value of the fraction and we get a new formula for the conditional probability:

$$\begin{aligned}\Pr(Y|X) &= [\#(X \text{ and } Y)/\#S]/[\#X/\#S] \\ &= \Pr(X \text{ and } Y)/\Pr(X).\end{aligned}$$

The $\Pr(Y|X)$ is read "the probability that the event Y occurs given that the event X has already occurred." The $\Pr(X|Y)$ is read "the probability that the event X occurs given that the event Y has already occurred." By interchanging X and Y, we get the following:

The conditional probability of X given Y is

$$\Pr(X|Y) = \Pr(X \text{ and } Y)/\Pr(Y).$$

The next example shows the calculations for conditional probabilities using both the definition and the formula.

Example 7.24.

a. Pr(A|HR) = #(HR and A)/#HR = 755/1506
   = Pr(HR and A)/Pr (HR) = (755/26360)/(1506/26360)
   = 755/1506.
b. Pr(HR|A) = #(HR and A)/#A = 755/13919
   = Pr(HR and A)/Pr(A) = (755/26360)/(13919/26360)
   = 755/13919.

Notice that Pr(A|HR) is not equal to Pr(HR|A).

In Section 7.8 we promised to give a probability formula for the compound event (X and Y). Here it is:

$$\begin{aligned}\Pr(X \text{ and } Y) &= \Pr(X) * \Pr(Y|X) \\ &= \Pr(Y) * \Pr(X|Y).\end{aligned}$$

Table 7.2 Batter A's results

Pitchers	H	O	Total ABs
Pitcher C	2	12	14
Rest of the pitchers	124	312	436
Totals	126	324	450

We return to the players mentioned in Exercise 7.1. They are batter A, batter B, and pitcher C. We summarize the results for batter A in Table 7.2.

The batting average, AVG, for player A for the season against *all* the pitchers he faced is the probability that player A gets a hit in an at bat. We express this as

AVG of player A = Pr(H) = #H/#AB = 126/450 = .280.

The conditional probability for player A getting a hit given that pitcher C is on the mound is

The AVG of player A against pitcher C = Pr(H | pitcher C) = 2/14.

Exercise 7.21. Duplicate the above for player B.

Some fans believe that in the National League, the Colorado Rockies have a home advantage because of Coors Field. The same is said of the Boston Red Sox because of Fenway Park. We analyze the results for Colorado for 2007; the Boston Red Sox for 2007 and 2008 are left as an exercise.

Table 7.3 Colorado Rockies' wins and losses for the 2007 season

Location	Wins	Loses	Totals
Home	51	31	82
Away	39	42	81
Totals	90	73	163

We begin by producing Table 7.3 for the wins and losses for the 2007 season. These data came from the website www.retrosheet.org.

For the 2007 season, the probability that the Colorado Rockies won a game was

Pr(Colorado Wins) = 90/163 = .55.

The probability that the Colorado Rockies won a home game was

Pr(Colorado Wins | a home game) = 51/82 = .62.

Exercise 7.22. Duplicate the above for the Colorado Rockies for the 2008 season. How similar is this result to the result in 2007?

Exercise 7.23. Repeat the above for the Boston Red Sox for the 2007 and 2008 seasons. How similar are the results for those two years?

7.9.1. Independence of Two Events and Two Variables

Is there a relationship between the location of the game (home or away) and the outcome of the game (win or loss) for the Colorado Rockies for the 2007 season?

This question leads us to the topic of the independence of two events. Two events X and Y are *independent* if

$$\Pr(X \text{ and } Y) = \Pr(X) * \Pr(Y).$$

If the above equation is satisfied, using the definition of conditional probability, we have

$$\Pr(X|Y) = \Pr(X \text{ and } Y)/\Pr(Y) = \Pr(X) * \Pr(Y)/\Pr(Y) = \Pr(X),$$
$$\Pr(Y|X) = \Pr(Y \text{ and } X)/\Pr(X) = \Pr(Y) * \Pr(X)/\Pr(X) = \Pr(Y).$$

Therefore, if X and Y are independent, we have the following alternate formulas:

$$\Pr(X|Y) = \Pr(X).$$
$$\Pr(Y|X) = \Pr(Y).$$

The equations above describe the independence of X and Y. These two equations say that the probability of one event occurring is not influenced by the fact that the other event already occurred.

The following four statements are equivalent. This means that if any one of the four statements is true, so are the other three; if any one of the four statements is false, so are the other three.

1. The events X and Y are independent.
2. $\Pr(X \text{ and } Y) = \Pr(X) * \Pr(Y)$.
3. $\Pr(X|Y) = \Pr(X)$.
4. $\Pr(Y|X) = \Pr(Y)$.

Two events are *theoretically independent* if any one of the above statements is true.

In classical probability, the *true* probability that an event occurs is known. In that case, we can use any one of the three equations above to determine if the two events are *theoretically independent*. If any one of the three equations is not satisfied, we conclude that the two events are dependent.

The next two examples involve probability experiments from classical probability.

Example 7.25. A coin is tossed twice. X is the event of getting a head on the first toss and Y is the event of getting a head on the second toss. From the sample space in Example 7.1, Pr(X and Y) = 1/4. We know that Pr(X) = 1/2 and *Pr*(Y) = 1/2. Therefore, Pr(X and Y) = Pr(X) * Pr(Y) and we can conclude that these two events are *independent*.

Example 7.26. First a coin is tossed, followed by the rolling of a die. X is the event that the toss of the coin results in a head and Y is the event that the roll of the die results in an even number. Using the sample space in Example 7.1, we see that Pr(X and Y) = 3/12. We know that Pr(X) = 1/2 and Pr(Y) = 3/6. Therefore, Pr(X and Y) = Pr(X) * Pr(Y) and we can conclude that the two events are *independent*.

For relative frequency probability, we deal with samples from a population. Thus, the *true* probabilities are unknown. Since we are dealing with *sample* probabilities, even if two events are independent, we do not expect any of the three equations above to be satisfied. The question is, how badly do they fail?

So far we have looked at whether two *events* are independent. We now define what is meant by two *variables* being independent.

Two variables X and Y are independent if *the combination of each simple event from X with each simple event from Y is independent. If the variables are not independent, we say that there is a relationship between the variables.*

Given that X is choice of player and Y is the result of a plate appearance, the variables X and Y are *theoretically independent* if each of the following equations is satisfied:

$$\Pr(1B) = \Pr(1B|A) = \Pr(1B|B),$$
$$\Pr(2B) = \Pr(2B|A) = \Pr(2B|B),$$
$$\Pr(3B) = \Pr(3B|A) = \Pr(3B|B),$$
$$\Pr(HR) = \Pr(HR|A) = \Pr(HR|B),$$
$$\Pr(BB) = \Pr(BB|A) = \Pr(BB|B),$$
$$\Pr(SF) = \Pr(SF|A) = \Pr(SF|B),$$
$$\Pr(HBP) = \Pr(HBP|A) = \Pr(HBP|B),$$
$$\Pr(O) = \Pr(O|A) = \Pr(O|B).$$

The next example calculates the above probabilities for the sample data in Table 7.1.

Example 7.27. The relative frequency probabilities below are calculated from Table 7.1 and rounded to two decimal places. Table 7.1 represents the sample results for the career combined plate appearances of Aaron and Bonds.

Pr(1B) = 3771/26360 = .14	Pr(1B|A) = 2294/13919 = .16	Pr(1B|B) = 1477/12441 = .12
Pr(2B) = 1221/26360 = .05	Pr(2B|A) = 624/13919 = .04	Pr(2B|B) = 597/12441 = .05
Pr(3B) = 175/26360 = .01	Pr(3B|A) = 98/13919 = .01	Pr(3B|B) = 77/12441 = .01
Pr(HR) = 1506/26360 = .06	Pr(HR|A) = 755/13919 = .05	Pr(HR|B) = 751/12441 = .06
Pr(BB) = 3922/26360 = .15	Pr(BB|A) = 1402/13919 = .10	Pr(BB|B) = 520/12441 = .20
Pr(SF) = 211/6360 = .01	Pr(SF|A) = 121/13919 = .01	Pr(SF|B) = 90/12441 = .01
Pr(HBP) = 137/26360 = .01	Pr(HBP|A) = 32/13919 = .00	Pr(HBP|B) = 105/12441 = .01
Pr(O) = 15417/26360 = .58	Pr(O|A) = 8593/13919 = .62	Pr(O|B) = 6824/12441 = .55

Given that the variable X consists of the simple events 1B, 2B, 3B, HR, BB, SF, HBP, and O, X is *theoretically independent* of the variable Y, whose outcomes are the two simple events Aaron (A) and Bonds (B), if the following equations are satisfied for each variable X:

$$\Pr(X) = \Pr(X|A) = \Pr(X|B).$$

Looking at Example 7.27, we can see that the above equations are not satisfied exactly for any plate appearance outcome. The events that come closest to satisfying the two equations above are 2B, 3B, HR, SF, and HBP. The events BB, O, and 1B are not close to satisfying the two equations.

However, since the probabilities in Example 7.27 are only *sample* probabilities, we cannot conclude whether the *true* probabilities are or are not equal.

To further analyze the independence of player choice and plate appearance result, we use the graphs in Figures 6.3–6.8 (Chap. 6). Since a probability is a proportion, for each plate appearance result X, $\Pr(X|A) = \Pr_X$ for Aaron and $\Pr(X|B) = \Pr_X$ for Bonds. Therefore, the heights of the rectangles, above each simple event X for a plate appearance in Figures 6.3 and 6.4, are the conditional probabilities $\Pr(X|A)$ and $\Pr(X|B)$, based on the combined sample plate appearances in Table 7.1.

Based on the assumption that the variables *choice of player* and *plate appearance result* are *theoretically independent* for their *theoretical population* (the combined plate appearances of Bonds and Aaron for an infinite number of plate appearances), $\Pr(X|A) = \Pr(X|B) = \Pr(X)$. The graphs in Figures 6.6–6.8 are based on the assumption of independence. The heights of the rectangles over each outcome of a plate appearance for Bonds and Aaron are equal.

The more Figures 6.3–6.5 differ in appearance from Figures 6.6–6.8, the more likely the variables *choice of player* and *plate appearance result* are *dependent* (not independent). If the variables are dependent, we say that there is a *relationship between the two variables.*

Figures 6.3–6.5 allow us to conjecture that the events 1B and O are more likely to occur for Aaron in a plate appearance than for Bonds and the event BB is more likely to occur for Bonds in a plate appearance than for Aaron. This would indicate that there is a relationship between choice of player and their plate appearance outcome. These conclusions, however, are nothing more than a subjective opinion concerning the sample data.

We now turn to Table 7.3 and ask if there is a relationship for the Colorado Rockies for the 2007 season between the location of the game (home or away) and the outcome of the game (won or lost). From Table 7.3 we have

$$\Pr(\text{W}) = 90/163 = .55 \qquad \Pr(\text{W}|\text{H}) = 51/82 = .62 \qquad \Pr(\text{W}|\text{A}) = 39/81 = .48,$$
$$\Pr(\text{L}) = 73/163 = .45 \qquad \Pr(\text{L}|\text{H}) = 31/82 = .38 \qquad \Pr(\text{L}|\text{A}) = 42/81 = .52.$$

Since the $\Pr(\text{W})$ is not close to $\Pr(\text{W}|\text{H})$ and $\Pr(\text{L})$ is not close to $\Pr(\text{L}|\text{H})$, it appears that there *is* a relationship between the result and the location of the game.

For our final examples on independence, we look at situations involving more than two events.

Example 7.28. Vincent Edward "Bo" Jackson was a Heisman Trophy–winning football player and an NFL running back for the Los Angeles Raiders. He also was an outfielder for the Royals and Angels, with a career batting average of .250. Using this as an estimate for his *true* batting average, find the probability that in five at bats he gets five hits. Assume that the at bats are independent.

Solution: The experiment consists of five at bats. Assuming that the outcome of each at bat is independent and each at bat results in a hit, the result of his next five at bats would be "HHHHH." Since the Pr(H) = .250, we have

$$\begin{aligned}\text{Pr(H and H and H and H and H)} &= \text{Pr(H)} * \text{Pr(H)} * \text{Pr(H)} * \text{Pr(H)} * \text{Pr(H)} \\ &= (.250)^5 = .0009.\end{aligned}$$

Example 7.29. Players Bo Jackson, Roberto Alomar, and Wally Backman have career batting averages of .250, .300, and .275, respectively. If they all played in a game, find the probability that they all went 0 for 4.

Solution: Assuming that each at bat is independent and player performance is independent, Pr(Bo[0 for 4]) = $(1-.250)^4$ = .3164, Pr(Roberto[0 for 4]) = $(1-.300)^4$ = .2401, and Pr(Wally[0 for 4]) = $(1-.275)^4$ = . 2763. As player performance is independent, the probability that each player goes hitless in four at bats is the product .3164 * .2401 * .2763 = .0210.

Exercise 7.24. In Example 7.28 find the probability that Bo gets five outs in his next five at bats.

Exercise 7.25. Using the contingency table for the combined career plate appearances for Joe DiMaggio (JD) and Ted Williams (TW) in Exercise 7.18, compute all the probabilities mentioned in Example 7.27.

Exercise 7.26. Discuss any relationships you can detect between choice of player (DiMaggio or Williams) and their plate appearance outcomes. Use the 100% stacked bar graph as an aid.

Exercise 7.27. Suppose, many years from now, that the Cubs take on the White Sox in the World Series. Assuming that the probability that the Cubs win each game is 1/2 and the results of the games are independent, find the probability that the Cubs win the series in five games (see Example 7.5). Find the probability that either the Cubs or the White Sox win the series in five games. Suppose that the probability is 3/5 that the Cubs win each game. Repeat this exercise. Then repeat this entire exercise for the Cubs to win the World Series in six and seven games.

7.10. Mutually Exclusive Events

For any two mutually exclusive events X and Y, Pr(X and Y) = 0. Therefore, we have the following definition:

For any two mutually exclusive events X and Y,

$$\Pr(X \text{ or } Y) = \Pr(X) + \Pr(Y).$$

Clearly, any two outcomes X and Y of a plate appearance are mutually exclusive. For example, the Pr(1B or 2B) = Pr(1B) + Pr(2B).

Given the sample space of career plate appearances for Aaron,

$$\text{OB} = (\text{1B or 2B or 3B or HR or BB or HBP}),$$
$$\text{OBP} = \Pr(\text{OB}) = \Pr(\text{1B}) + \Pr(\text{2B}) + \Pr(\text{3B}) + \Pr(\text{HR}) + \Pr(\text{BB}) + \Pr(\text{HBP}).$$

Given the sample space of career at bats for Aaron,

$$\text{H} = (\text{1B or 2B or 3B or HR}),$$
$$\text{BA} = \Pr(\text{H}) = \Pr(\text{1B}) + \Pr(\text{2B}) + \Pr(\text{3B}) + \Pr(\text{HR}).$$

The next example and the next exercise demonstrate the above for Aaron's career.

Example 7.30. Calculate the career OBP for Aaron, given that his sample space consists of his 13919 career plate appearances.

Solution:

$$\begin{aligned}\text{OBP} &= \Pr(\text{1B}) + \Pr(\text{2B}) + \Pr(\text{3B}) + \Pr(\text{HR}) + \Pr(\text{BB}) + \Pr(\text{HBP})\\ &= 2294/13919 + 624/13919 + 98/13919 + 755/13919 + 1402/13919\\ &\quad + 32/13919\\ &= 5498/13919 = .374.\end{aligned}$$

Exercise 7.28. Calculate the career BA (AVG) of Aaron, given that his sample space consists of his 12364 career at bats.

Example 7.31. As the first team to win four games wins the championship, the total number of games in a World Series can be four, five, six, or seven. Let the variable X be the total number of games in a World Series. The possible simple events are the outcomes four, five, six, and seven. Define the events E = series ends in four games, F = series ends in five games, G = series ends in six games, and H = series ends in seven games. The events E, F, G, and H are simple events and $\Pr(E \text{ or } F \text{ or } G \text{ or } H) = \Pr(E) + \Pr(F) + \Pr(G) + \Pr(H) = 1$. (In the 11 years from 1998 through 2008, five series were sweeps, three ended after five games, one after six games, and two times the series went to the seventh and final game.)

We now summarize our important probability formulas.

7.11. Summary of Important Probability Formulas

7.11.1. General Rules for Simple Events

For any simple event X, we have $0 \leq \Pr(X) \leq 1$. The sum of all the probabilities for all the simple events of an experiment is equal to 1.

7.11.2. Addition and Multiplication Formulas for Compound Events

Given any two events X and Y:

Addition Formula for Probability

$\Pr(X \text{ or } Y) = \Pr(X) + \Pr(Y) - \Pr(X \text{ and } Y)$.

If X and Y are mutually exclusive events, $\Pr(X \text{ or } Y) = \Pr(X) + \Pr(Y)$.

Multiplication Formula for Probability

$\Pr(X \text{ and } Y) = \Pr(X/Y) * \Pr(Y) = \Pr(Y/X) * \Pr(X)$.

If X and Y are independent events, $\Pr(X \text{ and } Y) = \Pr(X) * \Pr(Y)$.

Complement Formula for Probability

$\Pr(\text{not } X) = 1 - \Pr(X)$.

The three *P*'s (probability, percentage, and proportion) can be used interchangeably as a measure for the likelihood that an event occurs.

7.12. Constructing Models to Simulate a Player's Plate Appearances

We conclude the chapter by describing how to construct various types of models to simulate a plate appearance for a player.

7.12.1. Construction of Physical Models

We can construct a disk to represent a physical model for the plate appearances for a player. A disk consists of both the interior of a circle and the circle itself. We can think of a disk as a pie and a sector of the circle as a slice of the pie. The vertex of the angle of the sector is located at the center of the circle. In Figure 7.3, the symbol ψ (Greek letter *psi*) represents the angle of the sector. The arc length from A to B is also equal to ψ in radian measure. Any angle whose vertex is the center of the circle is called a central angle.

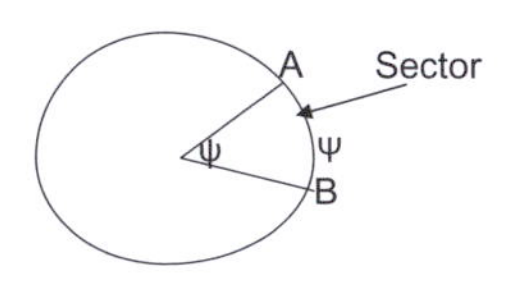

Figure 7.3 Central angle

For each plate appearance outcome of a player, we form the product of the probability of the outcome and 360 degrees, resulting in the central angle for the sector. Table 7.4 shows these calculations. The area of that sector

Table 7.4 Calculation for measures for central angles associated with sectors representing PA outcomes for Aaron and Bonds

PA outcomes	Probabilty*360	
	Aaron	Bonds
1B	.16*360=57.60	.12*360=43.20
2B	.04*360=14.40	.05*360=18.00
3B	.01*360=3.60	.01*360=3.60
HR	.05*360=18.00	.06*360=21.60
BB	.10*360=36.00	.20*360=72.00
SF	.01*360=3.60	.01*360=3.60
HBP	.00*360=0.00	.01*360=3.60
O	.62*360=223.20	.55*360=198.00

divided by the area of the circle is equal to the relative probability of the outcome. From this, we conclude that the relative probability of an outcome represents the proportion of the area of the circle determined by the area of its sector.

Since the sum of all the probabilities of the outcomes of a plate appearance for a player is equal to 1, the sum of all central angles representing all the outcomes of a plate appearance for a player will equal 360 degrees. Because of round-off error, the sum in Table 7.4 will only be close to 360 degrees. Using the central angles from Table 7.4 as the angles for our sectors, we can construct a probability disk for each player.

7.12.2. Four Methods for Constructing the Probability Disk (Spinning Disk)

There are four methods for the actual construction of the disk:

Method 1: Use any online spinner that produces accurate results or the spinner available at http://illuminations.nctm.org/activitydetail.aspx?ID=79. You may need to correspond numbers to colors for some online disks.

Method 2: Knowing the central angles that correspond to the outcomes of a plate appearance, a protractor can be used to construct the sectors.

Method 3: The disk can be divided into 36 sectors, each of which has an angle of 10 degrees. For each spin of the pointer the probability of the pointer landing on a given sector is 1/36. Since 1/36 equates to the decimal .0277, the probability of the pointer landed on any sector is 2.77%. The next step is to assign the correct number of sectors and partial sectors to the probability of the outcome of the plate appearance. For example, 1B for Aaron has a relative frequency probability of 0.16, so we would assign

5.75 sectors to the outcome of 1B. By creating 36 sectors each with the same area, we can associate the outcome 1B with 5.75 sectors. The same procedure can be applied to each plate appearance outcome. Since each of the 36 sectors is equally likely to occur, even though the plate appearance outcomes are not equally likely, we can consider the experiment of spinning a player's disk as a classical probability experiment.

Method 4: An easy method is to let Excel construct a pie graph based on the probabilities in Table 7.4 and use that pie graph as the disk.

Figures 7.4 and 7.5 contain the relative probability disks for Aaron and Bonds, respectively. The sectors of the disks correspond to their relative frequency probabilities of the corresponding outcomes.

Using the central angles to construct the sectors, we have a physical model that represents the probability of each outcome occurring. By adding a pointer to the disk, we create a process for generating plate appearances. The process of using probability theory to produce an artificial representation that mimics a real-life situation is called a *simulation*. Since the relative frequency probabilities for the outcomes of their career plate appearances are close to their *true* probabilities, we can use these disks to create samples of plate appearances. By the Law of Large Numbers, the more plate appearances there are, the closer the sample probabilities will be to the *true* probabilities.

The constructed disk enables us to create as many sample plate appearances for a player as we want. If the sample probabilities used to construct the disks are calculated from actual plate appearances for a given period of time (all postseason games, a particular year, a particular series of years, or a career), we can use this disk to create sample plate appearances for that same period of time.

In using the disk to generate plate appearances for a player, the following properties can be assumed:

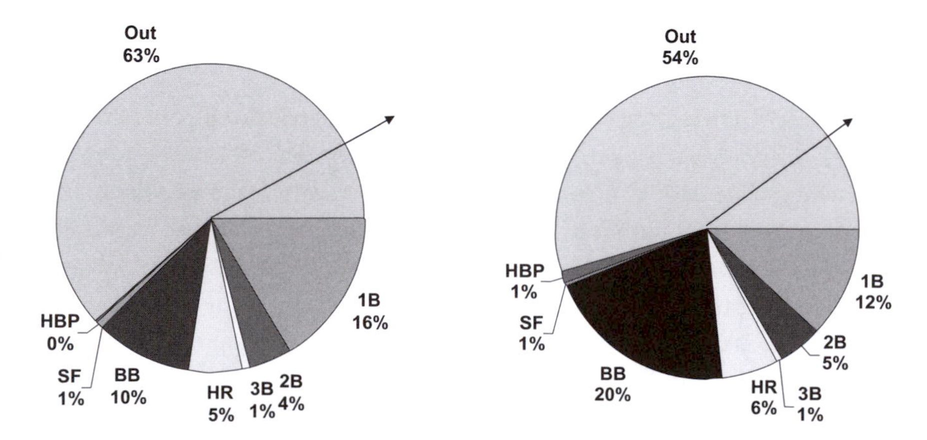

Figure 7.4 (left) Probability disk for Aaron's plate appearance outcomes

Figure 7.5 (right) Probability disk for Bonds's plate appearance outcomes

1. The sample probabilities represented by the sectors of the disk are close to the *true* probabilities. By the Law of Large Numbers, the larger the sample size, the closer the sample probability is to the *true* probability.
2. We can generate as many plate appearances as necessary.
3. The spins are independent of each other.
4. The probability of any outcome of a plate appearance will not change from spin to spin.
5. Since a disk is composed of 36 sectors each with an angle of 10 degrees, for each spin the probability of the pointer landing on any sector is 1/36. Therefore, the process of spinning a pointer on a disk results in 1 of 36 equally likely outcomes.

Property 5 above shows that the spin of a pointer on a disk is an experiment in classical probability. Therefore, the spin of a pointer is the same type of experiment as the tossing of a coin or the rolling of a die.

As discussed earlier, a baseball player's *true* batting average, or for that matter any *true* parameter for a given year or an entire career, is unknown. But we can use samples of plate appearances to estimate these parameters. The Law of Large Numbers states that as the sample size increases, we get closer to his *true* parameter. Thus, a career batting average based on many plate appearances will be an excellent estimate of a *true* career batting average.

7.12.3. Simulating Plate Appearances with a Mathematical Model

Another method of simulating plate appearances of a player is by using a mathematical model called the *random number generator.* A random number generator is a mathematical formula that generates in a completely random fashion an integer between any two given integers.

For this application, we select integers between 1 and 100. Based on Aaron's probability for each individual outcome, we assign that number of integers to the outcome. Using Aaron's probability disk, we would assign the numbers 1–16 (corresponds to the percentage of 16) to the outcome of 1B, the numbers 17–20 to the outcome 2B, the number 21 to the outcome of 3B, the numbers 22–31 to the outcome BB, the numbers 32–36 to the outcome HR, the number 37 to the outcome SF, and the numbers 38–100 to the outcome of an O. Each plate appearance would then be simulated by executing the random number generator. The easiest way to perform this simulation is to use the function RANDBETWEEN(1, 100) provided by Excel, which produces a number between 1 and 100. This number will correspond to one of the possible outcomes.

Therefore, we have two methods of simulating plate appearances: the physical model of the disk and the mathematical model of the random number function.

Exercise 7.29. Construct player batting disks for Ted Williams and Joe DiMaggio. Only use the seven plate appearance outcomes of 1B, 2B, 3B, HR, BB, HBP, and O. Treat the SF for Williams as outs. Use Tables 3.6 and 3.9 to assign probabilities for each of the plate appearance outcomes.

Exercise 7.30. Use the steps that follow to construct player disks for your two players.

a. Download from the website www.baseball-almanac.com the yearly baseball statistics for your two players. (You should have already done this.)
b. Construct two tables similar to Table 4.1. (You should have already done this.)
c. Construct two player disks.
d. Using the Excel function RANDBETWEEN(1, 100), assign the range of numbers needed to correspond to the probability for each plate appearance outcome occurring. For example, if the probability of a 1B was .15, we would assign the numbers 1–15 to the probability of a 1B occurring.

You will be asked to simulate plate appearances for your two players later in this book.

Chapter Summary

Probability is a numeric measure of the likelihood of the occurrence of a particular event. Three types of probability are defined.

Classical probability involves probability experiments whose outcomes occur strictly by chance and are equally likely to occur. The probability that an event occurs is calculated by dividing the number of outcomes in the event by the total number of possible outcomes (the sample space).

Relative frequency probability deals with outcomes of an experiment repeated a fixed number of times. The probability of an event occurring is calculated by dividing the number of times any outcome in the event occurs by the number of times the experiment is run.

Subjective probability involves experiments that cannot be repeated. The probability that an event occurs is assigned by a person who possesses a certain expertise in the area.

The evaluation of probabilities requires counting the number of objects in an event. Three methods for doing this were introduced. They include a *direct listing*, a *tree diagram*, and the use of a *formula*. The *Fundamental Law of Counting* (FLC) is a formula that can be used to count the number of outcomes in a multistage experiment.

Different ways of performing simple random sampling (SRS) were introduced. We looked at sampling *with or without replacement* and whether *the order of selection is important.*

A *permutation* (an arrangement of distinct elements in which the order of the elements is important) and a *combination* (an arrangement of distinct elements in which the order of the elements is not important) were used to count the number of possible different samples that could be selected from a finite population.

Relative frequency probability was applied to a contingency table consisting of the combined career plate appearances of Aaron and Bonds. The probabilities for *simple and compound events* were defined.

The concepts of *mutually exclusive* and *independent* events were defined. *Conditional probability* was defined and used to determine when two events are independent. The concept of the *independence of two variables* was defined.

The chapter concluded with the construction of a physical and a mathematical model to represent the career plate appearances for Aaron and Bonds. These models are used later in the book to *simulate* (mimic) the plate appearances of any player.

CHAPTER PROBLEMS

1. Suppose that the starting position players for an All-Star Team are to be selected by the vote of the fans. The ballot consists of 10 players for each of the eight positions. How many different starting lineups are possible for the All-Star Team? What counting principle did you use?
2. Suppose that 10 pitchers are listed on the ballot. The top 3 vote getters will be automatically placed on the All-Star Team. How many different possible selections of three pitchers are possible? What counting principle did you use?
3. Construct a tree diagram for Team A winning the World Series in seven games. List all the possible ways this can happen. (Hint: One way is AABABBA.)
4. Using the results of problem 3, calculate the probability that Team A wins the series in seven games (assume that each team has the same probability of winning each game and the outcomes of the games are independent of each other).
5. Suppose that everything in problem 4 is the same, except that for each game the probability that Team A wins is .60. Find the probability that Team A wins in seven games.
6. The probability that the series in problem 5 ends in seven games is ________.
7. Suppose that a manager uses four pitchers in a game: a starter, a middle reliever, a setup pitcher, and a closer. He has four pitchers in his starting rotation, and in the bullpen he has three middle relievers, two setup

pitchers, and one closer. How many different four pitchers can he use in that game?

8. Using the assumptions in problem 4, given that the series ends in five games, find the probability that Team A wins the series.
9. Answer the same question as problem 8, except that the series ends in six games.
10. We look at the number of runs scored in each inning by the visiting team in a nine-inning game. In how many ways can the visiting team score a total of (a) 0 runs, (b) 1 run, (c) 2 runs, (d) 3 runs, and (e) 4 runs?
11. A player has been assigned a locker and a lock. He was given a four-digit code for his lock; each digit is an integer from 0 to 9, and no two digits repeat. Unfortunately, he has forgotten the code.
    a. Suppose that he remembered that each of the four integers in his code was an odd number and the number 5 was not one of them. How many different permutations must he test to find his combination?
    b. Suppose that a friend gave him the following additional information. He told him that the integers 1 and 3 were grouped together in some order. Now, how many different permutations must he check?
    c. Another friend told him that not only were 1 and 3 grouped together in some order but the numbers 7 and 9 were also grouped together in some order. Now, how many different permutations must he check?
12. Construct a contingency table as described in Exercise 7.18 for your two chosen players. Construct a second contingency table as described in Exercise 7.19 for your two chosen players.
13. Sketch graphs similar to Figures 6.3–6.5 for your two players. Compare your two players with respect to their plate appearance results. Does this say anything about the type of hitter each player was?
14. Repeat Exercise 7.20 for the first contingency table in problem 12. Of course, you should use the initials of your two players.

Chapter 8

Sports Betting

The Colorado Rockies qualified for the playoffs in 2007 by winning 14 of their last 15 regular season games. In the divisional series they won three straight games against Philadelphia to advance to the league championship series against Arizona. They then proceeded to sweep the Diamondbacks. Therefore, Colorado won 21 of their last 22 games to make it to the World Series. The American League team they had to face was the Boston Red Sox.

What odds do you think the oddsmakers placed on Colorado winning the World Series? The odds were 5 to 12 that Colorado would win. If we place the larger number first, the odds were 12 to 5 that Boston would win. Many fans felt that these odds were inflated. Those fans who believed that it was worth losing $5 to win $12 bet on Colorado. What process is used to set odds on the outcome of a sporting event? What do the odds mean? Insight into these questions will be provided in this chapter.

Other questions explored in this chapter are as follows:

- How do casinos that run sports betting make their profits?
- What strategy should be used in deciding whether to place a bet on a certain sports outcome?
- What is meant by an expected gain or loss in a bet?
- What is meant by a fair bet?
- How is a betting line established?
- Using the betting line of a casino, how do we determine the amount to be won from a bet?

To understand sports betting, it is necessary to understand the concept of odds and the relationship between odds and probability. In this chapter we learn how to convert odds to probability and probability to odds. Because it's easy to go from odds to probability or from probability to odds, the advantage of using each one is discussed later in this chapter.

8.1. Odds and Probability

Suppose that the odds of event E happening are 2 to 1 (2:1). This implies that every three times the experiment is run, event E will occur two times. Note that the number 3 is the sum of 2 and 1. This leads to the definition of odds in terms of probability.

8.1.1. Converting Odds to Probability

If the odds of an event E occurring is a:b, then $\Pr(E) = a/(a+b)$.

Example 8.1. Converting odds to probability.

Odds	Calculation	Probability
3 to 2	3/(3+2)	3/5
3 to 4	3/(3+4)	3/7
1 to 1	1/(1+1)	1/2

8.1.2. Converting Probability to Odds

Suppose that $\Pr(E) = a/b$. To convert the probability that an event E occurs to the odds of event E occurring, we divide $\Pr(E)$ by $[1 - \Pr(E)]$:

$$\begin{aligned}\Pr(E)/[1-\Pr(E)] &= (a/b)/[(1-a/b)]\\ &= (a/b)/[(b-a)/b] = a/(b-a).\end{aligned}$$

We say that the odds of event E occurring are a to $(b - \text{a})$, or a:$(b - \text{a})$.

Example 8.2. Converting probability to odds.

Probability	Calculation	Odds
3/7	3 to (7 – 3)	3:4
1/2	1 to (2 – 1)	1:1
5/9	5 to (9 – 5)	5:4

Exercise 8.1. If a die is rolled once, the odds of the number 2 occurring are ______. The odds of getting an even number are ______.

Exercise 8.2. If the odds placed on a team winning are 6 to 5, then the probability assigned to that team winning is ______.

Exercise 8.3. Suppose that we spin the pointer on Aaron's disk (Fig. 7.4) one time. The odds of the simple event 1B occurring are ______. The odds of the simple event HR occurring are ______.

8.2. Expected Gain or Loss in Gambling

Before discussing betting, we must understand what is meant by the amount of money won and lost on a bet. In any bet, two amounts of money are involved. The first is the amount of money at risk that would be lost if the desired result of the bet does *not* occur. The other is the amount of money that would be won if the desired result *does* occur. To initiate a bet, the money at risk is considered to be held by a third party. If the bet is lost, the money is gone; if the bet is won, the money held is returned along with the amount won on the bet.

There are many ways to bet on baseball. We can bet on a game, on whether a particular result in a game occurs, or on a series of games. In betting on a baseball game, we can choose to bet on either team winning. The casino sets the odds on a team winning. Let us look at the following example.

Example 8.3. The Yankees are playing the Mets. The casino gives 5 to 4 odds on the Yankees winning. Let E be the event that the Yankees win. Let F be the event that the Mets win [F = (not E)]. Converting odds to probabilities, we have

$$\Pr(E) = 5/(5+4) = 5/9 \text{ and } \Pr(F) = \Pr(\text{not } E) = 1 - 5/9 = 4/9.$$

If you are willing to bet on the Yankees and the odds are 5:4 in favor of the Yankees winning, you are willing to risk \$5 to win \$4. This would be a *fair bet*, because if you made this bet nine times, you would expect to win five times and lose four times. If you win five times, the total amount won would be \$20; if you lose four times, the total amount lost would be \$20.

A bet is fair if the same bet made over and over again (theoretically an infinite number of times) results in a gain and loss of \$0.

We are ready to evaluate what is called the *expected gain* (a loss is interpreted as a negative gain) per bet. We denote the expected gain by $E(G)$.

Assuming that the same bet was made an infinite number of times, the $E(G)$ is the average amount won or lost [$E(G)$ is negative] per bet on a particular event E occurring. $E(G)$ is the theoretical average defined as follows:

$$\begin{aligned} E(G) &= W * \Pr(E) + -L * \Pr(\text{not } E) \\ &= W * \Pr(E) - L * \Pr(\text{not } E), \end{aligned}$$

where W = the amount won if event E occurs and L = the amount lost if event E does not occur. If $E(G) = 0$, then the bet is fair.

Example 8.4. The casino sets 4 to 3 odds of the Detroit Tigers defeating Tampa Bay in a given game. A bet of L = \$150 is placed on Detroit. The amount to be won is W = \$100. What is the $E(G)$ from the bet? Should we make the bet? What is the interpretation of $E(G)$?

Solution: If we bet the same game seven times, we would expect to win the bet four times and lose the bet three times. Converting odds to probabilities, we would get $\Pr(E) = 4/7$ and $\Pr(\text{not } E) = 3/7$, where E is the event of Detroit winning the game. Theoretically, if this bet was made an infinite number of times, the expected gain would be

$$E(G) = 100 * (4/7) - 150 * (3/7) = -50/7 = -7.14.$$

Since $E(G)$ is negative, we have an expected *loss* per bet. The interpretation is that if we made this bet 100 times, our expected loss would be $100 * (-7.14) = -\$714$. Clearly, we should not make this bet.

Example 8.5. The odds given by a casino for the Yankees beating the Mets is 5:4. A person bets $125 on a Yankee win. How much do you need to win for the bet to be fair? As the person loses $125 if the Yankees lose, we have the following: E = Yankees winning; $\Pr(E) = 5/9$; $\Pr(\text{not } E) = 4/9$; W = amount to be won; $L = 125$ = amount to be lost.

$$E(G) = (5/9) * W - (4/9) * 125 = 0,$$
$$5 * W - 4 * 125 = 0,$$
$$W = (4/5) * 125,$$
$$W = 100.$$

Thus, for the bet to be fair, $W = \$100$.

Exercise 8.4. The odds that Team A wins a game are 6 to 5. If $200 is bet on Team A, how much money should be won for the bet to be a fair bet?

8.3. Casino Betting on Baseball

A casino usually attaches a service charge for placing a bet, the amount varying from casino to casino. Without loss of generality, we arbitrarily use the amount of $5 as the service charge. The next two examples take into consideration the $5 service charge.

Example 8.6. If the bet on the Yankees in Example 8.5 is won, the winner receives only $95 because of the $5 service charge. The expected gain becomes

$$E(G) = (5/9) * 95 - (4/9) * 125 = -2.77.$$

Because of the service charge, the expected loss per bet would be –$2.77.

Example 8.7. If another person bets $100 on the Mets winning, then how much do we need to win for the bet to be fair? What is the expected gain if the service charge is applied?

Solution: As the person loses $100 if the Yankees win, we have the following: E = Mets win the game; $\Pr(E) = 4/9$; $\Pr(\text{not } E) = 5/9$; W = amount to be won; $100 = amount to be lost. So

$E(G) = (4/9) * W - (5/9) * 100 = 0,$
$4 * W - 5 * 100 = 0,$
$W = 500/4,$
$W = 125.$

Thus, if the bet is a *fair bet*, the person would win $125.

With the $5 service charge deducted from the winning amount, we have

$$E(G) = (4/9) * 120 - (5/9) * 100 = -2.22.$$

The above result means that with the service charge of $5, the bettor's expected loss per bet of $100 is –$2.22.

There is a problem exposed by these two examples. The expected gain changes depending on whether you bet on the favorite or the underdog. Clearly, this information would influence which team to wager on. A bettor having mathematical knowledge would always choose to bet on the result with the smaller expected loss.

From Ken Ross's *A Mathematician at the Ballpark*, the following method is used by some casinos to conduct betting on baseball. This method of betting corrects the inequity discussed above.

Many casinos will list the betting line in the following way: [Yankees –130, Mets +120]. This betting line means that a bet on the Yankees requires you to risk $130 in order to win $100; a bet on the Mets requires you to risk $100 in order to win $120. Translating the above to odds, the odds on the Yankees winning would be 130:100; the odds on the Mets winning would be 100:120.

Where do the numbers –130 and +120 come from? The process begins with the casino setting the odds of a team winning. For our example, the casino sets the odds of 5:4 on the Yankees winning. Multiplying both numbers by 25 yields the odds of 125:100. Since the odds are 5:4 for the Yankees winning, the odds are 4:5 (100:125 after multiplying both numbers by 25) for the Mets winning. The $5 service charge is added to the $125 at risk, giving the odds of 130:100 for the Yankees winning. At the same time, the $5 is subtracted from the amount to be won in betting on the Mets, giving the odds of 100:120 on the Mets winning.

- If you bet on the Yankees to win, $E(G) = (5/9) * 100 - (4/9) * 130 = -2.22.$
- If you bet on the Mets to win, $E(G) = (4/9) * 120 - (5/9) * 100 = -2.22.$

No matter which team you bet on to win, the expected gain is –2.22.

Can you explain why the above two expected gains are equal? Will this always be true?

- If $E(G) = 0$, we say that the bet is a fair bet.
- If $E(G) > 0$, we say that the bet favors the bettor.
- If $E(G) < 0$, we say that the bet favors the casino.

The service charge shifts the bet from a fair bet to a bet that favors the casino.

8.3.1. How Does a Casino Make Money on a Baseball Betting Line?

To establish a betting line, the casino first sets the odds for one of the two teams winning the game. To set these odds, the casino uses *subjective probability*. The betting line (Yankees –130, Mets +120) was established after the odds of 5:4 were set for the Yankees winning. Automatically, the odds on the Mets winning would be 4:5.

The casino makes its money as long as they can even out the money bet in such a way that for every $5 bet on the Yankees, $4 will be bet on the Mets. How does that work? Suppose $5000 is bet on the Yankees by various bettors and $4000 is bet on the Mets by various bettors. Since $5000 is collected for bets on the Yankees, we divide 5000 by 130 (the amount at risk), yielding 38.46 bettors. We take the $4000 bet on the Mets and divide that number by 100 (the amount at risk), yielding 40 bettors. If the Yankees win the game, the payout would be 38.46 times $100 (the amount to be won), giving a total payout of $3846. Since $4000 was collected from the bets lost on the Mets, the casino would make $154. On the other hand, if the Mets win the game, the total payout would be 40 times $120 (the amount won if the Mets win), yielding $4800. Since $5000 was collected from the bets lost on the Yankees winning, the casino would make $200. In either case, the casino makes close to 4% on the money collected. The money is made because of the service charge. If the service charge is increased, the profit the casino makes also increases. Suppose a bettor wishes to put at risk a different amount of money. How can the person decide the amount to be won? The next example addresses this question.

Example 8.8. Using the betting line above, suppose that a bettor puts $250 at risk in a bet on the Yankees. If the Yankees win, how much will the bettor win? Suppose that the bettor puts $250 at risk with a bet on the Mets. If the Mets win, how much will the bettor win?

Solution: If a bettor puts $250 at risk with a bet on the Yankees, we let W be the amount to be won. From the betting line, it takes $130 at risk to win $100.

Therefore,

$$(100/130) = (W/250),$$
$$W = (100 * 250)/130 = 25000/130 = 192.31.$$

A bet of $250 on the Yankees would enable the bettor to win $192.31.

If a bettor puts $250 at risk with a bet on the Mets, we have

$$(100/120) = (250/W),$$
$$W = (250 * 120)/100 = 300.00.$$

A bet of $250 on the Mets would enable the bettor to win $300.00.

Exercise 8.5. Given that a casino has a betting line of [Yankees –220, Mets +180], and assuming a service charge of $20, what is your guess at the odds the casino has set? Verify that your guess is correct.

Example 8.9. In this example, use the disk with pointer constructed for Aaron (Fig. 7.4). The experiment consists of one spin of the pointer on Aaron's disk. Let E be the event of getting a hit. The amount we put at risk is $L = \$100$. If the bet is fair, find the amount W that will be won if event E occurs.

Solution: We need to calculate the probability that E occurs. Using the fact that the events 1B, 2B, 3B, and HR are mutually exclusive with the data from Table 2.2, $\Pr(E) = \Pr(1B) + \Pr(2B) + \Pr(3B) + \Pr(HR) = 2294/13919 + 624/13919 + 98/13919 + 755/13919 = 3771/13919$. Since $\Pr(E) = 3771/13919$, $\Pr(\text{not } E) = 1 - \Pr(E) = 10148/13919$. For the bet to be a fair bet, the expected gain must be zero. We now calculate W as follows:

$$E(G) = W * \Pr(E) - 100 * \Pr(\text{not } E),$$
$$0 = W * (3771/13919) - 100 * (10148/13919),$$
$$0 = 3771 * W - 1014800,$$
$$W = 1014800/3771 = 269.11.$$

The spinning of the pointer on Aaron's disk is a classical probability experiment. The reason is that the outcome of a spin occurs strictly by chance (no skill is involved). Also, each spin is independent of any other spin. Theoretically, we can perform this experiment an infinite number of times. Notice that these are the same properties of the classical probability experiment of tossing a coin.

Exercise 8.6. A coin is tossed and a die is rolled. You decide to bet $100 on the event E = the coin results in a head and the die results in an even number. For this to be a fair bet, what would the payoff be if the event E occurs? What are the odds of event E occurring?

Exercise 8.7. The experiment consists of one spin of the pointer on Aaron's disk. The amount we put at risk is $100. The bet is won if the result of the spin is a HR. If the bet is won and it is a fair game, how much would we win?

Exercise 8.8. The experiment consists of two spins on Aaron's disk. Let E be the event of getting two hits. The amount we put at risk is $L = \$100$. If the bet is a fair bet, find the amount W that will be won if event E occurs. (Hint: The two spins are independent events.)

8.4. Strategy for Baseball Betting

There is a reasonable strategy for betting on multiple baseball games. First, choose a certain number of games that, based on your expertise, you feel you can assign the odds of a certain team winning. Next, decide the amount of money you are willing to risk. Then, calculate the amount you should win for the bet to be fair. Finally, check various gambling websites and see if any of these sites will pay out at least the amount you calculated for a fair bet. The next example illustrates this method.

Example 8.10. We choose three games. The games and odds, assigned by us, are presented below.

- Yankees against the Mets (5:4 Yankees win)
- Braves against Rockies (3:2 Braves win)
- Reds against Pirates (1:1 Reds win)

The events are E = the Yankees win, F = the Braves win, and G = the Reds win. We decide to risk \$100. The bet is won if the events E and F and G all occur. We assume that the three games are *independent* of each other. We now want to find the amount we need to win for the bet to be fair.

Let W = amount to be won, L = \$100. Since the events are independent, we have $\Pr(E \text{ and } F \text{ and } G) = \Pr(E) * \Pr(F) * \Pr(G) = [5/(5+4)] * [3/(3+2)] * [(1/(1+1)] = (5/9) * (3/5) * (1/2) = 15/90$.

Assuming that the bet is a fair bet, we have

$$\begin{aligned} 0 = E(G) &= W * \Pr(E \text{ and } F \text{ and } G) - 100 * \Pr(\text{not}[E \text{ and } F \text{ and } G]) \\ &= W * (15/90) - 100 * (75/90) \\ &= 15 * W - 7500, \\ W &= 500. \end{aligned}$$

The final step is to examine various betting sites to see if any site has a payout that makes this bet advantageous:

- If the casino pays out \$500, the bet is fair.
- If the payout exceeds \$500, the bet is in our favor.
- If the payout is less than \$500, the bet favors the casino.

Exercise 8.9. Using the information in Example 8.10, a bet is made that the Yankees win, the Rockies win, and the Pirates win. What amount should be won if the bet is a fair bet?

8.5. Step-by-Step Method to Establish a Betting Line

The betting line expression [Yankees −130, Mets +120] means that a person who bets on the Yankees to win must put $130 at risk to win $100, whereas a person who bets on the Mets to win must put $100 at risk to win $120. The service charge of $5 is included in the −130 and +120.

The reasons for a casino to use a betting line are twofold. First, it makes it easy for a person to place a bet. Second, the service charge is already included.

The advantage of using odds in betting is that it makes it easy for the person to see what will be won or lost on a given bet. Knowing how to calculate probabilities is necessary to evaluate $E(G)$.

A step-by-step method for going from either the probability of 5/9 or the odds of 5:4 for the Yankees winning against the Mets to the casino betting line of [Yankees −130, Mets +120] is provided next. The odds and probability of the Yankees losing are, respectively, 4:5 and 4/9. The service charge is $5.

8.5.1. Converting Odds to the Betting Line

Let E be the event that the Yankees win. Let (not E) be the event that the Mets win (the Yankees lose).

Step 1: We choose the event with the highest odds (in this case it is E).
Step 2: Assuming a fair bet, we evaluate the amount at risk L to win $100.

$$E(G) = (5/9) * 100 - (4/9) * L = 0 \text{ (L is the amount at risk).}$$

Step 3: Solving the above equation for L, we get $L = \$125$.
Step 4: We add the service charge of $5 to $125 and get $130.
Step 5: We wish to find the amount W to be won on a bet of $100 on the Mets winning (not E occurring).
Step 6: Assuming a fair bet, we get the following equation:

$$E(G) = (4/9) * W - (5/9) * 100 = 0 \text{ (W is the amount we will win).}$$

Step 7: Solving the above equation for W, $W = \$125$.
Step 8: We subtract the service charge of $5 from $125 and get $120.

The betting line is [Yankees −130, Mets +120].

In the next example, we use the method outlined above to go from the odds to the betting line.

Example 8.11. Suppose that the odds established by a casino for the Marlins to beat the Rockies are 3 to 2. We go through the steps necessary to create a betting line. The odds 3 to 2 correspond to a probability of 3/5 of the Marlins winning;

the probability that the Marlins lose is 2/5. For this bet a service charge of $10 is used.

Let E be the event that the Marlins win. Let (not E) be the event that the Marlins lose (Rockies win).

Step 1: We choose the event with the highest odds (in this case it is E).
Step 2: Assuming a fair bet, we evaluate the amount at risk L to win $100.

$E(G) = (3/5) * 100 - (2/5) * L = 0$ (L is the amount at risk).

Step 3: Solving the above equation for L, we get $L = \$150$.
Step 4: We add the service charge of $10 to $150 and get $160.
Step 5: We wish to find the amount W to be won for a bet on (not E) with a risk of $100.
Step 6: Assuming a fair bet, we get the following equation:

$E(G) = (2/5) * W - (3/5) * 100 = 0$ (W is the amount we will win).

Step 7: Solving the above equation for W, we get $W = \$150$.
Step 8: We subtract the service charge of $10 from $150 and get $140.

The betting line is [Marlins –160, Rockies +140].

Observe that the L computed in step 3 becomes the W in step 7. Why is this always true? Show how we can go directly from step 5 to step 8.

Exercise 8.10. Assuming that the odds of the Indians defeating the Blue Jays is 4 to 3, use the eight steps above to establish the betting line with a $20 service charge included.

Exercise 8.11. Given the betting line [Yankees –300, Mets +260] with the service charge of $20 included, find the original odds set by the casino for the Yankees winning the game.

Chapter Summary

This chapter introduced the concept of the odds of an event occurring. Formulas to convert odds to probabilities and probabilities to odds were given.

The expected gain, $E(G)$, of a bet on event F occurring was defined by

$$E(G) = W * \Pr(F) - L * \Pr(\text{not } F),$$

where W = amount to be won and L = amount at risk.

The expected gain $E(G)$ represents the average amount won or lost per bet. If the expected gain is negative, it is called an expected loss. Clearly, it is important and necessary to be able to calculate probabilities in order to calculate the expected gain.

A bet made by a player is a *fair bet* if $E(G)=0$. If $E(G)<0$, the bet favors the casino. If $E(G)>0$, the bet favors the player. If the same bet could be repeated a certain number of times n, multiplying $E(G)$ by n provides an estimate of the total amount won (if $E(G)>0$) or the total amount lost (if $E(G)<0$). The $E(G)$ gives the average amount won or lost per bet.

The two types of casino betting studied in this chapter were betting on baseball games and betting on games of chance. Casino betting on baseball games requires the use of subjective probability, whereas casino betting on games of chance (spinning a pointer on a player's disk) uses classical probability. The idea of a betting line was introduced. A step-by-step method for the creation of a betting line from initial odds was developed. A strategy for betting on multiple baseball games was given. A discussion of what is necessary for a casino to make money was presented.

CHAPTER PROBLEMS

1. Using the probabilities for the outcomes of a plate appearance for Aaron listed in Table 7.1 and Figure 7.4, find the odds for each of the simple events 1B, 2B, 3B, HR, BB, O, HBP, and SF occurring and the odds of each of the simple events not occurring. As an example, the odds of 1B occurring is $\Pr(1B)/[1-\Pr(1B)]=.16/.84=16/84=4/21$. This gives the odds as 4:21 of 1B occurring in a plate appearance for Aaron. Therefore, the odds of 1B not occurring (not 1B) is 21:4.
2. For each simple event in problem 1 express the betting line in the form of [favored outcome occurs $-x$, favored outcome does not occur $+y$]. We demonstrate obtaining the betting line for 1B using the eight steps in Section 8.5:

   Step 1: The outcome with the highest odds is (not 1B). The $\Pr(\text{not } 1B)=84/100=21/25$.

   Step 2: Assuming a fair bet, we evaluate the amount at risk L to win \$100.

   $$E(G)=(21/25)*100-(4/25)*L=0.$$

   Step 3: Solving the equation for L, we get $L=\$525$.

   Step 4: We add the service charge C (C can range from \$0 to any amount) to L. The result is $L+C$.

   Step 5: We wish to find the amount W to be won on a bet of \$100 that 1B occurs.

   Step 6: Assuming a fair bet, we get

   $$E(G)=(4/25)*W-(21/25)*100=0.$$

Step 7: Solving for W, we get $W = \$525$ ($W = L$).

Step 8: We subtract the service charge (the amount used in step 4) and get $W - C$.

The resulting betting line is [(not 1B) − $(L + C)$, 1B + $(W - C)$].

If the service charge is $C = \$10$, the betting line becomes [(not 1B) −535, 1B +515]. This betting line means that a bet on the event (not 1B) occurring requires a bettor to put \$535 at risk to win \$100. If a player bets that the event 1B occurs, the player would win \$515 with a bet of \$100.

3. A bet of \$200 is made on the following outcomes of a plate appearance for Aaron occurring: 1B, (not HR), 2B, (not O), and O. Assuming that the service charge $C = \$10$, how much would be won with a successful bet? We illustrate this with a bet of \$200 on 1B occurring. Since a bet of \$100 wins \$515, a \$200 bet would win \$1030.
4. Let the service charge $C = \$10$. Using the betting lines established in problem 2, compute the $E(G)$ for a bet on each of the simple events 1B, 2B, 3B, HR, BB, O, HBP, and SF occurring.

   We illustrate this for the simple event 1B. The $E(G)$ for betting on 1B is

   $$E(G) = (4/25) * 515 - (21/25) * 100 = -1.60.$$

   The expected gain would be −1.60 per bet.
5. Same problem as problem 4, except use the service charge $C = 0$.

Chapter 9

Baseball and Traditional Descriptive Measures

We revisit the relationships between the descriptive measures used in baseball (see Chap. 3) and those used in any traditional statistics course (see Chap. 1).

Most traditional statistics (descriptive measures) are amounts, means, proportions, or ratios.

From Chapter 3, we have the following baseball definitions:

Hit (H) = {1B, 2B, 3B, HR},
Out (O) = IPO + {SO}.

For this chapter, safe on error (SE) is included in the set O. This is justified since SE is treated as an out in the scorebook.

At bat (AB) = H + O,
Plate appearance (PA) = AB + {BB, HBP, SF},
In-play at bat (IPAB) = AB − {SO}.

In this chapter, Tables 2.2, 4.1, and 5.16 will be used. Table 2.2 provides the complete yearly baseball data for Henry Aaron. Table 4.1 provides the complete baseball data for Barry Bonds through 2007. Table 5.16 provides the complete baseball data for Alex Rodriguez through 2007.

9.1. Baseball Statistics That Are Means and Proportions

A player's batting average (AVG or BA) is formed by dividing the number of hits by the number of at bats (#H/#AB). The two outcomes of an at bat are the mutually exclusive events of "H" and "O." If we think like a manager and not a pitcher, we assign success to getting a hit and failure to making an out. We can then talk about the proportion of successes as the number of successes divided by the number of at bats. Consequently, AVG is the proportion of hits to at bats. Therefore, AVG is a proportion. But AVG can also be interpreted as a mean. To understand this, assign the number 1 to each at bat that resulted in a hit and the number 0 to each at bat that resulted in an out. Sum these numbers and divide that sum by the number of at bats. The result is the mean number of hits per at bat.

(AVG or BA) Formula = #H/#AB.

A player's on-base percentage or on-base average (OBP or OBA) is formed by dividing the sum of the number of hits, base on balls, and hit by pitch by the sum of the number of at bats, base on balls, hit by pitch, and sacrifice flies. The formula is (#H + #BB + #HBP)/(#AB + #BB + #HBP + #SF). OBP is the proportion of plate appearances that results in a player getting on base. OBP can also be interpreted as the mean number of times a player reaches base per plate appearance. This can be seen by assigning the number 1 to each plate appearance that results in getting on base and the number 0 to the other plate appearances. This sum is divided by the number of plate appearances. The result is the mean number of times on base per plate appearance.

(OBP or OBA) = (#H + #BB + #HBP)/#PA.

A player's home run average (HRA) is formed by dividing the number of home runs by the number of at bats. In the same way that AVG and OBP can be interpreted as a proportion and as a mean, the HRA is the proportion of at bats that result in a home run or the mean number of home runs per at bat.

HRA Formula = #HR/#AB.

The following formulas can also be interpreted as a proportion or as a mean:

SOA (strikeout average) = #SO/#AB,
BBA (base-on-balls average) = #BB/#PA,
IPBA (in-play batting average) = #H/#IPAB,
IPHR (in-play home run average) = #HR/#IPAB.

Why are all these baseball statistics proportions?
Use Tables 2.2 and 4.1 for the exercises below.

Exercise 9.1. For the year 1971, compute for Aaron his AVG, OBP, HRA, SOA, BBA, IPBA, and IPHR.

Exercise 9.2. For his career, compute for Aaron his AVG, OBP, HRA, SOA, BBA, IPBA, and IPHR.

Exercise 9.3. For the year 2001, compute the AVG, OBP, HRA, SOA, BBA, IPBA, and IPHR for Bonds.

Exercise 9.4. For his career, compute the AVG, OBP, HRA, SOA, BBA, IPBA, and IPHR for Bonds.

9.2. Baseball Statistics That Are Ratios and Not Proportions

The number of total bases (TB) equals the number of singles plus 2 times the number of doubles plus 3 times the number of triples plus 4 times the number of home runs. Alternately, the number of total bases is equal to the number of hits plus the number of doubles plus 2 times the number of triples plus 3 times the number of home runs. Both formulas appear below.

The statistic TB is a discrete variable that gives an amount.

$$TB = (1 * \#1B + 2 * \#2B + 3 * \#3B + 4 * \#HR),$$
$$TB = (\#H + \#2B + 2 * \#3B + 3 * \#HR).$$

Exercise 9.5. Show that the two formulas for TB give the same result.

A player's slugging percentage (SLG) is another type of average called a weighted mean. In this interpretation a single counts as 1 point, a double counts as 2 points, a triple counts as 3 points, and a home run counts as 4 points. For this statistic, all hits are not considered equal. More weight is given to hits with power. The power ranking of hits is from the home run, which is the highest, down to the single, which is the lowest.

$$\text{SLG Formula} = TB/\#AB.$$

The SLG formula is a ratio but not a proportion.

There is a difference between a proportion and a ratio. Both are formed in the same way by dividing a number representing one variable defined on a set by a second number representing a second variable defined on a second set. For a *proportion*, the set used in the numerator must be a *subset* of the set used in the denominator. A proportion will always be a number between 0 and 1. A proportion is one type of ratio. Ratios need *not* be proportions. SLG is a ratio that is not a proportion. SLG is a number between 0 and 4. The formulas presented next are all ratios and not proportions.

Exercise 9.6. Using four at bats for a player, give examples for the four at bats that would result in an SLG of

a. 4.00
b. 0.00
c. 1.00
d. .500
e. .400

A player's home run rate is formed by dividing the number of at bats by the number of home runs. This statistic tells us the average number of at bats it takes before the player hits his next home run.

HRR Formula = #AB/#HR.

A player's rate of strikeouts to base on balls is formed by dividing the number of strikeouts by the number of base on balls.

SO/BB = #SO/#BB.

To compare a discrete variable *X* between two players for a certain time interval (a portion of a year, a year, several years, or a career), we divide the #*X*, for each player, by their number of either outs, games, AB, or PA for that time period. These ratios are continuous variables (measurements).

An example of this is the ratio of the number of runs scored by a player divided by the number of games the player appeared in.

R/G Formula = #R/#G.

Since a player is credited with appearing in a game if he is a defensive replacement or a pinch hitter, an improved method for comparing a discrete variable *X* between two players for a certain time interval is obtained by dividing #*X* by their number of outs (#AB − #H). An example of this is the ratio of the number of runs scored by a player divided by the number of outs made by the player.

R/O Formula = #R/(#AB − #H).

Since most games are completed when a team makes 27 outs, we consider the number of runs scored by a player per 27 outs.

As discussed in Chapter 3, the formula below uses 25.5 instead of 27 because certain outs in a game are not caused by a player's at bat or an at bat can cause two outs.

R/27 Formula = 25.5 * #R/(#AB − #H).

Exercise 9.7. Compute Aaron's TB, SLG, SO/BB, HRR, R/G, and R/27 for 1971.

Exercise 9.8. Compute Aaron's TB, SLG, SO/BB, HRR, R/G, and R/27 for his career.

Exercise 9.9. Compute Bonds's TB, SLG, SO/BB, HRR, R/G, and R/27 for 2001.

Exercise 9.10. Compute Bonds's TB, SLG, SO/BB, HRR, R/G, and R/27 for his career.

9.3. Recent Baseball Statistics Derived Specifically for Baseball

Runs created (RC) was defined in the 1970s by Bill James. The original formula (there are now over 24 different RC formulas) is

$$RC = [(\#H + \#BB)/(\#AB + \#BB)] * TB.$$

Since HBP and SF are rare occurrences, $[(\#H + \#BB)/(\#AB + \#BB)]$ is a good approximation for OBP. Consequently, we have the following approximation for RC:

$$RC\ (\text{Approximation}) = OBP * TB.$$

Multiplying RC by 25.5 and dividing this product by the number of outs gives the number of runs created by a player per 27 outs.

$$RC/27 = 25.5 * RC/(\#AB - \#H).$$

The baseball statistic RC/27 can be used to measure hypothetically the number of runs produced by a player for a team where he is cloned to occupy all nine positions of the team.

The batter's run average or rate (BRA or BRR) is the product of OBP and SLG.

$$BRA = OBP * SLG = OBP * (TB/\#AB) = OBP * TB/\#AB.$$

On-base plus slugging (OPS) is defined to be OBP + SLG.

$$OPS = OBP + SLG.$$

BOP, bases over plate appearances, was introduced in the 1970s and is defined by replacing hits by TB in the OBP formula. Replacing H by TB puts more emphasis on slugging.

$$BOP = (TB + \#BB + \#HBP)/\#PA.$$

TA, the total average, measures the overall offensive contributions of a hitter. The TA was introduced by Thomas Boswell in the 1970s. Simply put, the TA is the ratio of the number of bases a player accumulates to the number of outs a player makes. In this formula SB is stolen base, CS is caught stealing, and GIDP is grounded into a double play.

$$TA = [(TB + \#HBP + \#BB + \#SB) - \#CS]/[(\#AB - \#H) + \#CS + \#GIDP].$$

A newer formula for evaluating a player's performance as a power hitter is called the total power quotient (TPQ). The numerator is $(TB + \#HR + \#RBI)$, while the denominator is #AB.

The formula for TPQ is

TPQ = (TB + #HR + #RBI)/#AB.

In Chapter 5, we used linear regression and correlation to find the baseball statistics that would be the best estimator of R/27 for Aaron and Bonds. For both Aaron and Bonds, the top four baseball statistics, each with a correlation coefficient r greater than 0.90, for predicting R/27 were OPS, BRA, SLG, and RC/27. OPS was the best predictor of R/27 for Aaron, and BRA was the best predictor of R/27 for Bonds. In the exercises that follow, you will be asked to use TA and TPQ as a predictor for R/27 for both Aaron and Bonds.

Exercise 9.11. Compute Aaron's RC/27, BRA, OPS, BOP, TA, and TPQ for 1971.

Exercise 9.12. For his career, compute Aaron's RC/27, BRA, OPS, BOP, TA, and TPQ.

Exercise 9.13. Compute Bonds's RC/27, BRA, OPS, BOP, TA, and TPQ for 2001.

Exercise 9.14. For his career, compute Bonds's RC/27, BRA, OPS, BOP, TA, and TPQ.

Exercise 9.15. For the career years of Aaron, find the equation for the regression line and the correlation coefficient r for X = TA and Y = R/27. Repeat for X = TPQ and Y = R/27. Compare the correlation coefficients with the ones obtained in Chapter 5 for X = BA, OBA, SLG, OPS, ISO, RC/27, and BRA.

Exercise 9.16. For the career years of Bonds, find the equation for the regression line and the correlation coefficient r for X = TA and Y = R/27. Repeat for X = TPQ and Y = R/27. Compare the correlation coefficients with the ones obtained in Chapter 5 for X = BA, OBA, SLG, OPS, ISO, RC/27, and BRA.

Exercise 9.17. For the career years of your two chosen players, find the equation for the regression line and the correlation coefficient r for X = TA and Y = R/27. Repeat for X = TPQ and Y = R/27. Compare the correlation coefficients with the ones obtained in Chapter 5 for X = BA, OBA, SLG, OPS, ISO, RC/27, and BRA.

Chapter Summary

This chapter related baseball statistics to the standard statistics of an amount, mean, proportion, and ratio. The standard statistics used in any statistics course were introduced in Chapter 1. Statistics used in baseball were introduced in Chapters 3 and 5.

Baseball statistics are used to evaluate a player's, a team's, and a league's batting performance. Batting performance for a player can be broken down into measuring a player's success in getting on base, hitting with power, creating runs for his team, making contact with the ball, and directing a batted ball to where fielders are not positioned.

In all areas outside of baseball, performance is also evaluated. In health care, the efficacy of drugs in the treatment of disease is measured. In business, the effect of an advertising campaign is measured. In finance, the performance of a stock is measured. Educators assess students. The descriptive measures of mean, median, standard deviation, variance, proportion, and ratio are used to measure performance in these areas.

To evaluate a drug treatment, the proportion of people in a sample that improve after taking the drug is calculated. To evaluate an advertising campaign, the mean amount of money that a sample of people will spend on a product is analyzed. To decide on whether to buy a stock, the variance of its price over a period of time is studied. To assess a new teaching technique, the proportion of students who improve because of the new technique is measured.

CHAPTER PROBLEMS

Many baseball people believe that the two best hitters from 2000 to 2008 were Alex Rodriguez and Albert Pujols.

1. Use the baseball data from Table 5.16 for Alex Rodriguez at the end of Chapter 5 to fill in Tables 9.1 and 9.2.
2. Download the yearly baseball data for Albert Pujols from www.baseball-almanac.com. Then complete the above tables for Albert Pujols (use his appropriate career years). You will use these tables to compare Rodriguez to Pujols in the next chapter.
3. Using the career years for your two chosen players, complete Tables 9.1 and 9.2.

Table 9.1 Batting statistics (proportions) for Alex Rodriguez

Year	AVG	OBP	HRA	SOA	BBA	IPBA	IPHR
1994							
1995							
1996							
1997							
1998							
1999							
2000							
2001							
2002							
2003							
2004							
2005							
2006							
2007							
Career							

Table 9.2 More batting statistics for Alex Rodriguez

Year	SLG	HRR	SO/BB	R/G	R/27	TA	RC/27	BRA	OPS	BOP	TPQ
1994											
1995											
1996											
1997											
1998											
1999											
2000											
2001											
2002											
2003											
2004											
2005											
2006											
2007											
Career											

Chapter 10

Final Comparison of Batting Performance between Aaron and Bonds

This chapter concludes our work with *descriptive statistics*, covered in Chapters 1–9. The goal of this chapter is to bring together the techniques of descriptive statistics that were covered in the preceding chapters. These techniques were used to compare the batting performance between two players.

At the end of this chapter, four tests, developed by Bill James to compare players, are presented. These are the *Black-Ink Test*, the *Gray-Ink Test*, the *Hall of Fame Standards Test*, and the *Similarity Test*. There are more tests that Bill James and others have developed to compare players. To see some, visit www.baseball-reference.com and look under the category "leaders."

Descriptive statistics provide techniques for the *collection*, *organization*, *presentation*, and *summarization* of data. The baseball data used were *collected* from the websites www.baseball-almanac.com, www.mlb.com, www.baseball-reference.com, www.retrosheet.org, and www.baseballhalloffame.org and from the books *The SABR Baseball List & Record Book*, by the Society for American Baseball Research, and *The Elias Book of Baseball Records*, by Seymour Siwoff. The downloaded baseball data were *organized* into special tables. Various types of graphs were used to *present* the baseball data. Finally, the *summarization* of the baseball data was done using descriptive measures.

Baseball scouts evaluate positional players on the basis of "five tools." These include hitting, hitting with power, fielding, running, and throwing. The only two tools we study here are hitting and hitting with power.

In Chapter 4, Aaron and Bonds were compared on the basis of three groups of batting statistics: BA, OBA, and SLG; RC/27 and RPR/27; and SOA, BBA, IPHR, and SO/BB.

The choice of those groups of baseball statistics was arbitrary. In Chapter 4, we gave the rationale for our choice of the statistics used in each of the above groups. We will see later in this chapter that for his *Black-Ink Test* Bill James used 12 baseball statistics.

The same techniques, used in this chapter to compare Aaron and Bonds, can be used to compare any two or more players, any two or more teams, and the National League with the American League. At the end of this chapter, you will be asked to compare the two players you chose.

We continue the comparison of the batting performances of Aaron and Bonds by looking at specific groups of baseball statistics. These groups are based on certain criteria and are studied over different time periods. The groups, criteria, and time periods are defined next.

The Seven Groups of Baseball Statistics

1. (GR-1) Getting on base: #BB, #H, BA, and OBP.
2. (GR-2) Hitting with power and advancing runners: #HR, #2B, TB, #XBH, SLG, ISO, TPQ, BOP, HRA, and IPHR.
3. (GR-3) Getting on base combined with slugging: BRA and OPS.
4. (GR-4) Special League Awards: MVP and selection to the All-Star Team. (For modern-day players, I would add the Silver Slugger Award, which started in 1980. This award is given to the best offensive player at each position.)
5. (GR-5) Creating runs for their teams: #R, #RBI, R/G, RC/27, and RPR/27.
6. (GR-6) Having a good batting eye: #SO, #BB, SO/BB, and SOA.
7. (GR-7) Directing a batted ball to where the fielders are not positioned: BA and IPBA.

We'll compare the batting performances of Aaron and Bonds by applying the following eight criteria to the individual baseball statistics housed in each of the groups GR-1 through GR-7.

The Eight Criteria

1. (CR-1) A comparison of each player's career statistics with the all-time career leaders for these statistics (only statistics in which at least one of the players is in the top 50 will be listed).
2. (CR-2) A comparison of each player's seasonal individual statistics with the individual statistics for all players in their league for that season (only player statistics that were league leading are listed).
3. (CR-3) A comparison of the two players, head to head, with respect to the number of times they led the league for a season for any of the statistics within a group.
4. (CR-4) A comparison of the two players, head to head, using their yearly statistics (these statistics are based on all plate appearances for a season).
5. (CR-5) A comparison of the two players, head to head, using their career statistics (these statistics are based on their cumulative plate appearances).

6. (CR-6) A comparison of the two players, head to head, based on their cumulative plate appearances for the period of time between the ages of 35 and 39.
7. (CR-7) A comparison of the two players, head to head, with respect to their best individual season in the majors.
8. (CR-8) A comparison of the two players with respect to the type of hitter they were.

The Four Time Periods

1. (T-1) Year by year (seasonal).
2. (T-2) Career performance, based on the cumulative totals for a player's career.
3. (T-3) Performance of a player from the age of 35 to the age of 39, based on the cumulative totals for those years only. (It should be noted that the time period T-3 was chosen because our initial studies showed that Barry Bonds had such a strong offensive performance late in his career. In many comparisons a time period consisting of each player's 5 or 10 best years of their careers is used.)
4. (T-4) Best year of their careers.

Many baseball statistics and their corresponding graphs will be used to compare Aaron and Bonds with respect to their batting performance. Almost all fields of study use statistical information to draw conclusions from data. But conclusions can be tricky. As Mark Twain said in his autobiography, "Figures often beguile me, particularly when I have the arranging of them myself; in which case the remark attributed to Disraeli would often apply with justice and force: 'There are three kinds of lies: lies, damned lies, and statistics.'"

Another consideration is the influence of outside factors on the provided statistics. The statistics provided in a study, based on the variables in the study, may indicate that certain statements are facts, but these could turn out not to be true. External factors, not mentioned in the study, may lead to such false conclusions. We show later in this chapter how certain rule changes in baseball could have affected Aaron's batting performance in the later years of his career. Also, it is alleged that the use of certain banned substances could have affected the batting performance of Bonds in the later years of his career.

In choosing the statistics to use for my comparison of the batting performances of Aaron and Bonds, I claim no bias. It is impossible to choose every statistic and every possible graph for my analysis. You are welcome to choose different individual statistics, different groups of statistics, and different criteria. These choices may lead to the same or different conclusions.

We must differentiate between a player's *batting performance* and *batting ability*. The statistics presented in the preceding chapters were used to measure the batting performance of a player. The player's batting performance was based solely on his actual plate appearances for a period of time. We considered these plate appearances as sample data. The descriptive measures used to summarize the batting performance of a player, such as AVG, OBP, and SLG, are considered statistics (descriptive measures of a sample).

Beginning in Chapter 13, techniques will be developed that use the sample proportions (e.g., AVG, OBP, BBA) to estimate the *true* (population) proportions (e.g., *true* AVG, *true* OBP, *true* BBA). The *true* proportions measure a player's batting ability. These techniques are part of the area of statistics called *inferential statistics.* The next discussion will illustrate these concepts.

Through the 2007 season, Bonds had a career batting average of .298. Aaron had a career batting average of .305. The statistic AVG is a sample proportion that measures one type of a player's batting performance. Is Aaron's ability with respect to batting average higher than that of Bonds? This question refers to the *true* AVG of each player, which is the population proportion. A person who answers "yes" to that question believes that the seven-point difference in their sample batting averages is significant. A person who answers "no" argues that if Bonds had just 69 more hits in his 9847 career at bats, his batting average would be .305. Since chance plays a large role in whether a batted ball is a hit or an out, 69 more or less hits is not unreasonable.

This raises the following question: How large should the difference between the *sample* batting averages be for us to conclude that their *population* batting averages, as a measure of their respective batting abilities, are different? The same question can be asked about any baseball statistic that is a proportion. This includes such baseball statistics as OBP, SOA, and HRA. The answers to these questions are left for the second half of this book. We now return to our discussion of comparing the batting performances of Aaron and Bonds.

To compare the batting performance between players, we must consider the talent of the era in which the players played, the home ballpark effect, any changes in baseball rules, the improvement in the equipment used, and the development in baseball strategy over the years.

Section 3.4 presented methods for adjusting a player's batting statistics (e.g., AVG, OBP, SLG) for the time he played in the league and for the effect of his home ballpark. New ballparks are built with uniform and smaller dimensions, which aid the hitting of home runs. The dimensions of the power alleys in the new stadiums are less than in the older stadiums, which may account for less triples being hit today.

The career statistics for Bonds includes his entire 2007 season. The discrete baseball statistics used include the number of home runs (#HR), extra-base hits (#XBH), RBIs (#RBI), hits (#H), runs scored (#R), games played (#G), at bats (#AB), plate appearances (#PA), total bases (TB), base on balls (#BB), intentional base on balls (#IBB), singles (#1B), and doubles (#2B) for their individual seasons and their careers. The continuous baseball statistics used include AVG (BA), OBP (OBA), SLG, OPS, HRA, BOP, BRA, ISO, RC/27, RPR/27, R/G, SOA, BBA, IPBA, IPHR, and SO/BB (see Chaps. 3 and 9 for their definitions).

We are now ready to begin our comparison of the batting performances of Aaron and Bonds with a brief biography of the two players.

10.1. Henry Aaron and Barry Bonds

10.1.1. Henry Aaron

Henry Aaron was born on February 5, 1934, in Mobile, Alabama. Hank made his Major League debut at the age of 20. A right-hander, he was known for his quick wrists and for hitting screaming line drives. Curt Simmons once said, "Trying to sneak a pitch past Hank Aaron is like trying to sneak the sunrise past a rooster." The famous Mets pitcher Tug McGraw was asked the best way to pitch to Aaron, and he said, "The same way as to anybody else, except don't let it go."

Usually a right fielder, Hank Aaron was a consistent batter. Helped by his 6' frame and 180 pounds, he hit 30 or more home runs in a season 15 times, but he never hit more than 50 home runs. He had more than 100 RBIs in 11 seasons. He scored at least 100 runs 15 times. His best year occurred at age 37, when he hit 47 home runs, had 118 RBIs, and led the league in slugging (.669). He was the first player to have 3000 hits and 500 home runs. Each of these two accomplishments has since become a benchmark for a player's election into the Hall of Fame.

Hank's playing career lasted from 1954 to 1976. Five years later, in 1982, he was inducted into the Hall of Fame. He received 97.80% of the votes, the sixth best percentage for all players inducted into the Hall of Fame.

10.1.2. Barry Bonds

Born in Riverside, California, on July 24, 1964, Barry Bonds made his Major League debut in 1986 at the age of 21. Standing 6' 1" tall and weighing in at 185 pounds, Barry combines power and speed. In his earlier years, the left fielder weighed around 190 pounds, but in his later years his weight noticeably increased to approximately 230 pounds. His father, Bobby Bonds, was a former Major Leaguer, and his godfather was the great "Say Hey" Willie Mays. Barry's best year arguably was in 2001, at the age of 36, when the lefty hit 73 home runs, had 137 RBIs, and had an SLG of .863.

In the sections that follow the eight criteria (CR-1 through CR-8) will be applied to the seven groups (GR-1 through GR-7) to compare the batting performance of Aaron with that of Bonds.

10.2. Comparison of Aaron and Bonds using CR-1 and CR-2

Tables 10.1 and 10.2 compile for each player their career statistics that rank in the top 50 and the number of seasons in which they led the league for various statistics.

10.3. Comparison of Aaron and Bonds with Respect to CR-3

Figures 10.1–10.4 provide side-by-side frequency bar graphs for the number of times each player led the league for a season for various baseball statistics. The statistics are grouped according to different categories of batting performance.

Exercise 10.1. Provide a side-by-side frequency bar graph for the number of times each player led the league for the OPS and BRA statistics.

After completing this exercise, make your own comparison of Aaron and Bonds based on CR-3.

10.4. Comparison of Aaron and Bonds with Respect to CR-4

We now proceed to criterion CR-4, a head-to-head comparison of Aaron and Bonds with respect to their yearly statistics.

In Chapter 4, the yearly comparison of various statistics between Aaron and Bonds was done through the use of various graphs, including the side-by-side bar graph, the box-and-whisker graph, the stem-and-leaf graph, and the time-series graph. The descriptive measures used to summarize the yearly results for these statistics included the mean, the median, the first and third quartiles, the variance, the standard deviation, the coefficient of variation, and the interquartile range. The mean, median, and mode were used to compare the *typical yearly value* for each statistic. The variance, the standard deviation, the coefficient of variation, and the interquartile range were used to compare the *variability* for each statistic. The box-and-whisker graph and the stem-and-leaf graph were used to compare the *shape of the yearly values* for each statistic. The year-by-year values for a statistic (beginning with year 1 of each player's career) were compared through the use of time-series graphs. I suggest that you review Chapter 4 at this time. Some of the results presented there are discussed next. Those statistics not covered

Table 10.1 **Batting accomplishments for the career of Henry Aaron**

Feat	Position on all-time list
Career home runs (755)	2
Career extra-base hits (1477)	1
Career RBIs (2297)	1
Career hits (3771)	3
Career runs scored (2174)	4
Career games played (3298)	3
Career at bats (12364)	2
Career plate appearances (13940)	3
Career total bases (6856)	1
Career base on balls (1402)	23
Career intentional walks (293)	2
Career SO/HR (1383/755 = 1.83)	11
Seasons with at least 100 runs (15)	1
Seasons with at least 100 RBIs (11)	6
Seasons with at least 100 hits (21)	4
Seasons batting at least .300 (14)	11
Seasons with .300 AVG, 30 HR,100 RBIs (7)	7
Consecutive seasons with 30+ HR (7)	18
Career singles (2294)	11
Career doubles (624)	10
Career AVG (.305)	—
Career SLG (.555)	25
Career OBP (.374)	—
Career OPS (.929)	38
Career RC (2552)	4
Seasonal achievements	**Number of times**
All-Star Team	21
Batting title	2
MVP title	1
Led the league in home runs	4
Led the league in RBIs	4
Led the league in total bases	8
Led the league in runs scored	3
Led the league in hits	2
Led the league in doubles	4
Led the league in SLG	4
Led the league in extra-base hits	5
Led the league in HRA	3
Led the league in IBB	1
Led the league in RC	3

Table 10.2 **Batting accomplishments for the career of Barry Bonds**

Feat	Position on all-time list
Career home runs (762)	1
Career extra-base hits (1440)	2
Career RBIs (1996)	4
Career hits (2935)	31
Career runs scored (2227)	3
Career games played (2986)	10
Career at bats (9847)	28
Career plate appearances (12602)	9
Career total bases (5976)	4
Career base on balls (2558)	1
Career intentional walks (688)	1
Career SO/HR (1539/762 = 2.02)	14
Seasons with at least 100 runs (12)	4
Seasons with at least 100 RBIs (12)	4
Seasons with at least 100 hits (17)	10
Seasons batting at least .300 (11)	14
Seasons with .300 AVG, 30 HR, 100 RBIs (9)	3
Consecutive seasons with 30+ HR (13)	1
Career singles (1495)	—
Career doubles (601)	14
Career AVG (.298)	—
Career SLG (.607)	6
Career OBP (.444)	6
Career OPS (1.051)	4
Career RC (2892)	1
Seasonal achievements	**Number of times**
All-Star Team	14
Batting title	2
MVP title	7
Led the league in total bases	1
Led the league in RBIs	1
Led the league in home runs	2
Led the league in OBP	8
Led the league in walks	12
Led the league in SLG	7
Led the league in RC	9
Led the league in extra-base hits	3
Led the league in HRA	8
Led the league in IBB	12

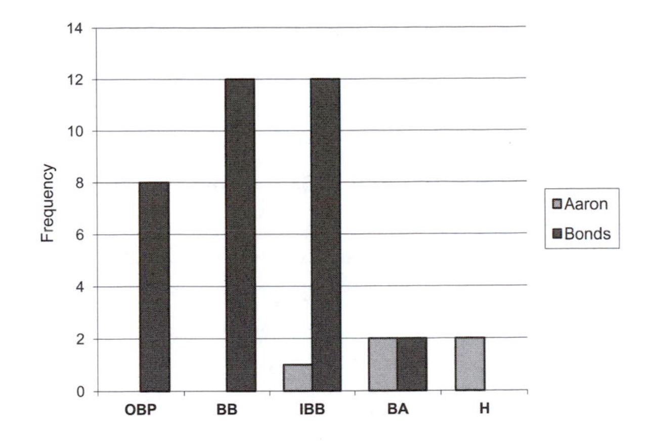

Figure 10.1 Number of times led league in group GR-1 (Aaron and Bonds)

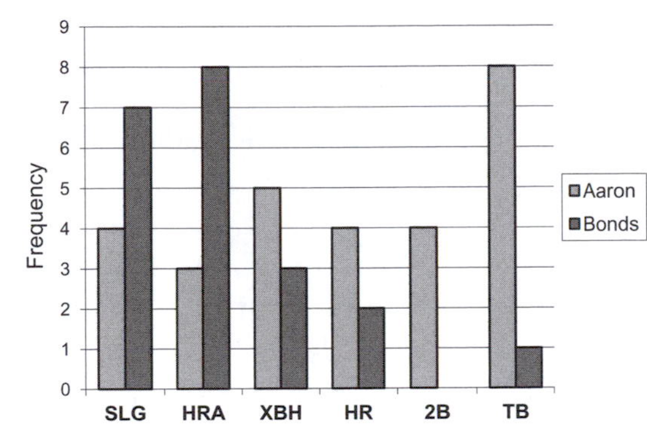

Figure 10.2 Number of times led league in group GR-2 (Aaron and Bonds)

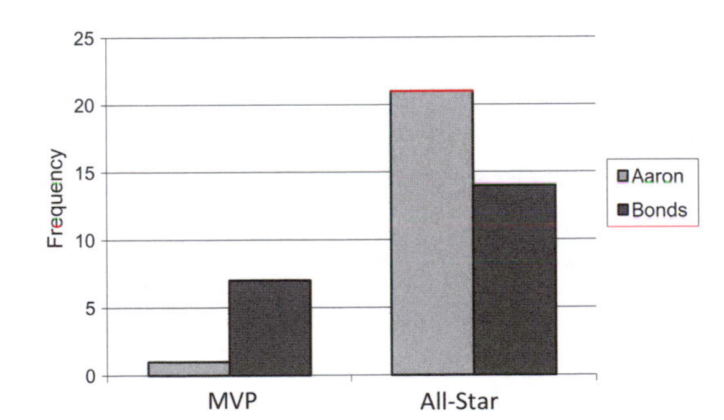

Figure 10.3 Number of times MVP Award and All-Star selection (Aaron and Bonds)

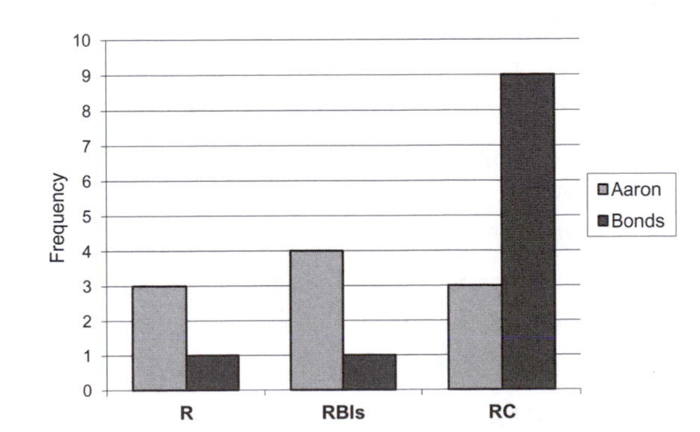

Figure 10.4 Number of times led league in group GR-5 (Aaron and Bonds)

will be left as exercises. We begin the year-by-year (seasonal) analysis with the statistic AVG.

10.4.1. AVG

From the summary statistics in Table 4.5, Aaron's mean and median yearly batting average were .301 and .307, respectively, whereas Bonds's mean and median yearly batting average were .299 and .298, respectively. The stem-and-leaf graph for AVG, in Figure 4.2, shows that Aaron had a batting average between .320 and .329 for seven seasons. Aaron was clearly more consistent with respect to batting average. The time-series graph in Figure 4.3 shows that in the early years of their respective careers Aaron's batting average was higher, in the middle years the averages were similar, and in the later years Bonds's average was higher. This trend will be repeated with many other baseball statistics.

Comparing AVG, Aaron has a slight edge over Bonds.

10.4.2. OBP

The summary statistics in Table 4.5 show that the mean and median yearly OBP for Aaron were .372 and .379, respectively, compared with .442 and .439 for Bonds. Bonds also had two extremely high outliers. The stem-and-leaf graph, in Figure 4.2, shows that Aaron's modal yearly OBP is .375, compared with Bonds's modal yearly OBP of .425. The time-series graph in Figure 4.3 shows that for the first seven years of their respective careers their OBPs are similar. From years 8 to 14, Bonds's OBP is slightly higher. From years 15 to 20, Bonds's OBP was much higher than Aaron's OBP. In fact, these later years contain his outliers. The coefficient of variation for Bonds's OBP was .1641, compared with .0705 for Aaron's OBP. This indicates twice the variability for Bonds's OBP. The stem-and-leaf graph also shows the increased variability for Bonds's yearly OBP.

Comparing OBP, Bonds has a large edge.

10.4.3. SLG

From Table 4.5 the mean and median yearly SLG for Aaron were .548 and .560, respectively, whereas the mean and median for Bonds were .615 and .612, respectively. The box-and-whisker graph in Figure 4.1 shows that the box for Bonds is shifted to the right of the box for Aaron, indicating higher results for Bonds. The time-series graph in Figure 4.3 shows that Aaron's yearly SLG is higher for the first six years. From years 6 to 14 a slight edge in yearly SLG goes to Bonds. Years 15 to 20 show a big advantage in yearly SLG for Bonds. The coefficient of variation shows that Bonds's yearly SLG is more variable than Aaron's yearly SLG.

Comparing SLG, Bonds has the edge.

10.4.4. BBA

Is there an explanation for the superiority of Bonds over Aaron for yearly OBP and yearly SLG? One possible explanation is the difference in walks (base on balls) between the two players. Remember, a walk is *not* considered an at bat. Since a walk, excluding an intentional walk, gives a player an opportunity to get an extra-base hit without increasing his total number of at bats, a walk can be considered a free chance of getting an extra-base hit that does not decrease the SLG. Since walks increase the OBP, walks have a positive effect on the OBP. Looking at Figure 4.11, we see an extreme edge in yearly BBA for Bonds. The time-series graphs for OBA, SLG, and BBA for Bonds show a similar large spike in the last third of his career.

Comparing BBA, Bonds has a large edge.

10.4.5. SO/BB

In looking at the yearly ratio of strikeouts to base on balls, a smaller number is better. From Table 4.14, the mean and median for Bonds's yearly SO/BB were .705 and .643, respectively; the mean and median for Aaron's yearly SO/BB were 1.021 and 1.050, respectively. The time-series graphs, in Figure 4.11, show that throughout their careers Bonds had more strikeouts and base on balls. Since Bonds's yearly base on balls were so much higher than Aaron's, his ratio was much better.

Comparing SO/BB, Bonds has the edge.

10.4.6. RC/27

This measures the runs a player produces for his team under the assumption that the player bats in all nine positions in the batting order. Table 4.10 shows that the mean and median yearly RC/27 for Bonds were 10.29 and 9.72, respectively, compared with Aaron's mean and median of 7.59 and 7.86, respectively. The time-series graph in Figure 4.7 follows the same pattern as the time-series graphs for the statistics OBA, SLG, and BBA. The box-and-whisker graph in Figure 4.5 shows Bonds's box shifted to the right of Aaron's box. The stem-and-leaf graph in Figure 4.6 shows more variability for Bonds. Table 4.10 shows that the coefficient of variation for Bonds is close to twice that of Aaron.

Comparing RC/27, Bonds has the edge.

10.4.7. Discussion of Comparison of Aaron and Bonds with Respect to CR-4

We have not looked at every yearly statistic. But those we did look at give the clear overall edge to Bonds.

Table 10.3 **Coefficient of variation comparison of statistics between Aaron and Bonds**

Player	AVG	OBP	SLG	SOA	BBA	IPHR	SO/BB	RC/27	RPR/27
Aaron	0.1035	0.0705	0.1433	0.2286	0.2686	0.3304	0.2733	0.2256	0.1872
Bonds	0.1187	0.1640	0.1925	0.1853	0.3429	0.3802	0.5201	0.3948	0.2324

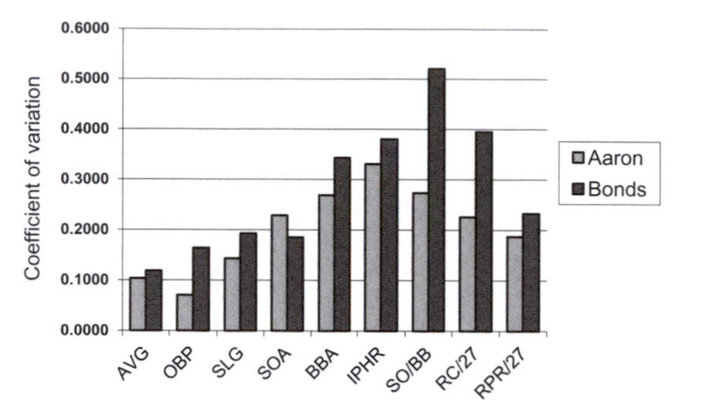

Figure 10.5 Comparison of consistency for baseball statistics (Aaron and Bonds)

Aaron was known for his consistency as a hitter. With this in mind, an analysis comparing the consistency of the statistics on a seasonal basis seems necessary. Table 10.3 and Figure 10.5 compare the coefficient of variation for the year-by-year results for various batting statistics. The smaller the coefficient of variation, the less variability there is. For every statistic except SOA, Aaron is more consistent than Bonds.

Comparing consistency, Aaron has the edge.

Exercise 10.2. Use CR-4 to compare Aaron and Bonds for the statistics OPS, BRA, RPR/27, HRA, BOP, IPBA, TA, and ISO. Your analysis should be based on the techniques introduced in Chapter 4.

Exercise 10.3. What aspect of batting performance does each of the statistics in Exercise 10.2 relate to?

In 1969, when Aaron was 35, two important rule changes occurred. The pitching mound was lowered 5 inches and the strike zone was shrunken to the area from the armpits to the top of the knees. When we look at the batting performance of Aaron between the ages of 35 and 39, we will see how these rule changes affected his batting results. Besides rule changes, the improvement of the equipment used by batters affected their performance. The construction of the baseball bat has improved from decade to decade. The baseball itself had been wound tighter in recent years. Bonds had the advantage of the 1969 rule changes throughout his career. Also, Bonds had the advantage of certain nutritional programs and supplements not available during Aaron's career.

10.5. Comparison of Aaron and Bonds with Respect to CR-5, CR-6, and CR-7

We now turn to criteria CR-5, CR-6, and CR-7 to compare Aaron and Bonds, head to head, with respect to their career performance, their performance between the ages of 35 and 39, and their best year in the majors. It turns out that Aaron and Bonds both had their best year in the majors at approximately the same age. In 1971, Aaron was 37, and in 2001, Bonds was 36. Table 10.4 contains the necessary statistics for each of these criteria. Figures 10.6–10.11 provide side-by-side bar graphs. These bar graphs are based on the groups GR-1, GR-2, GR-3, GR-5, GR-6, and GR-7 and time periods T-2, T-3, and T-4.

Figures 10.6–10.11 show the overall superiority of Bonds to Aaron with respect to all the groups for the time periods T-2, T-3, and T-4. The only statistic that appears to be close is BA. One conclusion is that the unbelievable performance of Bonds between the ages of 35 and 39 probably skewed all the other results in favor of Bonds. This same unbelievable performance for Bonds was shown at the beginning of this chapter when

Table 10.4 **Comparison of statistics between Aaron and Bonds for (career, ages 35–39, and their best years)**

	Career		35–39		Best year	
Statistic	Aaron	Bonds	Aaron	Bonds	Aaron (1971)	Bonds (2001)
BA	0.305	0.298	0.299	0.339	0.327	0.328
OBP	0.374	0.444	0.396	0.535	0.410	0.515
SLG	0.555	0.607	0.601	0.781	0.669	0.863
OPS	0.929	1.051	0.997	1.316	1.079	1.378
TB	6856	5906	1442	1658	331	411
SO	1383	1523	274	316	58	93
RC	2576.35	2615.21	572.98	881.61	136.26	209.59
HRA	0.061	0.077	0.085	0.122	0.095	0.153
BOP	0.596	0.680	0.653	0.843	0.705	0.899
BRA	0.208	0.269	0.238	0.418	0.274	0.444
ISO	0.250	0.307	0.302	0.442	0.342	0.535
RC/27	7.65	9.65	8.69	16.03	10.43	16.70
RPR/27	11.03	12.77	11.03	16.39	12.71	15.38
SOA	0.112	0.155	0.114	0.149	0.117	0.195
BBA	0.113	0.260	0.163	0.411	0.143	0.372
SO/BB	0.986	0.595	0.699	0.362	0.817	0.525
IPHR	0.069	0.092	0.096	0.143	0.108	0.191
IPBA	0.343	0.353	0.337	0.399	0.371	0.407

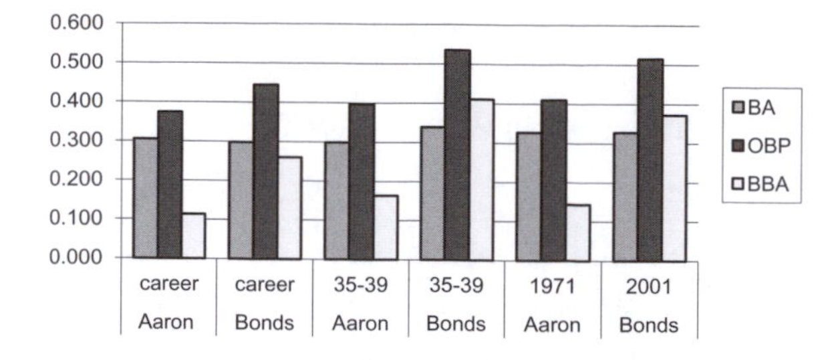

Figure 10.6 Comparison of some group 1 statistics for career, ages 35–39, and best year for Aaron and Bonds

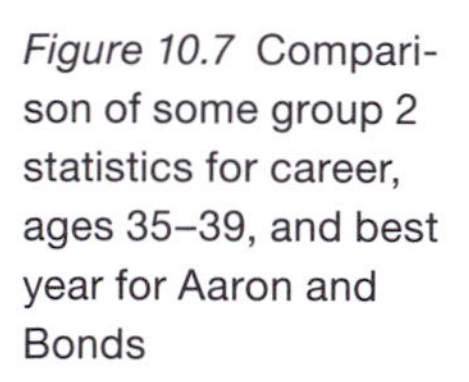

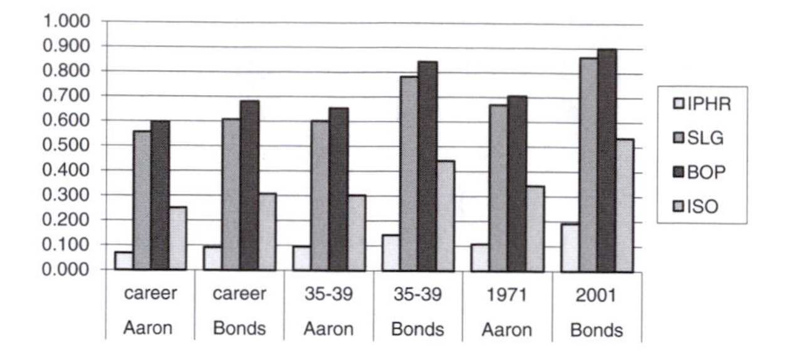

Figure 10.7 Comparison of some group 2 statistics for career, ages 35–39, and best year for Aaron and Bonds

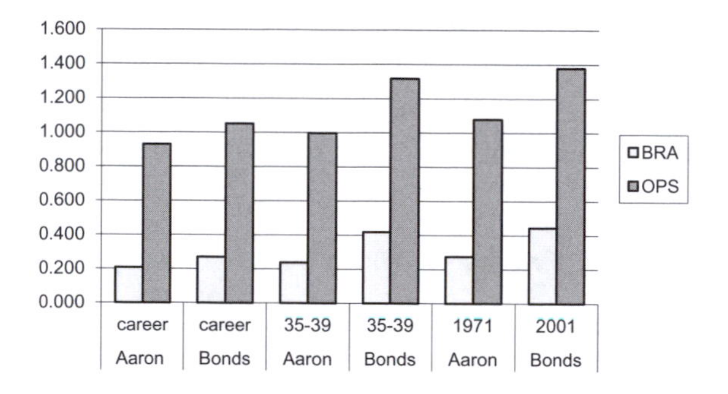

Figure 10.8 Comparison of some group 3 statistics for career, ages 35–39, and best year for Aaron and Bonds

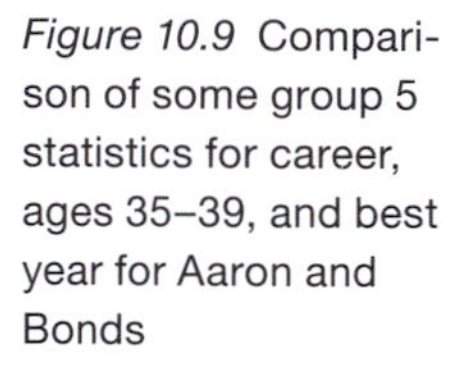

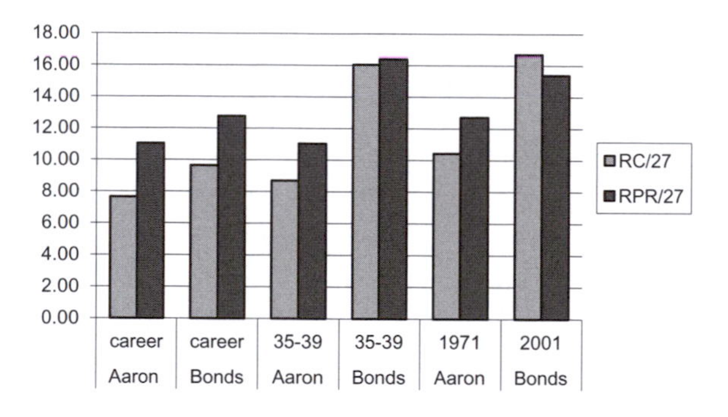

Figure 10.9 Comparison of some group 5 statistics for career, ages 35–39, and best year for Aaron and Bonds

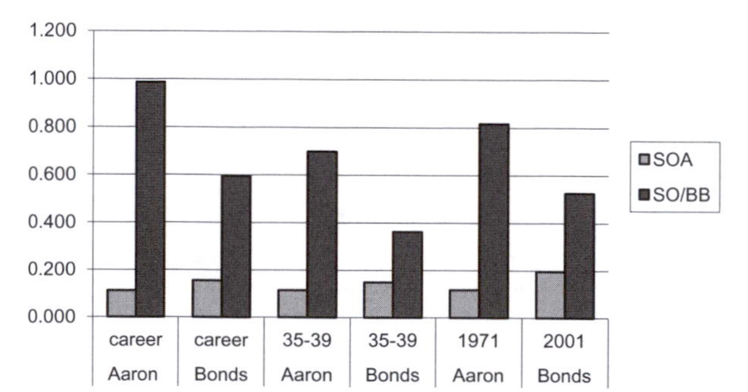

Figure 10.10 Comparison of some group 6 statistics for career, ages 35–39, and best year for Aaron and Bonds

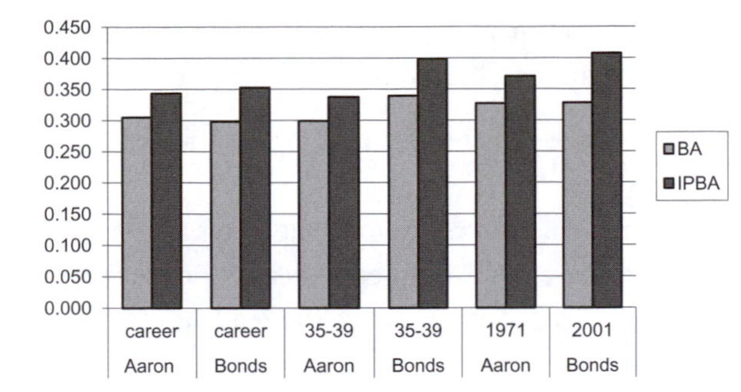

Figure 10.11 Comparison of some group 7 statistics for career, ages 35-39, and best year for Aaron and Bonds

we looked at the criterion CR-4, the yearly batting statistics of Bonds and Aaron.

Exercise 10.4. Use criteria CR-5, CR-6, and CR-7 to compare Aaron and Bonds with respect to the statistic TA = [(TB + #HBP + #BB + #SB) − #CS]/[(#AB − #H) + #CS + #GIDP]. Remember from Chapter 9 that TA stands for total average.

10.6. Comparison of Aaron and Bonds with Respect to CR-8

For our final comparison between Aaron and Bonds, we look at CR-8, the type of hitter each one was. For this comparison, we go to Chapter 6 and use Table 6.2 and Figures 6.3, 6.4, and 6.5. Based on their total career plate appearances, we will highlight those outcomes that showed a major difference. For Aaron, 16% of his plate appearances resulted in singles. For Bonds, the number was 12%. A slight edge in the percent of home runs favors Bonds. For Aaron, 10% of his plate appearances resulted in walks, but for Bonds, it was 20%. This base-on-balls percentage difference between Aaron and Bonds helps to account for their percentage difference of outs. For Aaron, 62% of his plate appearances resulted in outs, but for Bonds, it was only 55%. The results for the other outcomes of a plate appearance appear similar.

We conclude the chapter by looking at four tests used to evaluate a baseball player's career. The Black-Ink and Gray-Ink Tests look at the yearly accomplishments of a player. The Hall of Fame (Career) Standards Test is used to score a player based on certain statistics for his career. The last test discussed is the Similarity Test. This test is used to see which players are similar in batting performance.

10.7. The Black-Ink Test

This test was developed by Bill James, and its name "Black-Ink" comes from the fact that league-leading numbers are set in boldface type. This test can be used to compare players, and it rewards a player for leading his league in an important baseball statistic. The negatives are as follows:

- Players who do not lead the league for a statistic but are in the top 10 receive no credit.
- As modern players are in a league with 14 or 16 teams, compared to days when there were only eight teams per league, it is more difficult for a current player to have a league-leading statistic.

The breakdown of the test is as follows:

- Four points each are awarded for leading the league in home runs, runs batted in, and batting average.
- Three points each are awarded for leading the league in runs scored, hits, and slugging percentage.
- Two points each are awarded for leading the league in doubles, walks, and stolen bases.
- One point each is awarded for leading the league in games, at bats, and triples.

James also had a Gray-Ink Test, which awards the same point values as the Black-Ink Test to any player in the top 10 for that baseball statistic.

Exercise 10.5. Complete Table 10.5. The data can be found at www.baseball-reference.com.

Table 10.5 Black-Ink Test table for Aaron and Bonds (Exercise 10.5)

		Number of times led league		Points*player	
Categories	Points	Aaron	Bonds	Aaron	Bonds
AVG	4	2	2	8	8
RBI	4				
HR	4				
R	3	3	1	9	3
SLG	3				
H	3		0		
2B	2	4	0		0
BB	2				
Stolen bases	2	0	0		0
G	1	1	1		1
AB	1	0	0		0
3B	1	0	0		0
Totals					

Exercise 10.6. Using the "Black-Ink Test" with the table in Exercise 10.5, find the totals for Bonds and Aaron. What does this test say about the comparison between Bonds and Aaron?

Exercise 10.7. Use the "Black-Ink Test" to compare Joe DiMaggio and Ted Williams with Bonds and Aaron.

Exercise 10.8. Use the "Black-Ink Test" to compare your two players. Create a table similar to Table 10.5 for your two players.

Exercise 10.9. Repeat Exercises 10.5–10.8 for the Gray-Ink Test. Replace the column "Number of Times Led League" in Table 10.5 with the column "Number of Times in the Top 10."

Exercise 10.10. Create your own test by choosing your own baseball statistic categories and their associated point values. Use your test to compare your chosen two players. What conclusions can you draw from your comparison?

10.8. The Hall of Fame Career Standards Test

The test score for an average Hall of Famer is 50 points. Points awarded to a player, based on their career statistics, are as follows:

- One point for each 150 hits above 1500 (limit 10).
- One point for each .005 of batting average above .275 (limit 9).
- One point for batting over .300.
- One point for each 100 runs over 900 (limit 8).
- One point for scoring more than .500 runs per game.
- One point for scoring more than .644 runs per game.
- One point for each 100 RBIs over 800 (limit 8).
- One point for driving in more than .500 runs per game.
- One point for driving in more than .600 runs per game.
- One point for each .025 of slugging percentage above .300 (limit 10).
- One point for each .010 of on-base percentage above .300 (limit 10).
- One point for each 200 home runs.
- One point if home runs are more than 10% of hits.
- One point if home runs are more than 20% of hits.
- One point for each 200 extra-base hits over 300 (limit 5).
- One point for each 200 walks over 300 (limit 5).
- One point for each 100 stolen bases (limit 5).
- Defensive value: 20 points for catcher, 16—shortstop, 14—second base, 13—third base, 12—center fielder, 6—right fielder, 3—left fielder, 1—first baseman, 0—designated hitter.

Exercise 10.11. Use the Hall of Fame Career Standards Test to calculate a test score for Bonds, Aaron, DiMaggio, and Williams. What do the results indicate?

Exercise 10.12. Use the Hall of Fame Career Standards Test to compare your two chosen players. What do the results indicate?

Players who score high numbers in the Black-Ink Test, Gray-Ink Test, and the Hall of Fame Standards Test exhibit superior performance over a long period of time. Such players include Babe Ruth, Ty Cobb, and Lou Gehrig.

10.9. The Similarity Test

The final exercises in this chapter will be based on a fourth test called the "Similarity Test." This test was also introduced by Bill James to compare one player with another. This test starts at 1000 points. Points are then subtracted based on the statistical differences between the two players.

- One point for each difference of 20 games played.
- One point for each difference of 75 at bats.
- One point for each difference of 10 runs scored.
- One point for each difference of 15 hits.
- One point for each difference of 5 doubles.
- One point for each difference of 4 triples.
- One point for each difference of 2 home runs.
- One point for each difference of 10 RBI.
- One point for each difference of 25 walks.
- One point for each difference of 150 strikeouts.
- One point for each difference of 20 stolen bases.
- One point for each difference of .001 in batting average.
- One point for each difference of .002 in slugging percentage.

Finally, there is a positional adjustment. Each position has a value, and you subtract the difference between the two players' positions. If a player had more than one position, we will choose his primary position.

Table 10.6 **Similarity Test for Aaron and Bonds**

Player	G	AB	R	H	2B	3B	HR	RBI
Aaron (RF)	3298	12364	2174	3771	624	98	755	2287
Bonds (LF)	2986	9847	2227	2935	601	77	762	1996
Diff	312	2517	–53	836	23	21	–7	291
Diff(abs value)	312	2517	53	836	23	21	7	291
Points	15.6	33.6	5.3	55.7		5.3	3.5	29.1

- 240—Catcher
- 168—Shortstop
- 132—Second Base
- 84—Third Base
- 60—Center field
- 48—Right field
- 36—Left field
- 12—First Base
- 0—DH

The breakdown of the meaning of the final score is as follows:

950+ Unusually Similar
900+ Truly Similar
850+ Essentially Similar
800+ Somewhat Similar
700+ Vaguely Similar
<700 Not Similar

Exercise 10.13. Use the Similarity Test to compare Bonds and Aaron. Fill in the empty cells in Table 10.6. How would you classify their similarity?

Exercise 10.14. Use the Similarity Test to compare the similarity between your two chosen players. How would you classify their similarity?

Chapter Summary

How do the statistics of Aaron and Bonds compare with the all-time records in the history of baseball?

Both players rank in the top 5 for most career home runs, extra-base hits, RBIs, runs scored, total bases, and intentional walks; most seasons with at least 100 runs scored; and highest career RC.

In addition, Aaron ranked in the top 10 for most career hits (third), most seasons with at least 100 RBIs (sixth), most seasons with [.300 AVG, 30 HRs, 100 RBIs] (seventh), and most career doubles (tenth). Bonds ranked in

BB	SO	SB	BA	SLG	Pos	X	X
1402	1383	240	0.305	0.555	48	X	X
2558	1539	514	0.298	0.607	36	X	X
–1156	–156	–274	0.007	–0.052	12	Total	1000-
1156	156	274	0.007	0.052	12	Points	Total
46.2		13.7	7.000		12.0		

the top 10 for most seasons with at least 100 RBIs (fourth), most seasons with [.300 AVG, 30 HRs, 100 RBIs] (third), most career walks (first), most consecutive seasons with 30+ HRs (first), highest career SLG (sixth), highest career OBP (sixth), and highest career OPS (fourth).

Both Aaron and Bonds rank high for many important batting statistics. One major difference is with regard to SLG and OBP. Bonds has a rank of 6 in career SLG and 6 in career OBP; Aaron has a rank of 25 in career SLG and is not in the top 50 in career OBP.

For criterion CR-1, I give a slight edge to Bonds.

Turning to the two criteria CR-2 and CR-3, Aaron led the league more times than Bonds in hits, total bases, doubles, home runs, extra-base hits, RBIs, runs scored, and All-Star selections. In contrast, Bonds led the league more times than Aaron for SLG, HRA, OBP, BB, IBB, RC, and MVP selections.

What surprises me is that Aaron led the league in HRs, RBIs, and runs scored more times than Bonds. Aaron's advantage is 4 to 2 for home runs, 4 to 1 for RBIs, and 3 to 1 for runs scored. These results may indicate that within their respective eras, Aaron's batting performance was superior to Bonds. However, it should be pointed out that the National League, as a result of expansion, had more teams in Bonds's era than in Aaron's era.

For the two criteria CR-2 and CR-3, I give a very slight advantage to Aaron.

Within this chapter, a clear edge for criteria CR-4, CR-5, CR-6, and CR-7 was given to Bonds.

As mentioned before, external factors can affect batting performance. Both Aaron and Bonds displayed the unusual result of having their batting statistics increase dramatically during the time when they were between the ages of 35 and 39. Is there an explanation for this?

In the case of Aaron, he turned 35 in 1969. In 1969, the two rule changes of lowering the mound and tightening the strike zone were adapted. These two rule changes benefited a batter enormously.

In the case of Bonds, he turned 35 in 2000. It is alleged that Bonds took a combination of steroids and human growth enhancement drugs during the period of time between 2000 and 2004.

Finally, looking at criterion CR-8, the most striking difference was in the percentage of base on balls. About 20% of Bonds's plate appearances resulted in a walk, whereas about 10% of Aaron's plate appearances resulted in a walk.

The analysis of batting performance was based on the following seven groups: getting on base (GR-1), hitting with power (GR-2), getting on base combined with slugging (GR-3), special league awards (GR-4), creating runs for their team (GR-5), having a good batting eye (GR-6), and directing a batted ball (GR-7).

These seven groups were analyzed by using the eight criteria CR-1 through CR-8, mentioned at the beginning of the chapter.

These groups and criteria are not set in stone. You are invited to suggest your own groups and criteria. For example, you might want to choose, as a time period, a player's age from 27 to 33, which is believed by many baseball people to represent the peak time in a player's career. Another choice is a player's best 5 or 10 years in their careers. Other groups, such as the three groups described in Chapter 4, may be used. Another option for a group might be a collection of statistics chosen from a family of existing groups.

Some conclusions that can be drawn from this analysis are as follows:

- Bonds's much higher base-on-balls average might indicate that he was a more feared hitter than Aaron. It could also mean that Bonds's team did not have a strong hitter to protect him in the batting order. Finally, it could indicate that Bonds had a better batting eye.
- Since the majority of pitchers are right-handed, Bonds, being a lefty hitter, had an advantage over Aaron, a right-handed hitter.
- Since Bonds had a lower ratio of strikeouts to base on balls, it might be concluded that Bonds had a better batter's eye. It might also be the case that the result was influenced by the amount of intentional base on balls received by Bonds.
- Bonds had the advantage of the hitter-friendly rule changes implemented in 1969 for his entire career, while Aaron only had this advantage from age 35 on.
- Bonds hit for more power than Aaron.
- Aaron was a more consistent line-drive hitter.
- Bonds's best power numbers occurred between the ages of 35 and 39.
- The ball was livelier and the ballparks smaller in Bonds's era.
- Aaron was a better contact hitter.

Which player would I consider the better hitter? If the era they played in, the ballpark effect, and external factors were ignored, from the analysis done in Chapters 1–10, I would say Barry Bonds.

However, an argument can be made that both players as hitters are in the top 10 of all hitters in the history of baseball. This leads to another question. Who are the top 10 hitters of all time? A list of possible candidates besides Aaron and Bonds would include Babe Ruth, Ty Cobb, Joe Jackson, Honus Wagner, Rogers Hornsby, George Sisler, Paul Molitor, Joe DiMaggio, Lou Gehrig, Tony Gwynn, Ted Williams, Pete Rose, Robin Yount, Mickey Mantle, Wade Boggs, Alex Rodriguez, Albert Pujols, Stan Musial, Willie

Mays, and Rod Carew. Of course, this is only a partial list. This question will be revisited in Chapter 18. For those students interested in investigating this question now, I can recommend two books. The more technical is *Baseball's All-Time Best Hitters*, by Michael Schell; the less technical is *Who Is Baseball's Greatest Hitter?*, by Jeff Kisseloff.

The chapter concluded by using the four tests established by Bill James for evaluating a player's batting performance. These tests are the *Black-Ink Test*, the *Gray-Ink Test*, the *Hall of Fame Career Standards Test*, and the *Similarity Test*. In the exercises that followed, the student was asked to evaluate these tests for Aaron and Bonds and draw their own conclusions.

Remember, a positional baseball player is evaluated on five tools: hitting, hitting with power, fielding, running, and throwing. We have not looked at fielding, throwing, and running; we only looked at batting and batting with power. Their *batting performance* was evaluated from the results of their actual plate appearances. These plate appearances were considered sample data. Since our analysis was based on sample results, the conclusions drawn are based on opinion only.

Later in this book, we will compare Aaron and Bonds in terms of their *batting ability*. Theoretically, if Aaron and Bonds batted over and over again, what conclusions can we make about their *true* BA, *true* OBA, and other *true* offensive measurements?

CHAPTER PROBLEMS

1. Use the Black-Ink Test and Gray-Ink Test to compare the great post-2000 hitters: Alex Rodriguez, Manny Ramirez, Ichiro Suzuki, and Albert Pujols.
2. Create a new group by choosing one statistic from each of the groups GR-1, GR-2, GR-3, GR-5, GR-6, and GR-7. Use the seven criteria (CR-1 through CR-7) applied to your new group to compare Ted Williams with Joe DiMaggio.
3. Perform an analysis of batting performance for your two chosen players based on your choices in problem 2.
4. You are now ready to answer the question posed at the beginning of the book about your two chosen players. Should the potential Hall of Fame player be admitted into the sacred "Hall of Fame"? Base your argument on the descriptive statistics presented in the first 10 chapters.

Chapter 11

Probability Distribution Functions for a Discrete Random Variable

In Chapter 7, we explored the topic of probability. We looked at classical and relative frequency probability. A probability is a number between zero and one that measures the likelihood of a particular outcome occurring. In this chapter, we explore *probability distribution functions* for a *discrete random variable*. These distributions are based on outcomes that result from a probability experiment. A *probability experiment* is any process whose outcomes occur by chance. Theoretically, a probability experiment can be repeated an infinite number of times.

Example 11.1. The probability experiment is the process of tossing a coin. The set of all possible outcomes is the sample space. The sample space equals {H, T}. If we performed this experiment three times, the sample space is equal to {HHH, HTH, THH, HHT, TTH, THT, HTT, TTT}. The outcomes are the elements in the sample space.

Even though the concept of a *discrete random variable* was introduced in Chapter 1, we review the concept briefly here. A *variable* is a characteristic that assigns a value to each possible outcome of an experiment or each member of a population. These values, which can be the actual outcomes themselves, are either numbers or categories. If the values of the variable are numbers, not treated as categories, the variable is quantitative; categorical values come from qualitative variables. If a positive integer can be assigned to the number of possible values of the variable, we say that the variable is finite. If one cannot assign a positive integer to describe the number of possible values, we say that the variable is infinite. The term *random* means that the outcomes occur by *chance*. The term *discrete* means that the total number of possible values for the variable is a known finite number or, if infinite, a positive integer can be assigned to each value of the variable. We say that a discrete variable has a countable number of values. If the variable X is the number of career plate appearances for every player who played in the Major Leagues in the past, present, and future, it would be infinite and discrete. If the variable X is the distance of every home run hit in the Major Leagues in the past, present, and future, it would be infinite and continuous (not discrete). Remember from Chapter 1 that variables that are counted are discrete, whereas variables that are measured are continuous.

We assign a capital letter (X, Y, etc.) to name the variable; lowercase letters (x, y, etc.) represent the values of the variable. We present the following definition for a discrete probability distribution:

> A *probability distribution function* (*pdf*) for a *discrete variable* X is a function that sets up a correspondence between each distinct value of a discrete variable and its associated probability. The set of all distinct values of the discrete variable is the domain of the function, and the range or y-values are the corresponding probabilities. We express these y-values by the notation $y = \Pr(X = x)$.

A pdf for a discrete variable must satisfy the following two conditions:

1. Each y-value of the discrete random variable must be between 0 and 1 inclusive.
2. The sum of all the y-values must be equal to 1.

A pdf for a discrete variable X can be constructed independently of any physical model. Any function that satisfies the two conditions above is considered a pdf for X.

A pdf for a discrete variable can be represented (1) by a table, (2) by a graph in the form of a histogram, or (3) in a functional notation form ($f(x)$). The next example illustrates all three of these representations.

Example 11.2. Let the discrete random variable $X = \{0, 1\}$. We assign $x = 0$ to $y = 1/2$ and assign $x = 1$ to $y = 1/2$. This assignment is a pdf because each y-value is a probability and the sum of the probabilities is equal to 1.

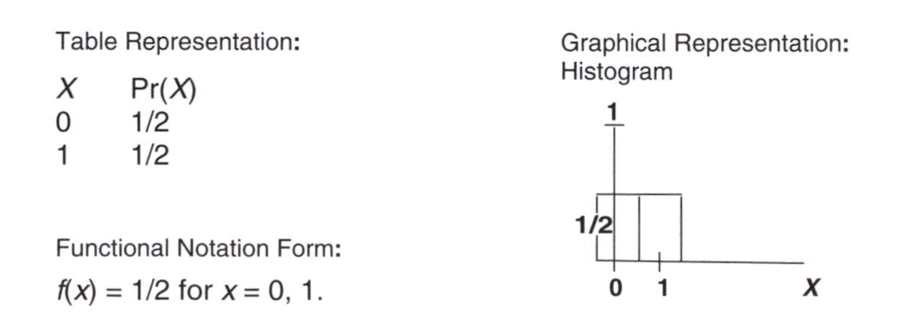

Table 11.1 **Career plate appearance outcomes for Aaron and Bonds**

Player	1B	2B	3B	HR	BB	SF	HBP	Out	PA total
Aaron	2294	624	98	755	1402	121	32	8593	13919
Bonds	1477	597	77	751	2520	90	105	6824	12441
Totals	3771	1221	175	1506	3922	211	137	15417	26360

Note: The plate appearances for Bonds include up to the first 81 games of 2007.

By converting frequencies to probabilities, any *frequency* distribution can be converted into a *probability* distribution function. The probability distribution has the same *x*-values as the frequency distribution, but the *y*-values are now probabilities. The next four examples use the data included in Table 11.1 to demonstrate this. This is the same table used in Chapter 7 to introduce basic probability.

Example 11.3. The experiment is a career plate appearance for either Aaron or Bonds. The experiment is repeated 26360 times. The sample space consists of the 26360 outcomes. Let the variable *X* be the outcome of a given plate appearance. The values of the discrete variable *X* are represented by *x* and consist of {1B, 2B, 3B, HR, BB, SF, HBP, and O}. We correspond to each *x* its relative frequency probability. Table 11.2 gives a representation for the pdf for *X*.

Table 11.2 **Probability distribution function for *X* = outcomes for career plate appearances for Aaron and Bonds**

X	Frequency	Probability
1B	3771	.1430
2B	1221	.0463
3B	175	.0066
HR	1506	.0571
BB	3922	.1488
SF	211	.0080
HBP	137	.0051
O	15417	.5849
Total PAs	26360	.9998

The sum of all the frequencies must equal the total number of plate appearances, and the sum of all the probabilities must equal 1.00. In the above example, the sum of all the probabilities is equal to .9998. This is due to a round-off error. If we represented each probability as a fraction (Pr(1B) = 3771/26360), the sum *would* equal 1.00. The variable *X* in this example is discrete, since there are only a finite number of values for the variable.

Example 11.4. The experiment is the same as in Example 11.3, but this time the sample space consists of those plate appearances that are considered an at bat. The sample space results in 22090 at bats. Table 11.3 represents the pdf for the discrete variable *X* (the outcome of a given at bat; BB, HBP, and SF are excluded).

Table 11.3 **Probability distribution function for *X* = outcomes of the combined at bats for Aaron and Bonds**

X	Frequency	Probability
1B	3771	.1707
2B	1221	.0553
3B	175	.0079
HR	1506	.0682
O	15417	.6979
Total ABs	22090	1.0000

Table 11.4 **Probability distribution function for X = (1) getting on base and (0) not getting on base for the career plate appearances for Aaron and Bonds**

X	Frequency	Probability
1	10732	.4071
0	15628	.5929
Total PAs	26360	1.0000

Example 11.5. The sample space is the same as in Example 11.3. Table 11.4 represents the pdf for the discrete variable X, where X is {0, 1}. The number "1" is assigned to the outcomes 1B, 2B, 3B, HR, BB, and HBP (getting on base), and the number "0" is assigned to the outcomes O and SF (not getting on base).

Example 11.6. The sample space is the same as in Example 11.4. Table 11.5 represents the pdf for the discrete variable X = the number of total bases from an at bat. An out results in 0 bases, a 1B results in 1 base, a 2B results in 2 bases, and so on.

Table 11.5 **Probability distribution function for X = total bases for the career at bats of Aaron and Bonds**

X	Frequency	Probability
0	15417	.6979
1	3771	.1707
2	1221	.0553
3	175	.0079
4	1506	.0682
Total ABs	22090	1.0000

Any frequency distribution can be changed into a discrete probability distribution. The probability distribution functions shown above were obtained by reducing a data set to a frequency table and then converting the frequencies into probabilities. Probability distributions obtained from empirical data are said to be *data dependent*. Clearly, if different players were selected, different probability distributions would occur. For our next example we look at Example 8.3.

Table 11.6 **Probability distribution function for X = {winning amount ($75), amount at risk (–$100)}**

X	Probability
75	5/9
–100	4/9

Example 11.7. This experiment involves betting on a baseball game between the Yankees and the Mets. The probability assigned to the Yankees winning is 5/9. A bet is made that the Yankees will win. The amount at risk is $100 and the amount to be won is $75.

Let the discrete variable X = {winning amount ($75), amount at risk (–$100)}. The pdf for the variable X is represented by Table 11.6.

The next three examples of pdfs come from classical probability.

Example 11.8. The experiment is rolling a die one time. The discrete variable X consists of the set of possible outcomes {1, 2, 3, 4, 5, 6}. Table 11.7 shows the pdf for X in table form. Following this table, the pdf is represented by a histogram and in functional notation form.

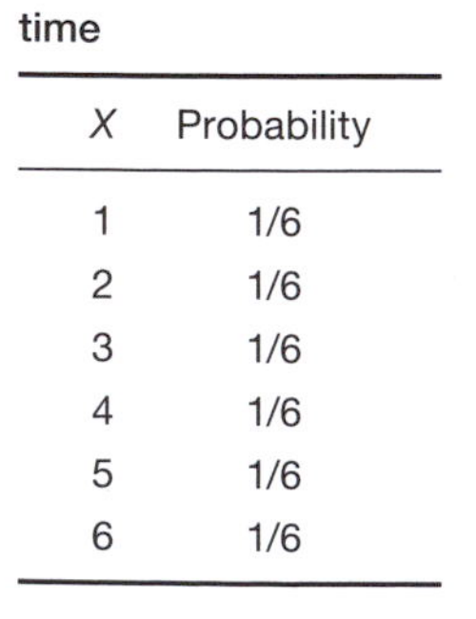

Table 11.7 **Probability distribution function for X = outcomes of rolling one die one time**

X	Probability
1	1/6
2	1/6
3	1/6
4	1/6
5	1/6
6	1/6

Histogram

Pr(X)
1/6
1 2 3 4 5 6 X

Functional Notation Form

$f(X) = 1/6$ for X = 1, 2, 3, 4, 5, 6

Example 11.9. The experiment is tossing one coin two times. The discrete variable X is equal to the number of heads. Since the possible outcomes are {HH, HT, TH, or TT}, the variable X has the values 0, 1, and 2. Table 11.8 and the histogram below both represent the pdf for X. Later in this chapter, we shall see that this pdf is an example of a theoretical discrete probability distribution called the *binomial distribution*. Because this pdf fits the binomial model, we will see later that there is a formula for the pdf.

Table 11.8 **Probability distribution function for X=number of heads resulting from tossing one coin two times**

X	Probability
0	1/4
1	1/2
2	1/4

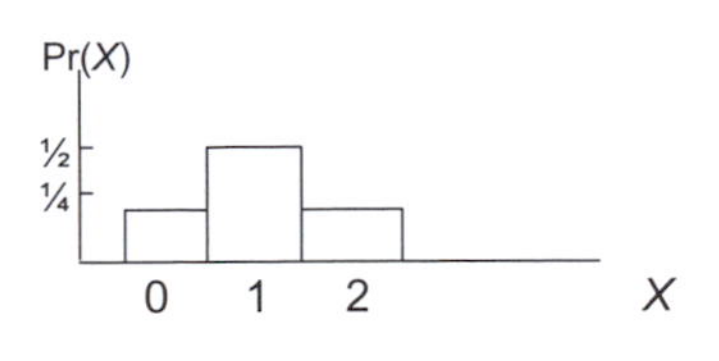

For our final example, we look at a pdf for a discrete variable X with an infinite number of possible values.

Example 11.10. The experiment is rolling a die repeatedly until the first "6" occurs. The discrete variable X is the roll number corresponding to the first occurrence of a "6." Since the rolls of a die are independent of each other, the $Pr(X=1)=1/6$, the $Pr(X=2)=(5/6)*(1/6)$ (the 5/6 comes from the fact that the first roll results in not getting a 6), the $Pr(X=3)=(5/6)*(5/6)*(1/6)$ (the first two rolls result in not getting a 6), and so on. The corresponding pdf for X is shown in Table 11.9.

Table 11.9 **Probability distribution function for X=trial number for the first 6 occurring after repeated rolls of one die**

X	Probability
1	1/6
2	5/36
3	25/216
↓	↓

Later in this chapter, we will see that this pdf is an example of a theoretical discrete probability distribution called a *geometric distribution*.

In Examples 11.3 and 11.4, the values of the discrete variable X were the actual possible outcomes.

In Examples 11.5 and 11.6, each possible outcome is assigned an integer.

Exercise 11.1. Construct a contingency table, similar to Table 11.1, for your two players.

Exercise 11.2. Construct a discrete probability distribution, similar to the one constructed in Example 11.3, for your two players.

Exercise 11.3. The sample space consists of all 56 games in Joe DiMaggio's 56-game hitting streak (see Table 1.1). Let X=the number of hits in each game. Construct the pdf for the discrete variable X.

Exercise 11.4. The sample space consists of all 56 games in Joe DiMaggio's 56-game hitting streak. Let X=the number of at bats in each game. Construct the pdf for X.

Exercise 11.5. The sample space consists of all 56 games in Joe DiMaggio's 56-game hitting streak. Let X = the number of home runs in each game. Construct the pdf for X.

Exercise 11.6. The sample space consists of all home runs hit by Barry Bonds in 2001. Let X = the inning number of a home run. Construct the pdf for X (See Table 1.2).

Exercise 11.7. The sample space consists of all home runs hit by Barry Bonds in 2001. Let X = the position of the home run. Construct the pdf for X (see Table 1.2).

Exercise 11.8. The experiment consists of repeatedly tossing a coin until the first "head" appears. Let X = the number of the toss that resulted in the first "head." Construct the pdf for the discrete variable X. Are the possible values of X finite or infinite? Which previous example does this experiment resemble?

Exercise 11.9. Bo Jackson had a lifetime batting average equal to .250. A new season begins. Let X = the number of the first at bat in which Bo gets his first hit for the season. Construct the pdf for X. What assumption must be made about the at bats? In the history of baseball, what do you think the largest possible value of X, for any Major League player, has been? See if you can use the Internet to find the answer.

11.1. The Mean and Standard Deviation for a Discrete Variable with a Discrete Probability Distribution

In Chapter 8, we defined the *expected gain* $E(G)$ on a bet as follows:

$$E(G) = W * \Pr(E) - L * \Pr(\text{not } E).$$

In the above equation W = the amount won if E occurs, L = the amount at risk (lost) if (not E) occurs, E = the team wins, and (not E) = the team loses.

From Example 11.6, we have E = Yankees win, (not E) = Yankees lose, W = \$75, L = –\$100, $\Pr(E) = 5/9$, and $\Pr(\text{not } E) = 4/9$. For this information, the expected gain on betting that the Yankees win is

$$E(G) = 75 * 5/9 - 100 * 4/9 = -25/9 = -2.77.$$

$E(G)$ represents the "theoretical average gain" of the experiment per bet. If the bet was made nine times and the theoretical probabilities were correct, we would expect the Yankees to win five times and lose four times. The mean for the amounts that correspond to five winning bets and four losing bets would be calculated as follows:

$$(75 + 75 + 75 + 75 + 75 + -100 + -100 + -100 + -100)/9 = -25/9 = -2.77.$$

The above calculation was based on nine bets, but the calculation of the expected gain is not limited to any particular number of bets. The value –2.77 is the mean amount lost per bet. If we bet the same game 100 times, our expected total loss would be –$277.

We substitute the symbol X for G and call $E(X)$ the *expected value* for the variable X. From the above we say that the expected value is –2.77.

We use the above to help motivate our definition for the expected value or mean of a discrete variable X whose values are numbers. We assume that each value x has the known probability $\Pr(X=x)$.

The mean or expected value of a discrete variable X is defined as

$$\mu = E(X) = \Sigma[x * \Pr(X=x)],$$

where $\Sigma[x * \Pr(X=x)]$ represents the sum of each x multiplied by the probability of the value x occurring.

The mean of a discrete random variable has this interpretation. If the experiment were performed many thousands of times, the observed mean of all the numerical outcomes would be close to the theoretical mean, $\Sigma[x * \Pr(X=x)]$, of the discrete variable X.

Exercise 11.10. The experiment is rolling a single die 100 times. Record the number resulting from each roll of the die. Calculate the mean of the 100 outcomes. This is the observed mean. The discrete variable $X=\{1, 2, 3, 4, 5, 6\}$ corresponds to the possible outcomes of rolling a die one time. Calculate the theoretical mean $\Sigma[x * \Pr(X=x)]$ of the discrete variable X. Compare the observed mean with the theoretical mean. Are the two means close to each other?

The definition of the mean of a discrete variable X is consistent with the definition of the mean for any quantitative data set. To see this, we look at the frequency distribution of the discrete variable X = the number of total bases in an at bat, where the sample space consists of the career at bats of Henry Aaron. The data for the career at bats for Aaron come from Table 2.2, and the frequency distribution appears in Table 11.10.

As defined in Chapter 1, the mean is calculated by summing the product of each value x with its frequency and then dividing the result by the sum of the frequencies:

$$\begin{aligned}
\mu &= \Sigma(x * f)/\Sigma f \text{ (definition of the mean for a quantitative data set)}\\
&= (0 * 8593 + 1 * 2294 + 2 * 624 + 3 * 98 + 4 * 755)/12364\\
&= \Sigma[x * (f/\Sigma f)]\\
&= 0 * (8593/12364) + 1 * (2294/12364) + 2 * (624/12364) + 3 * (98/12364)\\
&\quad + 4 * (755/12364)\\
&= 0 * \Pr(X=0) + 1 * \Pr(X=1) + 2 * \Pr(X=2) + 3 * \Pr(X=3) + 4 * \Pr(X=4)\\
&= \Sigma[x * \Pr(X=x)] \text{ (definition of the mean of a discrete variable)}\\
&= .555.
\end{aligned}$$

Table 11.10 **Frequency distribution for X = total bases for the career at bats of Henry Aaron**

X	Frequency (f)
0	8593
1	2294
2	624
3	98
4	755
Total ABs	12364

Table 11.11 Frequency distribution for $X=(1)$ getting a hit or (0) making an out for the career at bats of Henry Aaron

X	Frequency (f)
0	8593
1	3771

Observe that the mean of this discrete variable is the career SLG for Aaron.

We now create another discrete variable X for the career at bats for Aaron. Let $X=1$ (for an at bat resulting in a hit) and 0 (for an at bat resulting in an out).

The frequency distribution for X appears in Table 11.11.

The parameter $\mu=\Sigma[x*\Pr(X=x)]=1*(3771/12364)+0*(8593/12364)=$ .305. What baseball statistic does .305 represent?

The above examples show that the definition for the mean of a quantitative data set, introduced in Chapter 1, is consistent with our definition of the mean for a discrete variable X.

The *variance* and *standard deviation* for a population of quantitative data measure how spread out the data are around the mean. The *variance* and *standard deviation* for a frequency distribution are given by the following formulas:

$$\sigma^2=\Sigma[(x-\mu)^2*f]/\Sigma f,$$
$$\sigma=\sqrt{\Sigma[(x-\mu)^2*f]/\Sigma f}.$$

The function f is the frequency at each x.

Since for each x the $\Pr(X=x)=f/\Sigma f$, we have the following definitions for the *variance* and *standard deviation* for a discrete probability distribution:

$$\sigma^2=\text{Variance of } X=\Sigma[(x-\mu)^2*\Pr(X=x)],$$
$$\sigma=\text{Standard Deviation of } X=\sqrt{\Sigma[(x-\mu)^2*\Pr(X=x)]}.$$

The next example shows that the definitions for the variance and standard deviation for a discrete random variable are consistent with those of a frequency distribution.

Example 11.11. X is the discrete variable in Table 11.10.

Variance of $X=\sigma^2=\Sigma[(x-\mu)^2*f]/\Sigma f$ (definition of variance of data set)

$=[(0-.555)^2*8503+(1-.555)^2*2294+(2-.555)^2*624$
$+(3-.555)^2*98+(4-.555)^2*755]/12364$

$=\Sigma\,[(x-\mu)^2*(f/\Sigma f)]$

$=[(0-.555)^2*(8503/12364)+(1-.555)^2*(2294/12364)$
$+(2-.555)^2*(624/12364)+(3-.555)^2*(98/12364)$
$+(4-.555)^2*(755/12364)]$

$=\Sigma\,[(x-\mu)^2*\Pr(X=x)]$ (definition of variance of discrete variable X)

$=588.96.$

The standard deviation of X is $\sqrt{588.96}$, which is 24.27.

For another example, we look at the discrete variable in Example 11.8. The next example shows the calculations for the variance and standard deviation of the discrete variable X in Example 11.8.

Example 11.12. X is the discrete random variable in Table 11.7. Verify that the mean of the discrete probability distribution is 3.5.

$$\begin{aligned}\text{Variance of } X = \sigma^2 = \Sigma\,[(x-\mu)^2 * \Pr(X=x)] &= [(1-3.5)^2] * (1/6) + [(2-3.5)^2] * (1/6)\\ &+ [(3-3.5)^2] * (1/6) + [(4-3.5)^2] * (1/6) + [(5-3.5)^2] * (1/6)\\ &+ [(6-3.5)^2] * (1/6).\end{aligned}$$

The variance of $X = 70/24$.

Both the variance and standard deviation for a discrete random variable X measure how spread out the values x, weighted by their probabilities, are around the mean.

Exercise 11.11. Compute the mean, variance, and standard deviation for the discrete probability distribution in Exercise 11.3.

Exercise 11.12. Compute the mean, variance, and standard deviation for the discrete probability distribution in Exercise 11.4.

Exercise 11.13. Compute the mean, variance, and standard deviation for the discrete probability distribution in Exercise 11.5.

We now switch our attention from discrete probability distributions that are data dependent to theoretical discrete probability distributions that are based on axioms.

11.2. Binomial and Geometric Distributions

It is often possible to use a formula to define the y-values (probabilities) for the x-values of the pdf. If a formula exists, we can simply substitute the value of the discrete variable into the formula to obtain the probability.

Theoretical probability distributions have special properties or assumptions that allow us to obtain such a formula. These distributions are often called discrete probability models, and we say that these distributions are *axiom* not *data* dependent. Since theoretical distributions are not data dependent, they can be used repeatedly with many different data sets. The effectiveness of a mathematical model is measured by its ease of use and its ability to predict future outcomes with minimal error. Since these models are not dependent on the data, we are able to tabulate, in advance, all probabilities and provide tables for these probabilities. These tables can be used to reduce the amount of arithmetic generally associated with some of the formulas. We now consider two of the most important theoretical distributions.

The two theoretical discrete distributions we look at are the *geometric* and *binomial distributions.* Both these distributions are generated from special types of probability experiments. These two probability experiments are the binomial experiment and the geometric experiment. We define these experiments next.

A *binomial experiment* satisfies the following properties:

1. The experiment is *repeated a fixed number of times n*, where n can be any positive integer. Each repetition of the experiment is called a trial. All the repetitions are performed under identical conditions. The possible outcomes for each trial are identical.
2. Each trial results in two and only two mutually exclusive events E and F, where the compound set (E or F) contains all possible outcomes of a trial. Since the sets are mutually exclusive, (E and F) = { }. One of the events is called *success*; the other is called *failure*.
3. The probability of success is denoted by p and that of failure by q; $p+q=1$. For each trial, the probabilities p and q remain constant.
4. The trials are *independent.* That is, the outcome of one trial does not affect the outcome of another trial.

A *geometric experiment* replaces the first property in a binomial experiment with the following property:

1. The process is *repeated until the first success occurs.*

Properties 2, 3, and 4 for a binomial experiment also hold for a geometric experiment.

Example 11.9 is an example of a binomial experiment. Example 11.10 is an example of a geometric experiment. The difference between the two examples is that in Example 11.9 there are a fixed number of trials $n=2$. In Example 11.10 the trials will continue until the first "6" occurs. In Example 11.9, each trial results in either a head (E = {H}) or a tail (F = {T}). Success is getting a head with $p=1/2$, and failure is getting a tail with $q=1/2$. In Example 11.10, each trial results in either a "6" (E = {6}) with $p=1/6$ or not a "6" (F = {1, 2, 3, 4, 5}) with $q=5/6$. Once the first "6" occurs, the experiment ends.

The next two examples of probability experiments use *relative frequency probabilities.* In order to consider these experiments as either binomial or geometric, we must assume that the trials are independent and the probability of success remains constant from trial to trial. When models are applied to real-world experiments, we cannot expect the assumptions of a binomial or geometric experiment to be perfectly satisfied. For these models to work, all we need is that the assumptions are close to being satisfied.

Example 11.13. The probability experiment is the next five plate appearances for a particular player. Each trial results in the possible outcomes of a 1B, 2B, 3B, HR, BB, HBP, SF, and an Out. We assign success to E={1B, 2B, 3B, HR, BB, HBP} and failure to F={SF, Out}. Success corresponds to getting on base (OB). Unlike in classical probability, p=OBP and q=(1 – OBP) must be determined by looking at many previous plate appearances. For some players, their plate appearances will come close to the assumptions needed for a binomial experiment. For other players, the assumptions will not work. If we assume that the assumptions are almost satisfied, this would be an example of a binomial experiment.

Example 11.14. Everything is the same as in Example 11.13, except that there are not a fixed number of plate appearances. Instead, the experiment will be repeated until an outcome of a plate appearance belongs to the event E (getting on base). This is an example of a geometric experiment.

In each of the above examples, the event assigned to "success" and the event assigned to "failure" satisfy these two properties:

1. The union of the two events includes all the possible outcomes of a trial.
2. The two events are mutually exclusive.

In the next two sections, we look at the two theoretical probability distribution functions, the *geometric distribution* and the *binomial distribution.*

11.2.1. Geometric Distributions

For any geometric experiment, we can create a discrete probability distribution by defining the discrete random variable X=the trial number corresponding to the first "success." Why does this discrete random variable have an infinite number of possible values? This pdf for X, whose values can be any positive integer, is called a *geometric distribution.* Note that X cannot be 0 and that, in theory, X can have an infinite number of integer values. Consider the pattern shown below:

$$\underbrace{q\,q\,q\,q\,q \cdots q}_{x-1} \cdot \underset{x}{p}$$

x (x represents the trial number of the first "success")

The only way that the first p can occur at trial number x is if the first $(x-1)$ of previous trials were each q (failures). Therefore, we now have a formula for $\Pr(X=x)$ for a geometric distribution.

Geometric Formula

For a geometric experiment, the probability that the first success occurs at trial number x is

$\Pr(X=x)=q^{x-1}p; \qquad x=1, 2, 3, 4, \ldots$

The probability is equal to this product because of the assumption of independence. We now present two examples of geometric distributions. The first example involves a game of chance; the second involves a baseball experiment.

Example 11.15. A die is rolled until the first "6" appears, at which time the experiment terminates. For a given trial we assign success to getting a "6" and failure to not getting a "6"; $p=\Pr(\text{"success"})=1/6$ and $q=\Pr(\text{"failure"})=5/6$. Let X=trial number corresponding to the first success. Suppose that we were asked to find the $\Pr(X=4)$. Using our model, $x=4$, $p=1/6$, and $q=5/6$, so

$$\Pr(X=4)=[(5/6)^{(4-1)}]*(1/6)=125/1296=.0964.$$

Example 11.16. We look at Example 11.13. The player is Henry Aaron. Success is any outcome corresponding to getting on base, and failure is any outcome corresponding to not getting on base. Let $p=\Pr(\text{"success"})$. For the value p, we use Aaron's career on-base average of .374. Therefore, $q=.626$. Suppose that our trials are repeated spins of the pointer on the disk for Aaron, shown in Figure 7.4. The variable X is the spin number of the first occurrence of an outcome belonging to the event "success" (getting on base). The experiment consists of a spin of a pointer on Aaron's disk. Since the spinning of a pointer on a disk satisfies the properties of a geometric experiment (why?), the geometric distribution applies. What is the probability that we can stop after one spin, after two spins, after three spins, and after four spins? What is the probability that we will have more than four spins?

Solution: X is the trial number of the first success (X cannot equal 0). Using the formula for a geometric distribution, we have the following:

$$\Pr(X=1)=(.626)^{1-1}*(.374)=.3740,$$
$$\Pr(X=2)=(.626)^{2-1}*(.374)=.2341,$$
$$\Pr(X=3)=(.626)^{3-1}*(.374)=.1466,$$
$$\Pr(X=4)=(.626)^{4-1}*(.374)=.0917,$$
$$\Pr(X>4)=1-.3740-.2341-.1466-.0971=.1536.$$

Remember, the $\sum\Pr(X=x)=1$.

Table 11.12 Geometric distribution in Example 11.16

X=trial #	$\Pr(X=x)$
1	.3740
2	.2341
3	.1466
4	.0917
↓	↓

We construct Table 11.12 to represent the geometric distribution in Example 11.16.

We can see that the probability that we would have to spin more than four times is about 15%. We also see that the probability of obtaining the first success at trial number x decreases rapidly as x increases.

The geometric distribution always has an inverted J shape in that it starts off high and drops off quickly. Knowing the formula for a probability distribution allows us to obtain a formula for the mean and variance of the distribution. For any geometric distribution we have

mean $= \mu = E(X) = 1/p$,
variance $= \sigma^2 = q/p^2$.

For Example 11.16 with $p = .374$ and $q = .626$,

$\mu = E(X) = 1/.374 = 2.67$,
$\sigma^2 =$ variance of $X = .626/(.374)^2 = 4.48$.

The mean says that from any starting point, it takes on average 2.67 plate appearances before Aaron gets on base.

Example 11.17. Roberto Alomar had a career AVG of .300. Assuming that a new season begins and Roberto is close to satisfying the axioms of a geometric experiment, find the probability that his first hit occurred in his 11th at bat. On the average, how many at bats are needed before Alomar gets his first hit? Why must we assume that the conditions of a geometric experiment are satisfied?

Solution: Using the geometric model, $p = .30$, $q = .70$.

$\Pr(X = 11) = (.70)^{10}(.30) = .0084 = .84\%$.

The mean is equal to $(1/.300) = 3.33$.

Since a player is not a robot, his plate appearances are controlled by both skill and chance; therefore, perfect independence from plate appearance to plate appearance cannot be assumed. For example, in one plate appearance he could be facing a Hall of Fame pitcher, whereas in the next plate appearance the pitcher may be mediocre. Many other variables can also affect a plate appearance. Players less affected by outside influences will have results closer to those predicted by our models.

Example 11.18. The is the same problem as Example 11.17, except that we assume that the batter is "Mr. Baseball" Bob Uecker, a Milwaukee Braves catcher turned actor who was a .200 career hitter.

Solution:

$\Pr(X = 11) = (.80)^{10}(.20) = .0214 = 2.14\%$.

The mean is $(1/.200) = 5$.

Exercise 11.14. For Examples 11.17 and 11.18, explain why Uecker has a higher probability of obtaining his first hit in his 11th at bat. Explain why the mean for Uecker is greater than the mean for Alomar.

11.2.2. Binomial Distributions

The discrete random variable X for a binomial experiment counts the number of successes X for a fixed number of trials n. Unlike a geometric experiment, where the variable X can be theoretically any positive integer, the variable X for a binomial experiment is limited to nonnegative integers from 0 to n.

Let X represent the number of successes in a fixed number of trials n. We let $p=\Pr(\text{success})$ and $q=1-p=\Pr(\text{failure})$. The pattern below represents only one of many sequences with exactly x successes and $(n-x)$ failures.

$$\underbrace{pppp\ldots p}_{x}\underbrace{qqq\ldots q}_{n-x}$$

Since the trials are independent, the probability that the above sequence occurs is $p^x q^{n-x}$. In Section 7.5, we showed that the number of arrangements of x identical p's and $(n-x)$ identical q's is the number of combinations of n things taken x at a time $({}_nC_x)$.

Binomial Formula

For a binomial experiment, the probability of exactly x successes in n trials is given by

$$\Pr(X=x)={}_nC_x\, p^x q^{n-x},\ x=0, 1, 2, 3, \ldots, n\ [{}_nC_x=n!/x!(n-x)!].$$

Note that x can equal 0 but cannot be greater than n.

The above formula can be cumbersome. Four alternate methods for evaluating binomial probabilities are presented next.

11.2.2.1. Method 1: Using the Cumulative Binomial Table (Table C.1 in Appendix C)

For values of $n=3$, 4, 5, 10, and 20 and certain values of p, we can use the cumulative binomial table (Table C.1) to calculate probabilities for a binomial experiment. It should be noted that other books have tables ranging from $n=3$ to $n=25$. Table C.1 is called a *cumulative* probability table because a lookup of a value of x gives the *sum* of all the probabilities for the variable X from 0 to x. The function $A(x)$ below represents this cumulative probability.

$$A(x)=\Pr(X=0)+\Pr(X=1)+\Pr(X=2)+\cdots+\Pr(X=[x-1])+\Pr(X=x).$$

The lookup in Table C.1 involves knowing n, p, and x. The probability obtained from the table is $A(x)$, where $A(x)=\Pr(X\leq x)=$ lookup of x.

The rules shown below enable us to find other probabilities.

$\Pr(X=x)=A(x)-A(x-1)$	Pr(exactly x successes)
$\Pr(X\leq x)=A(x)$	Pr(of at most x successes)
$\Pr(X\geq x)=1-A(x-1)$	Pr(of at least x successes)
$\Pr(a\leq X\leq b)=A(b)-A(a-1)$	Pr(that X is between a and b inclusive)

$\Pr(X < x) = A(x-1)$ Pr(of at most $x-1$ successes)
$\Pr(X > x) = 1 - A(x)$ Pr(of more than x successes)

In many binomial experiments, each trial results in only two mutually exclusive outcomes. The two outcomes could be true-false, yes-no, on-off, male-female, hit-out. If we were using a binary system with the numbers 0 and 1, we could use these values as our outcomes. It is also possible to reduce a multinomial variable (one with more than two outcomes) to a binomial variable by selecting one or more of the outcomes as "success" and the others as "failure." For example, an at bat can have the five possible outcomes of 1B, 2B, 3B, HR, or an Out. If we consider the outcomes of the set {1B, 2B, 3B, HR} to be "success" and the outcome "Out" to be "failure," we have a binomial variable. Success would equate to getting a hit.

In the work that follows, the disk (displayed in Fig. 11.1) represents the career at bats for Barry Bonds. The binomial variable values are success = "Hit" and failure = "Out," where each at bat is a trial. The disk in Figure 11.1 was constructed using Bonds's career batting average of .298 as the Pr("Hit"). The central angle for the sector of the disk representing the outcome "Hit" is found by multiplying 360 times .298 = 107.28 degrees; the central angle for the outcome "Out" is equal to the product of 360 times (1 − .298) = 252.78 degrees.

The central angle for the outcome "Hit" is approximately 107 degrees and for the outcome "Out" is approximately 253 degrees.

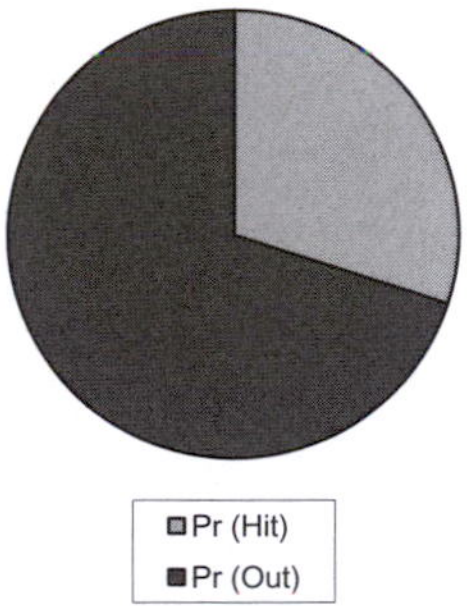

Figure 11.1 Probability disk for Bonds's at-bat outcomes "H" and "O"

Example 11.19. We found, in Chapter 2, that for Bonds the Pr(X = "Hit") = .298. Suppose that we were to spin the pointer on his disk (Fig. 11.1) 20 times.

1. Verify that the experiment of spinning a pointer on the disk is a binomial experiment.
2. Use the binomial table to find the probability of obtaining the following:
   a. Exactly five hits.
   b. No more than five hits.
   c. At least five hits.
   d. Between five and seven hits inclusive.
   e. Less than five hits.
   f. More than five hits.

Solution:

1. Each spin of the pointer is a trial. Each spin results in two possible outcomes; success is associated with a "Hit," and failure is associated with an "Out." The trials are independent. The probability of "success" is the same for each trial.
2. To use the binomial table, we need to find n, p, and x. Since $p = .298$ is not in the table, we use $p = .300$ (the closest p to .298). The number of spins is $n = 20$. Using the appropriate column, we can then calculate the above probabilities.

Table 11.13 **Small portion of the binomial table, $n = 20$**

$x \setminus p$	.21	.22	.23	.24	.25	.26	.27	.28	.29	.30
0										
1										
2										
3										
4										.2375
5										.4164

$$\Pr(X=5) = A(5) - A(4) = .4164 - .2375 = .1789,$$
$$\Pr(X \leq 5) = A(5) = .4164,$$
$$\Pr(X \geq 5) = 1 - A(4) = 1.0000 - .2375 = .7625,$$
$$\Pr(5 \leq X \leq 7) = A(7) - A(4) = .7723 - .2375 = .5348,$$
$$\Pr(X < 5) = A(4) = .2375,$$
$$\Pr(X > 5) = \Pr(x \geq 6) = 1 - A(5) = 1 - .4164 = .5836.$$

We illustrate the use of the binomial table in Table 11.13 for the first two probabilities above.

11.2.2.2. Method 2: Using Microsoft Excel

The function BINOMDIST(number, trials, probability, cumulative) is available in Microsoft Excel. In this formula,

- *number* refers to the number of successes x;
- *trials* refers to n;
- *probability* refers to the probability of success, p;
- *cumulative* is either True [$\Pr(X \leq x)$] or False [$\Pr(X = x)$].

$$\Pr(X=5) = \text{BINOMDIST}(5, 20, .298, \text{False}) = .1805,$$
$$\Pr(X \leq 5) = \text{BINOMDIST}(5, 20, .298, \text{True}) = .4241.$$

There is a slight difference in probabilities between using method 1 and using method 2. The reason is that in method 1 we used $p = .300$ as an estimate for $p = .298$ and in method 2 we did use $p = .298$.

11.2.2.3. Method 3: Using the Normal Curve as an Approximation

This method will be discussed in the next chapter.

11.2.2.4. Method 4: Using a Binomial Calculator

Located at the website http://stattrek.com/tables/binomial.aspx is a binomial calculator. Using this calculator, you will be able to input n = number of trials, p = the probability of success, and the x-value. Pressing the calculate button will return many probabilities, including the probability that X

is either (=, <, ≤, >, or ≥) to the x-value. Try the calculator with $n=20$, $p=.298$, and $x=5$.

11.2.2.5. The Mean and Variance for the Binomial Distribution

Since we have a formula for the binomial distribution, we can also obtain a formula for the mean and variance of a binomial distribution:

$$\text{mean}=\mu=E(X)=np,$$
$$\text{variance}=\sigma^2=npq=np(1-p).$$

Example 11.20. Find the mean and variance for a binomial distribution in Example 11.19.

Solution: We use $p=.298$. The mean $\mu=20*.298=5.96$, and the variance $\sigma^2=20*(.298)*(.702)=4.18$.

The mean tells us to expect that on the average for every 20 spins of his disk, Bonds will get close to six hits. This does make sense.

11.2.2.6. More Examples of Binomial Distributions

Example 11.21. For this example, we construct a binomial distribution. Using the spins of the pointer on Bonds's disk in Figure 11.1 as our binomial experiment, construct the binomial distribution for X=number of hits in five spins. Each spin is a trial. The number of trials is $n=5$. Success is getting a "Hit," $p=\Pr(\text{"Hit"})=.298$. Since $p=.298$ is not in the binomial table, we use $p=.300$.

Solution: The probabilities are calculated using the binomial table.

X	Pr(X)
0	.1681 (lookup $x=0$)
1	.3601 (lookup $x=1$ – lookup $x=0$)
2	.3087 (lookup $x=2$ – lookup $x=1$)
3	.1323 (lookup $x=3$ – lookup $x=2$)
4	.0284 (lookup $x=4$ – lookup $x=3$)
5	.0024 (lookup $x=5$ – lookup $x=4$)

Exercise 11.15. Construct the binomial distribution for the discrete random variable X in Example 11.19 using Microsoft Excel with $p=.298$.

Exercise 11.16. Repeat Exercise 11.15, but use 10 spins instead of 20.

Using the spins of the pointer on Bonds's disk as the binomial experiment, we wish to construct the binomial distribution based on $n=30$ spins of the pointer. Based on this disk (Fig. 11.1), the probability of a hit is $p=.298$. Since $n>20$, we cannot use the binomial table in Appendix C to calculate the probabilities. Table 11.14 shows the binomial distribution that

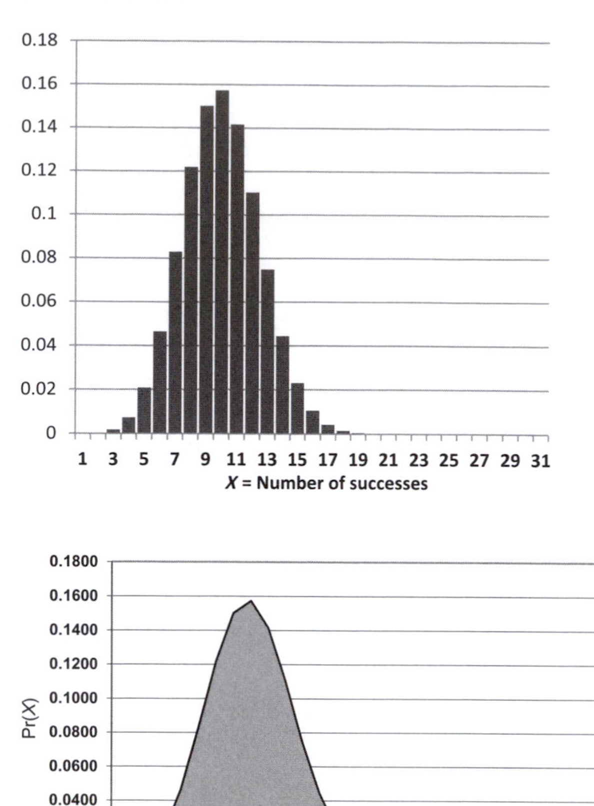

Figure 11.2 Histogram for binomial distribution generated from Bonds's disk

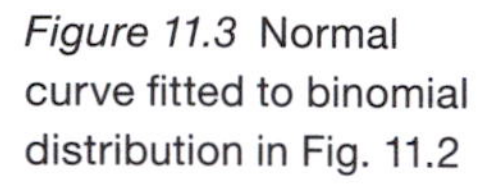

Figure 11.3 Normal curve fitted to binomial distribution in Fig. 11.2

was generated using Excel (method 2 discussed above). The Pr(X) column was calculated using the BINOMDIST function in Excel. Figure 11.2 shows the histogram for the binomial distribution, where the integers on the x-axis are the values for the discrete variable X = number of hits in 30 spins. Figure 11.3 shows a normal curve fitted to the histogram of the binomial distribution.

Exercise 11.17. Using the binomial experiment of 30 spins of the pointer on Bonds's disk, with X = number of hits in 30 spins, evaluate the following probabilities in two ways. First, use Table 11.14, where the probabilities are already evaluated, and then use the BINOMDIST function in Excel:

a. Pr($X > 10$)
b. Pr($X = 10$)
c. Pr($X \leq 10$)
d. Pr($X \geq 10$)
e. Pr($X < 10$)

Table 11.14 Binomial distribution for X=number of hits in 30 spins, p=.298

n	X	${}_nC_x$	Pr(X=x)
30	0	1	0.0000
30	1	30	0.0003
30	2	435	0.0018
30	3	4060	0.0072
30	4	27405	0.0208
30	5	142506	0.0464
30	6	593775	0.0829
30	7	2035800	0.1219
30	8	5852925	0.1501
30	9	14307150	0.1573
30	10	30045015	0.1416
30	11	54627300	0.1103
30	12	86493225	0.0749
30	13	119759850	0.0444
30	14	145422675	0.0231
30	15	155117520	0.0106
30	16	145422675	0.0042
30	17	119759850	0.0015
30	18	86493225	0.0005
30	19	54627300	0.0001
30	20	30045015	0.0000
30	21	14307150	0.0000
30	22	5852925	0.0000
30	23	2035800	0.0000
30	24	593775	0.0000
30	25	142506	0.0000
30	26	27405	0.0000
30	27	4060	0.0000
30	28	435	0.0000

Find the mean and variance for the binomial distribution.

Suppose that we wish to see how our binomial model compares with the actual career performance of Bonds for all games that he had exactly five at bats. To accomplish this, we would find all games in his career in which he actually had five at bats. For each of these games, we would record the number of hits and construct the frequency distribution for X. Finally, we would compare his observed frequencies with the expected frequencies, calculated using the binomial formula. In the next section, you will be asked to make a subjective comparison between Alex Rodriguez's observed and expected frequencies. If you feel that his observed frequencies are close to his expected frequencies, you may assume that his games with exactly five at bats fit the binomial model. On the other hand, if you feel that they are not close, you may conclude that his games with exactly five at bats do not fit the binomial model. Of course, any fixed number of at bats can be used.

We now apply the binomial distribution to the actual baseball data for Alex Rodriguez.

11.2.2.7. The Binomial Distribution and Alex Rodriguez

We ask the following question: For the year 2007, does the binomial distribution apply to those games where Alex had exactly three at bats? To help us answer this question, let's follow these steps:

Table 11.15 Frequency distribution for X = number of hits in those games in which Alex Rodriguez had exactly three at bats

X	Frequency
0	13
1	17
2	7
3	1

Step 1: We download, from the website www.retrosheet.org into Excel, all box scores for Rodriguez's games in 2007. In Excel, we sort on the field "number of at bats."

Step 2: We construct the frequency distribution table for those games with exactly three at bats. Table 11.15 shows the distribution.

Step 3: Table 11.16 shows the frequency distribution in step 2 converted into a discrete probability distribution for X, where the discrete variable X is equal to the number of hits in three at bats.

Step 4: We calculate the mean and variance for the pdf in step 3.

$$\mu = \Sigma[x * \Pr(X=x)] = .8947,$$
$$\sigma^2 = \text{variance} = \Sigma[(x-\mu)^2 * \Pr(X=x)] = .6204.$$

Table 11.16 Probability distribution function for the X in Table 11.15

X	$\Pr(X=x)$
0	13/38 = .3421
1	17/38 = .4474
2	7/38 = .1843
3	1/38 = .0263

Steps 1–4 dealt with Rodriguez's actual baseball data for 2007. Steps 5–7 are applied to the theoretical binomial distribution for $n=3$ trials and $p=.314$, the 2007 batting average for Rodriguez.

Step 5: The binomial distribution for $n=3$ trials and $p=.314$ is constructed in Table 11.17. The probabilities are calculated using the BINOMDIST function in Excel.

Step 6: We calculate the mean and variance for the binomial distribution with $n=3$ and $p=.314$.

$$\mu = np = 3 * .314 = .9420,$$
$$\sigma^2 = \text{variance} = np(1-p) = .6462.$$

Table 11.17 Probabilities calculated using the BINOMDIST function in Excel

X	$\Pr(X=x)$
0	BINOMDIST(0, 3, .314, False) = .3228
1	BINOMDIST(1, 3, .314, False) = .4433
2	BINOMDIST(2, 3, .314, False) = .2029
3	BINOMDIST(3, 3, .314, False) = .0309

Step 7: In comparing the binomial model for $n=3$ and $p=.314$ with the observed distribution of actual at bats for 2007, it appears that these two distributions and their corresponding means are similar.

What possible conclusions can be drawn from these similarities? One possible conclusion is that Rodriguez's actual at bats fit the axioms of the binomial model. That is, his at bats are independent of each other and his probability of getting a hit is close to .314 for each of his at bats. Can you think of any other possible reasons why the frequency distribution is so similar to the binomial distribution?

The problem with the analysis just done is that the final decision is based on a subjective opinion. This problem can be overcome by using inferential statistics, which you will learn about in Chapters 14 and 15.

In the chapter problems, you will be asked to repeat these steps for all games in which Alex had exactly four at bats and exactly five at bats in 2007.

Chapter Summary

In this chapter, we learned how to create a *probability distribution function* (pdf) for a *discrete random variable.* A *discrete random variable* associates a value with each outcome of a probability experiment. A pdf sets up a correspondence between each x-value of a discrete random variable and a y-value equal to the probability that the x-value occurs. Each y-value must be a number between 0 and 1 inclusive, and the sum of the y-values is equal to 1. A pdf can be represented by a table, by a histogram graph, or by a formula in functional form.

General formulas are provided that allow us to calculate the mean, variance, and standard deviation for a discrete random variable of a pdf.

Probability experiments such as the plate appearances of a player are said to be *data dependent.* A pdf for a data-dependent probability experiment requires that the experiment be executed and the data collected before the pdf can be constructed. The construction of its pdf is based on the relative frequency probability of each of its x-values.

Probability experiments that satisfy a certain set of properties are said to be *axiom dependent.* The two axiom-dependent probability experiments examined in this chapter were the geometric and binomial experiments. Since geometric and binomial experiments are based on axioms, we are able to provide formulas for their pdfs. These formulas allow us to calculate the probability of an x-value occurring without actually performing the experiment.

Although the general formulas provided for any pdf can be used to calculate the mean, variance, and standard deviation for a geometric or binomial distribution, the properties of these distributions give rise to simpler formulas.

The most used pdf is the binomial distribution. Four methods were provided to evaluate the probabilities for a binomial distribution. These are (1) a direct use of the formula, (2) the binomial table, (3) the binomial formula included in Microsoft Excel, and (4) an online binomial calculation. A fifth method uses a normal curve to approximate these probabilities. This method is explained in the next chapter.

We used the binomial distribution to look at two types of baseball experiments. First, we looked at the experiment involving a player's actual plate appearances. For this experiment, we assumed that a player's plate appearances were independent and the probability of success was the same for each plate appearance (of course, neither of these assumptions may be true). The second experiment consisted of spinning a pointer on a player's disk. For this experiment, the properties of both independence and the probability of success not changing are satisfied. The difference between the two experiments is that the former is based on *relative frequency* probabilities (both chance and skill), whereas the second example is based on *classical* probability (chance only).

The chapter ended with a study of all games in which Alex Rodriguez had exactly three at bats for the 2007 season. The discrete variable X was defined to be the number of hits in each of the games. Probabilities were then calculated using the probability distribution function for X formed from its frequency distribution. These probabilities were compared with the probabilities calculated using the binomial distribution. The results showed that the probabilities were close to each other. We conjectured that it is possible that Rodriguez's at bats satisfied the axioms for a binomial distribution.

There are other pdfs for a discrete variable that are not covered in this chapter. Two of these are the *Poisson distribution* and the *hypergeometric distribution*. Each distribution has its own set of axioms, which must be satisfied before it can be used. These distributions and their properties can be found in many textbooks on statistics.

We conclude with a summary of the axioms and formulas for the geometric and binomial distributions.

Geometric Distribution

Axioms

1. The probability experiment is repeated until the first success occurs.
2. The entire set of outcomes for each trial is separated into two events. One event is termed "success"; the other event is termed "failure."
3. The trials are independent of each other.
4. The probability of the event "success" occurring remains the same from trial to trial.
5. The discrete random variable X represents the number of the trial in which the first success occurs.

Formula

$$\Pr(X = x) = q^{x-1}p.$$

The formula gives the probability of the first success occurring in trial number x, where x = the number of the trial in which the first success occurs, p = probability of success in single trial, $q = 1 - p$, mean $= 1/p$, variance $= q/p^2$, and standard deviation $= \sqrt{q/p^2}$.

Binomial Distribution

Axioms

1. The probability experiment is repeated a fixed number of times (n).
2. The entire set of outcomes for each trial is separated into two events. One event is termed "success"; the other event is termed "failure."
3. The trials are independent of each other.
4. The probability of the event "success" occurring remains the same from trial to trial.
5. The discrete random variable X counts the number of trials out of the n fixed trials in which the event "success" occurs.

Formula

$$\Pr(X = x) = {}_nC_x p^x q^{n-x}.$$

The formula gives the probability of exactly x successes in n trials, where x = the number of successes in n trials, p = probability of success in single trial, $q = 1 - p$, n = the number of trials, ${}_nC_x = n!/x!(n-x)!$, mean $= np$, variance $= npq$, and standard deviation $= \sqrt{npq}$.

CHAPTER PROBLEMS

Table 11.18 is the frequency table for Alex Rodriguez for his number of hits in all games during the 2007 season in which he had exactly four at bats. Table 11.19 does the same thing for all games in which he had exactly five at bats in 2007. The data were retrieved from the website www.retrosheet.org.

Table 11.18 Frequency distribution for X = number of hits in each of Alex Rodriguez's games with exactly four at bats in 2007

X	Frequency
0	22
1	27
2	15
3	8
4	2

Table 11.19 Frequency distribution for X = number of hits in each of Alex Rodriguez's games with exactly five at bats in 2007

X	Frequency
0	1
1	5
2	11
3	4
4	2
5	0

1. Find Rodriguez's AVG for all games with four at bats and five at bats.
2. Repeat the seven steps in Section 11.2.2.7 for all games with four at bats and all games with five at bats.
3. Perform your own comparison on how well each of the three frequency distributions fit the binomial model.
4. Choose one of your players and a season for that player. For that season, prepare frequency tables for the discrete random variable X = number of hits in a game for all games with three at bats, four at bats, and five at bats. Using the player's batting average for that season for p, repeat the seven steps at the end of Section 11.2.2.7.

5. Research the Internet to find the player with the dubious record of the most at bats needed, from the beginning of a season, to get his first hit. For that player, calculate the probability of this actually happening. (Hint: Use the geometric distribution with p equal to his batting average for that year.)

Chapter 12

Probability Density Functions for a Continuous Variable

In Chapter 11, we studied discrete random variables and a probability distribution function (pdf) for them. The possible values of a discrete variable are countable. When each value of a discrete variable is coupled with the probability that it occurs, this correspondence is a pdf. We studied probability distribution functions that were data dependent and ones that were based on certain assumptions or axioms. The two discrete mathematical models, based on axioms, were the geometric and binomial distributions.

When a variable is a measurement, the possible values x of the variable X can be any real number in any interval. The entire real number line is an example of an infinite interval. When the values of a variable are measurements, we say that the variable is a continuous random variable.

The term *pdf* has a different meaning when applied to continuous variables. In the context of a continuous random variable, a *pdf* means a probability *density* function.

The next series of questions and answers will help to compare the probability distribution for a discrete variable with the probability density distribution for a continuous variable.

What is the difference between a discrete random variable and a continuous random variable?

A quantitative discrete random variable produces either a finite set of numbers or an infinite subset of integers. In either case, there are gaps between the numbers generated by a discrete variable. A continuous random variable is defined on an interval and includes any real number between any two numbers in the interval. Consequently, there are no gaps between the numbers. Often, we can distinguish between these two types of variables by referring to a discrete variable as a counting variable and a continuous variable as a measuring variable. In Example 12.1, several examples of each type of variable are provided.

Example 12.1.

Discrete Random Variables

- X = the number of hits a player has in a given season.
- X = the number of games won by each Major League team in the 2007 season.
- X = the number of players that hit at least 30 home runs for each year since 1900.

Continuous Random Variables

- X = the batting average of each player in 2007.
- X = the slugging average of each player in 2007.
- X = the distance of each of Barry Bonds's home runs in 2001.

Exercise 12.1. Classify each of the following random variables as discrete or continuous:

a. X = the number of base on balls for a player each year of his career.
b. X = the length of time for each Major League game in 2007.
c. X = the number of runs scored by a team for each game of a season.
d. X = the ratio BB/SO for each Major League team in 2007.
e. X = the proportion H/AB for each player in the 2007 season.
f. X = the OBA for each player in a season.

Why can't we assign probabilities for a continuous random variable in the same way as we did for a discrete random variable?

We learned in Chapter 11 that a probability distribution function sets up a correspondence between each x-value of a discrete random variable and the probability that the x-value occurs. This association was called a probability distribution function. This technique cannot be used for a continuous random variable. To understand why, we look at the following examples.

Example 12.2. Suppose that the population consisted of the heights of every person in the world. Choose a person at random from this population and ask the question, what is the probability that he or she is exactly 72 inches tall? Each person in the world theoretically has an exact height. An exact height of 72 inches would take the form of 72.00000000. . . . If we change the number at any decimal place from zero to another number, we get a different measurement. Since any measuring device can only measure to a certain number of decimal places, the exact height of any person will never be known. Therefore, we assign a probability of zero to a subject having an exact measurement.

Example 12.3. Suppose that a baseball player is picked at random. What is the probability that his batting average is exactly .333333 . . . ? Batting averages are displayed rounded to three decimal places. Taking into account the definition of batting average (#H/#AB), if one player was (200/600) = .33333 . . . and another player was (200/601) = .3327787022 (rounded to 10 decimal places), both players would have the same rounded batting average of .333. Clearly, even though their batting averages are displayed as the same number, they are different. Again, if we tried to assign a probability to a player having a particular exact batting average, the probability would have to be zero.

Examples 12.2 and 12.3 highlight one major difference between these two types of distribution. To summarize, for a probability distribution for a discrete variable, the probability of an individual x-value occurring is usually not zero, whereas for a probability density function for a continuous variable, by definition, the probability assigned to an individual x-value is always zero.

How then are probabilities evaluated for a continuous random variable?

Probabilities for a continuous random variable are evaluated by finding the probability that the x-values for the variable X lie between two numbers ($\Pr(a < X < b)$). This is done by finding the area of the region above the interval formed by two numbers and below a curve determined by a probability density function (pdf). These intervals can be infinite, such as $(-\infty < X < b)$ or $(a < X < \infty)$ or $(-\infty < X < \infty)$. As mentioned earlier, the probability that a continuous random variable assumes a single value is always zero. This is because the area of a line segment, which represents a single point, is zero ($\Pr(X = k) = 0$).

What is a probability density function for a continuous variable?

A probability density function for a continuous random variable X is a continuous function (a function without gaps) such that the $\Pr(a < X < b)$ is equal to the area of the region bounded on the sides by the lines $x = a$, $x = b$, bounded below by $y = 0$ (the x-axis), and bounded above by the probability density function (see Fig. 12.1). A probability density function is sometimes

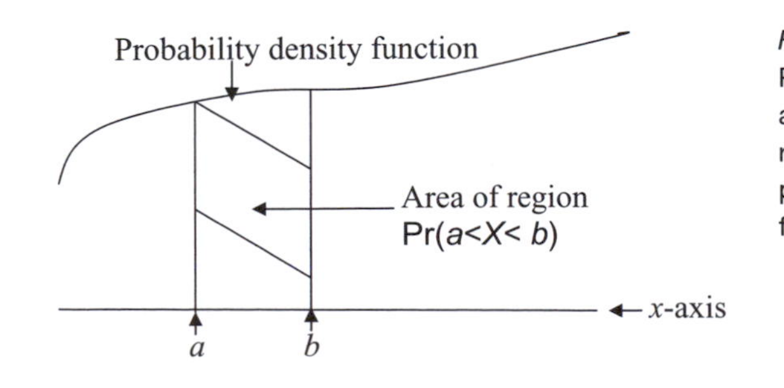

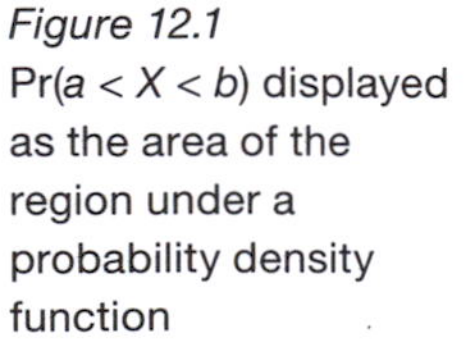

Figure 12.1
$\Pr(a < X < b)$ displayed as the area of the region under a probability density function

referred to as a continuous probability function. For any probability density function the total area under the curve and above the x-axis must equal 1 ($\Pr(-\infty < X < \infty) = 1$). This enables us to calculate probabilities as the area of a region of a probability density function. Remember, a probability must be a number between 0 and 1.

As in the case of discrete probability functions, there are many different probability density functions.

What is the difference between a pdf for a discrete random variable and a pdf for a continuous random variable?

The graph of a pdf for a discrete random variable is a histogram (see Examples 11.2, 11.8, and 11.9). The graph of a pdf for a continuous random variable is a continuous curve (a graph without breaks or gaps).The abbreviation pdf for a discrete random variable translates into probability *distribution* function. For a continuous random variable, it is a probability *density* function. The probability density function encountered most in nature has for its graph a normal curve. The y-value associated with an x-value for a pdf of a discrete random variable is the probability of that x-value occurring. The y-value associated with an x-value for a pdf of a continuous random variable is the y-value obtained by substituting the x-value into the probability density function and is *not* a probability.

What are normal distributions?

In Section 2.7, the properties were given for the quantitative data generated from a continuous random variable to be approximately normally distributed. The following checks should be made before we can assume that a continuous random variable has a normal distribution for its probability density function.

1. The histogram, polygon, and stem-and-leaf graphs for the data indicate a symmetric shape that resembles the shape of a bell.
2. The mean of the data is close to the median and the mode of the data (the mode for a small data set can be unreliable and may not even exist).
3. The 68-95-99.7 rule applies. The interval on the x-axis with endpoints equal to the [mean ± one standard deviation] contains approximately 68% of the data. The interval on the x-axis with endpoints equal to the [mean ± two standard deviations] contains approximately 95% of the data. The interval on the x-axis with endpoints equal to the [mean ± three standard deviations] contains approximately 99.7% of the data.
4. The graph of any normal curve approaches but never touches the x-axis as it extends farther away from the mean in either direction (we say that the graph is asymptotic to the x-axis).

5. There is a family of different normal curves. A particular normal curve is uniquely determined by the mean and standard deviation of the underlying population.
6. The mean μ lies directly under the peak of the normal curve.
7. The standard deviation σ determines the height of the peak. The smaller the standard deviation, the higher the peak. Consequently, the larger the standard deviation, the flatter the normal curve.
8. The points on the curve with x-values of $[\mu \pm 1\sigma]$ are inflection points (the shape of the curve changes).
9. The total area under any normal curve and above the x-axis is equal to 1.

Exercise 12.2. For the graphs of the two normal curves in Figure 12.2, labeled A and B, estimate the mean. If one of the curves has a standard deviation of 1.5 and the other has a standard deviation of 3.0, which curve has each of these standard deviations? What is the total area under each of these normal curves?

A continuous random variable, which has a normal curve for its probability density function, is called a normal random variable. Any normal curve is a bell-shaped curve with its highest point above the mean of the population. The values x of the normal variable X are unbounded ($-\infty < X < \infty$), and the curve is asymptotic to the x-axis. The bell shape occurs because most of the values of a normal variable are near the mean, and as you travel farther away from the mean, in either direction, the number of values diminishes rapidly and symmetrically.

Can we use a normal distribution to evaluate probabilities for a discrete variable?

Yes. If the histogram for our discrete variable takes a shape that resembles a normal curve, we will be able to obtain approximate probabilities for the discrete variable by using a normal curve. (At the end of this chapter, this will be demonstrated by using a normal curve to evaluate probabilities for a binomial distribution. This was method 3 discussed in the previous chapter.)

Many variables of interest are normal variables. In baseball, league batting averages (for players with a minimum of 200 at bats) are expected to be approximately normally distributed. A picture of a typical normal curve is shown in Figure 12.3.

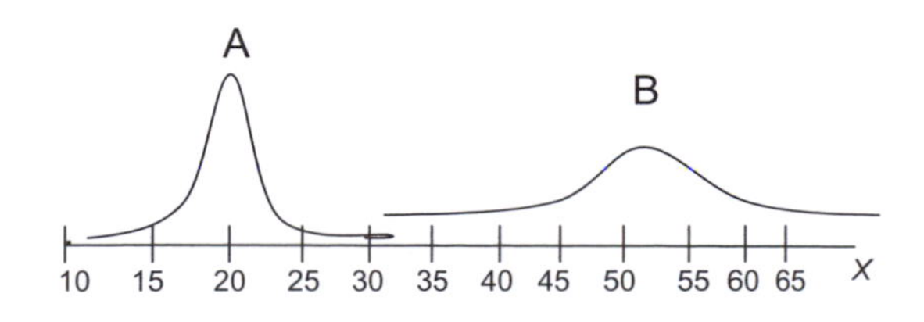

Figure 12.2 Two normal curves

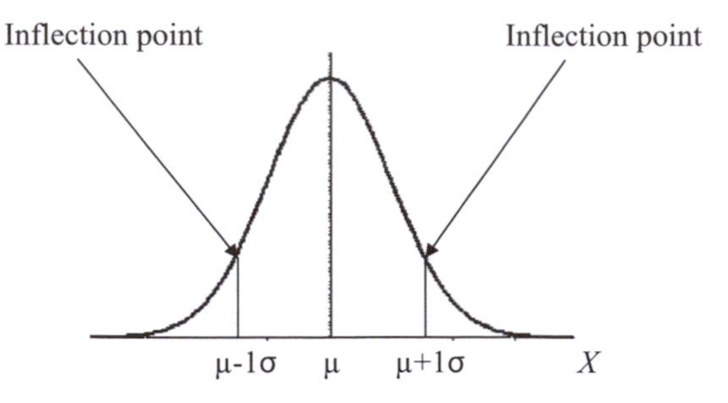

Figure 12.3 Typical normal curve

The normal curve model is the most important continuous distribution in statistics. It will be used in this chapter to predict the probability of a certain interval of outcomes occurring. In later chapters, we will see that the normal distribution plays a pivotal role in inferential statistics.

The theoretical formula for the graph of a normal curve depends solely on the mean and standard deviation of the underlying population.

Formula for the Probability Density Function for a Normal Distribution

$$f(x;\mu,\sigma)=\frac{1}{\sigma\sqrt{2\pi}}e^{-\frac{(x-\mu)^2}{2\sigma^2}}.$$

Using the methods from calculus, it can be shown that the total area between any normal curve and the x-axis is equal to 1. The variables μ (the population mean) and σ (the population standard deviation) depend on the underlying normal population. However, the area under the curve $f(x; \mu, \sigma)$ is always equal to 1. This allows us to calculate probabilities in terms of areas under the curve.

Both discrete and continuous probability models are used to predict probabilities for random outcomes occurring. As mentioned earlier, discrete probability models evaluate the probability that a certain outcome or collection of outcomes of a discrete random variable occur, whereas continuous probability models evaluate the probability that an interval of outcomes or several intervals of outcomes of a continuous variable occur. The probability for a continuous variable lying between $x=a$ and $x=b$ is equal to the area of the region bounded below by the x-axis, bounded above by the probability density function, and bounded on the sides by $x=a$ and $x=b$.

Section 12.1 provides the two methods for evaluating probabilities for a standard normal distribution, and Section 12.2 provides the two methods for evaluating probabilities for any normal distribution.

12.1. Evaluating Probabilities for the Standard Normal Distribution

The *standard normal distribution* is the normal distribution with $\mu = 0$ and $\sigma = 1$. The uppercase letter Z is the symbol used for the continuous variable for the standard normal curve. The lowercase z will represent the values for the Z-variable. The horizontal x-axis for the standard normal distribution is called the z-axis.

Formula for the Probability Density Function for the Standard Normal Distribution

$$f(z) = \left(e^{\frac{-z^2}{2}} \right) / \sqrt{2\pi}.$$

The probability that Z lies in an interval between $z = a$ and $z = b$ is equal to the area of a region under the standard normal curve and above the interval on the x-axis from a to b. Two methods for finding these areas are (1) to use a table and (2) to use a defined function provided in Excel.

Table C.2 in Appendix C provides the necessary areas under the standard normal curve. We refer to this as the Z-Table. The lookup of any z-value gives the area of the region under the standard normal curve to the left of the z-value on the z-axis ($-\infty < Z < z$). This area corresponds to $\Pr(Z < z)$. The Z-Table, like the Binomial Table, provides cumulative probabilities. A sample of cumulative probabilities obtained from the Z-Table by looking up some z-values is provided below. You should verify the probabilities shown in Table 12.1.

Table 12.1 **Sample of cumulative probabilities from the *Z*-Table**

z-lookup	Pr(*Z*<*z*)
−2.12	0.0170
−1.46	0.0721
−0.55	0.2912
0.00	0.5000
0.55	0.7088
1.46	0.9279
2.12	0.9830

To see how to use the Z-Table, we show the lookups of $z = 2.12$ and $z = -0.55$ in Figure 12.4.

The second method for finding $\Pr(Z < z)$ is to use a function provided in Excel, NORMSDIST(z). This function returns the area under the standard normal curve from $-\infty$ to z.

NORMSDIST(2.12) = .9830,
NORMSDIST(−0.55) = .2912.

Since $\Pr(Z < 3.89) = .9999$ and $\Pr(Z < -3.89) = .0001$, the $\Pr(Z < z)$ for any z-value greater than 3.89 will be rounded to 1.00 and the $\Pr(Z < z)$ for any z-value less than −3.89 will be rounded to 0.00.

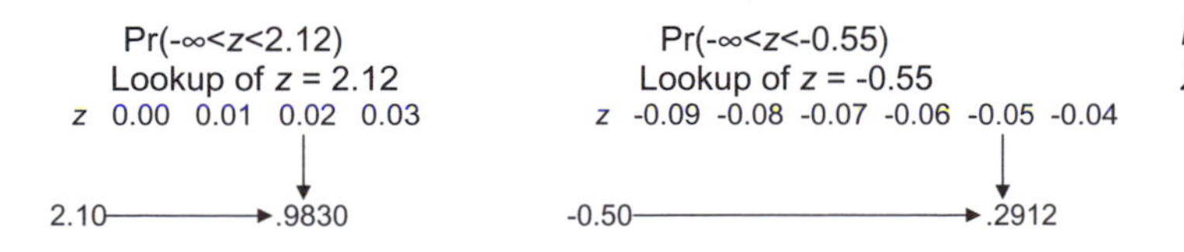

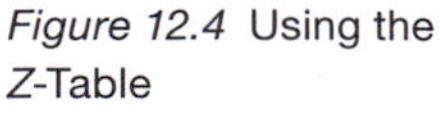

Figure 12.4 Using the Z-Table

12.1.1. Rules for Calculating Probabilities from the Standard Normal Distribution

The rules for finding areas (probabilities) for the standard normal distribution are similar to those used for finding the probabilities for the binomial distribution. One major difference is that for all continuous distributions, $\Pr(Z < z) = \Pr(Z \leq z)$ and $\Pr(Z > z) = \Pr(Z \geq z)$.

The function $A(z)$ gives the cumulative area of the region below the standard normal curve and above the z-axis in the interval from $-\infty$ to z.

- $A(z)$ = lookup of the z-value in the Z-Table.
- $A(z)$ = NORMSDIST(z) from Excel.

Rule 1: $A(a) = \Pr(Z < a) = \Pr(Z \leq a)$ = [left tail area].
Rule 2: $1 - A(a) = \Pr(Z > a) = \Pr(Z \geq a)$ = [right tail area].
Rule 3: $A(b) - A(a) = \Pr(a \leq Z \leq b) = \Pr(a < Z < b)$ = [area of the region between a and b].

The next two examples illustrate the calculation of probabilities using the Z-Table. The relationship between these probabilities and their corresponding regions is explored in Example 12.5.

Example 12.4. Find the following probabilities using the Z-Table:

a. $\Pr(z < 1.46) = A(1.46) = 0.9279$
b. $\Pr(z > -0.55) = 1 - A(-0.55) = 1 - .2912 = 0.7088$
c. $\Pr(0.55 < z < 2.12) = A(2.12) - A(0.55) = .9830 - .7088 = 0.2742$
d. $\Pr(z = \text{any constant}) = 0$

Example 12.5. Relating a probability to the area of its region, we have the following:

a. $\Pr(Z = \text{any constant } c) = 0$
b. $\Pr(0 \leq Z \leq 1.00)$ = (Lookup of $z = 1.00$) – (Lookup of $z = 0$) = .8413 – .5000 = .3413
c. $\Pr(0 < Z < 1.00)$ = same as (b)
d. $\Pr(Z < 1.28)$ = (Lookup of $z = 1.28$) = .8997
e. $\Pr(Z < -1.28)$ = (Lookup of $z = -1.28$) = .1003
f. $\Pr(Z > 1.28)$ = 1 – (Lookup of $z = 1.28$) = 1 – .8997
g. $\Pr(-1.28 < Z < 2.06)$ = (Lookup of $z = 2.06$) – (Lookup of $z = -1.28$) = .9803 – 1003 = .8800

The graphs in Figure 12.5 show the regions that are associated with the probabilities evaluated in (a) through (g) above.

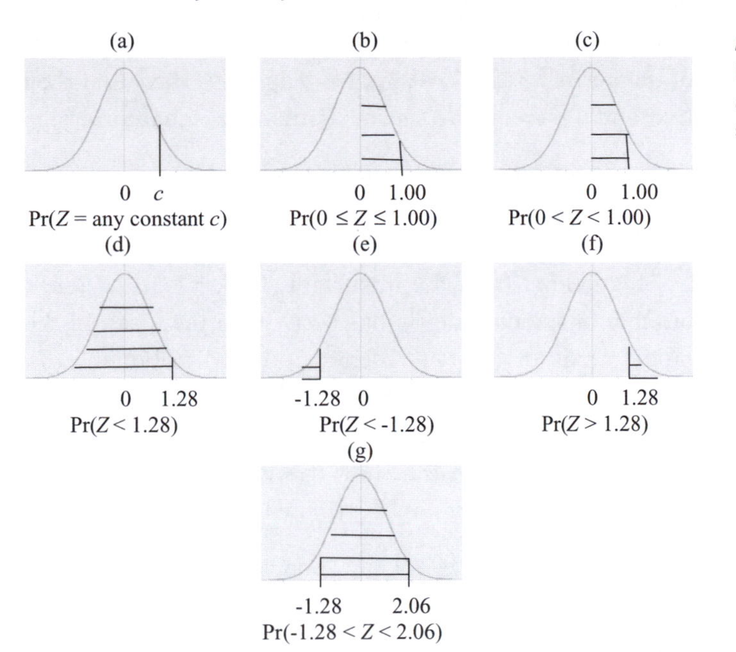

Figure 12.5 Graphical interpretations of the areas under the standard normal curve

Exercise 12.3.

1. Find the indicated areas under the standard normal curve:
   a. Left of $z = 1.76$
   b. Right of $z = 0.56$
   c. Between $z = 0.56$ and $z = 1.76$
   d. Left of $z = -1.38$
   e. Between $z = -0.67$ and $z = -2.09$
   f. Right of $z = -3.12$
   g. Between $z = -1.14$ and $z = 1.97$
   h. Left of $z = -1.12$
2. Find the indicated probabilities using the standard normal curve:
   a. $\Pr(Z < -2.36)$
   b. $\Pr(Z \leq -2.36)$
   c. $\Pr(Z > 1.47)$
   d. $\Pr(Z > -2.35)$
   e. $\Pr(-2.33 < Z < -0.87)$
   f. $\Pr(-1.24 < Z < 2.09)$
   g. $\Pr(Z = 0)$
   h. $\Pr(Z \leq 0)$

12.2. Finding Probabilities for any Normal Distribution

For any population of x-values which is normally distributed with mean μ and standard deviation σ, any x-value that is k standard deviations above or below the mean can be converted into $z=k$. If the x-value is to the right of the mean, k is positive; if the x-value is to the left of the mean, k is negative; and if x is equal to the mean, then k is equal to zero.

Formula for Converting an *x*-Value to a *z*-Value

$$z=(x-\mu)/\sigma.$$

This follows from the important property that if one travels the same number of standard deviations away from the mean in the same direction on the x-axis, for any two different normal distributions, the areas of the two regions under their respective normal curves will be equal. This property allows us to use the Z-Table to calculate probabilities for *any* normal curve. The next examples demonstrate how this is done. We refer to the x-values for a population that is normally distributed as the normal population X.

Example 12.6. Let X be a normal population with $\mu=100$ and $\sigma=10$. Find the following:

a. $\Pr(X<110)$
b. $\Pr(80<X<110)$

Solution: We convert the x-values to their z-values using the z-conversion formula.

$x=110$ converts to $z=(110-100)/10=1.00$.

This tells us that 110 is one standard deviation to the right of the mean.

$x=80$ converts to $z=(80-100)/10=-2.00$.

This tells us that the x-value 80 is two standard deviations to the left of the mean. These conversions are pictured in Figure 12.6.

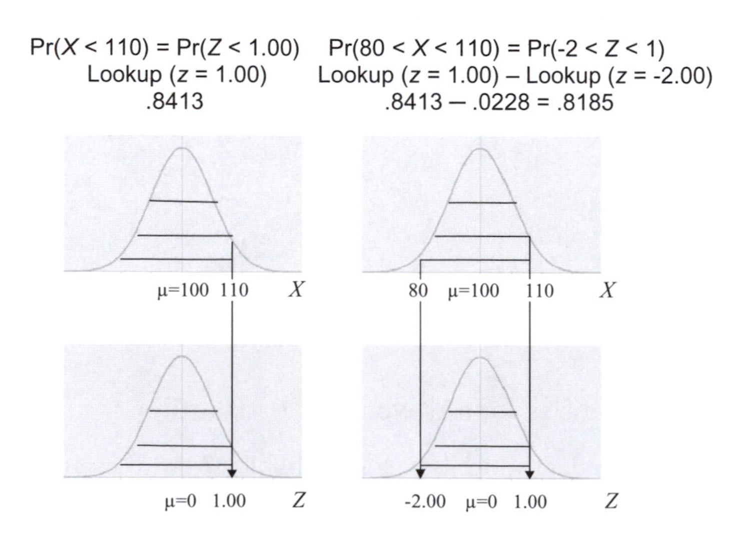

Figure 12.6 Converting an x-value from a normal distribution to a standard z-value

Alternately, the function NORMDIST(x, μ, σ, True) in Excel returns the cumulative area from $-\infty$ to x under the normal curve having mean μ and standard deviation σ.

NORMDIST(110, 100, 10, True) = .8413,
NORMDIST(80, 100, 10, True) = .0228.

Notice that NORMSDIST(1) = .8413 and NORMSDIST(−2) = .0228.

Exercise 12.4. The on-base averages (OBP) of all Major League players whose position is first base are assumed normally distributed with $\mu = .355$ and $\sigma = .039$. A first baseman is selected at random.

a. Find the probability that his OBP is greater than Larry Walker's, which was .400.
b. Find the probability that his OBP is less than that of the 2008 St. Louis Cardinals, .350.
c. Find the probability that his OBP is between .375 and .450.

Exercise 12.5. The batting averages (AVG) of all Major League players whose position is catcher, with at least 200 at bats, are assumed normally distributed. For the 2007 season, the following sorted data for Major League catchers' AVGs were obtained from the website www.retrosheet.org:

0.203	0.242	0.263	0.281
0.222	0.244	0.263	0.285
0.224	0.247	0.266	0.287
0.224	0.249	0.270	0.293
0.227	0.252	0.272	0.293
0.235	0.255	0.275	0.294
0.235	0.255	0.276	0.301
0.237	0.258	0.276	0.338
0.242	0.259	0.278	

1. Assuming that this represents population data, evaluate μ and σ.
2. Construct a frequency histogram using a width of .015, starting the first left endpoint of the first interval at .200. Make an argument that these data are approximately normal in shape.
3. Let the continuous random variable X = a Major League catcher's AVG. Evaluate the following probabilities:
   a. $\Pr(X < .300)$
   b. $\Pr(X > .250)$
   c. $\Pr(.250 < X < .300)$
4. Using the observed data, calculate the observed probabilities of (a), (b), and (c) above.
5. Compare your results in problems 3 and 4. Why are they different?

Exercise 12.6. Repeat Exercise 12.5 for all first basemen with at least 200 at bats for the 2007 season. These data can be retrieved from the website www.retrosheet.org.

The process, up to now, was to use for any normal population a given x-value to find the $Pr(X<x)$ or $Pr(X>x)$. We now reverse the process. That is, given that $Pr(X<x)$ = "a known probability," we are interested in finding the x-value. As before, we first look at the standard normal curve with variable Z.

The next example illustrates the process of finding a z-value for the area of a region under the standard normal curve.

Example 12.7. Find the z-value for $Pr(Z>z)=.0500$ in Figure 12.7.

Solution: To find the z-value, we must look up the area corresponding to $1-.0500=.9500$. If no area in the Z-Table matches the area we are looking up, use the area closest to the desired area. In our example, the area of .9500 is the midpoint of the two areas of .9495 and .9505. We choose to use .9495, which gives a z-value of 1.64. We could have chosen 1.65. Since 95% of the z-values are to the left of $z=1.64$, we say that the z-value of 1.64 represents the 95th percentile for the data.

Figure 12.7 Finding the z-value for a known probability

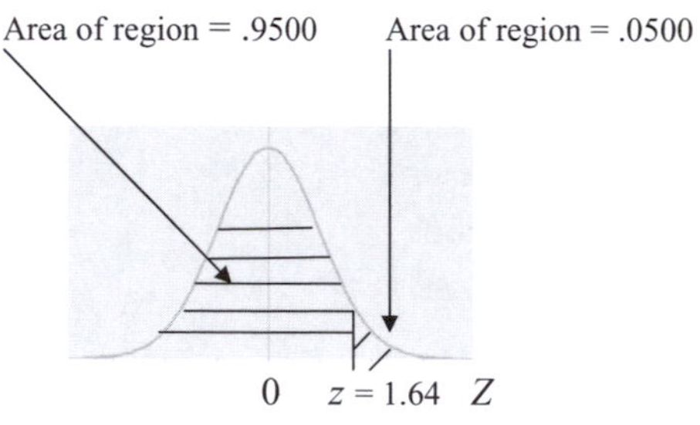

Exercise 12.7. Find the z-value that represents the 90th percentile, the 99th percentile, and the 80th percentile of the data.

Example 12.8. Find the z-value for $Pr(-z<Z<z)=.9500$ in Figure 12.8.

To find z, we observe that the area to the left of $-z$ is .0250. Looking up .0250 in the area portion of the Z-Table, we get $-z=-1.96$. Therefore, by symmetry, $z=1.96$.

Exercise 12.8. Find the z-value for $Pr(-z<Z<z)=.90$ and the z-value for $Pr(-z<Z<z)=.99$.

The next example illustrates the process of finding the x-value for the area of the region under any normal curve.

Example 12.9. The age of all Major League players for the 2008 season is assumed normally distributed with $\mu=30$ and $\sigma=3.5$. Find the age that corresponds to the 95th percentile.

Solution: From Example 12.7, a z-value of 1.64 corresponds to the 95th percentile. To find the correct x-value, we solve for the x-value in terms of the z-value as follows:

Figure 12.8 Finding the z-value for $Pr(-z < Z < z)$ (Example 12.8)

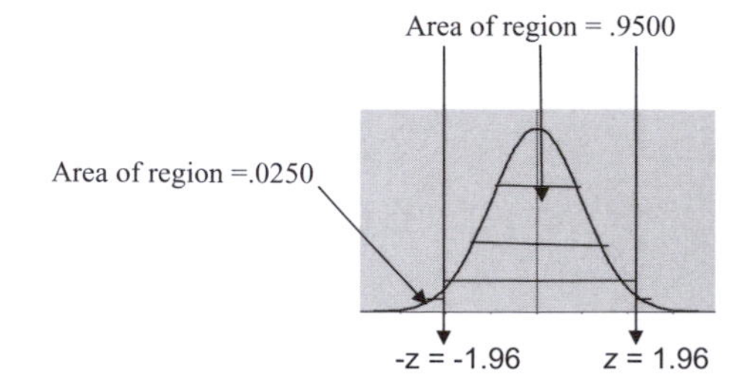

$$z=(x-\mu)/\sigma,$$

$$z*\sigma=x-\mu,$$

$\mu + z * \sigma = x,$

$x = \mu + z * \sigma.$

Since $\mu = 30$ and $\sigma = 3.5$ in our example, $x = 30 + 1.64 * 3.5 = 35.74$. The interpretation is that approximately 95% of the players have an age less than 35.74 years old.

Formula for Converting a *z*-Value to an *x*-Value

$x = \mu + z * \sigma.$

Exercise 12.9. For Exercise 12.5, find the AVG that corresponds to the 90th and 95th percentiles in two ways:

a. First, use only the observed data in Exercise 12.5.
b. Assuming that the data are normal, use the technique in Example 12.9.
c. How close are the results in (a) and (b)?

12.3. Using Normal Distributions to Approximate Probabilities for Binomial Distributions

As a binomial variable counts the number of successes x in n-trials and a normal variable involves measurements, the best we can do is use a normal curve to approximate binomial probabilities.

We now revisit the problem mentioned at the end of Chapter 11. Using the disk for Bonds shown in Figure 11.1, we saw in Figure 11.2 that the histogram for the discrete variable X = number of hits in 30 at bats resembled the shape of a normal curve.

In general, if $np > 5$ and $n(1-p) > 5$, the histogram for a binomial variable will resemble a normal curve and a normal-curve approximation can be used.

Since, for large values of n, the use of the binomial formula is time-consuming and the binomial table in our book only provides probabilities for n up to 20, another approach is necessary. Of course, if you have Excel available, the binomial formula supplied within that software can be used. For those binomial distributions whose shape resembles a normal curve, we can use a normal curve to approximate binomial probabilities.

As probabilities are evaluated differently for discrete and continuous distributions, if we wish to use a normal curve to approximate binomial probabilities, we need to make a certain adjustment. This adjustment is called a *continuity correction factor.*

The continuity correction factor has five cases. Assume that X is a discrete binomial variable. Case 1 is illustrated in Figure 12.9.

Figure 12.9 Case 1 for the continuity correction factor

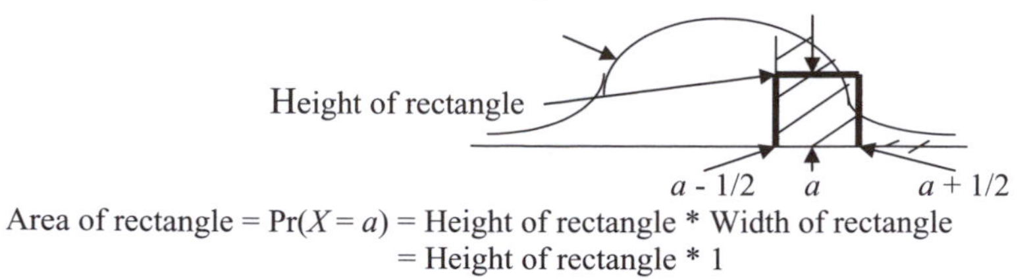

Area of rectangle = Pr($X = a$) = Height of rectangle * Width of rectangle
= Height of rectangle * 1

Binomial	Normal Correction
Case 1: $\Pr(X=a)$	$\Pr(a-1/2<X<a+1/2)$
Case 2: $\Pr(X\leq a)$	$\Pr(X<a+1/2)$
Case 3: $\Pr(X\geq a)$	$\Pr(X>a-1/2)$
Case 4: $\Pr(X<a)=\Pr(X\leq a-1)$	$\Pr(X<(a-1)+1/2)$
Case 5: $\Pr(X>a)=\Pr(X\geq a+1)$	$\Pr(X>(a+1)-1/2)$

The reason this correction factor gives us a better approximation is that the addition or subtraction of the number 1/2 picks up the entire area of the rectangle whose base extends from $(a-1/2)$ to $(a+1/2)$ and whose height is the $\Pr(X=a)$.

To choose the correct normal distribution, we need to know the mean and standard deviation for the underlying binomial population. We discovered in Chapter 11 that for a binomial population the mean is np and the standard deviation is $\sqrt{np(1-p)}$. The correct normal distribution is the one with mean $= np$ and standard deviation $= \sqrt{np(1-p)}$. The next example illustrates this approximation process.

Example 12.10. The experiment is 30 spins on Bonds's disk. The results appear in Table 11.14. Let X = the number of hits in 30 at bats. Use the normal curve to approximate the following probabilities:

a. $\Pr(X=10)$
b. $\Pr(X<10)$
c. $\Pr(X>10)$

Solution: For p we use 0.300 for Bonds's batting average. Since this is a binomial problem, we can compute the population mean and population standard deviation:

$$\mu = np = (30)(.30) = 9; \sigma = \sqrt{np(1-p)} = \sqrt{30(.30)(.70)} = 2.51.$$

We now use the normal distribution with mean = 9 and standard deviation = 2.51 to approximate the binomial probabilities:

a. $\Pr(X=10)=\Pr(9.5<X<10.5)=\Pr\left(\frac{9.5-9}{2.51}<Z<\frac{10.5-9}{2.51}\right)$
$=\Pr(.20<Z<.60)$
$=(\text{lookup of } z=.60)-(\text{lookup of } z=.20)$
$=.7257-.5793=.1464.$

b. $\Pr(X<10)=\Pr(X\leq 9)=\Pr(X<9.5)=\Pr\left(Z<\frac{9.5-9}{2.51}\right)$
$=(\text{lookup of } z=.20)=.5793.$

c. $\Pr(X>10)=\Pr(X\geq 11)=\Pr(X>10.5)=\Pr\left(Z>\frac{10.5-9}{2.51}\right)$
$=1-(\text{lookup of } Z=.60)=1-.7257=.2743.$

The reason we have $\Pr(X=10)=\Pr(9.5<X<10.5)$, $\Pr(X\leq 9)=\Pr(X<9.5)$, and $\Pr(X\geq 11)=\Pr(X>10.5)$ is to correct for using a continuous probability distribution to approximate a discrete probability distribution.

Recall that for a discrete variable the $\Pr(X\leq x)\neq \Pr(X<x)$.

Using Table 11.14, we see that the $\Pr(X\geq 10)=.4112$. This number was derived by summing the binomial probabilities from $x=10$ to $x=30$. Using the normal-curve approximation, we found $\Pr(X\geq 10)=\Pr(X=10)+\Pr(X>10)=.1464+.2743=.4207$. Observe how close the approximation is to the exact number.

Exercise 12.10. The experiment is 20 spins on Bonds's disk. Let X = the number of hits in 20 spins. Assume $p=.300$. Verify that a normal curve approximation can be used. Calculate the probabilities below by first using the binomial model and then using a normal curve approximation. Compare your results. You will need to use either the binomial table or the binomial formula provided in Excel. After completing your work, compare your answers. Are the results close to each other?

a. $\Pr(X=6)$
b. $\Pr(X<6)$
c. $\Pr(X>6)$

We conclude the chapter with another example of a probability density function for a continuous random variable, called a uniform distribution.

Example 12.11. The pdf is $F(x)=1/6$ for the continuous random variable X whose values lie between 0.5 and 6.5. The graph appears in Figure 12.10.

The area under the pdf is equal to 1.

$\Pr(1<X<5)=(5-1)*(1/6)=2/3$ = area of the rectangle whose height is 1/6 and whose base is the closed interval [1, 5].

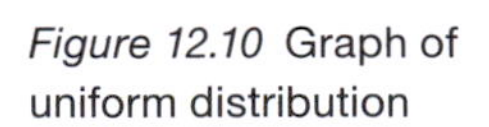

Figure 12.10 Graph of uniform distribution

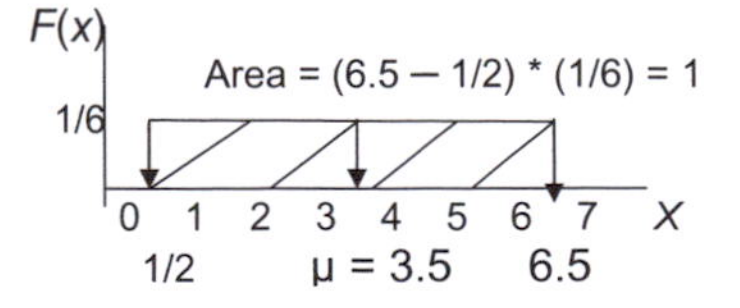

Exercise 12.11. Compare the pdf for the continuous variable in Example 12.11 with the pdf for the discrete random variable in Table 11.7. How are they different? Compare the means between the two pdfs.

Chapter Summary

This chapter introduced the concept of a probability distribution function (pdf) for a continuous variable. A probability distribution function for a continuous variable is called a probability density function. The graph of a probability density function has no holes or gaps and is said to be continuous. Continuous probability distributions are used to calculate probabilities for a continuous random variable.

We compared a probability distribution function for a discrete random variable and a probability density function for a continuous random variable. Both types of pdfs are used to calculate probabilities. For a discrete pdf, the y-value is the probability that the x-value occurs. The graph of the pdf for a discrete random variable is a histogram whose height is equal to the probability of x occurring ($y = \Pr(x)$). For a continuous random variable, the y-value is not a probability and the ordered pair (x, y) represents a point on the graph. Probabilities for a continuous pdf are calculated for an interval of x-values. The probability of an interval of x-values occurring is defined as the area of the region under the probability density function and above the interval on the x-axis. The probability of an individual x-value occurring is always zero.

The most common continuous probability distributions are normal distributions. Many characteristics in the real world are normally distributed. The graphs of normal distributions are called normal curves. Each normal curve is uniquely determined by the mean and standard deviation of the underlying population. The mean is the x-value under the peak of the normal curve, and the standard deviation determines the height of the peak.

Two methods for finding probabilities under any normal curve were given. The standard normal curve table (Z-Table) provides the cumulative probability for any z-value. By converting an x-value from any normal curve into a z-value, the Z-Table can be used to evaluate probabilities under any normal curve. The second method given was to use the formulas provided in Excel. The two formulas were NORMSDIST (for the standard normal curve) and NORMDIST (for any normal curve).

The other continuous probability distribution mentioned in this chapter was the uniform distribution.

A binomial distribution, which satisfies the two conditions that both np and $n(1-p)$ are greater than 5, has for its graph a histogram that resembles a normal curve and has a mean of np and a standard deviation of $\sqrt{np(1-p)}$. This allows us to use the normal curve with mean $= np$ and a standard deviation $= \sqrt{np(1-p)}$ to approximate probabilities for a binomial distribution. To compensate for the difference in the nature of variables (discrete or continuous) and improve the estimate, a continuity correction factor was introduced.

CHAPTER PROBLEMS

1. Use the standard normal curve to compute the following probabilities:
   a. $\Pr(Z=1.00)$
   b. $\Pr(Z>1.47)$
   c. $\Pr(Z>-2.34)$
   d. $\Pr(1.09<Z<2.87)$
   e. $\Pr(-2.67<Z<-0.97)$
   f. $\Pr(Z\leq-3.09)$
2. Use the standard normal curve to find the indicated areas:
   a. Left of $z=1.27$
   b. Right of $z=-2.96$
   c. Left of $z=-2.87$
   d. Right of $z=0.89$
   e. Between $z=-2.18$ and $z=1.49$
   f. Between $z=0.69$ and $z=3.01$
3. Find the z-value for the following probabilities:
   a. $\Pr(Z<z)=.2145$
   b. $\Pr(Z>z)=.2145$
   c. $\Pr(Z<z)=.7955$
   d. $\Pr(-z<Z<z)=.4657$
4. Use the standard normal curve to find the z-score that corresponds to the indicated percentiles:
   a. 88th percentile
   b. 50th percentile
   c. 99th percentile
5. From data downloaded from www.retrosheet.org, the mean and standard deviation for the on-base percentages (OBP) of all first basemen in the Major Leagues, with at least 200 at bats for the 2007 season, were .355 and .039, respectively. Assume that OBP is approximately normally distributed.
   a. How many standard deviations from the mean is an OBP = .375?
   b. How many standard deviations from the mean is an OBP = .330?
   c. Find the probability that a randomly selected first baseman has an OBP
      i. less than .300;
      ii. more than .400;
      iii. between .325 and .425.
   d. Find the OBP that corresponds to the 90th percentile.
   e. Find the minimum OBP that is required to be in the top 5% of all first basemen.

Chapter 13

Sampling Distributions

In previous chapters, we concentrated our efforts on *descriptive statistics*. In this chapter, we provide a bridge to the other major branch of statistics, *inferential statistics*. This bridge is provided by *sampling distributions*.

Inferential statistics involves techniques that allow us to use the result of a random sample from a population to make a decision about the entire population. One area of inferential statistics uses a statistic (descriptive measure of a *sample*) to estimate a parameter (descriptive measure of a *population*). *Confidence intervals* are used to make this estimate. Confidence intervals will be studied in the next chapter.

Sampling distributions provide important relationships between an original population and a population whose values are derived from a statistic calculated for all possible random samples of a fixed sample size selected from the original population.

Before defining a sampling distribution, we review definitions that were introduced in Chapter 1. A *population* consists of the complete set of data, derived from a variable (characteristic) applied to each subject in the entire group targeted for the study. A population can be finite or infinite (there is no list for the entire group). If a population is finite, the size is denoted by an uppercase N. A *sample* is a proper subset (excludes at least one member of the population), usually small in size relative to the population. The lowercase n denotes the sample size. A *parameter* is a descriptive measure of the *population*. A *statistic* is a descriptive measure of the *sample*. Some examples of parameters are population means (μ), population variances (σ^2), population standard deviations (σ), and population proportions (p). Some examples of statistics are sample means ($\bar{x}$), sample variances (s^2), sample standard deviations (s), and sample proportions ($\hat{p}$). Please do not confuse the data with descriptive measures of the data. Think of a descriptive measure such as a mean as a summary of the data.

In Chapter 11, *discrete random variables* were introduced. These are characteristics that assign a value to each member of a countable population. A population is countable if it is finite or (if infinite) an integer can be associated with each member of the population. A *probability distribution function* (pdf), for a discrete variable, sets up a correspondence between each value of the discrete variable and the probability that that value occurs.

Definition of a Sampling Distribution

A sampling distribution is a probability distribution function (pdf) whose discrete random variable X is a statistic that is calculated for all possible samples of the same sample size n, selected from the original population. Since a sampling distribution is a probability distribution function, each distinct value of the random variable is assigned to the probability of it occurring.

A sampling distribution is a special type of probability distribution function because the discrete variable X is a statistic. Each different statistic or different sample size gives rise to a different sampling distribution. For that reason, we don't just say "sampling distribution," we say a "sampling distribution of sample means, $\bar{x}$," or "a sampling distribution of sample standard deviations, s," or "a sampling distribution of sample proportions, $\hat{p}$."

13.1. Sampling Distributions for Three Types of Populations

The first type of population is a *finite* population of N objects. Since the population is finite, we are able to select all possible samples from the population in such a way that each sample of the same size has the same probability of being selected. As discussed in Chapter 7, sampling can be with or without replacement. Sampling *with* replacement guarantees that the selection of objects is independent. In either case, for each sample the statistic is calculated.

The second type of population is an *infinite* population (no complete list can be compiled or is available). In this case, several samples of the same sample size are selected without replacement. For each sample the statistic is calculated.

The statistic used in this chapter for the two population types above is the sample mean $\bar{x}$. The sampling distribution of sample means $\bar{x}$ is a function that associates each of the sample means with the probability of that sample mean occurring.

The third type of population is a *binomial* population. Binomial populations were presented in Chapter 11. Binomial populations, like tossing a coin, are generated from binomial experiments. A binomial experiment consists of a fixed number of trials, each of which results in two mutually exclusive events; one event is called "success," the other "failure." The trials are independent and the probability that each event occurs is the same from trial to trial. The outcomes for each of the n trials of the binomial experiment become a sample of size n. Each outcome in the sample belongs to one of the two mutually exclusive events. If the outcome belongs to the event "success," we assign a 1 to it. If the outcome belongs to the event "failure," we assign a 0 to it. The sample of size n consists of k successes and $(n-k)$ failures. For each sample, the statistic calculated is the sample proportion, $\hat{p}$, which is the quotient of the number of successes divided by the number of trials n (the sample size). The *sampling distribution of sample proportions* assigns each different value of $\hat{p}$ to its probability of occurring.

In this chapter, we study only two of the many possible sampling distributions. They are the *sampling distribution of sample means* and the *sampling distribution of sample proportions*. Remember that the *population* mean or *population* proportion can also be called the *true* mean or *true* proportion.

13.2. The Sampling Distribution of Sample Means

Suppose that we begin with a known finite population consisting of N numbers. The set of all such numbers is called the X-population. Since all the numbers in the population are known, both the population mean and population standard deviation can be calculated. The Greek letter μ (mu) symbolizes the population mean, and the Greek letter σ (sigma) symbolizes the population standard deviation. The shape of the population can be estimated by a histogram. In the next example, we present a finite population generated from a random variable X. The pdf of X will be constructed, and the population mean, variance, and standard deviation for the pdf will be calculated.

Example 13.1. Suppose that the X-population consists of the numbers 2, 4, 6, and 8. Assuming that one number is selected at random, the probability associated with each of the four possible outcomes is 1/4. The pdf for the X-population is given in Table 13.1. The $\Pr(X=x)$ represents the probability that each of the results in X is selected.

Table 13.1 Probability distribution function for the X-population in Example 13.1

X	$\Pr(X=x)$
2	1/4
4	1/4
6	1/4
8	1/4

We compute the mean and variance for the X-population in two ways. First, using the definitions of mean and variance for population data, we have

population size is $N=4$, $\Sigma x=20$, $\Sigma x^2=120$;

$\mu=(\Sigma x)/N=20/4=5$;

$\sigma^2=(\Sigma x^2-[(\Sigma x)^2/N])/N=[120-(20)^2/4]/4=5$, $\sigma=\sqrt{5}$.

From the definitions of the mean and variance for a pdf (Chap. 11) and using Table 13.1, we get

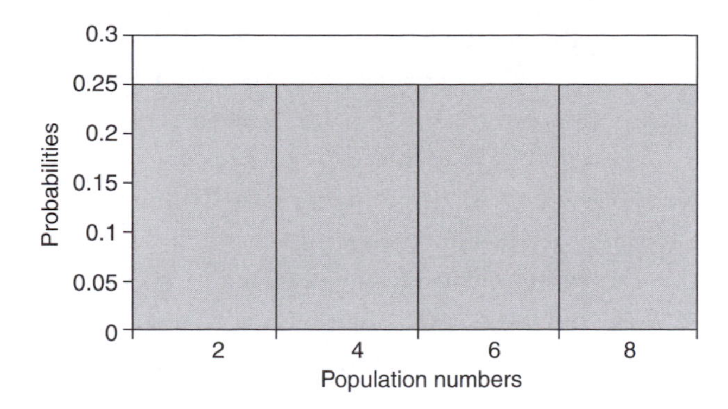

Figure 13.1 Histogram for the X-population in Example 13.1

$$\mu = \Sigma[x * \Pr(X=x)] = 2 * 1/4 + 4 * 1/4 + 6 * 1/4 + 8 * 1/4 = 5,$$
$$\sigma^2 = \Sigma[(x-\mu)^2 * \Pr(X=x)] = (2-5)^2 * 1/4 + (4-5)^2 * 1/4 + (6-5)^2 * 1/4 + (8-5)^2 * 1/4$$
$$= 20/4 = 5.$$

Notice that the mean and variance for the X-population correspond to the mean and variance of the pdf for the X-population. Do you think that this will always be true?

The histogram for the pdf for the X-population appears in Figure 13.1. The shape of this histogram is uniform. A uniform histogram occurs when the probabilities associated with different values of the random variable are equal.

13.2.1. Sampling from a Population (with Replacement)

In most cases, the total number of numbers N in the population is either very large or infinite, and it is not possible or too expensive to collect all the population data. If we can only observe n of the N values, we are forced to sample. Each time we select a sample of n of the values, we can calculate $\bar{x}$ (the sample mean) and s (the sample standard deviation). Suppose that we were able to select every possible sample, for a fixed size n, from the original population of size N. For each sample, we calculate $\bar{x}$. The collection of all sample means $\bar{x}$ for all possible samples of size n is called the $\bar{X}$-population. We can calculate the mean of the $\bar{X}$-population, symbolized by $\mu_{\bar{x}}$, and the standard deviation of the $\bar{X}$-population, symbolized by $\sigma_{\bar{x}}$. We can also describe the shape of the $\bar{X}$-population by constructing a histogram for the $\bar{X}$-population.

We now proceed to construct a sampling distribution of sample means for the population in Example 13.1. The sample size is $n=2$, and the sampling process is simple random sampling *with replacement* (refer to Chap. 7).

Figure 13.2 A disk divided into four sectors of equal area

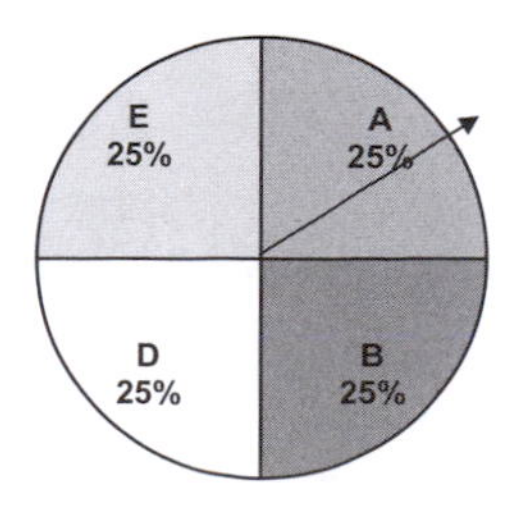

For the population in Example 13.1, the disk in Figure 13.2 will serve as a physical model for selecting samples. Each spin of the pointer on the disk generates an outcome for the sample. A sample of size n is generated by

spinning the pointer n times. The disk itself is divided into four sectors, each sector having the same area. Since each sector has the same area, each of the four outcomes has the same probability of occurring.

Each spin of the pointer will land on one of the four letters. We assign the number 2 to A, the number 4 to B, the number 6 to D, and the number 8 to E. The probability of landing on any of the four numbers is 1/4. We say that the four numbers are equally likely to occur.

If we spin the pointer two times and write down the number that results from spin number 1, followed by the number that results from spin number 2, we will have selected a random sample of size $n=2$ from the X-population. The sample (4, 6) is considered different from the sample (6, 4).

A sample created this way satisfies the following properties:

1. It is a simple random sample (every possible sample for a fixed sample size n has the same probability of being selected).
2. The sample is selected *with replacement.* The same number can be selected on both spins of the pointer.
3. The order in which the numbers are selected is important. If the same numbers are selected in a different order, the samples are considered different.

The different samples generated, from the experiment of spinning the pointer two times, are recorded in column 1 of Table 13.2. The $\bar{X}$-population of sample means is shown in column 2. Column 3 contains each sample mean squared. The $\bar{X}$-population consists of all possible samples of size 2.

Table 13.2 $\bar{X}$-population of sample means (with replacement)

Samples	$\bar{X}$-population sample means, $\bar{x}$	$\bar{x}^2$
2, 2	2	4
2, 4	3	9
2, 6	4	16
2, 8	5	25
4, 2	3	9
4, 6	5	25
4, 4	4	16
4, 8	6	36
6, 2	4	16
6, 4	5	25
6, 6	6	36
6, 8	7	49
8, 2	5	25
8, 4	6	36
8, 6	7	49
8, 8	8	64
$N=16$	$\Sigma\bar{x}=80$	$\Sigma\bar{x}^2=440$

Table 13.3 provides the sampling distribution of the sample means, for a sample size of $n=2$. Column 1 consists of the different sample means, column 2 shows the frequency for each of the sample means, and column 3 provides the probability of each sample mean being selected. Since each sample of size 2 has the same probability of $(1/4)*(1/4)=1/16$ of occurring, in Table 13.3 we can multiply the frequency (f) times (1/16) to obtain the $\Pr(\bar{X}=\bar{x})$. As an example of this, we look at the case where $\bar{x}=3$. There are two samples, (2, 4) and (4, 2), that have this sample mean. Therefore, the $\Pr(\bar{X}=3)=(1/16)+(1/16)=2/16$.

Figure 13.3 shows a histogram of the sampling distribution of sample means $\bar{x}$. Observe that for the sampling distribution of sample means each different sample mean is associated with its probability of occurring.

Using the individual $\bar{x}$'s from Table 13.2, we calculate the mean, variance, and standard deviation for the $\bar{X}$-population:

$$\Sigma\bar{x}=80 \text{ and } \Sigma\bar{x}^2=440,$$
$$\mu_{\bar{x}}=80/16=5,\ \sigma_{\bar{x}}^2=(\Sigma\bar{x}^2-[(\Sigma\bar{x})^2/16])/16$$
$$=(440-400)/16=2.5,\ \sigma_{\bar{x}}=\sqrt{2.5}.$$

Table 13.3 **Sample distribution of the sample means**

$\bar{X}$	Frequency (f)	$\Pr(\bar{X}=\bar{x})$
2	1	1/16
3	2	2/16
4	3	3/16
5	4	4/16
6	3	3/16
7	2	2/16
8	1	1/16
	$\Sigma f=16$	$\Sigma\Pr(\bar{X}=\bar{x})=1$

13.2.2. Relationships between the X-Population and the $\bar{X}$-Population When Sampling with Replacement

At this point, we wish to explore the relationships between the X-population in Example 13.1 and the $\bar{X}$-population shown in Table 13.2. From Example 13.1, the X-population had a population mean equal to 5 and a population variance equal to 5. We saw that the $\bar{X}$-population for sample size $n=2$ had a population mean equal to 5 and a population variance equal to 2.5 (the population variance for the X-population divided by the sample size). These relationships are summarized below.

(A1) The two population means are equal ($\mu_{\bar{x}}=\mu=5$).

(B1) The population variance of the $\bar{X}$-population is equal to the population variance of the X-population divided by the sample size ($\sigma_{\bar{x}}^2=\sigma^2/n=5/2$; n is the sample size 2).

(C1) Taking the square roots of both sides of the equation in (B1) gives the following relationship between the population standard deviations: $\sigma_{\bar{x}}=\sigma/\sqrt{n}=\sqrt{5/2}$.

(D1) The histogram for the $\bar{X}$-population resembles a normal curve.

The above experiment consisted of two spins of a pointer on a disk; each spin resulted in one of four equally likely outcomes. Using the *Fundamental Counting Principle*, the $\bar{X}$-population has a size of $N=4*4=4^2=16$ outcomes. If the experiment consisted of spinning the pointer three times, the

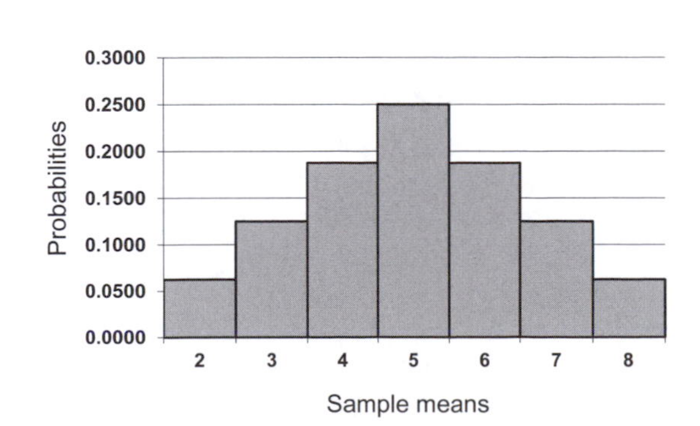

Figure 13.3 Histogram for sampling distribution of sample means in Table 13.3

sampling distribution of sample means for the sample size $n=3$ would have 64 outcomes. Therefore, the $\bar{X}$-population's size would equal 64.

For any finite population consisting of k outcomes, we can construct a disk divided into k sectors, where the area of each sector is determined by the probability that the outcome occurs. Each of the k outcomes would then have a known probability of occurring with each spin of the pointer. If we were to spin the pointer on this disk t times, the $\bar{X}$ -population would have a population size of $N=k^t$.

We now ask the following two questions: Will conditions A1, B1, and C1 hold for any X-population and any sample size? Will the histogram for the sampling distribution of sample means always be approximately normally distributed (condition D1)? These two questions will be answered later in this chapter.

The next example looks at a disk representing a baseball player's plate appearances for his career. (To construct such a disk, see Chap. 7.) The area of each of the sectors corresponds to a player's relative frequency probability for a particular outcome occurring. The relative frequency probability was based on the ratio of the number of occurrences of the outcome to the number of plate appearances. The set of all plate appearances of the player for his career can be considered a population for his actual plate appearances, or they can be considered a sample for his theoretical population consisting of an infinite number of plate appearances. Because of the *Law of Large Numbers*, whichever interpretation we use, we can be assured that, for a large number of plate appearances, the relative frequency probabilities are close to the *true* theoretical probabilities. When the spins of a pointer on the player's disk are used to simulate a player's batting performance, the player's batting performance satisfies the following axioms:

- The plate appearances (spins of pointer) are independent.
- The probability of any outcome occurring does not change from plate appearance to plate appearance (spin to spin).
- A probability can be assigned, in advance, to each outcome occurring.
- You can have as many plate appearances as you wish.

Of course, the *actual* plate appearance of a player is dictated by many outside variables, such as the pitcher he faces, the stadium he bats in, his age, the pressure of the situation, how fatigued he is, and his current injury status. These outside variables have no effect on the spinning of the pointer on the disk.

All Major League managers have a record of the plate appearances of a particular player against a particular pitcher. You can think of this as many disks, one for each pitcher. The probabilities used for these disks are conditional probabilities, based on the *given pitcher.* To decide which player to use in a batting situation, instead of using the overall player's disk, the manager uses the disk based on the pitcher the batter is to face. The problem is that usually the number of plate appearances against a given pitcher is small, which makes the relative probabilities on these disks less accurate.

Example 13.2. Aaron's disk (Fig. 7.4) contains the eight possible outcomes of a plate appearance. Each outcome resides in a sector, whose area is based on the probability of that outcome occurring. Aaron's disk differs from that constructed in Example 13.1. The disk in Example 13.1 had four equally like outcomes, whereas for Aaron's disk the outcomes have known but different probabilities associated with them. In this example, these eight outcomes are divided into two mutually exclusive events. The event "success" = {1B, 2B, 3B, HR, BB, HBP}, and the event "failure" = {SF, Out}. The outcomes belonging to the event "success" correspond to Aaron getting on base. Define a variable X as follows: $X=1$ for any spin whose outcome belongs to "success," and $X=0$ for any spin whose outcome belongs to "failure." Using the probabilities for each outcome of a plate appearance for Aaron (see Fig. 7.4), the pdf for the discrete variable X, based on one spin of the pointer on the disk, is shown in Table 13.4.

The above probabilities come from the fact that for Aaron the individual properties are Pr(1B) = .1648, Pr(2B) = .0448, Pr(3B) = .0070, Pr(HR) = .0542, Pr(BB) = .1007, Pr(HBP) = .0022, Pr(SF) = .0086, and Pr(O) = .6174.

As the outcomes are mutually exclusive, the probabilities can be added.

The mean for the pdf of X is

$$\mu = \Sigma[x * \Pr(X=x)] = 1 * .3737 + 0 * .6260 = .3737.$$

Of course, .374 is the career OBP for Aaron. The calculation of the variance for the pdf is left as an exercise.

Each result of three spins corresponds to one sample of size 3. By the *Fundamental Counting Law*, for three spins there would be $2^3 = 8$ possible samples of size 3. These are 000, 010, 001, 100, 110, 101, 011, and 111. If we calculated the sample mean for each of these samples and formed the sampling distribution of sample means, what do you think the mean $\mu_{\bar{x}}$ and the variance $\sigma_{\bar{x}}^2$ for the sampling distribution would be? To construct the sampling distribution of sample means for all samples of size 3, we would have to find the sample mean and the probability associated with each of the eight possible samples. For example, the outcome 000 would have a sample mean = (0 + 0 + 0)/3 = 0 and a probability of

Table 13.4 **Probability distribution function for the discrete variable *X***

Event	*X*	Probability
Success	1	1648 + .0448 + .0070 + .0542 + .1007 + .0022 = .3737
Failure	0	.0086 + .6174 = .6260

Pr(000) = .6260 * .6260 * .6260. The outcome 011 would have a sample mean = (0 + 1 + 1)/3 = 2/3 and a probability of Pr(011) = .6260 * .3737 * .3737. Why are we allowed to use the product of the three probabilities for Pr(011)? In the exercises that follow, you will be asked to construct the complete pdf for the $\bar{X}$-population.

Suppose that we increased the sample size to 5, 10, 100, and 400 spins; what would be the corresponding number of samples in the sampling distribution?

- For 5 spins, the $\bar{X}$-population would have a size of $N = 2^5 = 32$.
- For 10 spins, the $\bar{X}$-population would have a size of $N = 2^{10} = 1024$.
- For 100 spins, the $\bar{X}$-population would have a size of $N = 2^{100} = 1.267 * 10^{30}$.
- For 400 spins, the $\bar{X}$-population would have a size of $N = 2^{400}$ = calculator overflow.

Exercise 13.1. Using the population in Example 13.1, list all samples corresponding to *three spins* of the pointer on the disk in Figure 13.2.

Exercise 13.2. Construct the sampling distribution of sample means for the samples in Exercise 13.1. Calculate $\mu_{\bar{x}}$ and $\sigma_{\bar{x}}^2$ for the sampling distribution of sample means. Draw the histogram for the sampling distribution of sample means. Are the conditions A1, B1, and C1 above satisfied? Is the histogram for the sampling distribution approximately normal?

Exercise 13.3. Calculate the variance for the pdf for the variable X in Example 13.2. If you were to construct a sampling distribution of sample means, based on samples resulting from 3, 5, 10, 100, and 400 spins, what would you expect the mean, $\mu_{\bar{x}}$, and the variance, $\sigma_{\bar{x}}^2$, to equal?

Exercise 13.4. Consider a sample of three spins of the pointer on Aaron's disk. Using the two events and the discrete variable in Example 13.2, there would be $2^3 = 8$ different possible samples. For each sample, find the sample mean and the probability associated with that sample mean occurring. Use this to construct the sampling distribution of sample means. The table you construct should be similar to what was done in Table 13.3, with column 1 consisting of the different sample means, column 2 showing the frequency for each of the sample means, and column 3 providing the probability of each sample mean being selected. Finally, calculate the mean and variance for this sampling distribution. Sketch a histogram for the sampling distribution. How do your results compare with conditions A1, B1, C1, and D1 above?

For the next exercise you will need either a spinner device or the random number generator described in Section 7.12. You can also use the online spinner at http://illuminations.nctm.org/activitydetail.aspx?ID=79. A spinner device is included in the game *All Star Baseball* by Cadaco. You can adjust the radius of the pie graph for Aaron to fit the spinning device. Then you can print out the player's disk, cut it, and use it.

Exercise 13.5. Using the two events and discrete variable X from Example 13.2, perform the experiment of spinning the pointer on Aaron's disk three times. This will create a sample of size 3. Record your sample results. Now repeat this task 99 more times. When you are done, you will have 100 samples of size 3 (allow duplicates).

a. Create a table for the $\bar{X}$-population with the following three columns: column 1 will contain the 100 sample results, column 2 is the sample mean for each sample, and column 3 is the probability of that sample mean occurring.
b. For the $\bar{X}$-population consisting of the 100 sample means, calculate the mean ($\mu_{\bar{x}}$) and variance ($\sigma_{\bar{x}}^2$).
c. Construct the sampling distribution of sample means based on the population consisting of the 100 samples.
d. Sketch a histogram for the sampling distribution in (c).
e. What is the difference between Exercises 13.4 and 13.5? Compare the results in Exercise 13.4 with the results in Exercise 13.5. Why are they different?

Table 13.5 **$\bar{X}$-population of sample means (without replacement)**

Samples	Sample mean, $\bar{x}$	$\bar{x}^2$
2, 4	3	9
2, 6	4	16
2, 8	5	25
4, 2	3	9
4, 6	5	25
4, 8	6	36
6, 2	4	16
6, 4	5	25
6, 8	7	49
8, 2	5	25
8, 4	6	36
8, 6	7	49
$N=12$	$\Sigma\bar{x}=60$	$\Sigma\bar{x}^2=320$

Table 13.6 **Sample distribution of the sample means**

$\bar{X}$	Frequency (f)	$\Pr(\bar{X}=\bar{x})$
3	2	2/12
4	2	2/12
5	4	4/12
6	2	2/12
7	2	2/12

We now return to the X-population in Table 13.1. We change the sampling method from sampling with replacement to sampling without replacement to construct a sampling distribution of sample means for the population in Example 13.1.

13.2.3. Sampling from a Population (without Replacement)

The sample size is $n=2$, and the sampling process is simple random sampling without replacement (refer to Chap. 7). When sampling without replacement, the samples (2, 2), (4, 4), (6, 6), and (8, 8) cannot occur.

Table 13.5 lists all the possible samples of size 2 (selected without replacement) in column 1, all the sample means in column 2, and the sample means squared in column 3. Table 13.6 gives the sampling distribution of the sample means, for a sample size of $n=2$.

Using the individual $\bar{x}$'s from Table 13.5, the mean, variance, and standard deviation for the $\bar{X}$-population are calculated below.

$$\mu_{\bar{x}}=60/12=5, \sigma_{\bar{x}}^{2}=[\Sigma\bar{x}^2-(\Sigma\bar{x})^2/12]/12=(320-300)/12=5/3, \sigma_{\bar{x}}=\sqrt{5/3}.$$

13.2.4. Relationships between the X-Population and the $\bar{X}$-Population When Sampling without Replacement

We now explore the relationships between the X-population in Example 13.1 and its $\bar{X}$-population shown in Table 13.5:

(A1) The two population means are equal ($\mu_{\bar{x}} = \mu = 5$).

(B1) The population variance of the $\bar{X}$-population is equal to the product of the finite correction factor, $(N-n)/(N-1)$, and the quotient consisting of the population variance for the X-population in the numerator and the sample size in the denominator. $\sigma_{\bar{x}}^2 = [(N-n)/(N-1)] * (\sigma^2/n) = (2/3) * (5/2) = 5/3$; $N=4$ (population size), $n=2$ (sample size).

(C1) Taking the square roots of both sides of the equation in (B1) gives the following relationship between the population standard deviations:

$$\sigma_{\bar{x}} = \sqrt{(N-n)/(N-1)} * (\sigma/\sqrt{n}) = \sqrt{5/3}.$$

Exercise 13.6. Sketch the histogram for the $\bar{X}$-population in Table 13.5. What shape does the histogram take?

13.2.5. Sampling from a Binomial Population

A binomial population is generated from a binomial experiment. For the disk in Figure 13.2, we introduce a new Y-population by assigning the letter A to the number 1 and assigning B, D, and E to the number 0. The Y-population is (1, 0, 0, 0). Spinning the pointer on this disk represents a binomial experiment with the two mutually exclusive events {A} and {B, D, E}. The binomial experiment consists of one spin of the pointer on the disk. The two possible outcomes for one spin are either 0 or 1. The values of the binomial variable are 0 and 1. Since the four sectors of the disk have the same probability of occurring, the $\Pr(Y=0) = 3/4$ and the $\Pr(Y=1) = 1/4$. The pdf for the Y-population is displayed in Table 13.7.

Table 13.7 **Probability distribution function for the Y-population**

Y	Frequency	$\Pr(Y=y)$
0	3	3/4
1	1	1/4
		$\Sigma\Pr(Y=y) = 1$

13.2.5.1. The Sampling Distribution of Sample Means for the Y-Population in Table 13.7

The $\bar{Y}$-population can be derived from Table 13.2 by replacing $x=2$ by $y=1$ and $x=4$, 6, and 8 by $y=0$. Table 13.8 gives the $\bar{Y}$-population; Table 13.9 gives its pdf, and Figure 13.4 gives the histogram for the $\bar{Y}$-population.

Exercise 13.7. Show that the population mean for the Y-population is 1/4 and the population variance is 3/16.

Exercise 13.8. Compute the mean and variance of the pdf for the Y-population in Table 13.7.

Exercise 13.9. Compare the results in Exercise 13.7 and Exercise 13.8.

We now compute the mean and variance for the $\bar{Y}$-population in Table 13.8.

$N = 16$, $\Sigma\bar{y} = 4$, and $\Sigma\bar{y}^2 = 10/4 = 5/2$,
$\mu_{\bar{y}} = 4/16 = 1/4$,
$\sigma_{\bar{y}}^2 = \{\Sigma\bar{y}^2 - [(\Sigma\bar{y})^2/16]\}/16 = (10/4 - 16/16)/16 = 6/64$
$= 3/32$,
$\sigma_{\bar{y}} = \sqrt{3/32}$.

Exercise 13.10. Compute the mean and variance for the $\bar{Y}$-population by calculating the mean and variance for the sampling distribution of sample means $\bar{y}$ in Table 13.9.

13.2.5.2. Relationships between the Y-Population and the $\bar{Y}$-Population When Sampling with Replacement

The relationships between these two populations are summarized below.

(A1) The two population means are equal ($\mu_{\bar{y}} = \mu = 1/4$).
(B1) $\sigma_{\bar{y}}^2 = \sigma^2/2 = (3/16)/2 = 3/32$.
(C1) $\sigma_{\bar{y}} = \sigma/\sqrt{2} = \sqrt{3/32}$.

The histogram of the sampling distribution of $\bar{y}$ shown in Figure 13.4 has a shape that is *not close to normal.*

Table 13.8 $\bar{Y}$-population

Samples	$\bar{y}$	$\bar{y}^2$
1, 1	1	1
1, 0	1/2	1/4
1, 0	1/2	1/4
1, 0	1/2	1/4
0, 1	1/2	1/4
0, 0	0	0
0, 0	0	0
0, 0	0	0
0, 1	1/2	1/4
0, 0	0	0
0, 0	0	0
0, 0	0	0
0, 1	1/2	1/4
0, 0	0	0
0, 0	0	0
0, 0	0	0
$N = 16$	$\Sigma\bar{y} = 4$	$\Sigma\bar{y}^2 = 5/2$

Table 13.9 Probability distribution for the $\bar{Y}$-population

$\bar{Y}$	Frequency	$Pr(\bar{Y} = \bar{y})$
0	9	9/16
1/2	6	6/16
1	1	1/16
	$\Sigma f = 16$	$\Sigma Pr(\bar{Y} = \bar{y}) = 1$

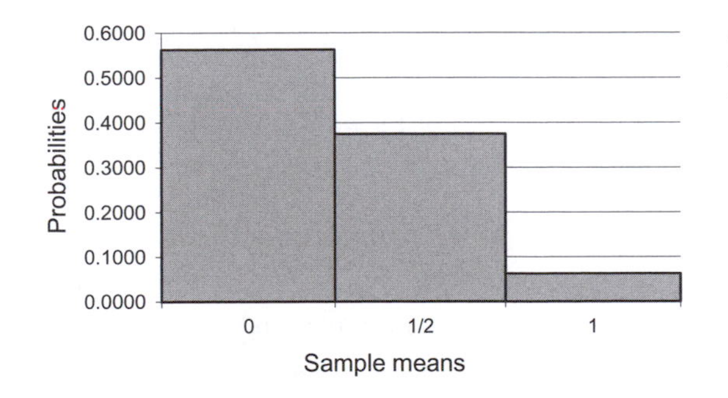

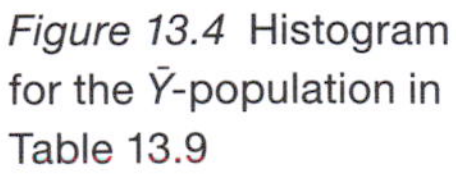

Figure 13.4 Histogram for the $\bar{Y}$-population in Table 13.9

13.2.6. Summary of Relationships between the Original Populations and Their Sampling Distributions of Sample Means

So far in this chapter we have looked at three sampling distributions of sample means. Table 13.1 contains the original X-population, and Table 13.7 contains the original Y-population. Two sampling distributions were constructed for the original X-population. Table 13.3 displays the sampling distribution of sample means, where the sampling was for a fixed size $n = 2$ (with replacement). Table 13.6 displays the sampling distribution of sample means, where the sampling was for a fixed size $n = 2$ (without replacement). Table 13.9 contains the sampling distribution of sample means for the Y-population, where the sampling was for a fixed sample size $n = 2$ (with replacement).

For all three sampling distributions of sample means, we observed the following:

(A) The population mean of the sampling distribution of sample means is equal to the population mean of the original population:

$$\mu_{\bar{x}} = \mu.$$

For those sampling distributions of sample means in which the sampling method was with replacement, we observed the following:

(B) The population variance of the sampling distribution of sample means is equal to the variance of the original population divided by the sample size:

$$\sigma_{\bar{x}}^2 = \sigma^2/n.$$

(C) The population standard deviation of the sampling distribution of sample means is the square root of the population variance:

$$\sigma_{\bar{x}} = \sigma/\sqrt{n}.$$

For those sampling distributions for sample means in which the sampling method was without replacement, we observed the following:

(B′) The population variance of the sampling distribution of sample means is equal to the product of the finite correction factor, $(N - n)/(N - 1)$, and the quotient consisting of the variance of the original population divided by the sample size:

$\sigma_{\bar{x}}^2 = [(N-n)/(N - 1)] * (\sigma^2/n)$; $n =$ sample size, $N =$ population size.

If $n < (.05) * N$, the correction factor can be dropped.

(C′) The population standard deviation for the sampling distribution of sample means is the square root of the population variance:

$$\sigma_{\bar{x}} = (\sigma/\sqrt{n}) * \sqrt{(N-n)/(N-1)}.$$

The next theorem summarizes these results.

Theorem 13.1. When all possible samples of a specific size are selected with replacement from a population, properties A, B, and C hold for the sampling distribution of sample means. Properties B′ and C′ only come into play when sampling is without replacement and the sample size is greater than 5% of the population.

When the original population is infinite or the finite population is large and it is either impossible or difficult to list all possible samples for a fixed sample size, many random samples can be collected and used for the sampling distribution. In this case, the relationships in properties A, B, and C are good estimates.

13.2.7. Special Definitions Used with Sampling Distributions

Below, the definition of the standard error of the mean and the definition of an unbiased estimator are introduced. These two definitions are then related to our examples.

- The standard deviation of the sampling distribution of sample means $\sigma_{\bar{x}}$ is also called the *standard error of the mean.*
- If the mean of the sampling distribution of a statistic for a population is equal to the parameter of that population, we say that the statistic is an *unbiased estimator* for that parameter.
- In our examples $\bar{x}$ was the statistic. The mean of the sampling distribution of sample means $\bar{x}$ was equal to the parameter μ. Therefore, we can say that $\bar{x}$ is an unbiased estimator for μ. The sample median would be a biased estimator for the population mean. This completes our discussion, started in Section 1.7, of types of biases.
- Looking at the histograms for the sampling distribution of sample means discussed in this section, we observed that only some of the histograms resembled a normal curve.

13.2.8. The Shape of the Sampling Distribution of Sample Means

What conditions or assumptions are necessary to conclude that the shape for the sampling distribution of sample means is approximately normal? The following two theorems address this question.

Theorem 13.2. If the original population X itself is normally distributed, the sampling distribution of sample means is normally distributed.

Theorem 13.3 (Central Limit Theorem). Given that the original population X has any functional form (not necessarily normal), the sampling distribution of sample means will be approximately normal, provided that the sample size n is at least 30 ($n \geq 30$). The greater the sample size, the better the approximation.

How do we interpret Theorems 13.2 and 13.3? They tell us that if the original population is normal or if the original population is not normal but the fixed sample size is at least 30, then the sampling distribution of sample means will be approximately normal in shape.

Be careful to understand what the above does not say: if the X-population is not normal and the sample size is *less* than 30, we cannot *automatically* conclude that the sampling distribution of sample means is not approximately normal. The above simply says that we cannot *guarantee* that the sampling distribution is approximately normal.

We saw for the X-population (2, 4, 6, 8) that the shape of the sampling distribution of $\bar{x}$ was close to normal, whereas for the Y-population (1, 0, 0, 0) the shape of the sampling distribution of $\bar{y}$ was not close to normal. In both cases the original X-populations were not normal and the sample size was 2.

Exercise 13.11. Explain why the result for the Y-population does not contradict the Central Limit Theorem.

We now turn our attention to the sampling distribution of sample proportions $\hat{p}$.

13.3. The Sampling Distribution of Sample Proportions

We return to the Y-population (1, 0, 0, 0). We assign the event "success" to getting the outcome 1 and the event "failure" to getting the outcome 0. One outcome of the Y-population is a "success," and three outcomes of the Y-population are "failures." To keep the three zeroes separate, we can call them 0_1, 0_2, and 0_3. The population proportion for the category "success" for the Y-population is 1/4. Table 13.10 provides all the possible samples (sampling with replacement) of size $n=2$ for the Y-population. The sample proportion of successes is the ratio x/n, where x is the total number of outcomes belonging to the category "success" and n is the sample size. Each sample mean $\bar{y}$ calculated in Table 13.10 corresponds to the sample proportion of successes. To help understand this, we look at two of the samples. The sample (1, 1) has sample mean $=(1+1)/2=1$ and sample proportion $=2/2$ (since both outcomes are successes). The sample (0, 1) has sam-

Table 13.10 All possible samples of size $n=2$ with their associated $\bar{y}$ and $\hat{p}$

Samples	$\bar{Y}$-population $\bar{y}$	$\hat{P}$-population $\hat{p}$
1, 1	1	2/2
1, 0_1	1/2	1/2
1, 0_2	1/2	1/2
1, 0_3	1/2	1/2
0_1, 1	1/2	1/2
0_1, 0_1	0	0/2
0_1, 0_2	0	0/2
0_1, 0_3	0	0/2
0_2, 1	1/2	1/2
0_2, 0_1	0	0/2
0_2, 0_2	0	0/2
0_2, 0_3	0	0/2
0_3, 1	1/2	1/2
0_3, 0_1	0	0/2
0_3, 0_2	0	0/2
0_3, 0_3	0	0/2
$N=16$	$\Sigma\bar{y}=5/2$	$\Sigma\hat{p}=4$

ple mean $=(0+1)/2=1/2$ and sample proportion $=1/2$ (since one outcome is a success and the other is a failure). Table 13.10 shows each sample mean with its corresponding sample proportion, $\hat{p}$. The set of all sample proportions is called the $\hat{P}$-population.

Since the numbers in the $\hat{P}$-population are identical to the numbers in the $\bar{Y}$-population, the mean and variance of the $\hat{P}$-population will equal the mean and variance of the $\bar{Y}$-population.

$$\mu_{\hat{p}}=\mu_{\bar{y}}=1/4,$$
$$\sigma_{\hat{p}}^{\ 2}=\sigma_{\bar{y}}^{\ 2}=3/32.$$

For the population (1, 0, 0, 0), the only outcome associated with the category "success" is 1. Since one out of four outcomes of the population corresponds to success, the population proportion of successes is $p=1/4$. Letting $n=2$ and $p=1/4$, we see that $p(1-p)/n=(1/4)*(3/4)/2=3/32$.

For the population (1, 0, 0, 0), we have shown that the following three properties are satisfied for its $\hat{P}$-population:

(A1) The mean of the $\hat{P}$-population $\mu_{\hat{p}}$ is equal to p (the population proportion of successes).
(B1) The variance of the $\hat{P}$-population $\sigma_{\hat{p}}^{\ 2}$ is equal to $[p(1-p)]/n$.
(C1) The standard deviation of the $\hat{P}$-population, $\sigma_{\hat{p}}$, is equal to $\sqrt{[p(1-p)]/n}$.

Since each sample mean is equal to a sample proportion, the histogram in Figure 13.4 also represents the histogram for the sampling distribution of $\hat{p}$. Clearly, the shape is not close to normal. We now look at another example to see if properties A1, B1, and C1 remain true.

For the disk in Figure 13.2, we introduce a new Z-population by assigning the letters A and B to the number 1 and assigning the letters D and E to the number 0. The Z-population $=$ (**1**, 1, **0**, 0). The use of bold keeps the outcomes separate. Spinning the pointer on the disk now represents a binomial experiment with the two mutually exclusive events of {A, B} and {D, E}. The values of the binomial variable are 0 and 1. The pdf for the Z-population is left as an exercise below. The binomial experiment consists of two spins of the pointer on the disk. The number 1 is assigned to "success," and the number 0 is assigned to "failure." For this population, the population proportion of successes is $p=2/4$. The $\hat{P}$-population for sample size $n=2$ appears in Table 13.11.

Table 13.11 $\hat{P}$-population for sample size $n=2$

Samples	$\hat{p}$	$\hat{p}^2$
1, **1**	2/2	1
1, 1	2/2	1
1, **0**	1/2	1/4
1, 0	1/2	1/4
1, **1**	2/2	1
1, 1	2/2	1
1, **0**	1/2	1/4
1, 0	1/2	1/4
0, **1**	1/2	1/4
0, 1	1/2	1/4
0, **0**	0/2	0
0, 0	0/2	0
0, **1**	1/2	1/4
0, 1	1/2	1/4
0, **0**	0/2	0
0, 0	0/2	0
$N=16$	$\Sigma\hat{p}=8$	$\Sigma\hat{p}^2=6$

Table 13.12 Sampling distribution of $\hat{p}$

$\hat{P}$	Frequency	$Pr(\hat{P}=\hat{p})$
0	4	4/16
1/2	8	8/16
1	4	4/16

The mean of the sampling distribution of $\hat{p}$ is $\mu_{\hat{p}} = \Sigma\hat{p}/16 = 8/16 = 1/2$.

The variance of the sampling distribution of $\hat{p}$ is $\sigma_{\hat{p}}^2 = (\Sigma\hat{p}^2 - [(\Sigma\hat{p})^2/16]/16) = (6-64/16)/16 = 2/16 = 1/8$.

We see that the same three properties are satisfied.

(A1) $\mu_{\hat{p}} = p = 1/2$.
(B1) $\sigma_{\hat{p}}^2 = p(1-p)/n = 1/2(1-1/2)/2 = 1/8$.
(C1) $\sigma_{\hat{p}} = \sqrt{[p(1-p)]/2} = \sqrt{1/8}$.

The sampling distribution of $\hat{p}$ is shown in Table 13.12.

This histogram for the sampling distribution of $\hat{p}$ is given in Figure 13.5.

Exercise 13.12. Compute the mean, variance, and standard deviation for the sampling distribution of $\hat{p}$ in Table 13.12.

Exercise 13.13. Construct the histogram for the pdf for the *Z*-population.

For the above two sampling distributions of sample proportions, we observed the following:

(A1) The population mean of the sampling distribution of sample proportions is equal to the population proportion of the original population:

$$\mu_{\hat{p}} = p.$$

(B1) The population variance of the sampling distribution of sample proportions is

$$\sigma_{\hat{p}}^2 = p(1-p)/n \qquad [n \text{ is the sample size}].$$

(C1) The population standard deviation of the sampling distribution of the sample proportion is

$$\sigma_{\hat{p}} = \sqrt{[p(1-p)]/n}.$$

The standard deviation of the sampling distribution of the sample proportion $\sigma_{\hat{p}}$ is called the *standard error of the mean.*

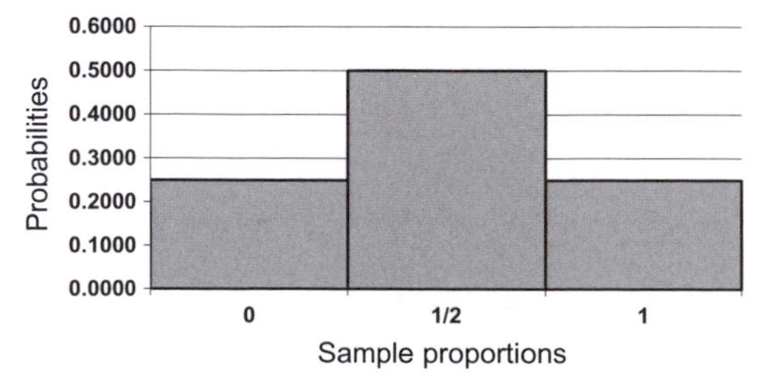

Figure 13.5 Histogram for sampling distribution of p̂ from Table 13.11

Since the mean of the sampling distribution of $\hat{p}$ is equal to the parameter p, we can say that $\hat{p}$ is an unbiased estimator for p.

Theorem 13.4. Given a binomial experiment consisting of n trials, a sample consists of the "successes" and "failures" in the n trials of the binomial experiment. A sample proportion $\hat{p}$, for the n trials of the binomial experiment, is equal to the number of successes divided by the number of trials. The sampling distribution of sample proportions satisfies the properties A1, B1, and C1.

Observe that the shape of the histogram in Figure 13.5 is closer to normal than the shape of the histogram in Figure 13.4. The reason is that the closer the population proportion is to 1/2, the closer the shape of the sampling distribution of $\hat{p}$ will be to a normal shape.

In general, the larger the sample size n and the closer the population proportion p is to 1/2, the closer the shape of the sampling distribution of $\hat{p}$ is to being normal. The next theorem gives a sufficient condition for the normality of the sampling distribution.

Theorem 13.5. If both np and $n(1-p)$ are greater than 5, we can assume that the sampling distribution of the sample proportion $\hat{p}$ is approximately normal in shape.

Example 13.3. We return to the spinning of a pointer on Aaron's disk described in Example 13.2. Assign the event "success" to {1B, 2B, 3B, HR, BB, HBP} and the event "failure" to {SF, Out}. The spinning of a pointer on Aaron's disk is a binomial experiment. For each spin of the pointer on the disk, a 1 is assigned to an outcome in "success" and 0 is assigned to an outcome in "failure." The probability of the event "success" occurring is equal to the sum of the probabilities of the outcomes corresponding to "success," which is equal to .374; the probability of the event "failure" occurring is equal to the sum of the probabilities corresponding to "failure," which is equal to .626. The Pr("success") = .374, which corresponds to Aaron's career OBP. Suppose that the experiment consists of spinning the pointer on the disk four times. One possible outcome of the four spins could be 1001. We say that this is a sample of size $n = 4$ with a sample proportion of successes equal to 2/4. The probability of the sample 1001 occurring is equal to .374 * .626 * .626 * .374. = .0548.

The possible sample proportions of successes, $\hat{p}$, for four spins include 0/4, 1/4, 2/4, 3/4, and 4/4. The sampling distribution of $\hat{p}$ will consist of two columns. The first column consists of the sample proportions 0/4, 1/4, 2/4, 3/4, and 4/4; the second column will be the probability associated with each of these sample proportions occurring.

Exercise 13.4.

a. Construct the sampling distribution of sample proportions for the four spins of the pointer on Aaron's disk by first completing Table 13.13 and then completing Table 13.14.

Table 13.13 **Binomial experiment for four spins on Aaron's disk**

Samples	Sample proportion ($\hat{p}$)	$Pr(\hat{P}=\hat{p})$
1111	4/4	.374 * .374 * .374 * .374 = .0196
1011	3/4	.374 * .626 * .374 * .374 = .0327
0000	0/4	.626 * .626 * .626 * .626 = .1536

Table 13.14 **Sampling distribution of $\hat{p}$**

$\hat{P}$	$\Sigma Pr(\hat{P}=\hat{p})$
0/4	
1/4	
2/4	
3/4	
4/4	

b. For Table 13.14, compute $\Sigma\hat{p}$ and $\Sigma Pr(\hat{P}=\hat{p})$.
c. Calculate the mean and variance for the binomial distribution with $p=$ OBP $=.374$, $n=4$ spins of the pointer.
d. Calculate the mean and variance of the sampling distribution given by Table 13.14.
e. Sketch a histogram for the sampling distribution in Table 13.14.
f. Does Theorem 13.4 apply? Explain.
g. Does Theorem 13.5 apply? Explain.

In the chapter problems at the end of this chapter, you will be asked to repeat Example 13.3 and Exercise 13.4 for one of your chosen players.

Chapter Summary

A sampling distribution of a chosen statistic is a special type of probability distribution function. In a sampling distribution, the random variable X is a statistic. Any statistic (descriptive measure of the sample) can be used as the random variable.

A sampling distribution is a probability distribution function (pdf) whose random variable is a statistic calculated for many random samples of the same sample size n, selected from the original population. Since a sampling distribution is a pdf, each distinct value of the statistic is associated with the probability of it occurring.

The two statistics used for the sampling distributions in this chapter were the sample mean and the sample proportion. The two sampling distributions studied in this chapter were the sampling distribution of sample means and the sampling distribution of sample proportions.

Both these sampling distributions have important relationships with their original populations. These relationships are summarized next.

If either the X-population is normal or the sample size $n \geq 30$, the sampling distribution of $\bar{x}$ satisfies the following conditions:

1. $\mu_{\bar{x}} = \mu$.
2. $\sigma_{\bar{x}} = \sigma/\sqrt{n}$.
3. Its shape is approximately normal.

If $np > 5$ and $n(1-p) > 5$, the sampling distribution of $\hat{p}$ satisfies the following conditions:

1. $\mu_{\hat{p}} = p$.
2. $\sigma_{\hat{p}} = \sqrt{p(1-p)/n}$.
3. Its shape is approximately normal.

Any statistic can be used to generate a sampling distribution. However, many statistics do not provide meaningful relationships with their original population.

What are the meaningful relationships between the sampling distributions for $\bar{x}$ and $\hat{p}$ and their original populations?

(A) Since $\mu_{\bar{x}} = \mu$ and $\mu_{\hat{p}} = p$, both $\bar{x}$ and $\hat{p}$ are unbiased estimators for μ and p.
(B) Under certain assumptions, both sampling distributions are approximately normally distributed.
(C) The standard deviations of both sampling distributions have known formulas. The standard deviation of the sampling distribution of $\bar{x}$ is $\sigma_{\bar{x}} = \sigma/\sqrt{n}$, and the standard deviation of the sampling distribution of $\hat{p}$ is $\sigma_{\hat{p}} = \sqrt{p(1-p)/n}$. These two standard deviations are called the *standard error of the mean.*

The standard error of the mean measures how spread out the population for the statistic is around the parameter being estimated. As the sample size n increases, the standard error gets smaller. The smaller the standard error of the mean, the more likely the statistic calculated from the sample will be closer to the parameter. This makes sense because a larger sample size looks more like the original population.

When these sampling distributions are approximately normal, we know that the parameter $\pm 1 *$ (standard error) forms an interval containing approximately 68% of the population values for the statistic. The parameter $\pm 2 *$ (standard error) forms an interval containing approximately 95% of the population values for the statistic. The parameter $\pm 3 *$ (standard error) forms an interval containing approximately 99.7% of the population values for the statistic.

Every statistic has a sampling distribution, but the appropriate sampling distribution may or may not be approximately normal. An example of a non-normal sampling distribution is the sampling distribution of the sample variances.

Since the sample proportion variable $\hat{p}$ is derived by dividing a binomial variable X by a fixed constant integer n, properties 1 and 2 of the sampling distribution of $\hat{p}$ can be justified by the observation mentioned in Section 2.4. This observation shows that the multiplication of a variable by a fixed positive constant results in a new variable, and that the mean and standard deviation of the new variable are equal to the products of the same constant with the mean and standard deviation of the original variable.

Applying the above to a binomial variable, X, whose mean is np and standard deviation is $\sqrt{np(1-p)}$, the variable $\hat{P} = X/n = (1/n) * X$ has a mean equal to $(1/n) * np = p$ and a standard deviation equal to $(1/n) * \sqrt{np(1-p)} = \sqrt{np(1-p)/(n * n)} = \sqrt{p(1-p)/n}$.

When dealing with many independent samples drawn from an infinite population, the corresponding results 1, 2, and 3 above become *approximations.*

Two major techniques used in inferential statistics are *confidence intervals* and *hypothesis testing.* Both of these techniques rely on sampling distributions. In Chapter 14 we will discuss confidence intervals, and in Chapter 15 we will discuss hypothesis testing.

CHAPTER PROBLEMS

For problems 1–5, you will use the disk for one of your players, consisting of the eight outcomes {1B, 2B, 3B, HR, BB, HBP, SF, O}. Using the procedure described in Example 13.3 and Exercise 13.4, you will be asked to construct different sampling distributions of sample proportions.

1. Using a sample size of $n = 5$ spins, construct tables similar to Table 13.13 and Table 13.14.

2. Using a sample size of $n = 10$ spins, create 100 samples of size $n = 10$. The samples are created by either spinning a pointer on your player's disk or using the random number generator in Excel or the online spinner. Construct tables similar to Tables 13.13 and 13.14. The sample proportions, $\hat{p}$, will be 0/10, 1/10, 2/10, 3/10, 4/10, 5/10, 6/10, 7/10, 8/10, 9/10, and 10/10. Repeat for $n = 20$ spins.
3. Compute $\mu_{\hat{p}}$, $\sigma_{\hat{p}}^2$, and $\sigma_{\hat{p}}$ for each of the above three sampling distributions of sample proportions, $\hat{p}$.
4. For the binomial experiment of spinning the pointer on your player's disk a fixed number of times, the population proportion $p =$ Pr("success") = the career OBP of your player. Compare $\mu_{\hat{p}}$ with p and $\sigma_{\hat{p}}^2$ with $p(1-p)/n$ for each of the three sampling distributions above. What conclusions can you make? What effect does the increase in sample size have on your conclusions?
5. Sketch a histogram for each of the three sampling distributions. Discuss the shape of the histograms. What effect does the increase in sample size have?

The next set of problems tests your knowledge of the theorems concerning sampling distributions.

6. Suppose that it is known for a certain population that $\mu = 240$ and $\sigma = 22$. The statements below refer to the sampling distribution of sample means with sample size $n = 100$.

   - The mean of the sampling distribution is ________.
   - The standard deviation of the sampling distribution is ________.
   - The standard error of the sampling distribution is ________.
   - The symbol used for the mean of the sampling distribution is ________.
   - The symbol used for the standard error of the sampling distribution is ________.
   - The shape of the sampling distribution is approximately ________.

7. Suppose that it is also known that the above population is normal. Fill in the same statements in problem 6 for the sampling distribution of sample means for the sample size $n = 16$.

Chapter 14

Confidence Intervals

14.1. Introduction to Inferential Statistics

Inferential statistics uses the result of a random sample, selected from a population, to estimate a result for that population. A common application uses a statistic (a descriptive measure of a sample), calculated from a random sample of a population, to estimate a parameter (a descriptive measure of a population) for the population. In particular, we use a sample mean to estimate a population mean, and we use a sample proportion to estimate a population proportion. The sample mean and sample proportion are both statistics. The population mean and population proportion are both parameters.

In Chapter 13, we showed that both statistics, the sample mean and sample proportion, are considered good estimators for their corresponding parameters, the population mean and population proportion. Both estimators satisfy the following properties:

- They are both *unbiased estimators*. The means of their sampling distributions are the parameters we wish to estimate.
- They are both *consistent estimators*. As their sample size increases, the variances of their sampling distributions are smaller.

Throughout the rest of the book the term *true* will have the same meaning as the term *population*. For example, the *true* proportion is the same as population proportion, and the *true* mean is the same as population mean.

Many examples and exercises in this and successive chapters will concentrate on samples and populations involving the plate appearances of a player. Each plate appearance results in one of the eight possible outcomes 1B, 2B, 3B, HR, BB, HBP, SF, and O (Out). The descriptive measures studied will be means and proportions such as batting averages (BA) and on-base percentage (OBP). The samples will consist of the observed outcomes for a finite number of plate appearances for a player. The samples will be acquired from either actual plate appearances or simulations.

For the remainder of this book two types of populations will be studied. For one, the population proportion will be known; for the second, the population proportion is unknown. The next two paragraphs describe these two populations.

In Chapter 7, we created disks for Aaron and Bonds, where the area of each sector was associated with the probability of a particular batting outcome occurring. The probability of a particular batting outcome occurring is calculated by dividing the total number of career plate appearances resulting in that outcome by the total number of career plate appearances. The spinning of a pointer on the player's disk simulates a plate appearance for that player. The experiment of spinning a pointer on a disk falls into the category of *classical probability*. When a disk is used for a player, each possible outcome of a spin has a known probability of occurring. Consequently, such population proportions as the OBP will be known. Since there is no limit on the number of spins, populations involving the spins on a disk create an *infinite population* of outcomes. Samples for a fixed size n are generated by performing n spins on the disk. A sample proportion, such as OBP, based on n spins on a disk is calculated by dividing the total number of spins that resulted in outcomes belonging to OB by the number n. We write $\hat{p}=\widehat{\text{OBP}}$ for the sample proportion. The population proportion is the player's career OBP, which is known. We write p for a player's population career OBP. Of course, there are many other proportions that can be studied, such as HRA and SOA. We use the same notation for these proportions.

The second population is the theoretical population based on an *infinite number* of actual plate appearances. This population does not use a disk. For this population, the population proportion is unknown. As mentioned earlier, a player's population career OBP can be referred to as his *true* career OBP. To estimate a player's *true* career OBP, his *actual* career OBP will be used as the sample, symbolized by $\widehat{\text{OBP}}$. The $\widehat{\text{OBP}}$ is calculated from his actual career plate appearances.

The major difference between using a disk to simulate a player's plate appearances and using the player's actual plate appearances is that, in the former case, the population proportion is known. In the latter case the population proportion is unknown.

In both cases, we use the result of a *sample* to estimate a *population* result. When using simulations with the disk, we can actually measure how good the sample estimate is for the population result. For example, the $\widehat{\text{OBP}}$ can be calculated from 100 spins of a player's disk. Since the population proportion is his known career OBP, we can compare the two results. When using a player's entire career plate appearances as the sample, the sample proportion, $\widehat{OBP}$, will be his career OBP. His *true* career OBP is based on an infinite number of actual career plate appearances and consequently is theoretical and hence unknown.

The *Law of Large Numbers* links these two approaches. Since a successful player's career plate appearances usually number in the thousands, his actual observed career OBP should be close to his *true* (population) career OBP. In the above discussion, we could have replaced a player's career years with a particular year or any collection of years.

14.2. Introduction to Confidence Intervals

The process begins by identifying the population (entire set of objects or subjects in the statistical study) and the variable (characteristic) to be studied. A random sample is then selected from the population. The variable is evaluated for each object or subject in the sample. This produces the sample data. The next example illustrates this process.

Example 14.1. We simulate Aaron's career plate appearances by using the disk in Figure 7.4. Each time we spin the pointer on Aaron's disk, a plate appearance is simulated. The eight possible outcomes of a given spin of the pointer are {1B, 2B, 3B, HR, BB, HBP, SF, O}. The objects of the population are the outcomes corresponding to each of the spins of the pointer. Since the spinning of the pointer can be done over and over, the population is considered infinite. The variable of interest is the outcome of each spin. Suppose that the random sample is the outcomes for the first 10 spins of the pointer. We perform the 10 spins of the pointer on Aaron's disk. The results appear in Table 14.1. Column 2 in the table displays the sample data. Notice that the sample data are qualitative.

Table 14.1 Outcomes for 10 spins of the pointer on Aaron's disk

Spin number	Outcome
1	1B
2	Out
3	1B
4	2B
5	SF
6	Out
7	SF
8	Out
9	BB
10	Out

We choose for our statistic the *on-base proportion* for the sample data. $\widehat{OBP}$ = #OB/#PA = 4/10 (see Table 3.2). The statistic $\widehat{OBP}$ is the sample proportion. By assigning the number 1 to each result corresponding to getting on base and the number 0 to each result corresponding to not getting on base, the sample data become {1, 0, 1, 1, 0, 0, 0, 0, 1, 0}. Since the sample data are now quantitative, we can talk about the *on-base average* ($\widehat{OBP}$) of the sample data.

Since $\widehat{OBP} = (1+0+1+1+0+0+0+0+1+0)/10 = 4/10$, $\widehat{OBP}$ can also be considered a sample mean. The above discussion shows that the concept of getting on base can be interpreted as either a proportion or a mean. Since the sample OBP is equal to the sample OBA, either notation can be used interchangeably.

Once the sample mean or sample proportion is found, the next step is to use this information to estimate the population mean or population proportion. For an experiment, the population mean or population proportion will usually be unknown. In the case of the experiment of spinning the pointer on Aaron's disk, the population proportion of Aaron getting on base will be known. This is because the sectors of the disk were constructed using probabilities (proportions) calculated from all the observed career plate appearances of Aaron. The population proportion of getting on base will equal the sum of the proportions corresponding to the six outcomes {1B, 2B, 3B, HR, BB, HBP}. Why is this true? The experiment of spinning the pointer on a disk is similar to the experiment of flipping a coin, where the *true* probability of getting a head is 1/2. Since both of these experiments satisfy the assumptions of a binomial experiment, their population mean ($\mu = np$) and their population variance ($\sigma^2 = np(1-p)$) are known.

We call a sample mean or a sample proportion a *point estimate* for the population mean or population proportion. Remember, the goal is to estimate an unknown mean or proportion for the population.

Why can't we just use the point estimate as our estimate? To answer this question, consider the following. Suppose that you bought a raffle ticket. If your ticket is selected, you win; otherwise, you lose. As no information is provided about your chances of winning, there is no meaningful way to make an estimate about your probability of winning. On the other hand, suppose that for the same raffle you were told that only 50 tickets were sold. Now, you would know that you had a 1 in 50 chance of winning. The difference is that in the first case you could not assign a probability to winning the raffle, whereas in the second case you can assign a probability of $1/50 = 2\%$ of winning the raffle. This is the problem with a point estimate. Since no probability can be assigned to the accuracy of the point estimate, the usefulness of a point estimate can be questioned. Because of this issue with point estimates, *confidence intervals* are used to make estimates.

An interval consists of a center point and a radius. A *confidence interval* is a special type of interval whose center is a point estimate and whose radius is called the *margin of error*. One endpoint of a confidence interval is equal to the margin of error added to the point estimate; the other endpoint of the confidence interval is equal to the margin of error subtracted from the point estimate. The *margin of error E* is a positive number determined by the *level of confidence L*. The *level of confidence* is the probability that the interval contains the population parameter. The default level of confidence is $L = .95$. Other often-used levels of confidence are $L = .90$, $L = .98$, and $L = .99$. The next section outlines the creation of a confidence interval with level of confidence L. Since any confidence interval depends on its level of confidence L, we name the confidence interval an $L\%$ confidence interval. For example, if $L = .95$, the confidence interval is called a 95% confidence interval (the decimal point is dropped). The terms level of confidence L and an $L\%$ confidence interval can be used interchangeably.

The *margin of error* for a *confidence interval* is also called the *precision* of the confidence interval. If the margin of error is small, the estimate of the parameter is more precise. To understand this, we look at two confidence interval estimates for p. One estimate says that, with a level of confidence of 95%, p lies between .40 and .60. Another estimate says that, with a level of confidence of 95%, p lies between .45 and .55. For the first confidence interval, the margin of error is .10. For the second confidence interval, the margin of error is .05. We can say that the second confidence interval gives a more precise estimate for p than the first confidence interval.

We discovered in Chapter 13 that, under certain assumptions, both the sampling distribution of sample means $\bar{x}$ and the sampling distribution of sample proportions $\hat{p}$ satisfy the following properties:

(A) They are both normally distributed.
(B) The means, μ and p, of their respective sampling distributions are the parameters of the population we wish to estimate.
(C) The standard deviations (standard error of the mean) for both sampling distributions are known. We use the notation SE for the standard error of the mean.

Properties A, B, and C above allow us to construct a confidence interval for the population mean and for the population proportion. Without these assumptions, the construction of confidence intervals would not happen. The construction process begins with the computation of the margin of error, the radius for the confidence interval.

Exercise 14.1. What assumptions are necessary to conclude that each of these sampling distributions is normally distributed?

14.2.1. The Calculation of the Margin of Error for an L% Confidence Interval

The margin of error is the radius of the confidence interval.

To calculate the margin of error, the *reliability coefficient* must be calculated first. The reliability coefficient depends on the level of confidence L. The next example demonstrates the process for finding the reliability coefficient for the standard level of confidence $L=.95$.

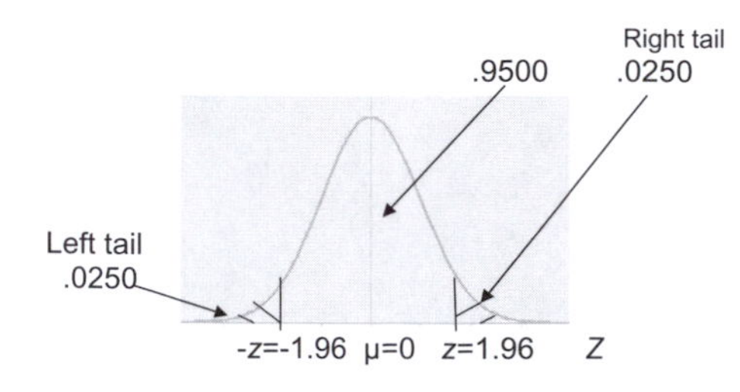

Figure 14.1 Standard normal curve with right and left tails having area equal to .0250

Example 14.2. Given $L=.95$, we find z such that the interval $(-z, z)$ contains 95% of the z-values. The area of the region from $-\infty$ to $-z$ is called the area of the left tail. The area of the left tail is calculated by subtracting .95 from 1 and then dividing the result by 2. We get $.0250=(1-.95)/2$. From the standard normal curve table, the Z-Table (Table C.2 in Appendix C), we find that the area .0250 corresponds to $-z=-1.96$. Therefore, $z=1.96$. Refer to Figure 14.1.

The interval (–1.96, 1.96) contains 95% of the z-values. The calculated z-value of 1.96 is the *reliability coefficient* for the level of confidence $L=.95$ and is symbolized by z_L. We have just shown that the *reliability coefficient* corresponding to *a level of confidence* $L=.95$, symbolized by z_L, is equal to 1.96. We write $z_{.95}=1.96$.

In general, given a level of confidence L, we can find the reliability coefficient by following these three steps:

Step 1: Calculate A = area to look up in the Z-Table. Let $A=(1-L)/2$.
Step 2: Locate in the body of the Z-Table the area closest to A.
Step 3: The reliability coefficient z_L is $-z$, where z corresponds to the lookup of the area A (z will be negative, so $-z$ will be positive).

The next example calculates $z_{.90}$.

Example 14.3.

Step 1: $A=(1-.90)/2=.0500$.
Step 2: Locating the area closest to .0500. Table 14.2 shows a small portion of the Z-Table.
Step 3: We can choose for z either –1.64 or –1.65. For no reason, we choose $z=-1.64$. Therefore, $-z=1.64$ and the reliability coefficient $z_{.90}=1.64$.

For the next exercise you will be asked to find a certain z_L.

Exercise 14.2. Find z_L for $L=.80$, $L=.98$, and $L=.99$.

We now use the reliability coefficient to calculate the margin of error. Recall that the margin of error, E, is the radius of a confidence interval.

We define the *margin of error E* as the product of the reliability coefficient z_L and the standard error of the mean SE (the standard deviation of a sampling distribution). Symbolically, the margin of error for the level of confidence L is

$$E=z_L * \text{SE}.$$

For the sampling distribution of sample proportions,

$$E=z_L * \sqrt{p(1-p)/n}.$$

Table 14.2 Portion of standard normal curve table

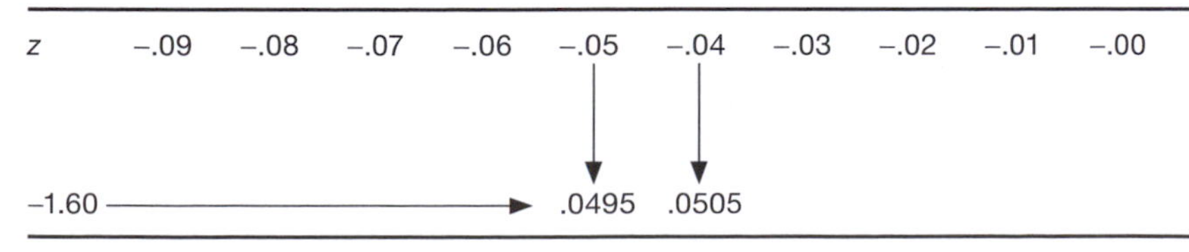

z	–.09	–.08	–.07	–.06	–.05	–.04	–.03	–.02	–.01	–.00
–1.60 →					.0495	.0505				

For the sampling distribution of sample means,

$E = z_L * (\sigma/\sqrt{n})$.

14.3. An $L\%$ Confidence Interval for the Population Proportion p

By adding E to and subtracting E from the mean p of the sampling distribution of sample proportions, an interval is formed having for its center the population proportion p and for its radius E. The left endpoint of the interval is $p - E$; the right endpoint of the interval is $p + E$. This interval will contain $L\%$ of the sample proportions for a fixed sample size n.

Why is L% of the sample proportions contained in this interval? The interval $(-z_L, z_L)$ contains $L\%$ of the z-values, where z_L is equal to the number of standard deviations away from the mean $z = 0$ for the standard normal curve. From properties A, B, and C above, both sampling distributions are normal with known standard deviations SE. Since the sampling distribution of sample proportions is normal and SE is its standard deviation, the interval $(p - z_L * \text{SE}, p + z_L * \text{SE})$ will also contain $L\%$ of the sample proportions. This follows from the important property satisfied by all normal curves: *If we move the same number of standard deviations away from the mean on any two normal curves, the two areas below the two normal curves and above the x-axis will be equal.*

Since $E = z_L * \text{SE}$ is equal to z_L standard deviations for the sampling distribution, the interval $(p - E, p + E)$, formed by adding and subtracting E from the mean of the sampling distribution, will contain $L\%$ of the values for the statistic $\hat{p}$.

Thus, $L\%$ of the sample proportions is contained within $(p - E, p + E)$, and each sample proportion within $(p - E, p + E)$ satisfies the following inequality:

$$p - E < \hat{p} < p + E,$$
$$p - z_L * \sqrt{p(1-p)/n} < \hat{p} < p + z_L * \sqrt{p(1-p)/n}.$$

Let $\hat{p}$ satisfy the above inequality. Using algebra, we transform the above inequality into an $L\%$ confidence interval. Adding $-p$ to each term,

$$-z_L * \sqrt{p(1-p)/n} < -p + \hat{p} < z_L * \sqrt{p(1-p)/n}.$$

Adding $-\hat{p}$ to each term,

$$-\hat{p} - z_L * \sqrt{p(1-p)/n} < -p < -\hat{p} + z_L * \sqrt{p(1-p)/n}.$$

Multiplying each term by -1 reverses the inequalities:

$$\hat{p} - z_L * \sqrt{p(1-p)/n} < p < \hat{p} + z_L * \sqrt{p(1-p)/n}.$$

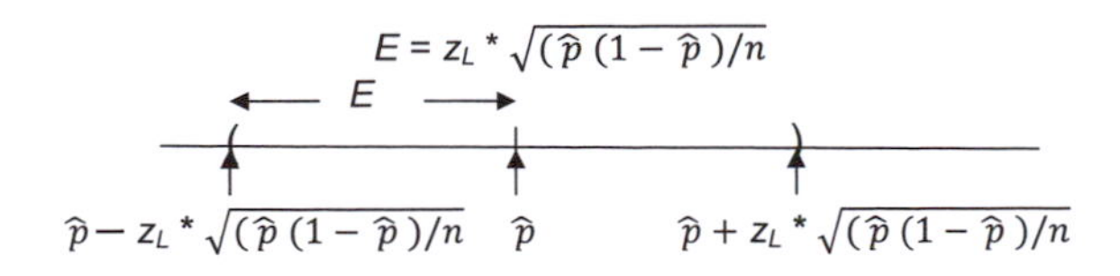

$E = z_L * \sqrt{(\hat{p}(1-\hat{p})/n}$
E
$\hat{p} - z_L * \sqrt{(\hat{p}(1-\hat{p})/n}$
$\hat{p}$
$\hat{p} + z_L * \sqrt{(\hat{p}(1-\hat{p})/n}$

Figure 14.2 L% confidence interval for *p*

Since p is usually unknown, we estimate the $SE = \sqrt{p(1-p)/n}$ by replacing p by $\hat{p}$, giving the final form for the L% confidence interval for p:

$$\hat{p} - z_L * \sqrt{\hat{p}(1-\hat{p})/n} < p < \hat{p} + z_L * \sqrt{\hat{p}(1-\hat{p})/n}. \tag{14.1}$$

The correct interpretation of an L% confidence interval for a population proportion is that if many L% confidence intervals are constructed, we expect L% of these confidence intervals to contain p. The practical interpretation (the one most used in books) is that with a probability of L% the population proportion p is contained in the confidence interval.

Figure 14.2 shows an L% confidence interval for p.

14.3.1. Rounding Rule for a Confidence Interval for the Population Proportion p

The endpoints of the confidence interval are rounded to three decimal places. E is rounded to four decimal places.

Remember, the L% confidence interval for p requires the sampling distribution of $\hat{p}$ to be normally distributed. For this to happen, both np and $n(1-p)$ must be greater than 5. Since usually p is unknown, we replace p by $\hat{p}$ and require $n * \hat{p} > 5$ and $n * (1-\hat{p}) > 5$.

Examples of L% Confidence Intervals Based on Spins of the Pointer on Aaron's Disk

Example 14.4. Find a 95% confidence interval for Aaron's *true* OBP using the experiment of the spins of the pointer on Aaron's disk. Suppose that the results of spinning the pointer on Aaron's disk 64 times are #1B = 11, #2B = 2, #3B = 1, #HR = 3, #BB = 7, #SF = 0, #HBP = 0, #O = 40.

Solution: The sample size is $n = 64$ spins. For those 64 plate appearances, the sample proportion $\hat{p} = \widehat{OBP} = (11+2+1+3+7+0+0)/64 = 24/64 = .375$. The level of confidence is $L = .95$, giving $z_L = 1.96$. The approximate $SE = \sqrt{\hat{p}(1-\hat{p})/n} = \sqrt{.375 * .625/64} = .0605$, and $E = 1.96 * .0605 = .1186$. The population proportion p is the *true* OBP.

A 95% confidence interval for the *true* OBP is

$$\hat{p} - E < p < \hat{p} + E,$$
$$.375 - .1186 < p < .375 + .1186,$$

$$.256 < p < .494,$$
$$.256 < true\ \text{OBP} < .494.$$

For the next example the number of spins is increased to 504.

Example 14.5. The results of spinning the pointer on Aaron's disk 504 times were #1B = 85, #2B = 19, #3B = 2, #HR = 20, #BB = 51, #SF = 1, #HBP = 1, #O = 325. For those 504 plate appearances, the sample proportion $\hat{p} = \widehat{OBP} = (85 + 19 + 2 + 20 + 51 + 1)/504 = 178/504 = .353$. The level of confidence is $L = .95$, giving a $z_L = 1.96$. The approximate $\text{SE} = \sqrt{p(1-p)/n} = \sqrt{.353 * .647/504} = .0213$, and $E = 1.96 * .0213 = .0417$. The population proportion p is the *true* OBP.

A 95% confidence interval for the *true* OBP is

$$\hat{p} - E < p < \hat{p} + E,$$
$$.353 - .0417 < p < .353 + .0417,$$
$$.311 < p < .395,$$
$$.311 < true\ \text{OBP} < .395.$$

14.3.2. Comparing the Results in Example 14.4 and Example 14.5

One difference in the two examples was the sample size. In Example 14.4, the sample size was 64 and E equaled .1186. In Example 14.5, the sample size was 504 and E equaled .0417. With the larger sample size, the margin of error was smaller. Since the margin of error is the radius of the confidence interval, we have a smaller radius as the sample size increases. Therefore, a larger sample size gives a more precise estimate.

What happens when the confidence is increased for a confidence interval? After completing the next two exercises, you should be able to answer this question.

Exercise 14.3. Using the information in Example 14.4, find a 90% and 99% confidence interval for the *true* OBP.

Exercise 14.4. Using the information in Example 14.5, find a 90% and 99% confidence interval for the *true* OBP.

Exercise 14.5. As we increase the confidence, what happens to the margin of error (radius) of the confidence interval?

Exercise 14.6. For Examples 14.4 and 14.5, why can we assume that the sampling distribution of $\hat{p}$ is approximately normally distributed?

Since the above examples involve binomial experiments and the disk was constructed based on Aaron's career plate appearances, the population OBP is known and is equal to .374. Observe that .374 lies in both confidence intervals. Since the *true* OBP is known, we could have used $\sqrt{p(1-p)/n}$ with $p = .374$ as the SE in the prior exercises and examples. We chose to use

$\hat{p}$ instead because in most applications the population proportion p will be unknown.

One can replace $L\%$ with such percentages as 80%, 90%, 95%, 98%, and 99%.

We now summarize the steps needed to construct any $L\%$ confidence interval.

14.3.3. Summary

The steps below are used to construct an $L\%$ confidence interval for any parameter.

1. The desired level of confidence L is declared.
2. A random sample is selected from the target population.
3. The desired statistic is calculated for the selected sample.
4. The assumptions needed to assume that the sampling distribution is normal are verified.
5. The margin of error E is calculated.
6. The $L\%$ confidence interval has the form

$$\text{statistic} - E < \text{parameter} < \text{statistic} + E.$$

In this section, the statistic was the sample proportion and the parameter was the population proportion. The sampling distribution of sample proportions was normal. Later in this chapter the statistic will be the sample mean and the parameter will be the population mean. The sampling distribution of sample means will be assumed normal.

14.3.4. Confidence Intervals for Baseball Descriptive Measures

We now turn our attention away from measuring Henry Aaron's ability to get on base through the spinning of a pointer on Henry Aaron's disk and toward Henry Aaron's ability to get on base by use of his actual plate appearances. The experiment of spinning the pointer on Aaron's disk is a binomial experiment, where a trial corresponds to a plate appearance found by spinning the pointer. The event "success" corresponds to outcomes that result in getting on base. A binomial experiment consists of a fixed number of identical trials that satisfy the following properties:

1. Each trial results in an outcome belonging to one of two mutually exclusive events.
2. The trials are independent.
3. The probability of success is the same for each trial.

Henry Aaron's actual career plate appearances cannot be assumed to satisfy the properties of a binomial experiment. Aaron's *true* OBP is unknown.

Batting success in baseball is a combination of a player's ability and the element of chance. A player's *batting ability* is divided into two parts: (1) to make contact with the ball (not striking out), and (2) when he does make contact, to direct the batted ball to where there are no fielders. Luck or chance comes into play, for a matter of inches can determine whether a batter gets a hit or makes an out. A player's *batting performance* is influenced by chance and can be measured by observing his actual plate appearances over a period of time. The period of time can be a portion of a year, an entire year, several years, or his entire career.

To measure a player's batting ability and performance, many descriptive measures are used. The descriptive measure studied in this section was the proportion. Some examples are on-base percentage (OBP), batting average (AVG or BA), home run average (HRA), strikeout average (SOA), base-on-balls average (BBA), and in-play batting average (IPBA). Each of these descriptive measures reflects some aspect of batting ability. Take the time now to review the aspect of batting measured by each of these descriptive measures (see Chap. 3). When using any of these measures as a statistic (a descriptive measure of a sample), we will use the symbol ^ above the measure; when using any of these measures as a parameter (a descriptive measure of a population), we will insert the words *true* or *population* in front of the measure. As a reminder, a *statistic* is used to measure the *batting performance for a sample*; a *parameter* is used to measure the *batting ability for a population.*

A sample will consist of observed plate appearances for a period of time. The population will be theoretical and based on an infinite number of plate appearances for that time period. The next example illustrates what was just discussed.

Example 14.6. The period of time is the 1964 season. The sample is all the plate appearances of Aaron for the 1964 season. The population is theoretical and based on Aaron batting over and over again in the 1964 season. The population is infinite. The variable is the outcome of a plate appearance. The statistic is his OBP for all observed plate appearances for the 1964 season, symbolized by $\widehat{OBP}$. The parameter is the unknown OBP, called the *true* OBP, for his theoretical infinite population of plate appearances for the 1964 season.

As stated throughout this book, performance is evaluated for a sample, while ability is evaluated for a population. Performance is measured by statistics (descriptive measures of a sample), while ability is measured by parameters (descriptive measures of a population). $\widehat{\text{OBP}}$ is a measure of Aaron's batting performance for the 1964 season; his *true* OBP is a measure of Aaron's batting ability for the 1964 season.

With these definitions, we are in a position to estimate Aaron's *true* OBP for the 1964 season and for his career. The next two examples illustrate the process of computing 95% confidence intervals for Aaron's *true* OBP for the 1964 season and for his career.

Example 14.7. We use for the sample the 634 actual plate appearances of Aaron in 1964. For those 634 plate appearances, Aaron successfully got on base 249 times (see Table 2.2). For 1964, his sample proportion ($\hat{p}=\widehat{OBP}$) was 249/634 = .393. The sample size was $n=634$. The reliability coefficient is $z_{.95}=1.96$. Using Formula (14.1), we get

$$\hat{p}-z_L*\sqrt{\hat{p}(1-\hat{p})/n}<p<\hat{p}+z_L*\sqrt{\hat{p}(1-\hat{p})/n},$$
$$.393-1.96*\sqrt{.393*.607/634}<p<.393+1.96*\sqrt{.393*.607/634},$$
$$.393-1.96*(.0194)<p<.393+1.96*(.0194),$$
$$.393-.0380<p<.393+.0380,$$
$$.355<p<.431,$$
$$.355<\textit{true}\ \text{OBP}<.431.$$

With 95% confidence, we can say that Aaron's *true* OBP for the year 1964 was between .355 and .431.

Example 14.8. We use for the sample the 13919 career plate appearances of Aaron (see Table 2.2). For those 13919 plate appearances, Aaron got on base successfully 5205 times. For his career, $\hat{p}=\widehat{OBP}=5205/13919=.374$. The 95% confidence interval for Aaron's true career OBP is calculated below:

$$.374-1.96*\sqrt{.374*.626/13919}<p<.374+1.96*\sqrt{.374*.626/13919},$$
$$.374-1.96*(.0041)<p<.374+1.96*(.0041),$$
$$.374-.0080<p<.374+.0080,$$
$$.366<p<.382,$$
$$.366<\textit{true}\ \text{OBP}<.382.$$

With 95% confidence, we can say that Aaron's *true* OBP for his career was between .366 and .382.

Exercise 14.7. For Examples 14.7 and 14.8, find the point estimate, the parameter being estimated, the margin of error, the reliability coefficient, and the standard error of the mean. Why can we assume that the sampling distributions are approximately normal? Which confidence interval is more precise? Why?

The OBP should be rounded to three decimal places, and the margin of error E should be rounded to four decimal places.

An L% Confidence Interval for p Expressed in Three Ways

- Using Example 14.8, we see the confidence interval expressed as an inequality in the form $\hat{p}-E<p<\hat{p}+E$.
- A shortcut form is $\hat{p}\pm E$.
- An interval form is $(\hat{p}-E,\ \hat{p}+E)$.

Another important descriptive measure of batting performance is a player's batting average. A player's batting average is symbolized by either $\widehat{\text{BA}}$ or $\widehat{AVG}$. A player's *population* batting average is his *true* BA or *true* AVG. For example, a player for a given year might have had 500 at bats. Of those 500 at bats, 100 resulted in hits. The quotient 100/500 is his *sample* batting average. Because of the chance factor in an at bat, we cannot consider this his *true* batting average for that year. In other words, if his batting average for the year was .200, it could have just as easily been .190 or .220. His *true* BA is unknown. In the next examples, we use L% confidence intervals to estimate his *true* batting average.

Example 14.9. Consider Henry Aaron's actual plate appearances for the year 1964, when he was 30 years old. That year, he had 570 at bats with 187 hits, giving him a yearly batting average of .328. Find a 95% confidence interval estimate for his *true* 1964 batting average.

Solution: Data: $\hat{p}=\#H/\#AB=.328$; $n=\#AB=570$; $L=.95$; $z_L=1.96$.

$$\hat{p}-z_L * \sqrt{\hat{p}(1-\hat{p})/n} < p < \hat{p}+z_L * \sqrt{\hat{p}(1-\hat{p})/n},$$
$$.328-1.96 * \sqrt{.328(1-.328)/570} < p < .328+1.96 * \sqrt{.328(1-.328)/570},$$
$$.328-1.96 * .0197 < p < .328-1.96 * .0197,$$
$$.328-.0386 < p < .328+.0386,$$
$$.289 < p < .367,$$
$$.289 < \textit{true}\ \text{BA} < .367.$$

The margin of error is .0386. With 95% confidence, we can say that Aaron's *true* BA for the year 1964 was between .289 and .367.

Example 14.10. The data are the same as in Example 14.9. Find a 90% confidence interval estimate for Aaron's *true* 1964 batting average.

Solution: Data: $\hat{p}=\#H/\#AB=.328$; $n=\#AB=570$; $L=.90$; $z_L=1.64$.

$$\hat{p}-z_L * \sqrt{(\hat{p}(1-\hat{p})/n} < p < \hat{p}+z_L * \sqrt{(\hat{p}(1-\hat{p})/n},$$
$$.328-1.64 * \sqrt{(.328(1-.328)/570} < p < .328+1.64 * \sqrt{(.328(1-.328)/570},$$
$$.328-1.64 * .0197 < P < .328-1.64 * .0197,$$
$$.328-.0323 < p < .328+.0323,$$
$$.296 < p < .360,$$
$$.296 < \textit{true}\ \text{BA} < .360.$$

The margin of error is .0323. With 90% confidence, we can say that Aaron's *true* BA for the year 1964 was between .296 and .360.

Observe that as the level of confidence increases, the margin of error increases. Since a smaller margin of error gives a more precise approximation of p, by lowering the level of confidence, we get a more precise estimate of p.

In the next two examples, we look at the corresponding results for Barry Bonds at age 30.

Example 14.11. Look at Barry Bonds's batting for the year 1995 when he was 30 years old. He had 506 at bats that year with 149 hits, giving him a yearly batting average of .294 (see Table 4.1). Find a 95% confidence interval for his *true* 1995 batting average.

Solution: Data: $\hat{p} = \#H/\#AB = .294$; $n = \#AB = 506$; $L = .95$; $z_L = 1.96$.

$$\hat{p} - z_L * \sqrt{\hat{p}(1-\hat{p})/n} < p < \hat{p} + z_L * \sqrt{\hat{p}(1-\hat{p})/n},$$
$$.294 - 1.96 * \sqrt{.294(1-.294)/506} < p < .294 + 1.96 * \sqrt{.294(1-.294)/506},$$
$$.294 - 1.96 * .0203 < p < .294 + 1.96 * .0203,$$
$$.294 - .0398 < p < .294 + .0398,$$
$$.254 < p < .334.$$

The margin of error is .0398. With 95% confidence, we can say that Bonds's *true* BA for the year 1995 is between .254 and .334.

Example 14.12. The data are the same as Example 14.11. Find a 90% confidence interval estimate for Bonds's *true* 1995 batting average.

Solution: Data: $\hat{p} = \#H/\#AB = .294$; $n = \#AB = 506$; $L = .90$; $z_L = 1.64$.

$$\hat{p} - z_L * \sqrt{\hat{p}(1-\hat{p})/n} < p < \hat{p} + z_L * \sqrt{\hat{p}(1-\hat{p})/n},$$
$$.294 - 1.96 * \sqrt{.294(1-.294)/506} < p < .294 + 1.96 * \sqrt{.294(1-.294)/506},$$
$$.294 - 1.64 * .0203 < p < .294 + 1.64 * .0203,$$
$$.294 - .0333 < p < .294 + .0333,$$
$$.261 < p < .327.$$

The margin of error is .0333. With 90% confidence, we can say that Bonds's *true* BA for the year 1995 is between .261 and .327.

Henry Aaron's most productive year occurred when he was 36 years old in 1971. Barry Bonds's most productive year occurred when he was 37 years old in 2001.

Exercise 14.8. Find both a 90% and 95% confidence interval for Aaron's *true* OBP, *true* BA, and *true* HRA for the year 1971.

Exercise 14.9. Find both a 90% and 95% confidence interval for Bonds's *true* OBP, *true* BA, and *true* HRA for the year 2001.

Exercise 14.10. For your two players pick their best year. For each player find both a 90% and 95% confidence interval estimate for their *true* OBP and *true* BA for that year.

Exercise 14.11. Repeat Exercise 14.11 for their careers.

Constructing an *L*% Confidence Interval Estimate for the *True* BA and *True* OBP

1. Declare the desired confidence level L.
2. Select one random sample from the population. Since for infinite populations simple random sampling is difficult, we must settle for a representative sample for that period of time. The representative sample will be all actual plate appearances for that time period.
3. For this sample, calculate the sample proportion ($\hat{p}$). For BA, $\hat{p}=$#H/#AB, and for OBP, $\hat{p}=$#OB/#PA.
4. As n is large, both $n\hat{p}$ and $n(1-\hat{p})$ are greater than 5. Therefore, the sampling distribution of $\hat{p}$ will satisfy the following properties:

   (A) It is normally distributed.
   (B) $\mu_{\hat{p}}=p$.
   (C) $\sigma_{\hat{p}}=\sqrt{p(1-p)/n}$.

   Since p is unknown, we need to approximate p in condition C above by using $\hat{p}$ as an estimate for p. Condition C becomes condition C′ below.

   (C′) $\sigma_{\hat{p}}=\sqrt{\hat{p}(1-\hat{p})/n}$.

5. Find z_L and let SE $=\sigma_{\hat{p}}$.
6. Set $E=z_L *$ SE.
7. The L% confidence interval is $\hat{p}\pm E$.

14.4. Comparing Two Confidence Intervals for Two Population Proportions

This section compares two population proportions for two populations. As usual, Hank Aaron is compared with Barry Bonds with respect to their career BAs and career OBPs. The career performances of these two players generate the statistics $\widehat{\text{BA}}$ and $\widehat{\text{OBP}}$, which are used to find 95% confidence intervals for their *true* career BA and their *true* career OBP.

In comparing two player's batting performances with respect to a given baseball statistic, how can we determine if one player's ability is better? At the age of 30, Bonds's batting average for that year was .294. When Aaron was 30, his batting average was .328. Can we conclude that Aaron's ability, with respect to batting average, was greater than Bonds's ability when both players were 30 years old? For age 30, we do know that Aaron's batting average performance was superior to that of Bonds. The confidence intervals constructed in Examples 14.9 and 14.11 will be used to compare their batting average abilities. Figure 14.3 shows both confidence intervals.

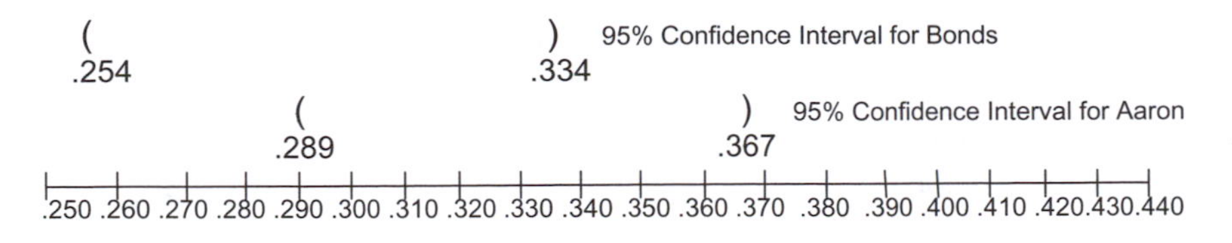

Figure 14.3 Confidence intervals for the BA of Aaron and Bonds when both were age 30

In comparing two confidence intervals, there are two possibilities. Either the confidence intervals overlap,

confidence interval 1 ($\hat{p}_1$)

confidence interval 2 ($\hat{p}_2$)

or they don't overlap,

confidence interval 1 ($\hat{p}_1$)

confidence interval 2 ($\hat{p}_2$)

Let the population proportion p_1 be associated with confidence interval 1 and population proportion p_2 be associated with confidence interval 2. Let $\hat{p}_1$ be the sample proportion that is the center of confidence interval 1 and $\hat{p}_2$ be the sample proportion that is the center of confidence interval 2. Without loss of generality, we can assume that both confidence intervals are 95% confidence intervals.

14.4.1. The Confidence Intervals Overlap

Suppose that the intervals overlap. Clearly, $\hat{p}_1$ is to the right of $\hat{p}_2$. However, it may be the case that p_1 is to the left of p_2. The reason is that both p_1 and p_2 can lie anywhere in their respective intervals. Therefore, statistically, there is no difference between the two population proportions. What does this mean when applied to batting averages or on-base percentages? It says that one player's higher BA or OBP for a year or a career does not necessarily imply that their *true* BA or *true* OBP is higher.

14.4.2. The Confidence Intervals Do Not Overlap

Suppose that the intervals do not overlap. If p_2 is contained in confidence interval 2 and p_1 is contained in confidence interval 1, p_2 will definitely be to the right of p_1. Since both samples are selected independently, we can conclude with confidence $.95 * .95 = .903$ that p_1 is in confidence interval 1 and p_2 is in confidence interval 2. What does this tell us about their *true* batting averages or *true* on-base percentages? It says that for a year or a career, with a confidence of at least 90%, the *true* BA or *true* OBP of the player associated with confidence interval 2 is higher than that of the player associated with confidence interval 1. When the two confidence intervals do not overlap, we say that the result is *statistically significant for a certain level of confidence.*

Figure 14.3 shows that the two confidence intervals overlap. It is possible that the batting average ability of Bonds for age 30 was really higher, but because of luck or chance variation, Aaron just happened to perform better. Thus, the 34-point difference in their respective season batting averages is not enough to say with a confidence of at least 90% that Aaron's batting average ability is higher than that of Bonds.

Warning: Many statisticians find a problem with using the logic just described to compare two population proportions. The problem is that the confidence level L has changed in the conclusion. When using confidence intervals, the level of confidence L must be declared before the study begins. If a 95% confidence level is desired at the onset, for the result to be statistically significant, the confidence level cannot be changed at the conclusion.

A statistically valid method for the testing of two population parameters can be found in Appendix A of this book, as well as in other statistics textbooks.

Since we are dealing with two sample proportions and two population proportions, we can't call them both p. To make things simpler, we use a letter from a player's name as a subscript of p or $\hat{p}$.

So, for Henry Aaron, sample proportions are, for example, $\hat{p} = \widehat{OBP} = \widehat{OBP}_A ; \hat{p} = \widehat{BA} = \widehat{BA}_A$.

Population proportions are likewise, for example, $p = true\ \mathrm{OBP}_A = \mathrm{OBP}_A$; $p = true\ \mathrm{BA} = \mathrm{BA}_A$.

For Bonds, we have sample proportions such as $\widehat{OBP}_B$ and $\widehat{BA}_B$ and population proportions such as OBP_B and BA_B.

Example 14.13. We now look at a comparison of their *true* career batting averages. We find a 95% confidence interval for p, the *true* career batting average (BA_A) for Henry Aaron.

Solution: Data: $\hat{p} = \#H/\#AB = .305$; $n = \#AB = 12364$; $L = .95$; $z_L = 1.96$.

$$\hat{p} - z_L * \sqrt{\hat{p}(1-\hat{p})/n} < p < \hat{p} + z_L * \sqrt{\hat{p}(1-\hat{p})/n},$$
$$.305 - 1.96 * \sqrt{.305(1-.305)/12364} < p < .305 + 1.96 * \sqrt{.305(1-.305)/12364},$$
$$.305 - 1.96 * .0041 < p < .305 + 1.96 * (.0041),$$
$$.305 - .0080 < p < .305 + .0080,$$
$$.297 < p < .313,$$
$$.297 < \mathrm{BA}_A < .313.$$

Example 14.14. We find a 95% confidence interval for p, the *true* career batting average (BA_B) for Barry Bonds up to and including his first 81 games in 2007.

Solution: Data: $\hat{p} = 2902/9726 = .298$; $n = \#AB = 9726$; $L = .95$; $z_L = 1.96$.

$$\hat{p} - z_L * \sqrt{\hat{p}(1-\hat{p})/n} < p < \hat{p} + z_L * \sqrt{\hat{p}(1-\hat{p})/n},$$
$$.298 - 1.96 * \sqrt{.298(1-.298)/9726} < p < .298 + 1.96 * \sqrt{.298(1-.298)/9726},$$
$$.298 - 1.96 * .0046 < p < .298 + 1.96 * .0046,$$
$$.298 - .0090 < p < .298 + .0090,$$
$$.289 < p < .307,$$
$$.289 < BA_B < .307.$$

The two confidence intervals overlap, so the difference between the two *true* career batting averages is not statistically significant.

Example 14.15. We now compare Aaron and Bonds with respect to their *true* career on-base percentages (OBP). We find a 95% confidence interval for p, the *true* career on-base proportion (OBP_A) for Henry Aaron.

Solution: Data: $\hat{p} = \#OB/\#PA = 5205/13919 = .374$; $n = 13919$; $z_L = 1.96$.

$$\hat{p} - z_L * \sqrt{\hat{p}(1-\hat{p})/n} < p < \hat{p} + z_L * \sqrt{\hat{p}(1-\hat{p})/n},$$
$$.374 - 1.96 * \sqrt{.374(1-.374)/13919} < p < .374 + 1.96 * \sqrt{.374(1-.374)/13919},$$
$$.374 - 1.96 * .0041 < p < .374 + 1.96 * .0041,$$
$$.374 - .0080 < p < .374 + .0080,$$
$$.366 < OBP_A < .382.$$

Example 14.16. We find a 95% confidence interval for p, the *true* career on-base proportion (OBP_B) for Barry Bonds, up to and including his first 81 games of 2007.

Solution: Data: $\hat{p} = \#OB/\#PA = 5527/12441 = .444$; $n = 12441$; $z_L = 1.96$.

$$\hat{p} - z_L * \sqrt{\hat{p}(1-\hat{p})/n} < p < \hat{p} + z_L * \sqrt{\hat{p}(1-\hat{p})/n},$$
$$.444 - 1.96 * \sqrt{.444(1-.444)/12441} < p < .444 + 1.96 * \sqrt{.444(1-.444)/12441},$$
$$.444 - 1.96 * .0045 < p < .444 + 1.96 * .0045,$$
$$.444 - .0088 < p < .444 + .0088,$$
$$.435 < OBP_B < .453.$$

Since the confidence intervals do not overlap and Bonds's interval is to the right of Aaron's interval, we can conclude with a confidence of at least 90% that Bonds's *true* OBP (OBP_B) is higher than Aaron's *true* OBP (OBP_A). The result is *statistically significant*.

As promised earlier in this chapter, we now consider confidence intervals for the population mean.

14.5. L% Confidence Intervals for the Population Mean, μ

Recall that, for the sampling distribution of sample means, the standard error (SE) is equal to $\sigma/\sqrt{n}$ and the margin of error (E) is equal to $z_L * (\sigma/\sqrt{n})$.

The interval formed by adding and subtracting E from the population mean of the sampling distribution of sample means provides an interval that is centered at μ and contains $L\%$ of the sample means for a fixed sample size n. Thus, $L\%$ of the sample means is contained in the interval $(\mu - E, \mu + E)$. For those sample means $\bar{x}$ contained in the $L\%$ confidence interval we have

$$\mu - E < \bar{x} < \mu + E,$$
$$\mu - z_L * (\sigma/\sqrt{n}) < \bar{x} < \mu + z_L * (\sigma/\sqrt{n}).$$

Using algebra, we can transform the above inequality into an $L\%$ confidence interval for μ, as we did for the confidence interval for p, giving Formula (14.2) below.

$$\bar{x} - z_L * (\sigma/\sqrt{n}) < \mu < \bar{x} + z_L * (\sigma/\sqrt{n}). \qquad (14.2)$$

Remember, the $L\%$ confidence interval for μ requires that the sampling distribution of sample means is normal. For this to happen, either the original X-population is normally distributed or the sample size is at least 30. Formula (14.2) also requires that σ is known. In the event that σ is not known and $n \geq 30$, we can use the sample standard deviation, s, as an estimate for σ. We then have Formula (14.3):

$$\bar{x} - z_L * (s/\sqrt{n}) < \mu < \bar{x} + z_L * (s/\sqrt{n}). \qquad (14.3)$$

Exercise 14.12. Derive Formula (14.2) using the same logic as in the derivation of Formula (14.1).

Exercise 14.13. Sketch a graph similar to Figure 14.2 for the $L\%$ confidence interval μ.

Example 14.17. This example provides a technique that allows any confidence interval for a population proportion to be expressed as a confidence interval for a mean. We use the sample data in Example 14.7 to construct a 95% confidence interval for the *true* OBA for Aaron for the year 1964. Notice that the *true* OBA (on-base average) has replaced the *true* OBP (on-base proportion). That is, we are now considering a confidence interval for the population mean.

For the sample of 634 plate appearances in Example 14.7, we assign the number 1 to those plate appearances for which Aaron got on base and the number 0 to those plate appearances for which he did not get on base. Since he got on base 249 times, we have $\hat{p} = \bar{x} = 249/634 = .393$.

We can use $\sqrt{\hat{p}(1-\hat{p})}$ as an estimate for the unknown population standard deviation σ. Why? You will be asked to verify this relationship in an exercise below. Using Formula (14.2), we construct a 95% confidence interval for μ.

$$\bar{x} - z_L * (\sigma/\sqrt{n}) < \mu < \bar{x} + z_L * (\sigma/\sqrt{n}),$$
$$.393 - 1.96 * \sqrt{\hat{p}(1-\hat{p})}/\sqrt{n} < \mu < .393 + 1.96 * \sqrt{\hat{p}(1-\hat{p})}/\sqrt{n}.$$

Since $\sqrt{\hat{p}(1-\hat{p})}/\sqrt{n} = \sqrt{\hat{p}(1-\hat{p})/n}$, we have

$$.393 - 1.96 * \sqrt{.393 * .607/634} < \mu < .393 + 1.96 * \sqrt{.393 * .607/634},$$
$$.393 - 1.96 * (.0194) < \mu < .393 + 1.96 * (.0194),$$
$$.355 < \mu < .431,$$
$$.355 < \textit{true}\ \text{OBA} < .431.$$

Looking back at Example 14.7, we see the same confidence interval.

It should be noted that many confidence intervals involving a population mean have nothing to do with a confidence interval for a population proportion.

Exercise 14.14. Given the data (1, 0, 0, 0), with 1 being success and 0 being failure, compute the population standard deviation for these data and show

a. $\sigma = \sqrt{p(1-p)}$, where $p = 1/4$ and $n = 4$.
b. $\sigma/\sqrt{n} = \sqrt{p(1-p)/n}$, where $p = 1/4$ and $n = 4$.

Exercise 14.15. Use the data in Example 14.8 to construct a 95% confidence interval for the *true* OBA.

A confidence interval for the population mean is used to estimate the population mean for an *X*-population whose data are quantitative.

Examples of Confidence Intervals for the Population Mean

Example 14.18. A study was done on the time it took to complete a nine-inning Major League baseball game in 2007. A random sample of 100 games had a mean time of 190 minutes with a sample standard deviation of 25 minutes. Find a 95% confidence interval for the population mean time to complete a Major League game in the 2007 season. Assume that σ is unknown.

Solution: The population consists of all Major League games played in 2007. The random variable is the time it takes to complete a game. Since the sample size is 100 and σ is unknown, we can use Formula (14.3). For the sample of 100 games, the sample mean is $\bar{x} = 190$ minutes and the sample standard deviation is $s = 25$ minutes. Using Formula (3),

$$\bar{x} - z_L * (s/\sqrt{n}) < \mu < \bar{x} + z_L * (s/\sqrt{n}),$$
$$190 - 1.96 * (25/\sqrt{100}) < \mu < 190 + 1.96 * (25/\sqrt{100}),$$
$$190 - 4.9 < \mu < 190 + 4.9,$$
$$185.1 < \mu < 194.9,$$
$$185 < \mu < 195 \text{ (rounded)}.$$

With a confidence of 95%, we can conclude that the population mean time to complete a Major League game in 2007 was between 185 minutes and 195 minutes.

Exercise 14.16. For Example 14.18, find L, z_L, SE, and E.

Exercise 14.17. Use the data in Example 14.18 to find a 90% and 99% confidence interval for μ.

Exercise 14.18. Royals reliever Joakim "The Mexicutioner" Soria is credited with a 96 mph heater and a 90 mph slider. Suppose that Joakim threw 40 fastballs in a game. The sample mean speed of the 40 fastballs was 92 mph, as measured by a speed gun, with a standard deviation of 2 mph.

a. Describe the population and variable of interest.
b. Why can we assume that the sampling distribution is approximately normal?
c. Which confidence interval formula should we use?
d. Find a 90% confidence interval for Soria's population mean fastball speed.
e. Write your interpretation of the 90% confidence interval.
f. For the 90% confidence interval, find SE and E.
g. If we increase the confidence to 95%, what is the effect on E?
h. With 95% confidence, can we conclude that Joakim does not have a 96 mph heater?
i. If we had seen 50 fastballs instead of 40 with a sample mean of 92 mph, what is the effect on E?

14.5.1. Rounding Rule for a Confidence Interval for a Mean

The endpoints of the confidence interval should have the same number of decimal places as the sample mean. The margin of error, E, is rounded to one more decimal place.

14.6. L% Confidence Intervals for the Population Mean μ (for Sample Size < 30)

As we saw in the previous section, if the sample size is ≥30, then two important consequences occur. First, we can assume that the sampling distribution of the sample mean is approximately normally distributed. Secondly, if σ is unknown, we can use the sample standard deviation, s, as an estimate for σ.

In many real-world situations, because of availability, cost, or time involved, the sample collected has a size less than 30. One of the reasons for finding a confidence interval for the population mean is that the population mean is unknown. If the population mean is unknown, it is safe to assume, in most cases, that the population standard deviation will also be unknown.

When (1) the sample size $n < 30$, (2) the population standard deviation is unknown, and (3) the original X-population is normal, the sampling distribution of sample means cannot be assumed to be normally distributed.

William S. Gosset (1876–1937) developed a new continuous distribution called the t-distribution. Gosset was employed at the Guinness Brewing Company in Dublin, Ireland, where he was in charge of sampling each batch of ale. He discovered that it was impractical to sample many glasses of ale, so he had to sample just a small number. Gosset saw that the standard normal curve (z-distribution) did not work for these sampling distributions.

Gosset developed the t-distribution. In his paper published in 1908 under the pseudonym Student, the distribution was called the Student t-distribution. He showed in his paper that if the original X-population is approximately normally distributed and s is used in place of σ, then the sampling distribution of sample means would follow a t-distribution.

14.6.1. Properties of the t-Distribution

The t-distribution consists of a family of curves, called t-curves, each of which is a function of a special number, called the degrees of freedom (df), which is calculated by subtracting 1 from the sample size. We write $\text{df} = (n - 1)$.

Each t-curve is *similar* to the standard normal curve in the following ways:

1. Each t-curve is bell shaped and symmetric about the mean.
2. The total area under each t-curve and above the x-axis is equal to 1.
3. The mean, median, and mode of each t-curve are equal to 0.
4. The t-curves are asymptotic to the x-axis (they get close to but never touch the x-axis as the x-values travel farther away from 0 in both directions).

Each t-curve *differs* from the standard normal curve in the following ways:

5. The variance of each t-curve is greater than 1.
6. Compared with the standard normal curve, the t-curves have a lower peak and higher tails.

The following relationships exist between different t-curves:

7. As the degrees of freedom increase, the t-curves approach the shape of the standard normal curve (the variances approach 1).
8. For samples of size at least 30, many statisticians allow the use of the z-distribution as an approximation for the t-distribution. In this book,

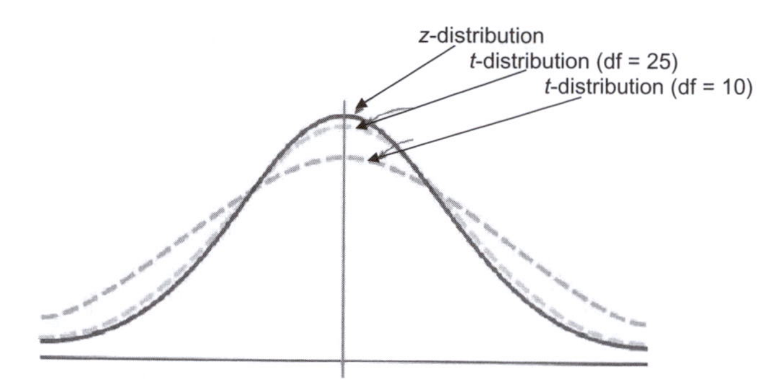

Figure 14.4 Comparison of a standard normal distribution and two *t*-distributions

we have already used the *z*-statistic when the samples size is ≥30 and σ is unknown (Formula (14.3)).

Figure 14.4 shows the standard normal distribution and two *t*-distributions with different degrees of freedom. Observe how close the shape of the *t*-distribution is to the *z*-distribution when the degrees of freedom is 25.

Formula (14.4) gives an $L\%$ confidence interval with a reliability coefficient taken from the *t*-distribution.

$$\bar{x} - t_L * (s/\sqrt{n}) < \mu < \bar{x} + t_L * (s/\sqrt{n}). \qquad (14.4)$$

Comparing Formula (14.4) with Formula (14.3), the only difference is that the reliability coefficient t_L replaces z_L. In general, given a confidence level L and a sample size n, we can find the t_L using Table C.3 in Appendix C.

Steps for Finding the Reliability Coefficient t_L

Step 1: Let $A = (1 - L)/2$ (A gives the area in either tail).
Step 2: Form the sum $A + L$.
Step 3: Choose the column whose subscript corresponds to $A + L$.
Step 4: Choose the row that corresponds to $\text{df} = n - 1$.
Step 5: The reliability coefficient corresponds to the intersection of that row and column.

The reliability coefficient t_L corresponds to the lookup from Table C.3 with column heading $(A + L)$ and row heading $(n - 1)$. The column headings are labeled $t'_{(A+L)}$. From the above, we have $t_L = t'_{(A+L)}$.

Example 14.19. This example demonstrates how to find the correct reliability coefficient from the *t*-distribution table, Table C.3. For this example, we are given a sample of size $n = 10$ and a confidence level of $L = .90$.

Steps for Finding the Reliability Coefficient t_L

Step 1: $A = (1 - .90)/2 = .05$.
Step 2: $A + L = .05 + .90 = .95$.
Step 3: Choose the column $t'_{.95}$.

Table 14.3 **Portion of Table C.3 (standard *t*-table)**

df	$t'_{.90}$	$t'_{.95}$	$t'_{.975}$	$t'_{.99}$	$t'_{.995}$
1		6.3138			
9 ⟶		1.8331			
30		1.6973			
200		1.6525			
∞	1.282	1.645	1.96	2.326	2.576

Step 4: Choose the row $= 10 - 1 = 9$.

Step 5: $t_{.90} = t'_{.95} = 1.8331$ (the reliability coefficient for a 90% confidence interval).

Table 14.3 shows a small part of Table C.3 (the *t*-distribution).

Looking at the portion of Table C.3, the reliability coefficients across the row ∞ correspond to the reliability coefficients $z_{.80}$, $z_{.90}$, $z_{.95}$, $z_{.98}$, and $z_{.99}$.

Exercise 14.19. Fill in the correct subscripts for the table below:

Confidence	z_L	$t_L = t'_{(A+L)}$
80%	$z_{.80}$	$t_{.80} = t'_{.90}$
90%		
95%		
98%		
99%		

Exercise 14.20. Using a sample size $n = 30$, fill in the reliability coefficients for the table in Exercise 14.19. Do the reliability coefficients z_L depend on the sample size?

Examples to Demonstrate the Construction of Confidence Intervals for the Population Mean (Sample Size $n < 30$, the Population Is Approximately Normal, and the Population Standard Deviation σ is Unknown)

Example 14.20. Find a 95% confidence interval for Aaron's *true* mean yearly batting average for the theoretical population consisting of his yearly batting averages. We use for our sample data the 23 years from 1954 to 1976. Assume that his population of yearly batting averages is approximately normal.

Solution: Because the sample size is less than 30, σ is unknown, and the population of yearly batting averages is approximately normal, we can use Formula (14.4) with the *t*-distribution. The population is theoretical and based on

Aaron playing year after year. The variable of interest is his yearly BA. The sample data can be acquired from Table 2.2. The sample mean is $\bar{x} = .301$, the sample standard deviation is $s = .031$, the df $= 23 - 1 = 22$, $L = .95$, and $t_L = t'_{(.95+.025)}$. Using Formula (14.4) with Table C.3, we have

$$\bar{x} - t_L * (s/\sqrt{n}) < \mu < \bar{x} + t_L * (s/\sqrt{n}),$$

$$.301 - 2.0739 * .031/\sqrt{23} < \mu < .301 + 2.0739 * .031/\sqrt{23},$$

$$.301 - .0134 < \mu < .301 + .0134,$$

$$.2877 < \mu < .3144,$$

$$.288 < \mu < .314.$$

We can conclude with 95% confidence that Aaron's *true* mean yearly batting average is between .288 and .314.

Exercise 14.21. Using the information in Example 14.20, find a 90% confidence interval for Aaron's *true* mean yearly batting average. Compare the results of Example 14.20 with the results in this exercise.

Exercise 14.22. Find a 95% confidence interval for Bonds's *true* mean yearly batting average for the theoretical population consisting of his yearly batting averages. For our sample data, use the yearly batting averages for the 22 years from 1986 to and including 2007 (use Table 4.1). Assume that the population of yearly batting averages is approximately normal.

Exercise 14.23. Compare the 95% confidence intervals in Example 14.20 and Exercise 14.22. What can you say, or not say, about their *true* mean yearly batting averages?

Exercise 14.24. A speed gun was used to measure Brewers relief pitcher David Riske's fastball speed. Suppose that David threw 16 fastballs in the game. The sample mean speed was 88 mph, and the sample standard deviation was 3 mph.

a. In order to construct the confidence interval, what assumption is necessary?
b. Find a 95% confidence interval for his *true* mean fastball speed.
c. Find z_L, SE, and E for a 95% confidence interval.

Later it was discovered that the pitch counter was in error and David actually threw 25 fastballs in the game. Suppose that the sample mean and sample standard deviation remained the same.

d. Find a 95% confidence for his *true* mean fastball speed.
e. Which confidence interval gave a more precise measurement for Riske's *true* mean fastball speed? Why?

14.7. Making a Decision on the Distribution (z, t, or Neither One) to Use for the Construction of the L% Confidence Interval for μ

We begin this section with an explanation of why we can use the z-distribution for any confidence interval where the sample size $n \geq 30$, whether σ is known or not known. Looking at Table C.3 for the t-distribution, for degrees of freedom ≥ 30, the reliability coefficients for each of the levels of confidence are very close to the reliability coefficients provided by Table C.2 (the Z-Table). Therefore, if df ≥ 30, the confidence interval is affected minimally by using z_L instead of t_L. For example, if df $= 30$ and $L = .95$, the reliability coefficient from Table C.3 for the t-distribution would be 2.04 compared to a lookup of 1.96 from the Z-Table. As the degree of freedom increases, the lookup from Table C.3 gets closer to the lookup from the Z-Table.

For the same level of confidence L, z_L is smaller than t_L. Thus, the use of z_L instead of t_L will provide a more precise estimate for μ. Since a more precise estimate is desirable, it is important to use the z-distribution whenever the necessary assumptions are satisfied.

Exercise 14.25. Complete the table below for a sample size equal to 28.

Confidence	L	z_L	t_L
80%			
90%			
95%			
98%			
99%			

To help you decide on which distribution to use, we provide Figure 14.5, which is a flowchart based on answering the following three questions:

1. Is $n \geq 30$?
2. Is the X-population approximately normal?
3. Is σ known?

Exercise 14.26. Fill in the boxes in the flowchart with the appropriate L% confidence interval formula.

Exercise 14.27. In 1988, Wrigley Field became the last ballpark to install lights. Suppose that the Cubs look to change their 1500-watt lightbulbs from Brand X to Brand Y. Using the assumptions below, decide on which formula from this chapter to use or that no formula from this chapter can be used. If a formula can be used, find a 95% confidence interval for the mean length of time the theoretical population of Brand Y bulbs will last.

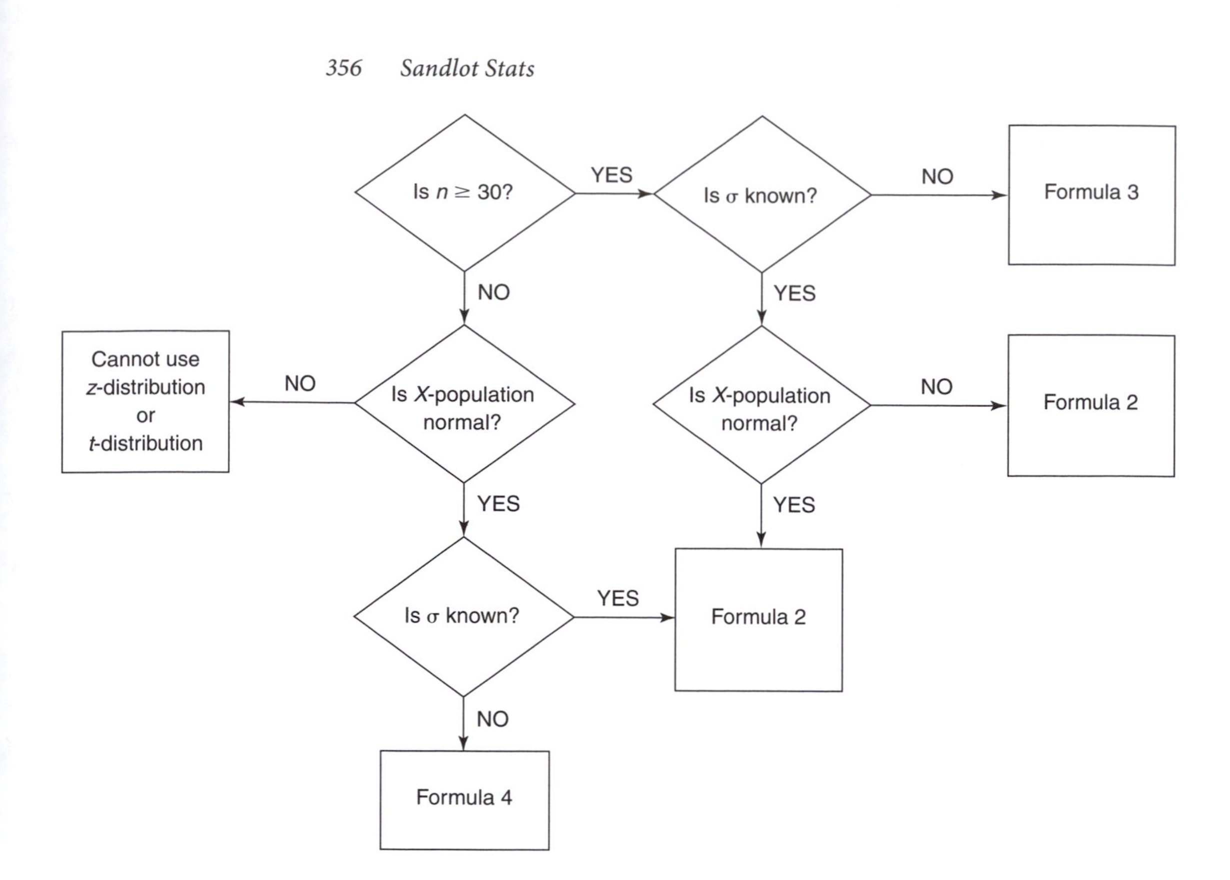

Figure 14.5 Flowchart for deciding on which distribution to use

a. Brand Y sends 25 bulbs to be tested. They include the specification that $\sigma = 20$ hours and the length of time for their bulbs is normally distributed. The mean length of time in hours for the 25 bulbs was 961 hours.
b. Brand Y sends 100 bulbs to be tested, providing no specifications. The 100 bulbs had a mean length of time of 1020 hours with a standard deviation equal to 30 hours.
c. Brand Y sends 25 bulbs to be tested, providing no specifications. The 25 bulbs had a mean length of time of 970 hours with a standard deviation equal to 22 hours.
d. Brand Y sends 25 bulbs to be tested, providing the specification that the length of time for their bulbs is normally distributed. The 25 bulbs had a mean length of time of 990 hours with a standard deviation equal to 35 hours.

14.8. Sample Size Determination

The sample size n and the level of confidence L determine the width, from endpoint to endpoint, of the confidence interval. The width of a confidence interval is equal to $2 * E$ (margin of error). From Formula (14.1), the *margin of error* for a confidence interval for p is $z_L * \sqrt{\hat{p}(1-\hat{p})/n}$. From Formulas

(14.2), (14.3), and (14.4), the margin of error for a confidence interval for μ is given by $z_L*(\sigma/\sqrt{n})$, $z_L*(s/\sqrt{n})$, and $t_L*(s/\sqrt{n})$, respectively. The margin of error is called the *precision* of a confidence interval because as it gets smaller, our estimate for p or μ becomes more exact. We have no control over $\hat{p}$, $\bar{x}$, s, or σ. We do have control over L, z_L, t_L, and n. A decrease in the level of confidence, L, results in a smaller z_L or t_L. A smaller z_L or t_L decreases the width of the confidence interval. Decreasing the level of confidence L makes the confidence interval more precise (a smaller width for the confidence interval), but at the expense of reducing the confidence. If we want to maintain a certain confidence, the other choice for decreasing the width of the confidence interval is to increase the sample size. Since n is in the denominator of every margin of error formula, an increase in n will decrease the margin of error, without affecting the confidence.

If the margin of error of a confidence interval is too wide, the usefulness of the confidence interval is greatly diminished. In the design of the research project, a desired sample size is determined before the study is begun. Assuming the standard 95% confidence interval, researchers decide on how wide of a confidence interval can be tolerated. After the margin of error is determined, the minimum sample size needed to accomplish the desired margin of error can be calculated. The rest of this section presents formulas to achieve this minimum sample size.

14.8.1. Determination of the Minimum Sample Size Needed for Estimating a Population Proportion with a Level of Confidence L

We use $L = .95$ in this section. Clearly, what follows will work for any L. Recall that at the time the sample size is chosen, the sample proportion $\hat{p}$ is unknown. We write the formula for the margin of error and then solve for n.

$$E = z_{.95} * \sqrt{\hat{p}(1-\hat{p})/n},$$
$$\sqrt{n} * E = z_{.95} * \sqrt{\hat{p}(1-\hat{p})},$$
$$\sqrt{n} = [z_{.95} * \sqrt{\hat{p}(1-\hat{p})}]/E,$$
$$n = \{[z_{.95} * \sqrt{\hat{p}(1-\hat{p})}]/E\}^2,$$
$$n = [(z_{.95})^2 * \hat{p}(1-\hat{p})]/E^2,$$
$$n = [(1.96)^2 * \hat{p}(1-\hat{p})]/E^2.$$

We now present the final form for the minimum sample size needed to construct a 95% confidence interval with a predetermined margin of error (radius of the confidence interval).

$$n = [(1.96)^2 * \hat{p}(1-\hat{p})]/E^2. \tag{14.5}$$

The sample size is determined *before* the sample is selected, which means that $\hat{p}$ is unknown. So, how can we use Formula (14.5), which requires us to know $\hat{p}$? There are two cases to consider. In the first, we have a preliminary estimate of $\hat{p}$. In the second, we do not.

If a preliminary estimate for $\hat{p}$ exists, we can use this as an estimate for $\hat{p}$ in Formula (14.5). If no estimate is available for $\hat{p}$, we use 1/2 for $\hat{p}$. From the study of calculus, we know that $\hat{p}=1/2$ produces a sample size that will work for any $\hat{p}$ between 0 and 1. Assuming a 95% confidence interval, using ½ for $\hat{p}$ and 2 as an estimate for $z_{.95}$, Formula (14.5) is transformed into Formula (14.6). The necessary calculations follow.

$$n=[(1.96)^2 * \hat{p}]/E^2,$$
$$n=(2^2 * 1/2 * 1/2)/E^2,$$
$$n=1/E^2 \text{ (based on 95\% confidence interval).} \qquad (14.6)$$

Formula (14.6) is used often because it is simple to use and easy to remember. Clearly, Formula (14.6) would change for a different level of confidence.

Remember, when the probability experiment is spinning a pointer on a player's disk, the descriptive measure for the population of plate appearances is known. In the examples that follow, the descriptive measure will be the on-base percentage (OBP). However, any descriptive measure that is a mean or a proportion could also be used in the examples.

Example 14.21. Suppose that we are given a disk, for a certain player, divided into eight sectors corresponding to the eight possible outcomes of a plate appearance. The area of each sector divided by the area of the circle is equal to the probability that that outcome occurs. In 2008, Florida's Hanley Ramirez had an OBP of .400. Using this as an estimate for $\hat{p}=\widehat{OBP}$, find the minimum sample size needed to estimate, with a confidence of 95%, that the sample proportion is within a margin of error of .0200 of p (the population OBP).

Solution: A sample size n corresponds to n spins of the disk. The reliability coefficient, $z_{.95}=1.96$, corresponds to 95% confidence and $\hat{p}=.400$. Using Formula (14.5), we have

$$n=[(z_{.95})^2 * \hat{p}(1-\hat{p})]/E^2$$
$$=[(1.96)^2 * (.400) * (.600)]/(.0200)^2$$
$$=2304.96.$$

Since the sample size must be an integer, we must decide how to round. You must always round *up* to the next integer. In our example, 2304 would fail because it is less than 2304.96.

By selecting a sample of size 2305 (2305 spins of the disk), we can conclude with a confidence of 95% that the interval $(\hat{p}-.0200, \hat{p}+.0200)$ contains Ramirez's *true* OBP. The calculation of $\hat{p}=\widehat{OBP}$ is based on the 2305 outcomes of the spins.

Example 14.22. A disk for an unknown player is now used, so that we have no estimate for his $\hat{p} = \widehat{OBP}$. Find the minimum sample size needed to estimate, with a confidence of 95%, that the sample proportion is within a margin of error of .0200 of his population OBP.

Solution: The reliability coefficient, $z_{.95} = 1.96$, corresponds to 95% confidence. Since we have no estimate for $\hat{p}$, we must use Formula (14.6).

$$n = 1/E^2 = [1/(.0200)^2] = 2500.$$

The reason the sample size in Example 14.22 exceeds that in Example 14.21 is because the sample size of 2500 must work for all possible $\hat{p}$.

Exercise 14.28. A survey is planned to find out what proportion of people, based on the promotion of free hot dogs discussed in Chapter 1, would buy a ticket to the right field stands at Dodger Stadium. In previous promotions to encourage people to buy tickets to the right field stands, about 35% answered yes. Find the minimum sample size required to estimate with a confidence of 95% that the sample proportion is within a margin of error of .0500 of the population proportion that would buy a ticket.

Exercise 14.29. Same problem as Exercise 14.28, except no preliminary estimate for $\hat{p}$ exists.

Exercise 14.30. Choose one of your player disks. Assuming no preliminary estimate for $\hat{p} = \widehat{OBP}$, find the minimum sample size needed to estimate, with a confidence of 95%, that the sample proportion is within .0400 of his career population OBP.

Exercise 14.31. Using the sample size found in Exercise 14.30, spin your pointer for your player's disk that many times or use the online spinner mentioned in the introduction, recording the outcome for each spin. Calculate his sample OBP. Compare his sample OBP to his population OBP (his career OBP). Is his career population OBP within the interval $(\hat{p} - .0400, \hat{p} + .0400)$?

14.8.2. Determining the Minimum Sample Size Needed for Estimating a Population Mean

Starting with the formulas for the margin of error $E = z_L * (\sigma/\sqrt{n})$, one can solve for the sample size n as follows:

$$E = z_L * (\sigma/\sqrt{n}),$$
$$\sqrt{n} * E = z_L * \sigma,$$
$$\sqrt{n} = (z_L * \sigma)/E,$$
$$n = [(z_L)^2 * \sigma^2]/E^2.$$

A problem with this formula for n is that, in many cases, σ is unknown. If so, we attempt to estimate σ. One such estimate of σ uses the sample standard deviation s obtained from a previous similar study. The two formulas are shown below. One formula uses the population standard deviation σ, and the other uses the sample standard deviation s.

$$n = [(z_L)^2 * \sigma^2]/E^2. \quad (14.7)$$
$$n = [(z_L)^2 * s^2]/E^2. \quad (14.8)$$

Example 14.23. The Yankees and Mets have some kosher concession stands, the Brewers have vegetarian options, and the Rockies have some gluten-free items on the menu. Suppose that the owner of a Major League team asks the head of concessions to estimate the average amount an adult spends at the concession stands during a game. The head of concessions wants to be 95% confident that the estimate is within \$1.25 of the population mean. A previous study showed a sample standard deviation equal to \$4.25. What is the minimum sample size needed?

Solution: The reliability coefficient is $z_{.95} = 1.96$, $E = 1.25$, and $s = 4.25$.

$$n = (1.96)^2 * (4.25)^2/(1.25)^2 = 44.41.$$

By rounding up to the next integer, the minimum sample size is 45. Therefore, to be 95% confident that the estimate is within \$1.25 of the population mean amount spent, a sample size of at least 45 is needed.

Exercise 14.32. TV stations want to schedule programming and generate ad revenue. Suppose that they want to estimate the average time, in minutes, it takes to complete a nine-inning game in the National League. The TV station wants to be 90% confident that its estimate is within 10 minutes of the population mean. A previous study revealed a sample standard deviation equal to 45 minutes. What is the minimum sample size needed?

Chapter Summary

This chapter concentrated on estimation, an important aspect of inferential statistics. Estimations are made for the two parameters p and μ. Estimations are accomplished by choosing a random sample from the population and computing a statistic as an estimator of the parameter. The statistic $\bar{x}$ was used to estimate μ. The statistic $\hat{p}$ was used to estimate p. The two statistics $\bar{x}$ and p are called point estimates.

Two types of estimates were discussed, a point estimate and a confidence interval estimate. A point estimate is a specific value. The problem with using a point estimate as a predictor for the population descriptive measure (parameter) is that no probability can be assigned to the prediction. For this reason, statisticians prefer using a confidence interval. A confidence interval estimate assigns a probability that the interval contains the population descriptive measure.

The construction of a confidence interval for the parameters p and μ depends on having a sampling distribution of a statistic whose mean is the parameter we wish to estimate. The sampling distribution must be approximately normal with a known standard deviation, called the standard error, SE. If certain assumptions are satisfied, the sampling distribution of $\hat{p}$ is normally distributed with mean equal to p and the sampling distribution of $\bar{x}$ is normally distributed with mean equal to μ. Since the mean of each of their sampling distributions is the parameter being estimated, both estimators are said to be *unbiased*.

The researcher must decide, before collecting the sample, the level of confidence L to assign to the interval. The standard level of confidence is 95%. Other common levels of confidence are 90%, 98%, and 99%. The level of confidence L is the probability that if a confidence interval is formed for many random samples, L% of the confidence intervals will contain the parameter being estimated.

The general form of a confidence interval is

$$\text{estimator} \pm [(\text{the reliability coefficient}) * (\text{the standard error})].$$

The confidence interval for p has the following form:

$$\hat{p} - z_L * \sqrt{\hat{p}(1-\hat{p})/n} < p < \hat{p} + z_L * \sqrt{\hat{p}(1-\hat{p})/n},$$

where $\hat{p}$ is the estimator (sample proportion), p is the parameter to be estimated (population proportion), n is the sample size, L is the level of confidence, z_L is the reliability coefficient from the z-distribution, $\sqrt{\hat{p}(1-\hat{p})/n}$ is an estimate for the standard error, and $z_L * \sqrt{\hat{p}(1-\hat{p})/n}$ is the margin of error.

The confidence interval for μ has two possible forms:

$$\bar{x} - z_L * (\sigma/\sqrt{n}) < \mu < \bar{x} + z_L * (\sigma/\sqrt{n}), \qquad \text{(Form 1)}$$

or, if σ is unknown, we have

$$\bar{x} - t_L * (s/\sqrt{n}) < \mu < \bar{x} + t_L * (s/\sqrt{n}), \qquad \text{(Form 2)}$$

where $\bar{x}$ is the estimator (sample mean), μ is the parameter to be estimated (population mean), n is the sample size, z_L is the reliability coefficient from the z-distribution, t_L is the reliability coefficient from the t-distribution, $\sigma/\sqrt{n}$ is the standard error, and $s/\sqrt{n}$ is an estimate for the standard error. If $n \geq 30$ and σ is unknown, we can use z_L as an estimate for t_L.

Multiplying the reliability coefficient times the standard error gives the radius for the confidence interval. The radius of the confidence interval is called the margin of error. The width of the confidence interval is equal to 2 times the margin of error.

Form 1 is used when σ is known and either $n \geq 30$ or the original population is approximately normal. Form 2 is used for small samples ($n < 30$), where σ is unknown and the *X*-population is normal. If $n \geq 30$ and σ is unknown, Form 2 can be used or Form 2 can be used with z_L replacing t_L. For the same level of confidence, Form 2 gives a wider confidence interval than Form 1. The reason for this is that when the sample size is small and σ is unknown, less is known about the *X*-population.

A higher level of confidence results in a wider confidence interval. A larger sample size results in a narrower confidence interval. A narrower confidence interval gives a more precise estimate for the parameter.

The last section of the chapter deals with finding the minimum sample size required to guarantee for a specified level of confidence that the estimate for the parameter is within a certain margin of error.

Throughout this chapter, various baseball descriptive measures are used to evaluate a player's batting performance and ability. Many of these descriptive measures are proportions, such as batting average, on-base proportion, strikeout average, base-on-balls average, and home run average. For each proportion, we can define a player's performance and ability. Performance is based on his observed plate appearances for a period of time. A player's performance is the sample proportion. Ability is theoretical and is based on a player batting an infinite number of times for a time period. A player's ability is the population proportion, which is unknown. A confidence interval for the population proportion gives an interval estimate for a player's ability with respect to a certain descriptive measure for his batting performance.

Confidence intervals can be used to compare two players with respect to certain descriptive measures of their batting ability. If the two confidence intervals overlap, we cannot conclude that the ability of one player exceeds that of the other player. If the two confidence intervals did not overlap, we can conclude that the player whose confidence interval is to the right has the greater ability for that descriptive measure of batting. As was pointed out in the chapter, since the level of confidence changes, this is not the best method to use to compare the batting ability between two players.

CHAPTER PROBLEMS

1. Harris, in January 2008, polled 1562 people who follow one or more sports and found that 30% say that professional football is their favorite. Second

place goes to baseball, with 15%. College football is third, with 12%, and 10% of sports enthusiasts say auto racing is their favorite.

a. What is the sample in the survey?
b. What is the population?
c. What is the variable of interest?
d. What is the sample size?
e. What is the measurement scale of the variable?
f. Find the sample proportion of people who chose baseball as their favorite sport.
g. Find a 95% confidence interval for the population proportion of people who would choose baseball as their favorite sport.
h. Find the radius of the confidence interval in (g).
i. Find the margin of error for the confidence interval in (g).
j. Find the minimum sample size needed to ensure with a confidence of 95% that the radius of the confidence interval is equal to .0100. Use the sample proportion from (f) as your estimate for the sample proportion.

2. The USA Baseball Medical & Safety Advisory Committee commissioned the American Sports Medicine Institute to study pitch limits in youth baseball. Surveys were sent to 85 baseball experts, consisting of orthopedic surgeons and coaches, about pitch limits. Twenty-eight of these experts responded. Table 14.4 was included in the study.

   There are two variables of interest. One variable is the opinion of baseball experts on the maximum number of pitches per game for a given age group. The second variable is the opinion of baseball experts on the maximum number of games per week for a given age group. Assume that the two populations are approximately normally distributed.

   a. Identify the sample in the study. What is the sample size?
   b. Find a 95% confidence interval for the population mean maximum number of pitches that baseball experts believe a child between the ages of 8 and 10 should throw per game.
   c. Same as (b), except change the age group to 13–14.
   d. In comparing the confidence intervals for (b) and (c), what conclusion can be drawn?
   e. Find the minimum sample size required to have 95% confidence that the sample mean maximum number of pitches per game is within three pitches of the population mean maximum number of pitches per game. Use the age group 8–10. Use $s = 15$ from the study above.

Table 14.4 **Maximum number of pitches recommended (sample mean ± sample standard deviation)**

Age	Maximum pitches/game	Maximum games/week
8–10	52 ± 15	2 ± 0.6
11–12	68 ± 18	2 ± 0.5
13–14	76 ± 16	2 ± 0.4
15–16	91 ± 16	2 ± 0.4
17–18	106 ± 16	2 ± 0.6

Note: Survey sent to 85 baseball experts (28 responded).

f. Find a 95% confidence interval for the population mean maximum number of games per week that baseball experts believe a child between the ages of 8 and 10 should play.
g. Same as (f), except change the age group to 13–14.
h. In comparing the confidence intervals for (f) and (g), what conclusion can be drawn?

3. Choose one of your players' disks.
   a. Spin the pointer 25 times or use the online spinner. Record the results of the 25 spins. What is the sample size? Calculate his sample batting average ($\widehat{BA}$). What is the population? Find a 90% and 95% confidence interval for his *true* (population) BA. What assumption is necessary about the population?
   b. Spin the pointer 50 times. Answer the questions in (a) above.
   c. Which sample batting average is a better estimate of his career batting average?
   d. Find the minimal sample size needed to estimate, with a confidence of 95%, that his sample batting average is within .010 (10 points) of his career batting average.

The next set of questions tests your knowledge of the theorems concerning confidence intervals.

Suppose that the mean of a certain population is unknown. One random sample of size 36 is selected. The sample mean is 70, and the sample standard deviation is 8.

4. The standard error is ______.
5. Find a 90% confidence interval for the population mean.
6. Find a 95% confidence interval for the population mean.

Suppose that we also know that the above population is normal. One random sample of size 16 is selected. The sample mean is 70, and the sample standard deviation is 8.

7. The standard error is ______.
8. Find a 90% confidence interval for the population mean.
9. Find a 95% confidence interval for the population mean.
10. The margin of error in problem 6 is ______ and in problem 9 is ______.

The next two questions ask you to compare the confidence interval in problem 6 with the confidence interval in problem 9.

11. Which confidence interval is more precise?
12. Explain the connection between the precision of the confidence interval and the margin of error.

The next two questions ask you to compare the confidence interval in problem 5 with the confidence interval in problem 6.

13. Which confidence interval is more precise?
14. Explain the connection between the precision of the confidence interval and the reliability coefficient.

Chapter 15

Hypothesis Testing for One Population

In Chapter 14, we studied one technique of inferential statistics called confidence intervals. A confidence interval uses a statistic (a descriptive measure of a sample) to create an interval estimate for a parameter (a descriptive measure of a population). In particular, a sample mean was used to create an interval estimate for a population mean, and a sample proportion was used to create an interval estimate for a population proportion.

In this chapter, we introduce a second technique used in inferential statistics. This technique is called *hypothesis testing*. Hypothesis testing is based on two hypotheses. A hypothesis is a statement about a population, which is either true or false. This statement can be about a parameter of the population or about a property of the population. In this chapter, our hypotheses will be about parameters. When the hypotheses are concerned with the parameters of a population, we say that the hypothesis testing is *parametric hypothesis testing*.

15.1. Introduction to Hypothesis Testing

Hypothesis testing is used in fields such as medicine, business, education, and psychology. In medicine, a pharmaceutical company wants to see if a new drug increases the survival time of a cancer patient. In business, a marketing manager needs to know if more advertising will increase the total sales. An educator wishes to explore whether a new technique will raise students' test scores. A psychologist may like to assess if a change in the workplace environment will reduce absenteeism.

In baseball, would a certain rule change decrease the time of a nine-inning game, or would it increase the league batting average?

Hypotheses concerning means and proportions, for one population, are investigated in this chapter. As with confidence intervals, the sampling distribution of sample proportions and the sampling distribution of sample means are used.

Three methods are used to perform hypothesis testing. They are the *classical* or *original* method, the *p-value* method, and the *confidence interval* method.

15.1.1. Statement of Null and Alternate Hypotheses

Parametric hypothesis testing, as it is known, begins with a research question that makes a claim about a parameter for a population. The research question leads to a pair of statements about a population parameter. These statements are called statistical hypotheses. These two hypotheses are complements of each other. When one of the statements is true, the other is false. The claim represented in the research question is linked directly to one of the two hypotheses.

The first hypothesis, symbolized by $H_{0,}$ is called the *null* hypothesis. There are three choices for H_0. If we let c be a constant, the three choices are as follows:

H_0: parameter $= c$; H_0: parameter $\leq c$; H_0: parameter $\geq c$.

In hypothesis testing, H_0 is assumed true *before* the study begins. The statement in H_0 contains a form of equality, either equal to (=), less than or equal to (≤), or greater than or equal to (≥).

The second hypothesis, written $H_{A,}$ is the *alternate* hypothesis. The alternate hypothesis is the complement of the null hypothesis and does *not* contain a form of equality. The three possible pairs of hypotheses are shown below:

H_0: parameter $= c$	H_0: parameter $\leq c$	H_0: parameter $\geq c$
H_A: parameter $\neq c$	H_A: parameter $> c$	H_A: parameter $< c$.

Only one of these pairs can be chosen. Regardless of which pair is chosen, H_0 is always assumed true at the beginning of the research study. A preponderance of evidence is necessary to reject H_0; otherwise, we will continue to assume that H_0 is true. The claim made by the researcher determines which pair of hypotheses to use. The claim can be made for either one of the two hypotheses in the pair selected. The next examples illustrate the choice of hypotheses.

Example 15.1. The health department of New Jersey claims that the population proportion of high school varsity athletes using steroids, without a doctor's prescription, is equal to 6%. The parameter being tested is the population proportion. Since the claim contains a form of equality, it appears as H_0. We have the following pair:

H_0: $p = .06$ (claim)
H_A: $p \neq .06$.

Example 15.2. Major League Baseball claims that the mean time of a baseball game since 2000 is greater than 3 hours. The parameter being tested is the population mean. Since the claim does not contain a form of equality, it appears as H_A. We have the following pair:

H_0: $\mu \leq 3$
H_A: $\mu > 3$ (claim).

If the claim was that the mean time is less than 3 hours, the pair of hypotheses would be

H_0: $\mu \geq 3$
H_A: $\mu < 3$ (claim).

Example 15.3. The areas of the sectors on a player's disk correspond to probabilities calculated from the player's career plate appearance. Spinning a pointer on the disk simulates a plate appearance for that player. We are not allowed to see the disk. Can we conclude from the results of several spins that this player's *true* career batting average is less than .300? Since a batting average can be considered a proportion or a mean, we can use for the parameter either μ or p. Since the claim does not contain a form of equality, it will appear in H_A.

H_0: $\mu =$ *true* BA $\geq .300$ H_0: $p =$ *true* BA $\geq .300$
H_A: $\mu < .300$ (claim) H_A: $p < .300$ (claim).

The reason for two pairs of hypotheses is that a batting average can be considered a mean or a proportion.

Exercise 15.1. The same study as in Example 15.3, except that the question is, can we conclude that the player's *true* career batting average is greater than .300? Construct the pair of hypotheses for this study.

Exercise 15.2. The same study as in Example 15.3, except that the question is, can we conclude that the player's *true* career batting average is not equal to .300? Construct the pair of hypotheses for this study.

15.1.2. Two Types of Errors and Their Associated Probabilities

At the end of the process, a decision must be made about the null hypothesis H_0. Remember, before the process begins, H_0 is assumed true. But do we continue to accept the statement in H_0 as true, or declare it false?

Since the decision is based on the result of a sample and not the entire population, there is a possibility that our declaration concerning H_0 is wrong. There are two possible errors that can occur. These are called the *type 1 error* and the *type 2 error.*

- The type 1 error occurs when we reject a *true* H_0.
- The type 2 error occurs when we accept a *false* H_0.

To understand these two hypotheses more fully, we digress and turn our attention to our legal system. Suppose that a person is arrested for murder. The two hypotheses to consider are as follows:

- H_0: The person is innocent of murder.
- H_A: The person is guilty of murder.

Unlike the previous examples, there are no parameters involved. The hypotheses are statements. In our legal system, the defendant is assumed innocent until proven guilty. Therefore, H_0 is assumed true at the beginning of the trial. The strength of the evidence presented by the prosecutor determines the decision of the jury. The judge instructs each member of the jury not to find the defendant guilty unless it is beyond any reasonable doubt. The jury returns a verdict of either not guilty or guilty as charged. Table 15.1 shows the jury's decision and the theoretical truth about the defendant.

The type 1 error results in putting an innocent person in prison. The type 2 error results in a guilty person being set free. Both errors are bad, but in our legal system, the type 1 error of finding an innocent person guilty is considered the more severe.

The statement that resides in the null hypothesis (H_0) is considered the status quo. Status quo means that the statement is assumed true at the beginning of the process.

Table 15.1 **Decision table for the null hypothesis H_0**

Jury's decision concerning H_0: person is innocent	Reality	
	Innocent	Guilty
Accept H_0	Correct decision	Type 2 error
Reject H_0	Type 1 error	Correct decision

Utilized in football, and soon to be added to the rules of baseball, is "instant replay." What statement would be the null hypothesis, and what statement would be the alternate hypothesis? For "instant replay," the two hypotheses are as follows:

- H_0: The ruling on the field is correct.
- H_A: The ruling on the field is wrong.

To overturn the ruling on the field, there must be indisputable evidence that it was wrong. The type 1 error in this situation is that we overrule the decision on the field when it was in fact correct. The type 2 error is to accept the ruling on the field when in reality it was wrong. Since "the ruling on the field is correct" is the null hypothesis, the type 1 error is the one that must be kept small. Therefore, the decision on the field should not be overturned without indisputable evidence.

Exercise 15.3. A baseball player's career statistics clearly qualify him for the Hall of Fame. If the baseball writer believes that the player used banned substances, he will not include him on his ballot. Before voting, a baseball writer looks at these two hypotheses:

H_0: The player did not use banned substances during his career.
H_A: The player did use banned substances during his career.

For these two hypotheses,

a. State the type 1 error. What are the consequences of making this error?
b. State the type 2 error. What are the consequences of making this error?
c. Which error does the sportswriter want to keep small? Why?

We are now interested in assigning a probability to each of these errors. The probability of committing a type 1 error is denoted by the Greek letter α (alpha) and is called the α-error. The probability of committing a type 2 error is denoted by the Greek letter β (beta) and is called the β-error. Both these errors will be displayed as a decimal.

- The α-error is the probability of committing a type 1 error.
- The β-error is the probability of committing a type 2 error.

Figure 15.1 Seesaw effect applied to two errors

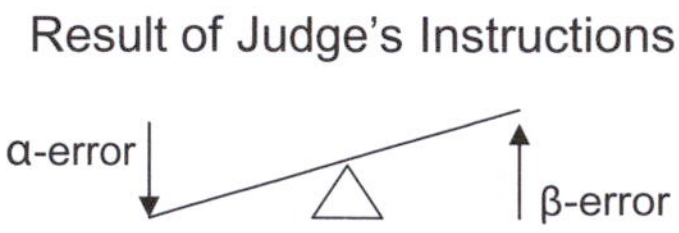

Result of Judge's Instructions

α-error

β-error

Since the judge's instructions to the jury make it much more difficult to find a person guilty, they have the effect of reducing the α-error. The unintended consequence of reducing the α-error is to *increase* the β-error. As the α-error decreases, the β-error increases. Since forcing one of the errors down pushes the other error up, we call this the seesaw effect (see Fig. 15.1).

A not-guilty verdict means that we assumed at the beginning of the trial that the person was innocent (H_0 is assumed true) and the evidence presented was not strong enough to find the person guilty. So at the end of the trial, we will continue to assume that H_0 is true. *A not-guilty verdict does not mean that we have proven that the person is innocent.* What it does mean is that there was not enough evidence to find the person guilty. A guilty verdict means that we reject the statement that the person is innocent. The result of rejecting that a person is innocent is accepting the fact that the person is guilty.

15.1.3. Relating Our Previous Discussion to Classical Hypothesis Testing in Statistics

The null hypothesis H_0 is assumed true at the beginning of the research process. The α-error is also set at the beginning. The sample data are then collected. If the statistic evaluated from the sample data is so extreme that the probability of it occurring strictly by chance, under the assumption that H_0 is true, is less than the α-error, our statistical decision is to reject the null hypothesis. If the statistic is not that extreme, we accept H_0. The acceptance of H_0 as true means that we will continue to assume that H_0 is true. However, since the β-error is not controlled, even though we accept H_0 as true, we have no control over the probability that our decision to accept H_0 as true was the wrong decision.

If H_0 is rejected, we conclude that the statement in H_0 is false, with the understanding that our conclusion could be wrong $(\alpha * 100)\%$ of the time. On the other hand, the acceptance of H_0 means that we will *continue to assume* that the statement in H_0 is true.

Exercise 15.4. For the hypotheses in Exercise 15.3,

a. State the α-error. What probability should the baseball writer assign to the α-error?
b. State the β-error. What probability should the baseball writer assign to the β-error?

15.2. Classical Hypothesis Testing

In this section, we break down classical hypothesis testing into eight steps.

The Eight Steps for the Classical Hypothesis Testing Method
Step 1: State the research question.
Step 2: State the two hypotheses (H_0 and H_A).

Step 3: State the level of significance (this is the maximum allowable α-error).

Step 4: Calculate the test statistic (ts) for the collected data and provide the necessary assumptions for the *X*-population. (In this chapter, the test statistic will be either $\bar{x}$ or $\hat{p}$.)

Step 5: Convert the test statistic (ts) to the standard test statistic (sts) ($\bar{x}$ to z or t, or $\hat{p}$ to z).

Step 6: Find the critical value or values (cv) that separate the acceptance and rejection region or regions and then locate the result of step 5 with respect to these regions.

Step 7: Make your statistical decision about H_0 (accept or reject H_0).

Step 8: Conclude by answering the research question.

Before embarking on the explanation for the eight steps, we use Theorem 15.1 to summarize our work with sampling distributions which was studied in Chapter 13.

Theorem 15.1.

a. If both $np > 5$ and $n(1-p) > 5$:
   i. The sampling distribution of $\hat{p}$ is normally distributed.
   ii. The formula $(\hat{p}-p)/\sqrt{p(1-p)/n}$ converts a *test statistic* $\hat{p}$ into the *standard test statistic z*.
b. If the *X*-population is normally distributed or $n \geq 30$ and σ is known:
   i. The sampling distribution of $\bar{x}$ is normally distributed.
   ii. The formula $(\bar{x}-\mu)/(\sigma/\sqrt{n})$ converts a *test statistic* $\bar{x}$ into the *standard test statistic z*.
c. If the sample size $n \geq 30$ and σ is unknown:
   i. The sampling distribution of $\bar{x}$ takes the shape of a *t*-curve with $(n-1)$ degrees of freedom.
   ii. The formula $(\bar{x}-\mu)/(s/\sqrt{n})$ converts a *test statistic* $\bar{x}$ into the *standard test statistic t* with $(n-1)$ degrees of freedom.
   iii. Since the *z-statistic* is a good estimate for the *t-statistic* when $n \geq 30$, we use the *z-statistic* instead of the *t-statistic*.
d. If the sample size $n < 30$, σ is unknown, and the *X*-population is normally distributed:
   i. The sampling distribution of $\bar{x}$ takes the shape of a *t*-curve with $(n-1)$ degrees of freedom.
   ii. The formula $(\bar{x}-\mu)/(s/\sqrt{n})$ converts a *test statistic* $\bar{x}$ into the *standard test statistic t*.

Remember, $\bar{x}$ and $\hat{p}$ are test statistics (abbreviated by ts) and z and t are standard test statistics (abbreviated by sts).

Example 15.4. This example provides an explanation of each of these eight steps associated with the research question posed in Example 15.3. For what follows, a batting average will be considered a proportion.

Step 1: State the Research Question. Can we conclude, with a level of significance of $\alpha = 0.05$, that the population of infinite spins on the disk will result in this player having a *true* batting average less than .300?

Step 2: Statement of Two Hypotheses. Since the claim of "less than" is a strict inequality, it resides in H_A.

H_0: $p \geq .300$
H_A: $p < .300$ (claim).

Step 3: State the Level of Significance.

$\alpha = 0.05$.

The type 1 and type 2 errors were discussed in the legal example. Recall that the type 1 error corresponds to rejecting H_0 when H_0 is actually true. The type 2 error corresponds to accepting H_0 when H_0 is false. The probability of committing a type 1 error is called the α-error; the probability of committing a type 2 error is called the β-error. From our previous discussion of the legal example, the seesaw effect showed that we cannot control both the α-error and the β-error. The error we choose to control and make small is the α-error. The level of significance is defined below.

The level of significance is the maximum allowable α-error. The standard level of significance is 0.05. Other possible values used for the level of significance are 0.01, 0.02, and 0.10.

Notice that the standard level of significance $\alpha = 0.05$ is equal to 1 minus the standard level of confidence of $L = 0.95$, studied in Chapter 14. Steps 1, 2, and 3 are declared before the sample data are collected.

At this point in a research study, a decision is made on the sample size and how to collect a random sample (or at least a sample that is representative of the population). After the sample is collected, the appropriate statistics must be calculated. Also, assumptions (if any) concerning the original X-population must be stated. This leads to step 4.

Step 4: The Calculation of the Test Statistic and Presentation of Any Assumptions about the X-Population. The sample data are obtained by spinning the pointer on the disk 500 times. The sample size consists of the 450 spins that resulted in at bats (50 of the outcomes were BB). Of the 450 at bats, 130 resulted in hits. The sample size is $n = 450$, and the sample proportion $\hat{p} = 130/450 = .289$. The X-population consists of the outcomes for an infinite number of spins. The statement in H_0 is always assumed true. Thus, $p \geq .300$

is assumed true. Unlike for confidence intervals, where the parameter is unknown, we can use our assumed p for two purposes. It can be used to show that the sampling distribution of $\hat{p}$ is approximately normal, and it can be used to calculate the standard error SE. Since $np = 450 * .300$ and $n(1-p) = 450 * .700$ are both greater than 5, we can assume that the sampling distribution is approximately normal.

Step 5: Convert the Test Statistic $\hat{p}$ to the Standard Test Statistic z. The standard deviation of the sampling distribution of $\hat{p}$ for $n=450$ is $\text{SE} = \sqrt{p(1-p)/n} = \sqrt{.300(1-.300)/450} = .0216$. The z-transformation, $z = (.289 - .300)/.0216 = -.51$, converts the test statistic $\hat{p} = .289$ to the standard test statistic $z = -.51$. The interpretation is that, for the sampling distribution, .289 is .51 standard deviations to the left (because of the negative sign) of the mean. The sampling distribution and the standard normal curve are shown in Figure 15.2.

Step 6: Find the Critical Value or Values (cv) That Separate the Acceptance from the Rejection Regions. The level of significance from step 3 and the alternate hypothesis, H_A, determine the critical value or critical values that separate the acceptance region from the rejection region. In Figure 15.3, α is the level of significance. We look at the three possible cases for H_A. We represent the acceptance region by A and the rejection region by R.

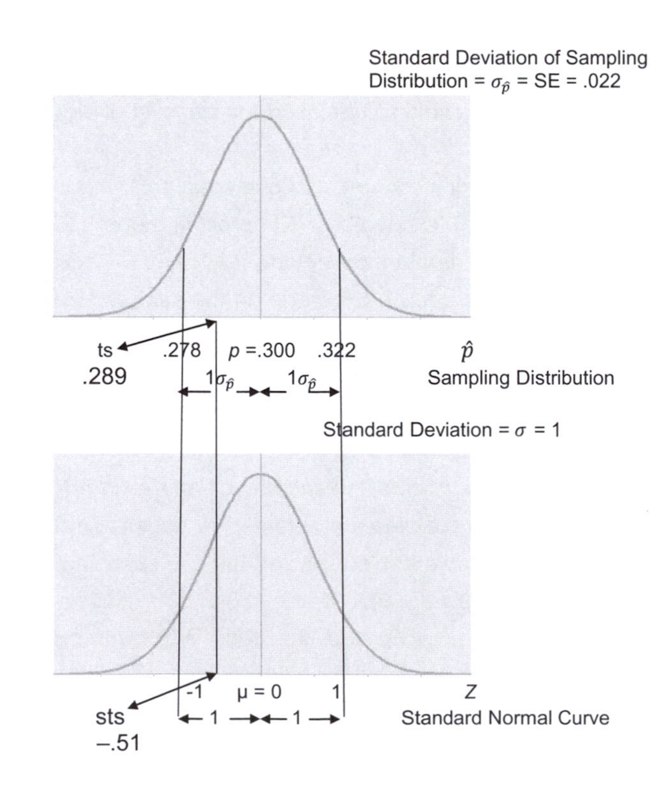

Figure 15.2 Sampling distribution for Example 15.4

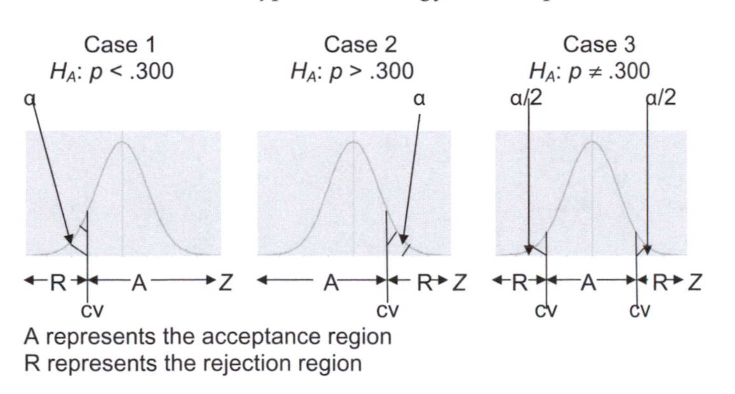

Figure 15.3 Acceptance and rejection regions for the three cases

The probability that the converted *z*-value lies under the rejection region in all three cases is α. Since α is the probability of rejecting a *true* H_0, the error in rejecting a true H_0 will not exceed α. Thus, we have controlled the α-error to a predetermined small probability.

Case 1: The one critical value is the number in the column headed by *z* in Table C.2, which corresponds to the area closest to α in the body of the table. In this case *z* is negative.

Case 2: The one critical value is the number in the column headed by *z* in Table C.2, which corresponds to the area closest to $(1 - \alpha)$ in the body of the table. In this case *z* is positive.

Case 3: There are two critical values. The easiest way to find them is to use Case 1 above with $\alpha/2$ to find the critical value on the left side. The critical value on the right side is then equal to the negative of the critical value on the left side.

What leads to the rejection of H_0? In Case 1, $\hat{p}$ must be less than .300 and far enough away from .300 so that the *z*-value lies in the rejection region. In Case 2, $\hat{p}$ must be greater than .300 and far enough away from .300 so that the *z*-value lies in the rejection region. In Case 3, $\hat{p}$ can be on either side of .300, but it is far enough away from .300 so that the *z*-value lies in the rejection region. Case 1 is called a *left-tailed* hypothesis test, Case 2 is called a *right-tailed* hypothesis test, and Case 3 is called a *two-tailed* hypothesis test.

For Example 15.4, Case 1 above applies. Since $\alpha = 0.05$, the *z*-value that separates the acceptance and rejection region, called the critical value, is $z = -1.64$ or $z = -1.65$. The lookup of the area closest to $\alpha = 0.05$ in Table C.2 produces $z = -1.64$ or $z = -1.65$.

Under the assumption that H_0 is true, the probability that $\hat{p}$ converts into a *z*-value less than –1.64 is less than or equal to 0.05. We reject H_0 only if the converted *z*-value is to the left of –1.64. The converted *z*-value is called the *standard test statistic* (sts). Since the probability of the *z*-value lying in the rejection region is less than 0.05, the error we would make in rejecting a

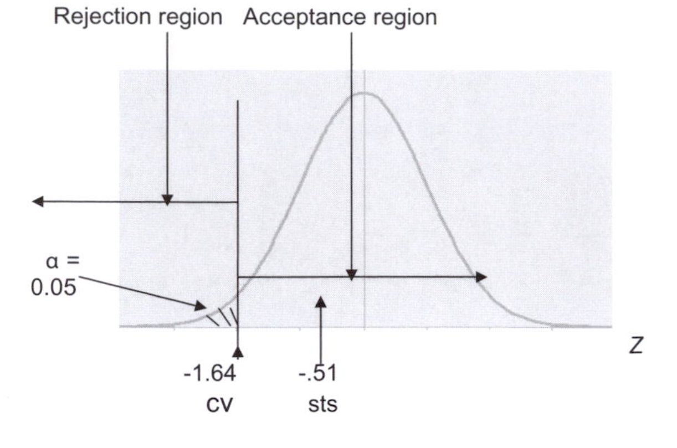

Figure 15.4 Acceptance and rejection regions for Example 15.4. The converted z-value is called the standard test statistic. The z-value that separates the acceptance region from the rejection region is called the critical value.

true H_0 has a probability less than 5%. Figure 15.4 shows the acceptance and rejection region for Example 15.4. The rejection region is to the left of $z = -1.64$. Since -1.64 separates the acceptance and rejection regions, it is called the *critical value*. Clearly, $-.51$ lies under the acceptance region.

Step 7: Statistical Decision on H_0. We have to make a decision either to accept H_0 as true or to reject H_0 and accept H_A. The decision is based on whether the converted test statistic lies under the acceptance or rejection region. Since $-.51$ lies under the acceptance region, the statistical decision is to accept H_0.

Since H_0 was accepted, we say that the sample result was not statistically significant. If H_0 had been rejected, we would have said that the sample result was statistically significant.

Step 8: Answer the Research Question. Since H_0 was accepted, we cannot conclude that this player's *true* batting average is less than .300.

Exercise 15.5. For each part (a–e) below, you are given the pair of hypotheses, the α-error, and the standard test statistic (sts). For each part, complete step 6 and step 7. Step 6 should include a graph showing the critical value(s), the standard test statistic, and the acceptance and rejection region. Step 7 should include your statistical decision to accept or reject H_0.

a. Given: H_0: $p \leq .40$; H_A: $p > .40$; $\alpha = 0.10$, $z = 1.47$.
b. Given: H_0: $p \geq .40$; H_A: $p < .40$; $\alpha = 0.10$, $z = -1.05$.
c. Given: H_0: $\mu \leq 6$; H_A: $\mu > 6$; $\alpha = 0.01$, $z = 2.48$.
d. Given: H_0: $\mu = 6$; H_A: $\mu \neq 6$; $\alpha = 0.01$, $z = -2.88$.
e. Given: H_0: $\mu \geq 10$; H_A: $\mu < 10$; $\alpha = 0.01$, $z = -2.08$.

15.2.1. Hypothesis Testing for One Population Proportion p (the Other Two Cases)

Example 15.4 showed hypothesis testing for the population proportion p in which the alternate hypothesis was the strict inequality "<." In the next two examples, we apply the eight steps of classical hypothesis testing to the other two possible cases for H_A. The eight steps are provided with minimal explanation.

Example 15.5. A disk of a player with sectors representing the probabilities of his plate appearance outcomes is supplied. We are not allowed to see the disk. A sample of 1000 spins produced 950 at bats with 314 hits. Can we conclude, with a level of significance of $\alpha = 0.05$, from the sample results, that this player's *true* career batting average is greater than .299?

Step 1: Can we conclude, with a level of significance of 0.05, that this player's *true* batting average is greater than .299?

Step 2: H_0: $p \leq .299$; H_A: $p > .299$ (claim).

Step 3: $\alpha = 0.05$.

Step 4: Sample data: $n = 950$, $\hat{p} = 314/950 = .331$. Since np and $n(1-p)$ are both greater than 5, the sampling distribution can be assumed normal.

Step 5: Convert test statistic $\hat{p}$ to the standard test statistic z:

$$z = (\hat{p} - p)/\sqrt{p(1-p)/n}$$
$$= (.331 - .299)/\sqrt{.299 * .701/950}$$
$$= 2.15.$$

Step 6: Set up acceptance and rejection region (see Fig. 15.5).

Step 7: Statistical decision: Since the test statistic lies under the rejection region, we reject H_0.

Step 8: We can conclude that the player's *true* career batting average *is* greater than .299.

Example 15.6. A disk of a player with sectors representing the probabilities of his plate appearance outcomes is supplied. We are not allowed to see the disk. A sample of 800 spins produced 725 at bats with 216 hits. Can we conclude, with a level of significance of $\alpha = 0.05$, from the sample results, that this player's *true* career batting average is different from .300?

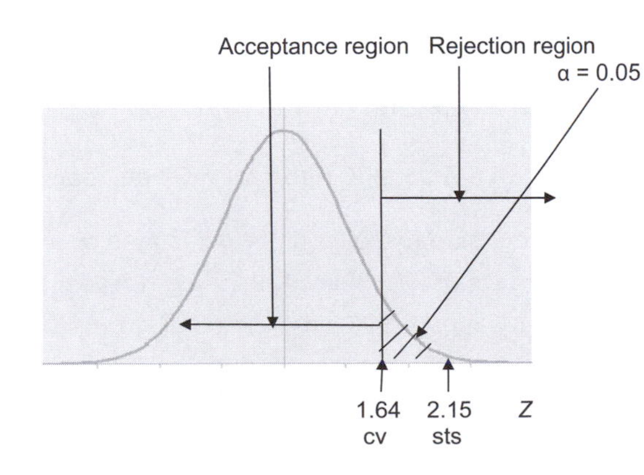

Figure 15.5 Acceptance and rejection regions for Example 15.5

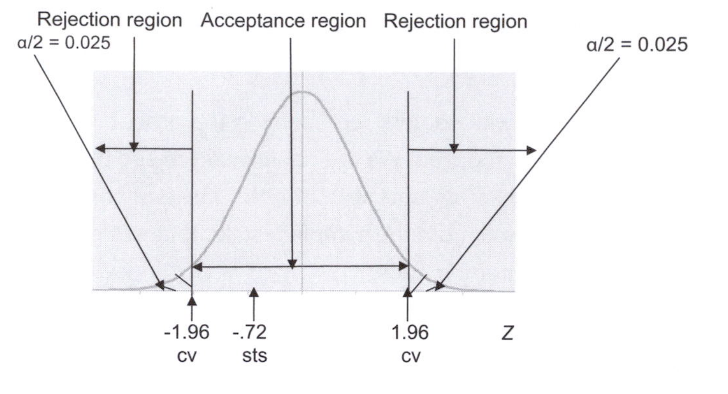

Figure 15.6 Acceptance and rejection regions for Example 15.6

Step 1: Can we conclude, with a level of significance of 0.05, that this player's *true* batting average is different from .300?

Step 2: H_0: $p=.300$; H_A: $p \neq .300$ (claim).

Step 3: $\alpha=0.05$.

Step 4: Sample data: $n=750$, $\hat{p}=216/750=.288$. Both np and $n(1-p)$ are greater than 5 (sampling distribution can be assumed normal).

Step 5: Convert test statistic $\hat{p}$ to the standard test statistic z:

$$\begin{aligned} z &= (\hat{p}-p)/\sqrt{p(1-p)/n} \\ &= (.288-.300)/\sqrt{.300*.700/750} \\ &= -.72. \end{aligned}$$

Step 6: See Figure 15.6 for the acceptance and rejection regions. This has a two-tail rejection region. The lookup of 0.025 in the Z-Table produces $z=-1.96$. The two critical values are $z=1.96$ and $z=-1.96$. The standard test statistic is $-.72$.

Step 7: Since the test statistic lies under the acceptance region, we accept H_0.

Step 8: We cannot conclude that the player's *true* career batting average is different from .300.

In baseball, there are certain magic numbers. Three of these magic numbers for a player's career are having a batting average of at least .300, hitting at least 500 home runs, and getting at least 3000 hits. A player who accomplishes at least two of these three milestones is almost guaranteed admission to the Hall of Fame.

Definition of a *True* .300 Hitter for a Given Year or for a Career

If we can conclude, at a level of significance of $\alpha=0.05$, that a player's *true* batting average is greater than .299 for a given year or for a career, we will say that that player is a *true* .300 hitter for the given year or for a career.

In the above definition, the population is assumed infinite and consists theoretically of an infinite number of at bats for that player for the given year or his career. The sample is his actual at bats for that year or his career.

In Chapter 17, a *true* .400 hitter for a given year is defined by substituting .399 for .299 and .400 for .300 in the definition for a *true* .300 hitter.

Exercise 15.6. Can we conclude that Henry Aaron was a *true* career .300 hitter? Use the data from Table 2.2. (Hint: H_A: $p > .299$.) Show the eight steps for hypothesis testing.

Exercise 15.7. Was Barry Bonds a *true* career .300 hitter? Use the data from Table 4.1. Show the eight steps for hypothesis testing.

Exercise 15.8. Write out the definition for a player being a *true* .400 hitter for a given year.

The topic of .400 hitters will be studied in Chapter 17.

15.2.2. Hypothesis Testing for One Population Mean, μ (for a Large Sample Size $n \geq 30$)

Since a batting average can be interpreted as both a proportion and a mean, we use as our first example for hypothesis testing of the population mean the information in Example 15.4. In Example 15.7, we present the eight steps of hypothesis testing for the population mean.

Example 15.7. Spinning a pointer on a disk for a player 500 times results in 50 base on balls and 130 hits. Can we conclude, at a level of significance of 0.05, that the player's *true* career batting average is less than .300?

Step 1: State the Research Question. Can we conclude, with a level of significance of 0.05, that the player's *true* batting average is less than .300?

Step 2: State the Hypotheses. H_0: $\mu \geq 300$; H_A: $\mu < 300$ (claim).

Step 3: State the Level of Significance. $\alpha = 0.05$.

Step 4: Sample Data and Population Assumptions. The sample consists of 450 at bats. Of the 450 numbers, 130 of the numbers are 1s and 320 of the numbers are 0s.

The test statistic $\bar{x} = \Sigma x/n = 130/450 = .289$,

The sample variance $s^2 = [\Sigma x^2 - (\Sigma x)^2/n]/(n-1)$

$= [130 - (130)^2/450]/449 = .206.$

The sample standard deviation $s = \sqrt{.206} = .4538$. There are no assumptions about the underlying X-population of spins. The sampling distribution of $\bar{x}$ is assumed normally distributed, since the sample size exceeds 30.

Step 5: Convert the Test Statistic $\bar{x}$ to the Standard Test Statistic z. Since the sample size is at least 30 and σ is unknown, from Theorem 15.1(c) we have

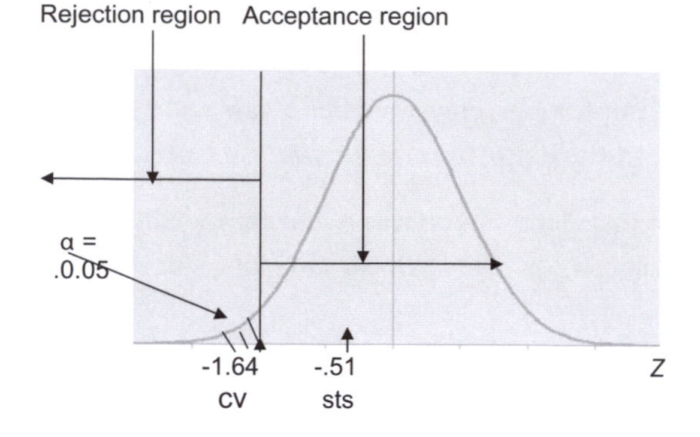

Figure 15.7 Acceptance and rejection regions for Example 15.7

$$z = (\bar{x} - \mu)/(s/\sqrt{n})$$
$$= (.289 - 300)/(.4538/\sqrt{450})$$
$$= -.51$$ (the z-statistic is used as a good estimate for the t-statistic).

Step 6: Set Up the Acceptance and Rejection Region. See Figure 15.7.

Step 7: Statistical Decision on H_0. Since the test statistic lies in the acceptance region, we accept H_0.

Step 8: Answer the Research Question. We cannot conclude that his *true* batting average is below .300.

This result is consistent with that of Example 15.4. Take a minute to look back at the discussion of Example 15.4.

Exercise 15.9. Using the information in Example 15.5, perform the eight-step hypothesis test for the population mean.

Exercise 15.10. Using the information in Example 15.6, perform the eight-step hypothesis test for the population mean.

Exercise 15.11. Compare the results in Exercise 15.9 with the results in Example 15.5.

Exercise 15.12. Compare the results in Exercise 15.10 with the results in Example 15.6.

We now look at examples of hypothesis testing for a population mean which do not deal with proportions.

Example 15.8. Baseball games seem to be taking longer. Suppose a baseball researcher wishes to establish whether a certain set of rule changes has changed the mean time in hours for the length of a nine-inning baseball game. From past records, it is assumed that $\mu = 3.00$ hours. A random sample of size $n = 100$ games is selected since the new rules have been implemented. For this sample $\bar{x} = 2.75$ hours and $s = 2.00$.

Can we conclude, with a level of significance $\alpha = 0.05$, that the population mean length of a baseball game now differs from 3.00 hours?

Solution:

Step 1: Can we conclude that the *true* mean game time, under the new rules, will be different from 3.00 hours?

Step 2: H_0: $\mu = 3.00$; H_A: $\mu \neq 3.00$ (claim).

Step 3: $\alpha = 0.05$.

Step 4: Sample data: $n = 100$, $\bar{x} = 2.75$, $s = 2.00$. There are no assumptions for the X-population.

Step 5: Convert the test statistic $\bar{x}$ to the standard test statistic z. Since $n > 30$, by Theorem 15.1(c) using the z-statistic as an estimate for the t-statistic,

$$\begin{aligned} z &= (\bar{x} - \mu)/(s/\sqrt{n}) \\ &= (2.75 - 3)/(2/\sqrt{100}) \\ &= -.25/.20 \\ &= -1.25. \end{aligned}$$

Step 6: Set up the acceptance and rejection regions (see Fig. 15.8).

Step 7: Statistical decision: Since –1.25 lies under the acceptance region, we accept H_0.

Step 8: Answer the research question: We cannot conclude that the mean length of time is different from 3.00 hours.

Exercise 15.13. What if the researcher was interested in whether the rule change *decreased* the time? Using the information in Example 15.8, with a level of significance of $\alpha = 0.05$, can we conclude that the rule change decreased the time of a nine-inning game? Show the eight steps for classical hypothesis testing.

Exercise 15.14. What if the researcher was interested in whether the rule change *increased* the time? Using the information in Example 15.8, with a level of significance of $\alpha = 0.05$, can we conclude that the rule change increased the time of a nine-inning game? Show the eight steps for classical hypothesis testing.

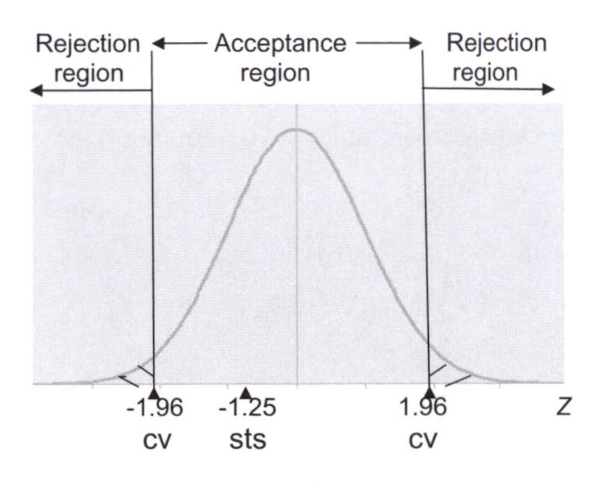

Figure 15.8 Acceptance and rejection regions for Example 15.8

Exercise 15.15. For which of the two previous exercises was it not necessary to do hypothesis testing to answer the question? Why?

Exercise 15.16. Redo Example 15.8, but change $\bar{x}$ to 3.50 and the level of significance to $\alpha = 0.10$. Keep all the other information the same.

Exercise 15.17. Redo Exercise 15.13, but change $\bar{x}$ to 3.50 and the level of significance to $\alpha = 0.10$. Keep all the other information the same.

Exercise 15.18. Redo Exercise 15.14, but change $\bar{x}$ to 3.50 and the level of significance to $\alpha = 0.10$. Keep all the other information the same.

Exercise 15.19. For which of the previous two exercises was it not necessary to do hypothesis testing to answer the question? Why?

15.2.3. Hypothesis Testing for the Population Mean, μ (Small Samples, $n < 30$)

If the sample size is less than 30, the population standard deviation σ is unknown, and the X-population is assumed normally distributed, then, by Theorem 15.1(d), the formula $(\bar{x} - \mu)/(s/\sqrt{n})$ converts the *test statistic* $\bar{x}$ into the *standard test statistic t*.

Example 15.9. Suppose that a speed gun was used to measure the velocity of relief pitcher Mariano Rivera's fastball. During a game, Mariano threw only 16 fastballs. Their mean speed was 94 mph, with a standard deviation of 3 mph. At a level of significance of $\alpha = 0.05$, can we conclude that the population mean speed for Rivera's fastball is greater than 90 mph? What assumption is needed?

Solution:

Step 1: Can we conclude, with a level of significance of 0.05, that Mariano's population mean fastball speed is greater than 90 mph?

Step 2: H_0: $\mu \leq 90$; H_A: $\mu > 90$ (claim).

Step 3: $\alpha = 0.05$.

Step 4: Sample data: $n = 16$, $\bar{x} = 94$, $s = 3$. We must assume, for the infinite X-population of fastballs thrown by the Yankee closer, that the speed of his fastball is normally distributed.

Step 5: Convert the test statistic $\bar{x}$ to the standard test statistic t. Since $n < 30$, σ is unknown, and the X-population is normally distributed, by Theorem 15.1(d) we have

$$\begin{aligned} t &= (\bar{x} - \mu)/(s/\sqrt{n}) \\ &= (94 - 90)/(3/\sqrt{16}) \\ &= 4/.75 \\ &= 5.33. \end{aligned}$$

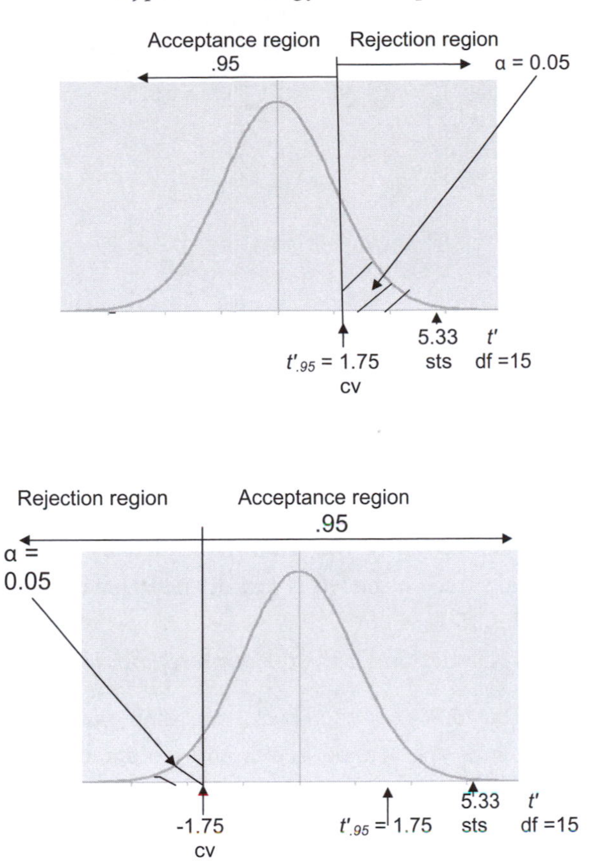

Figure 15.9 Acceptance and rejection regions for Example 15.9

Figure 15.10 Acceptance and rejection regions for $\mu < 90$ in Example 15.9

Step 6: See Figure 15.9 for the acceptance and rejection regions. The *t*-distribution will be used instead of the *z*-distribution. The degrees of freedom is $16-1=15$. The critical value (cv) is found by looking at the intersection of the column headed by $t'_{.95}$ and the row df $=15$. The subscript of t' provides the total area to the left of the cv. The area to the right corresponds to the level of significance. The critical value $= t'_{.95} = 1.75$.

Step 7: Since 5.33 lies under the rejection region, we reject H_0.

Step 8: We can conclude that the *true* mean speed of Mariano's fastball is greater than 90 mph.

If, in Example 15.9, H_A were the statement that $\mu < 90$, the rejection region would be one tail to the left. To find the critical value, we would repeat step 6 in Example 15.9, which would give us $t'_{.95} = 1.75$. The critical value would be $t = -1.75$ (see Fig. 15.10).

Since 5.33 lies under the acceptance region, we would accept H_0.

If, in Example 15.9, H_A were the statement that $\mu \neq 90$, the rejection region would be a two-tail region. Each tail will have an area of $\alpha/2$. Since

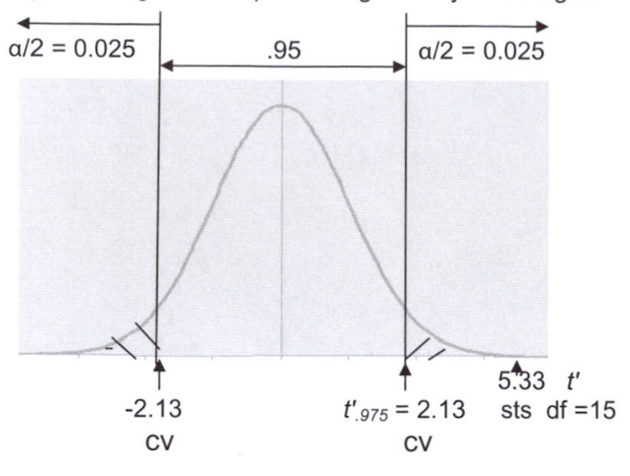

Figure 15.11 Acceptance and rejection regions for $\mu \neq 90$ in Example 15.9

$\alpha = 0.05$, $\alpha/2 = .025$. The critical value on the right is found by looking at the intersection of the column headed by $t'_{.975}$ and the row df=15. The critical value on the left is just the negative of the critical value on the right (see Fig. 15.11).

As 5.33 lies under the rejection region, H_0 will be *rejected* for this case.

Exercise 15.20. For each part (a–c) below, you are given a pair of hypotheses, the α-error, the sample size n, and the standard test statistic (sts). For each part, complete step 6 and step 7. Step 6 should include a graph showing the critical value(s), the standard test statistic, and the acceptance and rejection regions. Step 7 should include your statistical decision to accept or reject H_0.

a. Given: H_0: $\mu \geq 4$; H_A: $\mu < 4$; $\alpha = 0.05$, $n = 18$, $t = -1.45$.
b. Given: H_0: $\mu \leq 4$; H_A: $\mu > 4$; $\alpha = 0.10$, $n = 12$, $t = 2.76$.
c. Given: H_0: $\mu = 6$; H_A: $\mu \neq 6$; $\alpha = 0.01$, $n = 23$, $t = -2.88$.

Exercise 15.21. In the 2009 season, Mets shortstop José Reyes pulled a hamstring. Suppose that a survey was sent to 45 trainers and doctors who specialize in the treatment of such an injury. One of the questions dealt with the number of days it takes a player with such an injury to fully recover. The survey was returned by 22 of those surveyed. The sample mean and sample standard deviation, based on the returned surveys, were 26 days and 3.5 days, respectively. The researcher claims that the population mean number of days is 25 days. At a level of significance of $\alpha = 0.05$, can we reject the researcher's claim? What assumption is necessary? Would you place a player with such an injury on the 15-day or 60-day disabled list?

15.3. The *p*-Value Method for Hypothesis Testing

Current research literature favors the *p*-value method over classical hypothesis testing. We begin with the definition of the *p*-value.

Definition of the *p*-Value

The *p-value* is the probability of obtaining a test statistic with a value *as extreme as or more extreme than* the one evaluated from the chosen random sample, under the assumption that the statement in the null hypothesis is true. The phrase "with a value as extreme as or more extreme than the one evaluated from the chosen random sample" means farther away from the mean of the sampling distribution in the direction of the alternate hypothesis. In this chapter, the mean of the sampling distribution is the value of the parameter assumed true in the null hypothesis. The *p*-value is often referred to as the chance error.

The calculation of the *p*-value depends on three factors:

1. The sampling distribution being used, whether it takes the shape of the *z*-distribution, the *t*-distribution, or some other distribution.
2. The nature of the alternate hypothesis ($\neq$, $<$, or $>$).
3. The calculation of the standard test statistic.

15.3.1. The Three Cases for the Calculation of the p-Value for the Z-Distribution

The three cases that follow refer to Figure 15.12. In Cases 1 and 2 above, the *p*-value is equal to the area in the one-tail shaded region in the direction of the statement in H_A. In Case 3, the *p*-value is equal to the sum of the areas in the two shaded regions. The evaluation of the areas will use Table C, the table for the standard normal curve.

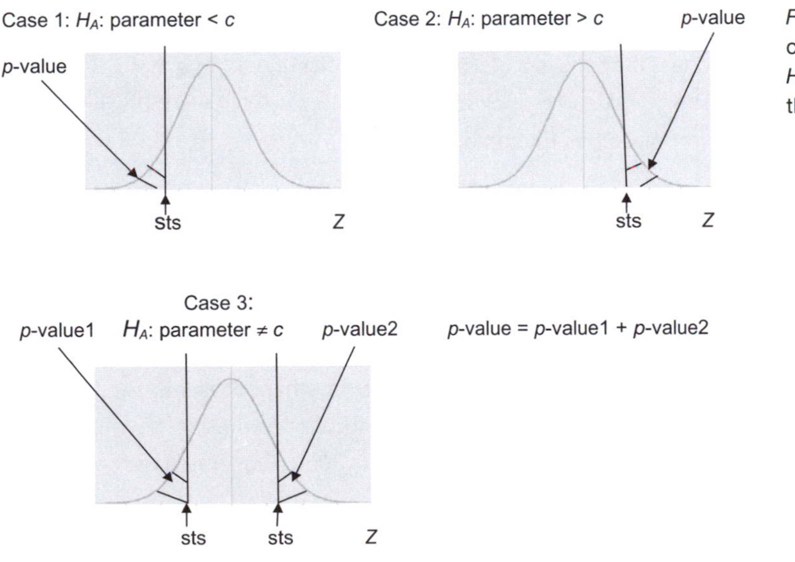

Figure 15.12 Three cases, depending on H_A, for the *p*-value with the *z*-distribution

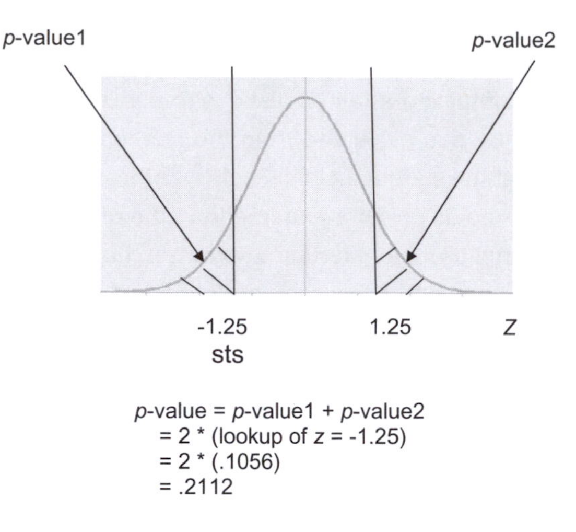

Figure 15.13 The *p*-value for Example 15.8

Example 15.10. We calculate the *p*-value for Example 15.8. Since $n > 30$, the sampling distribution can be assumed approximately normal in shape. The statement in H_A is $\mu \neq 3$, and the standard test statistic is $z = -1.25$. Figure 15.13 shows Case 3.

The *p*-value = .2112. Observe that this is *greater* than the α-error of .0.05 in Example 15.8.

Exercise 15.22. Add to Figure 15.13 in Example 15.10 the regions determined by the α-error in Example 15.8. What conclusion can you make concerning the *p*-value and the α-error?

The next two examples look at Cases 1 and 2.

Example 15.11. We find the *p*-value for Example 15.5. In this example the sampling distribution is normal, the alternate hypothesis is $p > .299$, and the standard test statistic is $z = 2.15$. Figure 15.14 shows Case 2.

The *p*-value = (1 − lookup of $z = 2.15$) = 1 − .9842 = .0158. This is *less* than the α-error of 0.05 in Example 15.5.

Exercise 15.23. Add to Figure 15.14 in Example 15.11 the region determined by the α-error in Example 15.5. What conclusion can you make concerning the *p*-value and the α-error?

Example 15.12. We find the *p*-value for Example 15.7. In this example, the sampling distribution is normal, the alternate hypothesis is $\mu < .300$, and the standard test statistic is $z = -.51$. Figure 15.15 shows Case 1.

The *p*-value = (lookup of $z = -.51$) = .3050. Observe that the *p*-value is *greater* than the α-error of 0.05.

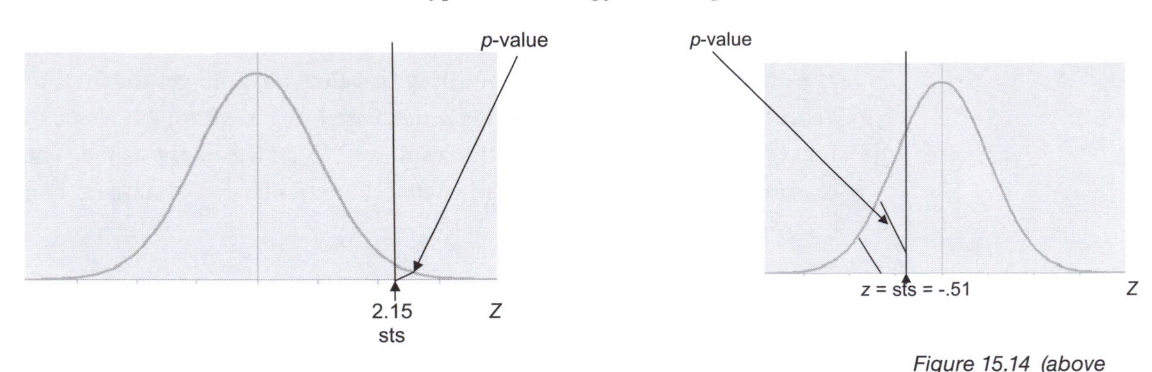

Figure 15.14 (above left) The p-value for Example 15.5

Figure 15.15 (above right) The p-value for Example 15.7

Exercise 15.24. Add to the picture in Figure 15.15 the region determined by the α-error in Example 15.7. What conclusion can you make concerning the p-value and the α-error?

15.3.2. The Calculation of the p-Value for the t-Distribution

Assuming that the sampling distribution is the t-distribution, the standard test statistic will be a t-value. We look at the calculation of the p-value for all three cases of H_A. The correct t-curve is found by subtracting 1 from the sample size, df $=(n-1)$.

Since Table C.3 for the t-distribution only provides a partial table for each t-curve, we are forced to approximate the p-values. If a complete table were provided for each t-curve, we could calculate exact p-values in the same way as we did for the z-distribution.

In the discussion that follows t_0 is the standard test statistic. We examine the three cases for H_A: $\mu > c$, $\mu < c$, $\mu \neq c$.

Case 1: H_A: $\mu > c$. We now look at the three possibilities for the standard test statistic t_0.

a. t_0 lies between two lookups from Table C.3 ($t_1 < t_0 < t_2$). Assume, without loss of generality, that $t_1 = t'_{.95}$ and $t_2 = t'_{.975}$. From Figure 15.16, we can say that $.025 < p\text{-value} < .05$.
b. t_0 lies to the right of $t'_{.995}$. From Figure 15.17, we can say that the p-value $< .005$.
c. t_0 lies to the left of $t'_{.90}$. From Figure 15.18, we can say that the p-value $> .10$.

Case 2: H_A: $\mu < c$. If t_0 is greater than or equal to 0, there is no need to proceed any further. Why? If $t_0 < 0$, we can calculate the p-value for $-t_0$, using the logic discussed in Case 1 for H_A: $\mu > c$. Because each t-curve is symmetric, the p-value for t_0 when $\mu < c$ is the same as the p-value for $-t_0$ when $\mu > c$.

Figure 15.16 The p-value for the t-distribution, case 1 part (a)

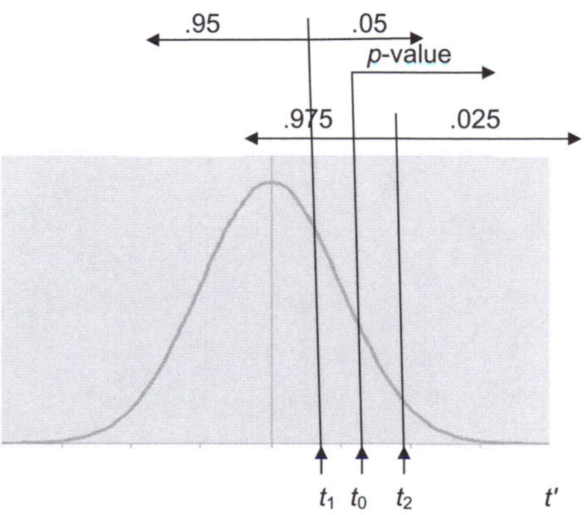

Case 3: H_A: $\mu \neq c$. If t_0 is positive, use the logic discussed in Case 1 for H_A: $\mu > c$ to find (p-value1). Then multiply (p-value1) by 2 to get the final p-value. If t_0 is negative, use the logic discussed in Case 2 for H_A: $\mu < c$ to find (p-value1). Then multiply the result by 2 to get the final p-value. The reason for multiplying by 2 is because the standard test statistic can occur on either side.

Exercise 15.25. Given that the sampling distribution is the t-distribution, H_A: $\mu \neq c$, and the df = 20, find the p-value for each of the following standard test statistics, t_0:

a. $t_0 = 1.89$.
b. $t_0 = -3.02$.
c. $t_0 = -1.15$.

Exercise 15.26. Given that the sampling distribution is the t-distribution, H_A: $\mu > c$, and the df = 20, find the p-value for each of the following standard test statistics, t_0:

a. $t_0 = 1.89$.
b. $t_0 = 3.02$.
c. $t_0 = 1.15$.

Figure 15.17 The p-value for the t-distribution, case 1 part (b)

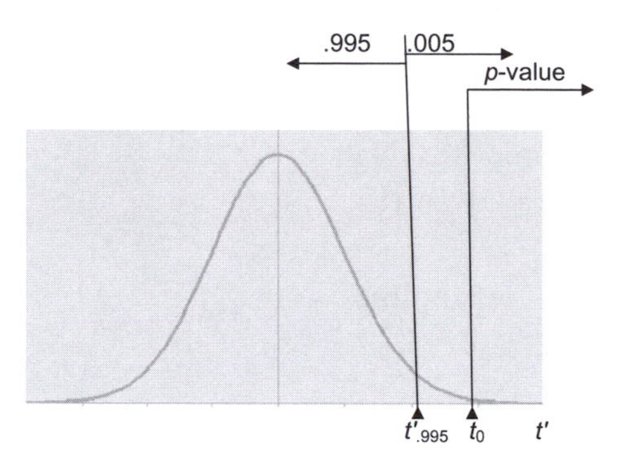

Exercise 15.27. Given that the sampling distribution is the t-distribution, H_A: $\mu < c$, and the df = 20, find the p-value for each of the following standard test statistics, t_0:

a. $t_0 = -1.89$.
b. $t_0 = -3.02$.
c. $t_0 = -1.15$.

Figure 15.18 The p-value for the t-distribution, case 1 part (c)

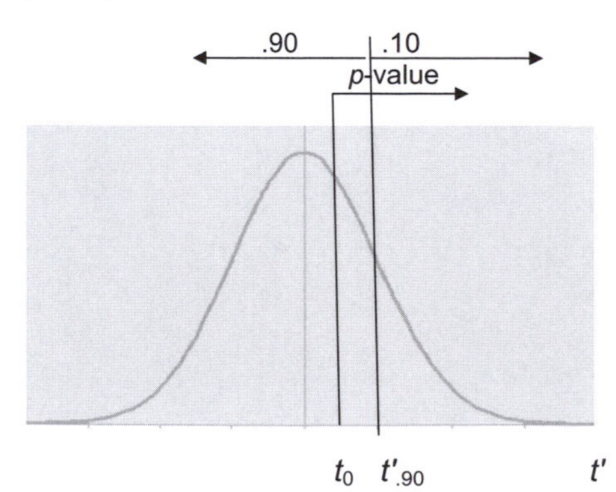

15.3.3. The Statistical Significance of the p-Value

Remember that the p-value represents the probability that the test statistic evaluated from the representative sample, under the assumption that the null hypothesis is true, occurs strictly by chance. Therefore, the smaller the p-value, the more likely H_0 is false. We use the following guidelines for the interpretation of the statistical significance of the p-value:

- If p-value $\leq .01$, the result is "highly significant."
- If $.01 < p$-value $\leq .05$, the result is "significant."
- If $.05 < p$-value $\leq .10$, the result is "mildly significant."
- If p-value $> .10$, the result is "not statistically significant."

15.3.4. The Seven Steps in the p-Value Method for Hypothesis Testing

For *classical hypothesis testing*, the process consisted of eight steps. The *p*-value method for hypothesis testing uses only seven steps:

Step 1: State the research question.
Step 2: State the two hypotheses.
Step 3: Calculate the test statistic for the collected data and provide any assumptions for the *X*-population.
Step 4: Convert the test statistic to the standard test statistic (z or t).
Step 5: Evaluate the *p*-value.
Step 6: Interpret the *p*-value.
Step 7: Answer the research question.

The next few examples demonstrate using the *p*-value method.

Example 15.13. The information in Example 15.5 is used.

Solution: Based on the seven-step *p*-value method:

Step 1: Can we conclude that this player's *true* batting average is greater than .299?

Step 2: H_0: $p \leq .299$; H_A: $p > .299$ (claim).

Step 3: Sample data: $n = 950$, $\hat{p} = 314/950 = .331$. There are no assumptions about the *X*-population. Since $n * p$ and $n * (1 - p)$ are both greater than 5, the sampling distribution can be assumed normally distributed.

Step 4: Convert the test statistic $\hat{p}$ to the standard test statistic z:

$$z = (\hat{p} - p)/\sqrt{p(1-p)/n}$$
$$= (.331 - .299)/\sqrt{.299 * .701)/950}$$
$$= 2.15.$$

Step 5: Evaluate the *p*-value. Figure 15.19 shows the evaluation of the *p*-value. The area to the right of 2.15 is the *p*-value. It is (1 – lookup of $z = 2.15$) = (1 – .98 42) = .0158. Thus, the *p*-value = .0158.

Step 6: Interpretation of the *p*-value. Since the *p*-value lies between .01 and .05, the *p-value is statistically significant*.

Step 7: Answer the research question. Since the probability of getting such an extreme value for the test statistic, under the assumption that $p \leq .299$ is true (the *p*-value), is so small, we conclude that the reason we got this extreme value for the test statistic is that this player's *true* batting average is not $\leq .299$. Therefore, we can say that this player was a *true* .300 hitter.

Figure 15.19 Graph showing *p*-value for Example 15.5

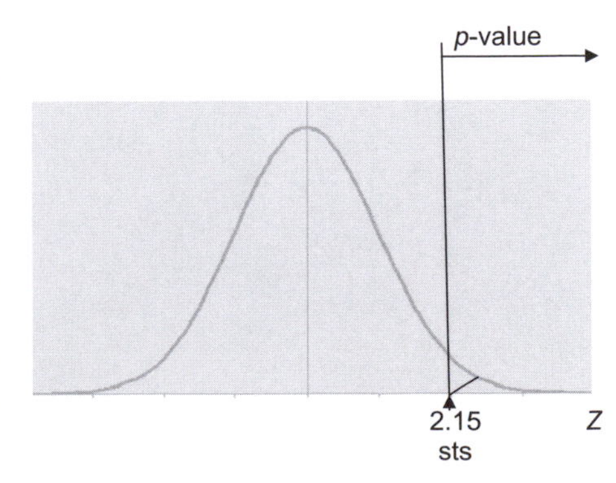

Example 15.14. Suppose that during the exhibition season baseball experiments with a new rule to reduce the time for a game. The new rule requires the batter to stay in the batter's box throughout his time at bat. The new rule was tried in 25 games. For the sample of 25 games, the sample mean time for a nine-inning game was 2.30 hours with a sample standard deviation equal to 2.10 hours. Can we conclude that for the population of all future games, using this new rule, the population mean time would be less than 3.00 hours? What must be assumed about the X-population?

Solution: Based on the seven-step p-value method:

Step 1: Can we conclude that the population mean time of all games, using the new rule, will be less than 3.00 hours?

Step 2: H_0: $\mu \geq 3.00$; H_A: $\mu < 3.00$ (claim).

Step 3: Sample data: $n = 25$, $\bar{x} = 2.30$, $s = 2.10$. Since the sample size is less than 30, we must assume that the X-population is normally distributed.

Step 4: Since $n < 30$ and σ is unknown, we must use the t-distribution. We convert test statistic $\bar{x}$ to the standard test statistic t:

$$t = (\bar{x} - \mu)/(s/\sqrt{n})$$
$$= (2.30 - 3.00)/(2.10/\sqrt{25})$$
$$= -.70/.42$$
$$= -1.67.$$

Step 5: Evaluate the p-value: Since the alternate hypothesis is one tail to the left, we multiply the standard test statistic by −1. We get $t_0 = -1 * (-1.67) = 1.67$. We now proceed as if H_A were $\mu > 3.00$. The degrees of freedom is $(25 - 1) = 24$. Since $1.318 = t'_{.90} < 1.67 < t'_{.95} = 1.7109$, the p-value lies between 0.05 and 0.10, $0.05 < p\text{-value} < 0.10$. Figure 15.20 shows the p-value.

Step 6: Interpretation of the p-value: The sample result is mildly significant.

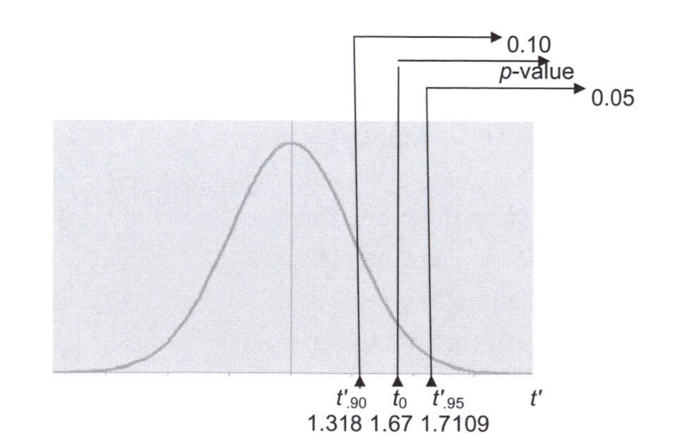

Figure 15.20 Graph showing p-value for Example 15.14

Step 7: Answer the research question: Since the probability of getting such an extreme value for a test statistic, under the assumption that $\mu \geq 3.00$ is true, is small (mildly significant), the researcher has the option of concluding that the rule change will reduce the mean time of a game to less than 3 hours.

Exercise 15.28. Perform the seven-step p-value method for hypothesis testing for Example 15.4.

Exercise 15.29. Perform the seven-step p-value method for hypothesis testing for Exercise 15.9.

Exercise 15.30. Perform the seven-step p-value method for hypothesis testing for Example 15.6.

Exercise 15.31. Perform the seven-step p-value method for hypothesis testing for Example 15.9.

Exercise 15.32. Perform the seven-step p-value method for hypothesis testing for Example 15.9, using the pair of hypotheses H_0: $\mu = 90$; H_A: $\mu \neq 90$.

Exercise 15.33. Perform the seven-step p-value method for hypothesis testing for Example 15.9, using the pair of hypotheses H_0: $\mu \geq 90$; H_A: $\mu < 90$.

15.3.5. The Relationship between the p-Value and the α-Error (Used in Classical Hypothesis Testing)

In doing the exercises, the connection between the p-value and the α-error should have become clear. We formally state this relationship next.

- If p-value $\leq \alpha$-error, we rejected H_0.
- If p-value $> \alpha$-error, we accepted H_0.

The *advantage* of the p-value method is that it gives a precise numeric measurement of the sample results occurring by chance. For example, if a p-value $= .06$ and the α-error $= .05$, we would fail to reject H_0. However, the p-value $= .06$ indicates a mildly significant result. The p-value method allows the researcher the opportunity to assign significance to the sample result, based on the precise probability of the test statistic occurring strictly by chance.

The *disadvantage* of the classical method is that it is hit or miss. If the standard test statistic lies under the rejection region, we reject H_0. If the standard test statistic lies under the acceptance region, we accept H_0.

As with confidence intervals, if σ is unknown, $n < 30$, and the X-population is normal, our reliability coefficient will come from the t-distribution.

Exercise 15.34. Compare the p-value and the α-error for Exercises 15.28–15.33. Show by way of a graph that if the $p\text{-value} \le \alpha$, we would reject H_0 in classical hypothesis testing.

15.4. The Confidence Interval Method for Hypothesis Testing

In Chapter 14, we studied the estimation technique called confidence intervals. Confidence intervals provide an interval estimate for an unknown parameter. In the *hypothesis testing* used in this chapter, a known parameter was assumed and appeared in both hypotheses.

The level of significance $\alpha = 0.xy$ used in hypothesis testing corresponds to the level of confidence $L = (1 - \alpha)$ used in confidence intervals. To see this, if we let $\alpha = 0.05$, $L = (1 - 0.05) = 0.95$. Therefore, the standard α-error of 0.05 corresponds to the standard level of confidence of $(1 - 0.05) = 0.95$.

Exercise 15.35. For each level of confidence, L, find the corresponding level of significance, α, and for each α-error find the level of confidence, L.

a. A level of confidence $L = 0.90$ corresponds to a level of significance $\alpha =$ ______.
b. A level of confidence $L = 0.99$ corresponds to a level of significance $\alpha =$ ______.
c. A level of significance $\alpha = 0.01$ corresponds to a level of confidence $L =$ ______.
d. A level of significance $\alpha = 0.10$ corresponds to a level of confidence $L =$ ______.

Even though these two techniques are considered different, under certain conditions, a confidence interval can be used to reach a decision on whether to accept or reject the null hypothesis. The only time a confidence interval can be used to make a decision about H_0 is when we have a two-tail hypothesis test. This situation is described next.

15.4.1. The Confidence Interval Method

Given a two-tail hypothesis test with a level of significance α, let $L = (1 - \alpha)$ be the level of confidence. Given the pair of hypotheses H_0: parameter = c; H_A: parameter ≠ c, *if the parameter lies in the confidence interval, we will accept H_0; if the parameter lies outside the confidence interval, we will reject H_0.*

We now revisit two earlier examples from this chapter. For these two examples, we construct their corresponding confidence intervals with level of confidence L.

Example 15.15. Refer to the information from Example 15.8. The assumed μ used in H_0 is $\mu = 3.00$. As the level of significance is $\alpha = 0.05$, the level of confidence that will be used is $L = 0.95$. The sample size is $n = 100$, $\bar{x} = 2.75$, and $s = 2.00$. Since n is greater than 30, the reliability coefficient can be taken from the z-distribution, $z_L = z_{.95} = 1.96$. The L% confidence interval is

$$\bar{x} - 1.96 * (2/\sqrt{n}) < \mu < \bar{x} + 1.96 * (2/\sqrt{n}),$$
$$2.75 - 1.96 * (2/10) < \mu < 2.75 + 1.96 * (2/10),$$
$$2.75 - .3920 < \mu < 2.75 + .3920,$$
$$2.36 < \mu < 3.14.$$

Since $\mu = 3.00$ lies within the 95% confidence interval, we accept H_0. Looking back at Example 15.8, we did accept H_0.

Example 15.16. Since Example 15.9 is a one-tail test, we cannot use the confidence interval method. But we can illustrate the confidence interval method if we use the information in Example 15.9 with the pair of hypotheses H_0: $\mu = 90$; H_A: $\mu \neq 90$.

The assumed μ used in H_0 is $\mu = 90$. Since the level of significance is $\alpha = 0.05$, the level of confidence that will be used is $L = 0.95$. The sample size is $n = 16$, $\bar{x} = 94$, and $s = 3$. Since n is less than 30, the reliability coefficient is from the t-distribution, $t_L = t_{.95} = t'_{.975}$. The intersection of the column $t'_{.975}$ and the row corresponding to df $= (16 - 1) = 15$ gives a reliability coefficient of 2.13. The 95% confidence interval is

$$\bar{x} - 2.13 * (3/\sqrt{n}) < \mu < \bar{x} + 2.13 * (3/\sqrt{n}),$$
$$94 - 2.13 * (3/4) < \mu < 94 + 2.13 * (3/4),$$
$$94 - 1.60 < \mu < 94 + 1.60,$$
$$92.40 < \mu < 95.60.$$

Since $\mu = 90$ lies outside of the interval, we would reject H_0.

Exercise 15.36. Given the pairs of hypotheses below and a level of significance α, use the confidence interval method to accept or reject H_0.

a. H_0: $p = .40$; H_A: $p \neq .40$; $\alpha = 0.10$. Sample data: $\hat{p} = .34$, $n = 81$.
b. H_0: $\mu = 60$; H_A: $\mu \neq 60$; $\alpha = 0.01$. Sample data: $\bar{x} = 64$, $n = 100$, $s = 6$.
c. H_0: $\mu = 25$; H_A: $\mu \neq 25$; $\alpha = 0.05$. Sample data: $\bar{x} = 31$, $n = 25$, $s = 7$; assume that the X-population is normal.

15.5. Should a One-Tail or Two-Tail Test Be Used in Hypothesis Testing?

One question that students often ask is, for hypothesis testing should I use a one-tail or two-tail test?

To help motivate the answer, we provide two examples.

Example 15.17. A manufacturer supplies the players' lounge with a coffee machine. If operating properly, the machine fills a cup with 8 ounces of coffee. If the machine underfills the cup or overfills the cup, it is malfunctioning. The manufacturer assumes that the machine is working properly at the onset but will repair the machine if it malfunctions. The manufacturer takes a sample of 36 cups. If the sample mean is 7.3 ounces and the sample standard deviation is 1.5 ounces, can we conclude that the machine is not working properly?

In this example, a *two-tail* hypothesis test should be used, for an extreme value on *either side* is bad.

Example 15.18. The key to success for a general manager is to pick young talent. In 2009, there were 10 players under 25 years of age who batted .300 or better. Suppose that a general manager will offer a player a long-term contract if he believes that that player is a *true* .300 hitter. The player has played for just 4 years. For those 4 years his career batting average is .316 based on 1440 at bats. Can the GM conclude that this player is a *true* .300 hitter?

In this example, a *one-tail* test should be used, because the general manager is only interested in one direction.

The default in hypothesis testing is the two-tail test. If there is a rationale for a researcher to be interested in only one direction, a one-tail test should be used.

Use a level of significance of $\alpha = 0.05$ for the next two exercises.

Exercise 15.37. Complete the eight-step hypothesis test for Example 15.17.

Exercise 15.38. Construct a 95% confidence interval for the information in Example 15.17. Use the confidence interval to accept or reject the H_0 in Example 15.17. Does this agree with the results in Exercise 15.37?

Chapter Summary

This chapter presented the second technique used in statistical inference, hypothesis testing. In Chapter 14, we studied the first technique of statistical inference called confidence intervals. The basic difference between the technique of confidence intervals and the technique of hypothesis testing is that, in confidence intervals, the population parameter is unknown and we wish to form a confidence interval to estimate it; in hypothesis testing, we assume that a population parameter is known and decide to continue to accept it as true or reject it as false.

Since we are dealing with one sample and not the population, the decision we make can be wrong. The type 1 error is the mistake we would make in rejecting a true null hypothesis. The type 2 error is the mistake we would make in accepting a false null hypothesis. We would like to make both errors small, but this is impossible. The probability of committing a type 1 error is called the α-error. The probability of committing a type 2 error is called the β-error. The error we choose to control is the α-error. The maximum allowable α-error is called the level of significance.

Hypothesis testing begins with two hypotheses about a population. The null hypothesis is assumed true from the onset. The alternate hypothesis is usually the hypothesis we want to show is true. From the research question, a claim is made that either the statement in H_0 or the one in H_A is true. If the claim is that the statement in H_0 is true, we want to continue to assume that H_0 is true. If the claim is that the statement in H_A is believed to be true, we want to reject the H_0 hypothesis. Since the α-error is the error we control and make small, the strongest result occurs with the rejection of H_0. The null hypothesis has a form of equality in it ($=, \leq, \geq$). The alternate hypothesis is the complement of the null hypothesis ($\neq, >, <$).

The hypothesis testing covered in this chapter was based on one parameter for one population. The two parameters studied in this chapter were μ and p. The statistical decision to accept or reject the null hypothesis is based on how extreme (far away from the mean of the sampling distribution) the test statistic is. The α-error is used to set up the rejection region. If the test statistic lies in the rejection region, we conclude that the sample result is *statistically significant*. This leads to the rejection of the null hypothesis. We say that, under the assumption that the null hypothesis is true, the probability of obtaining such an extreme sample result by chance is less than α. The standard α-error is 0.05. The rejection of the null hypothesis leads to the belief that the alternate hypothesis is true.

The two test statistics used in this chapter were $\hat{p}$ and $\bar{x}$. To measure how extreme the test statistics $\hat{p}$ and $\bar{x}$ are, the sampling distribution of $\hat{p}$ and the sampling distribution of $\bar{x}$ are used. In order to use these sampling distributions, it was necessary to assume that both sampling distributions are approximately normal and that the mean of the sampling distribution of sample means was μ and the mean of the sampling distributions of sample proportions was p. The test statistic $\hat{p}$ was converted to the standard test statistic z. The test statistic $\bar{x}$ was converted to the standard test statistic z or the standard test statistic t. The following describes what is needed to use the two sampling distributions.

- In the case of the sampling distribution of sample proportions, both np and $n(1-p)$ must exceed 5.
- In the case of the sampling distribution of sample means, a choice must be made between using the z-distribution and using the t-distribution for the standard test statistic. If σ is known and the X-population is normally distributed or if σ is unknown and the sample size is greater than

or equal to 30, the z-distribution is used. When the population standard deviation is unknown, s is used in place of σ.

- If σ is unknown, the sample size is less than 30, and the X-population is normally distributed, the t-distribution is used.

Three methods of performing hypothesis testing were introduced. They are the *classical method*, the *p-value method*, and the *confidence interval method*.

In the *classical method* a level of significance α is used to set up the acceptance and rejection regions. The standard level of significance is 0.05. Other levels commonly used are 0.01, 0.02, and 0.10. If the standard test statistic lies under the acceptance region, we continue to accept H_0 as true. If the standard test statistic falls under the rejection region(s), we reject H_0. The classical method involves an eight-step process:

Step 1: State the research question.
Step 2: Provide a statement for the two hypotheses (H_0 and H_A).
Step 3: State the level of significance (the largest allowable α-error).
Step 4: Calculate the test statistic for the collected data and provide any necessary assumptions for the X-population.
Step 5: Convert the test statistic (ts) to the standard test statistic (sts) ($\bar{x}$ to z or t, or $\hat{p}$ to z).
Step 6: Find the critical value or values (cv) that separate the acceptance and rejection regions and locate in which region the standard test statistic lies.
Step 7: Make a statistical decision to accept H_0 if the standard test statistic lies under the acceptance region and to reject H_0 if the standard test statistic lies under the rejection region.
Step 8: Base the answer to the original research question on the result in step 7.

In the *p-value method*, the probability of getting a sample statistic with a value as extreme as or more extreme than the one determined from the acquired sample, based on the assumption that the statement in H_0 is true, is calculated. The smaller the p-value, the more convincing is the argument that the statement in H_0 is not true. The p-value gives the probability of the sample result occurring strictly by chance. Guidelines for the significance of the p-value were given. In general, for a p-value to have *statistical significance* its value must be less than 0.10. The p-value method involves a seven-step method:

Step 1: State the research question.
Step 2: Provide a statement for the two hypotheses.
Step 3: Calculate the test statistic for the collected data and provide any necessary assumptions for the X-population.

Step 4: Convert the test statistic to the standard test statistic (z or t).
Step 5: Evaluate the p-value.
Step 6: Interpret the p-value.
Step 7: Base the answer to the original research question on the result in step 7.

The *confidence interval method* is limited to a two-tailed hypothesis test. If the confidence interval contains the parameter in the null hypothesis, we would accept H_0. If the parameter in the null hypothesis is outside the confidence interval, we would reject H_0.

The last section of the chapter looked at how to choose between a one-tail test and a two-tail test.

CHAPTER PROBLEMS

Through the 2008 season, Albert Pujols had 4578 at-bats and 1531 hits. Can we conclude, with a level of significance of $\alpha=0.05$, that the *true* batting average for Albert Pujols, through the 2008 season, is not equal to .300? Use this information to do problems 1 and 2 below.

1. Interpreting a batting average as a proportion:
   a. Perform the eight-step classical hypothesis testing method.
   b. Perform the seven-step p-value method for hypothesis testing.
   c. Perform the confidence interval method for hypothesis testing.
2. Interpreting a batting average as a mean:
   a. Perform the eight-step classical hypothesis testing method.
   b. Perform the seven-step p-value method for hypothesis testing.
   c. Perform the confidence interval method for hypothesis testing.
3. For the 2007 season, Alex Rodriquez had the following results: AB=583, H=183, BB=95, SF=9, HBP=21. Can we conclude, with a level of significance of $\alpha=0.05$, that Alex Rodriguez's *true* on-base percentage is greater than .400?
   a. Perform the eight-step classical hypothesis testing method.
   b. Perform the seven-step p-value method for hypothesis testing.
4. For the 2007 season, with a level of significance of $\alpha=0.05$, can we conclude that Alex Rodriguez's *true* batting average was greater than .299?
   a. Perform the eight-step classical hypothesis testing method.
   b. Perform the seven-step p-value method for hypothesis testing.
5. For each of your two players, using for your samples their career at bats, can we conclude that either one of your players was a *true* .300 hitter for their careers? Use both classical hypothesis testing and the p-value method.

Chapter 16

Streaking

16.1. Introduction

The year 1941 was a special one in baseball. Two remarkable feats occurred: Joe DiMaggio's 56-game hitting streak and Ted Williams having a batting average of at least .400 for a season, the last player to do so.

This chapter is devoted to the study of hitting streaks in baseball. In particular, we look at Joe DiMaggio's 56-game hitting streak and the 84-game on-base streak of Ted Williams. Other lesser-known streaks are also discussed. The next chapter will be devoted to the study of what it takes to be the next .400 hitter.

Four models are presented to evaluate the probability of a player having a 56-game hitting streak. The first two models involve formulas based on the rules of probability; one is a general function developed by Charles Blahous, while the second formula is a *recursive* function developed by Michael Freiman. A recursive function is a sequence of terms where the value of a beginning term is known and the values for succeeding terms are defined through the values of previous terms. The third formula is my own: a piecewise function of polynomials that can be used for streaks of at least 54 games. This would include the two principal streaks covered in this chapter. The last method is based on a simulation. Remember, a simulation is a probability experiment that mimics a real-world situation. The probability experiment can be performed through spinning a pointer on a disk, rolling a die, using the online spinner, or using the RANDBETWEEN function provided in Microsoft Excel.

We'll evaluate these models mathematically for validity and accuracy. Before undertaking the work for this chapter, we review some of the concepts of probability, first introduced in Chapter 7.

The events $E_1, E_2, E_3, \ldots, E_n$ are *mutually exclusive* if

$\Pr(E_1 \text{ or } E_2, \ldots \text{ or } E_n) = \Pr(E_1) + \Pr(E_2) + \ldots + \Pr(E_n)$;

they are *independent* if

$\Pr(E_1 \text{ and } E_2, \ldots \text{ and } E_n) = \Pr(E_1) * \Pr(E_2) * \ldots * \Pr(E_n)$.

If $E = E_1 = E_2 = \ldots = E_n$ and the events are *independent*, then

$\Pr(E \text{ and } E, \ldots \text{ and } E) = \Pr(E) * \Pr(E) * \ldots * \Pr(E) = [\Pr(E)]^n$.

The event (Not E) is the *complement* of the event E, and if

$\Pr(E) = k$, then $\Pr(\text{Not } E) = 1 - k$.

For this chapter, the definition of an official plate appearance (PA), which does not include sacrifice hits (SH), will be extended to *include* them. Therefore, a plate appearance will include all of the following: at bats (AB), hit by pitch (HBP), base on balls (BB), sacrifice flies (SF), and sacrifice hits (SH).

Our streaks are limited to one season. Before 1961, a season consisted of 154 games; thereafter, a season consists of 162 games. We will not consider a streak extending to another season. A streak longer than the one being asked for will also satisfy the smaller number. If a player gets a hit in 57 consecutive games, he will satisfy a 56-game hitting streak.

Finally, much of the information for this chapter comes from five articles, which appeared in the *Baseball Research Journal.* Since they will be referred to throughout the chapter, we assign a number to each of the articles:

1. "56-Game Hitting Streaks Revisited," by Michael Freiman;
2. "The DiMaggio Streak: How Big a Deal Was It," by Charles Blahous;
3. "DiMaggio's Hitting Streak," by Joe D'Aniello;
4. "Calculating the Odds," by Bob Brown and Peter Goodrich; and
5. "Ted Williams' On-Base Performances in Consecutive Games," by Herm Krabbenhoft.

Two articles from the *New York Times* will be used:

6. "A Journey to Baseball's Alternate Universe," by Samuel Arbesman and Steven Strogatz, published March 30, 2008; and
7. "In Defense of Joe DiMaggio," by Carl Bialik, published April 3, 2008.

Finally, one article from *Baseball Digest* will be used:

8. "Hitting Streaks: Complete List of Major League Players with Batting Skeins of 20 or More Games in One Season since 1900," by Bob Kuenster.

16.2. Joe DiMaggio's 56-Game Hitting Streak

To analyze Joe DiMaggio's 56-game hitting streak, we use a formula from article 1, a formula from article 2, a simulation, and a formula by Stanley Rothman.

16.2.1. The Charles Blahous Formula to Calculate the Probability of a Player Achieving a 56-Game Hitting Streak at Some Point in a Season

We develop the Blahous Formula step by step using a fictitious player called player A. The input information needed for the entire season is the number of games played (#G), number of plate appearances (#PA = #AB + #BB + #HBP + #SF + #SH), and number of hits (H). For player A's season, #G = 100, #PA = 440, and #H = 150. Here are the steps.

Step 1: The probability, p, of getting a hit (H) in a plate appearance (PA) is equal to #H/#PA. Therefore, $(1-p)$ is equal to the probability of him *not* getting a hit in a PA. For player A, $p = 150/440 = .341$, and so $(1-p) = .659$. We assume that this p is good for each plate appearance in the season.

Step 2: Each plate appearance of a player in a game is considered a simple event. If the plate appearances in a game are *independent* events, the probability of him not getting a hit in a game is $(1-p) * (1-p) * \ldots * (1-p) = (1-p)^n$, where n is the number of plate appearances in the game. Therefore, the probability of him getting *at least* one hit in a game is $1-(1-p)^n$. The probability that player A gets at least one hit in a game is equal to $1-(1-.341)^n = 1-.659^n$. Now we need to find n.

Step 3: If we had the box score for each of his games for the season, we would have the exact number of plate appearances for each game. Instead, we estimate the number of plate appearances for a typical game for a given season. Player A had 440 plate appearances in 100 games, so his average number of plate appearances for a typical game for that season was #PA/#G = 440/100 = 4.40. From step 2, the probability that he got at least one hit in four plate appearances is $[1-(1-.341)^4]$; the probability of getting at least one hit in five plates appearances is equal to $[1-(1-.341)^5]$. Since 4.40 plate appearances can never happen, we need to use a weighted average for Player A. If he had 60 games with four plate appearances and 40 games with five plate appearances, his average number of plate appearances per game for the season is $[(4 * 60) + (5 * 40)]/100 = 4 * .60 + 5 * .40 = 4.40$, so we use the 60:40 weighting. The probability U of player A getting at least one hit in 4.40 plate appearances (a typical game for that season) is equal to

U = [Pr(at least one hit in four PA)] $*$ .60 + [Pr(at least one hit in five PA)] $*$ .40
$= [1-(1-.341)^4] * .60 + [1-(1-.341)^5] * .40.$

The weighting factor of $.60 = 1 - .40$ is assigned to the number of four plate appearances, and the weighting factor of .40 is assigned to the number of five plate appearances.

Step 4: Assuming that the player's performance for each game is independent, the probability that he has a 56-game hitting streak in any particular 56-game span is equal to

$U^{56} = U * U * U * U * \ldots * U.$

Step 5: The probability of *not* getting a 56-game hitting streak in any particular 56-game span is equal to $1 - U^{56}$.

Step 6: We now want to calculate the probability that the player has a 56-game hitting streak *at some point* in a given season. For simplicity, assume that he plays in 65 games in that season. The possible 56-game hitting streaks can occur from games 1 to 56, 2 to 57, 3 to 58, 4 to 59, 5 to 60, 6 to 61, 7 to 62, 8 to 63, 9 to 64, and 10 to 65.There are $65 - 55 = 10$ possible streaks. We generalize this to x number of games, where x is greater than or equal to 56 and x represents the number of games the player was active in. The formula for the number of possible 56-game hitting streaks for a season is $x - 55$. Since player A's season consisted of 100 games, there are $100 - 55 = 45$ possible 56-game hitting streaks for that season. Assuming that the possible 56-game hitting streaks are independent events, the probability of *not* having a 56-game hitting streak for player A for any point in that season is

$(1 - U^{56})^{45} = (1 - U^{56}) * (1 - U^{56}) * \ldots * (1 - U^{56}).$

Step 7: The probability that player A has a 56-game hitting streak at some point in the season is equal to $1 - (1 - U^{56})^{45}$.

We now wish to generalize this formula so that it can be used to find the probability of any player achieving a 56-game hitting streak at some point in any season. To do so, the following information is needed for a player:

- The number of games played in the season (#G).
- The number of plate appearances in the season (#PA = #AB + #BB + #HB P + #SF + #SH).
- The number of hits in the season (#H).

Once the above information is known, we perform the following seven steps:

Step 1: Let $p = \#H/\#PA$.
Step 2: Let $j.kl = \#PA/\#G$.
Step 3: The weighted average for the probability of a player getting at least one hit in $j.kl$ plate appearances for a typical game in that season is equal to

$$U = [1-(1-p)^j] * (1-.kl) + [1-(1-p)^{(j+1)}] * .kl.$$

Step 4: The probability of a player having a 56-game hitting streak in any particular 56-game span is equal to U^{56}.
Step 5: The probability of a player not having a 56-game hitting streak in any particular 56-game span is equal to $1 - U^{56}$.
Step 6: The probability of not having a 56-game hitting streak at any point in that season is equal to $(1 - U^{56})^{(\#G-55)}$.
Step 7: The Charles Balhous Formula, hereafter called the CB Formula, is the probability that a player has a 56-game hitting streak at some point in his season.

The CB Formula

$$1-(1-U^{56})^{(\#G-55)}, \text{ where } U = [1-(1-p)^j] * (1-.kl) + [1-(1-p)^{(j+1)}] * .kl.$$

This version of the formula is based on a 56-game hitting streak. The Generalized CB Formula can be used for many s-game hitting streaks.

The Generalized CB Formula

$$1-(1-U^s)^{[\#G-(s-1)]}.$$

Here s is the length of the streak and U is the probability of at least one success in an event. An event can be a game, a plate appearance, or an at bat. A success could be, for example, getting on base, hitting a home run, or getting an extra-base hit.

Some examples of U are given:

1. U is getting on base at least one time in a game, where $p = \#OB/\#PA$, $j.kl = \#PA/\#G$, and $s = 84$.
2. U is hitting at least one home run in a game, where $p = \#HR/\#PA$, $j.kl = \#PA/\#G$, and $s = 8$.
3. U is getting at least one hit in a game, where $p = \#H/\#PA$, $j.kl = \#PA/\#G$, and $s = 56$. In this case, we would have the CB Formula.

Both versions of the CB Formula are based on the following assumptions:

1. The probability p is based on all the games in a season in which a player was active.

2. The probability p is used for each plate appearance in the season.
3. The plate appearances in a game are independent events.
4. The player has the same number of plate appearances in each game for the season.
5. A player's batting performance in each game is independent of his performance in any other game in the season.
6. The possible 56-game batting streaks are independent events.

Mathematically, assumptions 1-5 are good. As pointed out in article 1, the problem with the CB Formula is assumption 6. As many of the possible 56-game streaks *overlap*, the assumption of *independence* is invalid. Without the assumption of independence, the multiplication rule for independent events cannot be used.

Assumptions 1–4, taken together, describe a binomial experiment where each plate appearance is a trial and success is getting a hit. See if you can restate the CB Formula by using the Binomial Formula.

Example 16.1. This example illustrates the CB Formula. Suppose that during a season player A played in 66 games with #AB = 322, #H = 138, #BB = 20, #HBP = 0, #SF = 0, and #SH = 0. What is the probability that at some point in his season he had a 56-game hitting streak?

Solution: We follow the steps for the CB Formula.

Step 1: $p = \#H/\#PA = 138/(322+20+0+0+0) = 138/342 = .404$, where p is the probability of a hit in a plate appearance.

Step 2: The average number of plate appearances per game expressed as a decimal is $j.kl = \#PA/\#G = 342/66 = 5.18$. (So $j = 5$, $.kl = .18$.)

Step 3: The probability of a player getting at least one hit in $j.kl$ plate appearances per game is

$$\begin{aligned} U &= [1-(1-p)^j] * (1-.kl) + [1-(1-p)^{(j+1)}] * .kl \\ &= [1-(1-.404)^5] * (1-.18) + [1-(1-.404)^{(5+1)}] * .18 \\ &= [1-(.596)^5] * (.82) + [1-(.596)^6] * (.18) \\ &= .9248 * .82 + .9552 * .18 = .9302. \end{aligned}$$

Step 4: The probability of a player having a 56-game hitting streak in any particular 56-game span in his season of 66 games is equal to

$$U^{56} = .9302^{56} = .0174.$$

Step 5: The probability of a player not having a 56-game hitting streak in any particular 56-game span is equal to

$$1 - U^{56} = 1 - .0174 = .9826.$$

Step 6: The probability of player A not having a 56-game hitting streak at some point in his 66-game season is equal to

$(1 - U^{56})^{(\#G-55)} = .9826^{(\#G-55)} = .9826^{(66-55)} = .9826^{11} = .8244.$

The overlapping streaks can occur in games 1-56, 2-57, 3-58, 4-59, 5-60, 6-61, 7-62, 8-63, 9-64, 10-65, and 11-66.

Step 7: The probability that player A has a 56-game hitting streak at some point in his 66-game season is equal to

$1 - (1 - U^{56})^{(\#G-55)} = 1 - .8244 = .1756.$

Exercise 16.1. Suppose that a player played in 126 games, with #AB = 544, #H = 190, #BB = 37, #HBP = 7, #SF = 0, and #SH = 13; complete the following:

a. Calculate p = the probability of the player getting a hit in a plate appearance (sacrifice hits are included in the plate appearance totals).
b. Calculate #PA/#G, the player's average number of plate appearances per game.
c. Express #PA/#G as the decimal $j.kl$.
d. Find the weighted average, U, for the probability of getting at least one hit in $j.kl$ plate appearances for a typical game.
e. Find the probability of having a 56-game hitting streak in any particular 56-game span (U^{56}).
f. Find the probability of not having a 56-game hitting streak in any particular 56-game span ($1 - U^{56}$).
g. Find the probability of not having a 56-game hitting streak at any point of the season ($(1 - U^{56})^{(\#G-55)}$).
h. Find the probability that he has a 56-game hitting streak at some point in the season ($1 - (1 - U^{56})^{(\#G-55)}$).
i. List the games included in all the possible overlapping streaks.

Exercise 16.2. Use the CB Formula to find the probability of Joe DiMaggio having a 56-game hitting streak at some point in the 1941 season. Show all nine steps covered in Exercise 16.1.

We move to a second formula for calculating the probability that a player has at least one 56-game hitting streak in a season.

16.2.2. The Michael Freiman Formula to Calculate the Probability of a Player Achieving a 56-Game Hitting Streak at Some Point in the First n Games of a Season

The Michael Freiman Formula, hereafter called the MF Formula, for the probability of a player achieving a 56-game hitting streak at some point in the first n games of a season is given by the recursive function $D(n)$.

The MF Formula

$$D(n) = D(n-1) + [1 - D(n-57)] * (1-U) * U^{56}.$$

Here $n > 56$ is the first n games a player played in a season. Since a season in the Major Leagues consists of 162 games, the maximum n can be is 162.

$$U = [1-(1-p)^{j}] * (1-.kl) + [1-(1-p)^{(j+1)}] * .kl; \; p = \#H/\#PA \text{ and } j.kl = \#PA/\#G,$$

$$D(0) = D(1) = D(2) = \ldots = D(55) = 0; \; D(56) = U^{56}.$$

This is a recursive formula because the value of $D(n)$ is dependent on $D(n-1)$ being calculated first with $D(0) = 0$ being the starting point. The meaning of U, p, and $j.kl$ is the same as in the CB Formula. The average number of plate appearances per game is given by $j.kl$. U is equal to the probability of getting at least one hit in a typical game for a given season, and p is the probability of getting a hit in any plate appearance. U^{56} is the probability of having a 56-game hitting streak in any particular 56-game span.

The next formula, called the Generalized MF Formula, extends the MF Formula to other s-game batting streaks. The logic is the same as used to extend the CB Formula to the Generalized CB Formula. U is defined as it was for the Generalized CB Formula.

The Generalized MF Formula

$$D(n) = D(n-1) + \{1 - D[n-(s+1)]\} * (1-U) * U^{s}, \text{ where } n > s.$$

Define $D(0) = D(1) = D(2) = \ldots = D(s-1) = 0$, $D(s) = U^{s}$.

The first five assumptions for the MF Formulas are the same as for the CB Formulas. The MF Formula *differs* from the CB Formula in that it does not consider overlapping 56-game hitting streaks as independent events.

16.2.2.1. Explanation of the Logic for the MF Formula

For a player to have a 56-game hitting streak at some point in the first n games ($n > 56$), he must either have a 56-game hitting streak in the first $(n-1)$ games or have his first hitting streak in the last 56 games. The probability of having a 56-game hitting streak at some point in the first $(n-1)$ games is $D(n-1)$. To have his first hitting streak in the last 56 games, three events must all occur: (1) he must not have a 56-game hitting streak in the first $(n-57)$ games, (2) he must not have a hit in game number $(n-56)$, and (3) he must get a hit in each of the games $(n-55)$ through n. We define the three events that must occur for him to have his first 56-game hitting streak in the last 56 games.

- E_1 = Not having a 56-game hitting streak in the first $(n-57)$ games.
- E_2 = Not having a hit in game number $(n-56)$.

- E_3 = Getting a hit in each of the games from game number $(n-55)$ through game number n.

  Since the three events are independent, the following is true:

$$\Pr(E_1 \text{ and } E_2 \text{ and } E_3) = \Pr(E_1) * \Pr(E_2) * \Pr(E_3)$$
$$= [1 - D(n-57)] * (1-U) * U^{56}.$$

For $n = 120$ games, calculate $n-57$, $n-56$, and $n-55$. Then use the above logic to see why the probability of having the first 56-game hitting streak in the last 56 games is equal to $\Pr(E_1) * \Pr(E_2) * \Pr(E_3)$.

The event of having a 56-game hitting streak at some point in the first $(n-1)$ games and the event of (E_1 and E_2 and E_3) are mutually exclusive (they both cannot occur at the same time). Therefore,

$$D(n) = D(n-1) + [1 - D(n-57)] * (1-U) * U^{56}.$$

Example 16.2. To understand how to calculate the MF Formula, we present the following example. We use player A's information from Example 16.1. With this information, we use the MF Formula to compute the probability of player A having a 56-game hitting streak at some point in his 66-game season.

Solution: The first four steps are the same as in the CB Formula.

Step 1: $p = \#H/\#PA = 138/(322+20+0+0+0) = 138/342 = .404$.
Step 2: $j.kl = \#PA/\#G = 342/66 = 5.18$. (So $j = 5$, $.kl = .18$.)
Step 3: The probability of a player getting at least one hit in $j.kl$ plate appearances per game is

$$U = [1-(1-p)^j] * (1-.kl) + [1-(1-p)^{(j+1)}] * .kl$$
$$= [1-(1-.404)^5] * (1-.18) + [1-(1-.404)^{(5+1)}] * .18$$
$$= [1-(.596)^5] * (.82) + [1-(.596)^6] * (.18)$$
$$= .9248 * .82 + .9552 * .18 = .9302.$$

Step 4: The probability of player A having a 56-game hitting streak in any particular 56-game span in his season of 66 games is equal to

$$U^{56} = .9302^{56} = .0174.$$

Step 5: A recursive formula begins with a known term. The known beginning term is $D(56) = U^{56}$. Our goal is to reach $D(n)$, where $n = 66$. To reach this goal, we must find $D(57)$, $D(58)$, and so on.

$$D(0) = D(1) = D(2) = \ldots = D(55) = 0,$$
$$D(56) = .0174,$$
$$D(57) = D(56) + [1 - D(n-57)] \ (1-U) * U^{56}$$
$$= D(56) + [1 - D(0)] * (1 - .9302) * (.9302)^{56}$$
$$= .0174 + .0012 = .0186,$$

$D(58) = D(57) + [1 - D(n-57)] * (1-U) * U^{56} = D(57) + [1 - D(1)] * (1-U) * U^{56}$
$= .0186 + .0012 = .0198,$
$D(59) = D(58) + [1 - D(n-57)] * (1-U) * U^{56} = D(58) + [1 - D(2)] * (1-U) * U^{56}$
$= .0198 + .0012 = .0210.$

You will be asked to do $D(60)$ through $D(66)$ as an exercise. For the following exercises it is recommended that you use Excel.

Exercise 16.3. Complete Example 16.2 by finding $D(60)$ through $D(66)$.

The next three exercises show the difficulty in using a recursive formula. It is almost mandatory to do these problems by using a spreadsheet.

Exercise 16.4. Find the probability, using both the CB and MF Formulas, of Ted Williams having a 56-game hitting streak at some point in the 1941 season. Compare the results. How does the MF Formula change at $D(113)$?

Clearly, the recursive nature of the MF Formula provides a need for a spreadsheet to evaluate the probabilities. Next, I present my own formula to calculate the probability of a player achieving a 56-game hitting streak. This is the SR Formula. The SR Formula is a split function composed of polynomials for the variable U. U has the same meaning as in the MF and CB Formulas.

16.2.3. The SR Formula to Calculate the Probability of a Player Achieving a 56-Game Hitting Streak at Some Point in a Season

The SR Formula

Let n be the number of games played by a player in a season.

$R(n)$ is the probability of at least one 56-game hitting streak at some point in the n games.

For $0 \le n \le 55$, $R(n) = 0$.

For $56 \le n \le 112$, $R(n) = U^{56} + (n-56) * (1-U) * U^{56}$.

For $113 \le n \le 162$, $R(n) = U^{56} + (n-56) * (1-U) * U^{56} - .5(n-112)(1-U) * U^{112} * [(n-112)(1-U) + 1 + U]$.

The Generalized SR Formula

Let n be the number of games played by a player in a season.

$R(n)$ is the probability of at least one s-game batting streak at some point in the n games.

For $0 \le n \le (s-1)$, $R(n) = 0$.

For $s \leq n \leq 2s$, $R(n) = U^s + (n-s) * (1-U) * U^s$.

For $(2s+1) \leq n \leq (3s+1)$, $R(n) = U^s + (n-s) * (1-U) * U^s - .5(n-2s)(1-U) * U^{2s} * [(n-2s)(1-U) + 1 + U]$.

The Generalized SR Formula is valid only when $n \leq 3s + 1$. An error occurs for shorter streaks. For streaks where $n > 3s + 1$, new polynomial pieces, starting at $3s + 2$, would have to be added to the piecewise polynomial. How many additional polynomial pieces would be needed for a streak $s = 8$, assuming that the player was active in all 162 games? For those students interested in understanding the error analysis for shorter streaks, you are referred to Chapter 4 of the book *Mathematics and Sports* published by the Mathematics Association of America.

The meaning of U is the same as in the CB and MF Formulas. Since the number of games in a season has always been less than or equal to 162, the SR Formula can be applied to Joe DiMaggio's 56-game hitting streak and to Ted Williams's 84-game on-base streak. However, an error would occur in an eight-game batting streak.

Using the data from Example 16.1, we have $n = \#G = 66$ and $U = .9302$:

$$R(66) = .9302^{56} + (66-56)(1-.9302) * .9302^{56} = .0295.$$

For $s = 56$, the SR Formula gives the same result as the MF Formula (see your result for Exercise 16.3). The advantage of the SR Formula is that it gives a *closed*, not *recursive*, formula for calculating the probability.

Exercise 16.5. Apply the SR Formula and the MF Formula to the data in Exercise 16.1. Compare the two probabilities obtained.

Exercise 16.6. Apply the SR Formula and the MF Formula to the data in Exercise 16.2. Compare the two probabilities obtained.

16.2.4. A Comparison between the CB, MF, and SR Formulas

The data #G = 66, #H = 138, #PA = 342 produce the probability of .1738 from the CB Formula and .0295 from the MF Formula (you should get a probability close to this upon completion of Exercise 16.3). These results differ by a factor of 10. Put in perspective, in 1000 seasons, player A will have a 56-game hitting streak in 174 seasons if the CB Formula is correct, but only in 30 seasons if the MF Formula is correct.

Which formula should we use? The answer is, the one that gives the correct result! The next simple example helps answer this question. As both formulas depend on U, the probability of a player getting at least one hit in a typical game for that season, a comparison can be made by choosing a different value of s for U.

Example 16.3. Let $U=1/2$. Suppose that a player's season consisted of three games $n=\#G=3$. Find the probability of this player having a two-game hitting streak.

Solution: We provide four solutions to this problem: the CB Formula, the MF Formula, the SR Formula, and a formula from probability theory. For all four solutions, we assume that the events "having at least one hit in n consecutive games" are independent.

Solution 1: The CB Formula for an s-game hitting streak is

$$1-(1-U^s)^{[\#G-(s-1)]}.$$

$U=1/2$, so the probability of having at least one two-game hitting streak in the three-game season is

$$1-[1-(1/2)^2]^{[3-(2-1)]}=1-(3/4)^2=7/16.$$

Solution 2: The MF Formula for an s-game hitting streak is

$$D(n)=D(n-1)+\{1-D[n-(s+1)]\}*(1-U)*U^s,$$

where $n>s$.

Define $D(0)=D(1)=D(2)=\ldots=D(s-1)=0$, $D(s)=U^s$.

Since $U=1/2$, $s=2$, and $n=3$, $D(0)=0$, $D(1)=0$, $D(2)=(1/2)^2$, and the probability of having at least one two-game hitting streak in the three-game season is

$$\begin{aligned} D(3) &= D(2)+\{1-D[3-(2+1)]\}*[1-(1/2)]*(1/2)^2 \\ &= (1/2)^2+[1-D(0)]*(1/2)*1/4 \\ &= 1/4+(1-0)*(1/2)*(1/4) \\ &= 1/4+1/8=3/8. \end{aligned}$$

Solution 3: Since $n=3$ and $s=2$, we have $n<2s$; the Generalized SR Formula for an s-game hitting streak is

$$R(3)=U^s+(n-s)*(1-U)*U^{56}=R(3)=(1/2)^2+(3-2)(1-1/2)*(1/2)^2=3/8.$$

As expected, the MF Formula gives the same result as the SR Formula. However, the MF and CB Formulas disagree. To distinguish between them, we turn to probability theory.

Solution 4: The following three events are given: E=getting at least one hit in game number 1, F=getting at least one hit in game number 2, G=getting at least one hit in game number 3. Using the Fundamental Counting Theorem (Chap. 7), there are $2*2*2=8$ possible outcomes resulting from the three-game season. The following is a list of the eight possible outcomes along with the probabilities that each occurs. The events are assumed independent, so the probability of all three occurring is the product of their respective probabilities. Remember that $U = Pr(E) = Pr(F) = Pr(G) = 1/2$.

G#1	G#2	G#3	Probability
E	*F*	*G*	$Pr(E) * Pr(F) * Pr(G) = (1/2)^3 = 1/8$
E	Not *F*	Not *G*	$Pr(E) * Pr(\text{Not } F) * Pr(\text{Not } G) = (1/2)^3 = 1/8$
E	Not *F*	*G*	$Pr(E) * Pr(\text{Not } F) * Pr(G) = (1/2)^3 = 1/8$
E	*F*	Not *G*	$Pr(E) * Pr(F) * Pr(\text{Not } G) = (1/2)^3 = 1/8$
Not *E*	*F*	*G*	$Pr(\text{Not } E) * Pr(F) * Pr(G) = (1/2)^3 = 1/8$
Not *E*	Not *F*	*G*	$Pr(\text{Not } E) * Pr(\text{Not } F) * Pr(G) = (1/2)^3 = 1/8$
Not *E*	*F*	Not *G*	$Pr(\text{Not } E) * Pr(F) * Pr(\text{Not } G) = (1/2)^3 = 1/8$
Not *E*	Not *F*	Not *G*	$Pr(\text{Not } E) * Pr(\text{Not } F) * Pr(\text{Not } G) = (1/2)^3 = 1/8$

Of these eight outcomes, only the outcomes (*E* and *F* and *G*), (*E* and *F* and Not *G*), and (Not *E*, *F*, *G*) have two-game hitting streaks. Since the eight outcomes are mutually exclusive and equally likely to occur, the probability of having a two-game hitting streak is $1/8 + 1/8 + 1/8 = 3/8$.

Example 16.3 suggests that the MF Formula is the correct one, but this is *not* a proof. We have just shown that for the case of three games and a streak of two games with $U = 1/2$, the MF Formula is the one that works.

This brings up an interesting point: do we consider a three-game hitting streak a two-game hitting streak? The answer is yes.

But what would happen if Example 16.3 was done for a different value for *U*?

Exercise 16.7. Repeat Example 16.3 for $U = 1/4$ and $U = 2/5$. What conclusions can you draw about the four formulas?

Examples 16.1, 16.2, and 16.3 showed that the probability obtained from the CB Formula was greater than the probability obtained from the MF Formula.

As you must have discovered from the exercises, it is time-consuming to use the MF Formula by hand. The calculations for both formulas can be done with the use of an Excel spreadsheet.

Table 16.1 provides a list of all players with hitting streaks of more than 35 games (as of the writing of this book).

Table 16.2 provides the probabilities of various players having at least one 56-game hitting streak in a season, calculated using both the CB Formula and the (MF or SR) Formula. The probability *U*, of getting at least one hit in a game, is the same for both the CB and MF Formulas. The value of *U* is also displayed in Table 16.2.

Table 16.1 List of all players with hitting streaks of more than 35 games

Rank	Year	Player	Team	League	Length of streak
1	1941	Joe DiMaggio	New York	AL	56
2	1897	Willie Keeler	Baltimore	AL	44
2	1978	Pete Rose	Cincinnati	NL	44
4	1894	Bill Dahlen	Chicago	NL	42
5	1922	George Sisler	St. Louis	AL	41
6	1911	Ty Cobb	Detroit	AL	40
7	1987	Paul Molitor	Milwaukee	AL	39
8	1945	Tommy Holmes	Boston	NL	37

General Observations concerning Table 16.2

1. In most cases the closer that U (the probability of getting at least one hit in a typical game for that season) is to 1, the higher the probability of achieving the 56-game streak. Can you find any exceptions to this in Table 16.2? If so, why do you think this happened?
2. The probabilities calculated from the CB Formula are approximately 3 to 10 times higher than the probabilities calculated using the MF Formula. Are there any examples in the table where the MF probability is larger than the CB probability?
3. For 1941, when Joe DiMaggio had his 56-game hitting streak, the MF Formula revealed a probability of .00010 for Joe achieving his streak. This translates into a 1 in 10000 chance of the streak happening.
4. Pete Rose, whose 44-game hitting streak is second best, had a probability of .00001, which translates into a 1 in 100000 chance of having a 56-game hitting streak, whereas Willie Keeler, who tied Rose with a 44-game hitting streak, had a probability of .02500, a 2500 in 100000 chance of having a 56-game hitting streak. What do you think accounts for such a large difference?
5. Of the eight players listed with hitting streaks of at least 37 games, only Willie Keeler appears in the list of players with the best chance of achieving the streak.
6. In 1941, Joe DiMaggio's probability was .00010 and Ted Williams's probability was .00002. DiMaggio had a 1 in 10000 chance, and Williams had a 2 in 100000 chance. This equates to DiMaggio having 5 times the chance of accomplishing the streak. Despite having a batting average of .406, Ted's chance of performing the feat was small. This is explained by looking at his ratio of hits to plate appearances, which was only .305. The big difference between his 1941 season and DiMaggio's 1941 season was their number of base on balls: DiMaggio had 76, but Williams had

Table 16.2 **Probabilities of various players having at least one 56-game hitting streak in a season, calculated using the CB Formula and the (MF or SR) Formula**

#G	#AB	#H	#PA	#H/#AB	#H/#PA	*U*	CB Form.	MF Form.	Player	Year	Actual streak length
66	322	138	342	0.429	0.404	0.93003	0.17382	0.02925	A		
76	322	138	342	0.429	0.404	0.89895	0.05251	0.00775	B		
86	322	138	342	0.429	0.404	0.87141	0.01384	0.00218	C		
66	322	148	342	0.460	0.433	0.94589	0.39297	0.06837	D		
66	322	158	342	0.491	0.462	0.95871	0.66358	0.13322	E		
66	322	168	342	0.522	0.491	0.96896	0.87293	0.22410	F		
66	352	138	372	0.392	0.371	0.92477	0.12946	0.02195	G		
66	422	138	442	0.327	0.312	0.91718	0.08352	0.00958	H		
143	456	185	606	0.406	0.305	0.78397	0.00011	0.00002	T. Williams	1941	
146	601	215	677	0.358	0.318	0.82700	0.00218	0.00040	A. Rodriguez	1996	
147	373	135	617	0.362	0.219	0.64364	0.00000	0.00000	B. Bonds	2004	
154	629	223	693	0.355	0.322	0.82247	0.00175	0.00032	H. Aaron	1959	
Players with hitting streaks of at least 37 games											
139	541	193	621	0.357	0.311	0.80716	0.00052	0.00010	J. DiMaggio	1941	56
159	655	198	729	0.302	0.272	0.76323	0.00003	0.00001	P. Rose	1978	44
121	502	179	591	0.357	0.303	0.82708	0.00159	0.00030	B. Dahlen	1894	42
142	586	246	654	0.420	0.376	0.88304	0.07885	0.01043	G. Sisler	1922	41
146	591	248	654	0.420	0.379	0.87848	0.06227	0.00842	T. Cobb	1911	40
118	465	164	542	0.353	0.303	0.80589	0.00036	0.00007	P. Molitor	1987	39
154	636	224	713	0.352	0.314	0.82253	0.00175	0.00033	T. Holmes	1945	37
Best chances for a 56-game hitting streak by the MF Formula											
125	539	237	606	0.440	0.391	0.90812	0.27222	0.03320	H. Duffy	1894	22
66	322	138	342	0.429	0.404	0.93003	0.17382	0.02925	R. Barnes	1876	
129	564	239	618	0.424	0.387	0.90180	0.20314	0.02500	W. Keeler	1897	44
124	517	225	572	0.435	0.393	0.89722	0.14706	0.01839	T. O'Neill	1887	
133	586	240	647	0.410	0.371	0.89364	0.13386	0.01690	J. Burkett	1896	
131	544	232	584	0.426	0.397	0.89203	0.11892	0.01513	N. Lajoie	1901	
Other DiMaggio seasons											
138	637	206	668	0.323	0.308	0.83051	0.00252	0.00045	J. DiMaggio	1936	
151	621	215	692	0.346	0.311	0.81512	0.00102	0.00020	J. DiMaggio	1937	
120	462	176	524	0.381	0.336	0.82943	0.00184	0.00034	J. DiMaggio	1939	

147. Unfortunately, a base on balls has the same negative impact as an out does when it comes to a hitting streak.

7. Barry Bonds, in 2004, had an AVG of .362. His ratio of hits to plate appearances was .219. That year Bonds had 232 base on balls. His probability of having a 56-game hitting streak was .00000.
8. Using the MF Formula, the six players most likely to achieve the 56-game hitting streak all played before 1902. Why do you think this is so?
9. Joe DiMaggio had a higher probability of achieving his streak for the years 1936, 1937, and 1939 than he did in 1941.
10. Of the six players given the best chance of having a 56-game hitting streak for a specific season (from the MF Formula), only two had more than a 20-game hitting streak for that season. The two players were Hugh Duffy, with a 22-game hitting streak, and "Wee Willie" Keeler, with a 44-game hitting streak.

There are many more observations that can be made from this table. A later exercise will ask you to list your own observations. The next three exercises deal with Table 16.2. For those three exercises, the effect of the variables #G, #H, and #PA on the value of U will be examined. You will be asked to compare the effect on U when two of the variables have the same value and the third variable has a different value.

Exercise 16.8. The first three rows increase the number of games, leaving the other two variables constants. What is happening to the value of U and to the outcomes for the two formulas? Can you explain why this is happening?

Exercise 16.9. Rows 4 through 6 increase the number of hits for the season, leaving the other two variables constant. What is happening to the value of U and to the outcomes for the two formulas? Can you explain why this is happening?

Exercise 16.10. If we increase the number of plate appearances, leaving the other two variables constant, what happens to the value of U and to the outcomes for the two formulas? Can you explain why this is happening?

Exercise 16.11. Choose a player who is still active which you believe has the best chance of having a 56-game hitting streak. For that player, pick his best year and for that year use the (MF or SR) Formula to evaluate his probability of having a 56-game hitting streak. It would be useful to use either a spreadsheet or the SR Formula. What criteria would you use to decide on his best year?

Exercise 16.12. Give your own five observations about the results in Table 16.2.

16.3. Simulations

In this section, we compare the MF Formula with various simulations.

In article 4, "Calculating the Odds," a simulation was done to mimic 1000 seasons, where each season consisted of 10 games, for a player having $U=2/3$ as his probability of getting at least one hit in a game. The simulation was performed by rolling a six-sided die 10000 times. A result of a 1, 2, 3, or 4 meant that the player had at least one hit in the game; a 5 or 6 meant that the player went hitless in the game. One could have created a disk with six sectors, numbered 1 through 6, each having a central angle equal to 60 degrees. Then, spinning the pointer on the disk would give the same result as rolling the die. The function RANDBETWEEN (1, 6) supplied by Microsoft Excel would also work.

16.3.1. The Performance of a Simulation Based on 1000 Seasons with Each Season Consisting of 10 Games for a Player with a Probability of $U=2/3$

Since $U=2/3$ is used for each game, the probability of getting at least one hit in a game is the same for each game. Further, the use of a die to determine the result guarantees that the event of getting at least one hit in one game is independent of getting at least one hit in any other game.

Since a season for this player is 10 games, 10 tosses of a die represents one season. For each 10 tosses of the die, we let s equal the length of the longest hitting streak. Clearly, the streak can range from $s=0$ to $s=10$. An $s=0$ would mean that the 10 tosses of the die resulted in 10 hitless games. An $s=10$ would mean that all 10 tosses resulted in getting at least one hit in each game. What do you think a streak of $s=1$ would mean?

The experiment of tossing the die was carried out 10000 times (1000 seasons of 10 games), resulting in 1000 values for s, one for each season. Table 16.3 shows the observed results taken from article 4.

Table 16.3 Streak length (s) for the simulation of 1000 seasons of 10 games each taken from article 4 (tossing a die 10000 times)

Streak length s	Occurrences	Probability of streak
0	0	.000
1	22	.022
2	157	.157
3	248	.248
4	208	.208
5	146	.146
6	100	.100
7	54	.054
8	30	.030
9	19	.019
10	16	.016

Exercise 16.13. For each of the following 10-game seasons find the length of the longest hitting streak s. We use the letter H for getting at least one hit in a game and the letter N for (Not H). Assuming independence between games and $U=2/3$, find the probability of each sequence occurring.

a. NHNHNHNNNH
b. NNNNNNNNNN
c. HNNHHHNHHN

d. HHHHHHHHHH
e. HNHHHHHHHH

Exercise 16.14. Perform your own simulation for 50 seasons, with each season consisting of eight games for a player with $U = 2/3$. Display the observed results in a table similar to Table 16.3. For your simulations, you can use either a six-sided die, a disk, the online spinner, or the RANDBETWEEN function in Excel.

16.3.2. A Simulation of Joe DiMaggio's 1941 Season

The simulation for Joe DiMaggio's 1941 season of 139 games provided in article 4 was based on $U = .817$. The U we used in the MF Formula shown in Table 16.2 was $U = .807$. The difference is only .010. The difference is because article 4 used plate appearances for the seasons 1936 through 1940 to evaluate U, whereas we used only the plate appearances for the 1941 season to evaluate U. The simulation employed the RANDBETWEEN(0, 999) function. If the number generated from the function falls between 000 and 816, DiMaggio "hit safely" in the game. If it falls between 817 and 999, he went "hitless" in that game. The simulation involved 100000 seasons, each of which consisted of 139 games. Table 16.4 gives the results from article 4.

From Table 16.2, we saw that $U = .807$ gave us a probability of .0001 of Joe DiMaggio having at least a 56-game hitting streak in 1941.

What would happen if we used $U = .817$ instead? The MF Formula gives the probability of having a hitting streak of 56+ games in a season of 139 games as $.0002 = 2/10000$, while the probability of having a hitting streak of 41+ games in a season of 139 games is $.00476 = 476/100000$. We summarize these results in Table 16.5.

It appears that the frequency distribution for the length of the streak generated by the simulation fits the frequency distribution expected using the MF Formula.

Table 16.4 **Streak length (*s*) for the simulation of 100000 seasons, each of which consisted of 139 games taken from article 4**

Streak length *s*	Occurrences	Probability of streak
10 or less	2806	.02806
11–15	32005	.32005
16–20	36262	.36262
21–30	24957	.24957
31–40	3478	.03478
41–55	469	.00469
56+	23	.00023
Total	100000	

Table 16.5 **MF Formula applied for $U = .817$ to hitting streaks of 56+ and 41+ games**

Streak Length *s*	Probability	
	Simulation	MF Formula
41+	.00469 + .00023 = .00492	.00476
56+	.00023	.00020

16.3.3. A Simulation for the History of Baseball

The article "A Journey to Baseball's Alternate Universe" by Samuel Arbesman and Steven Strogatz was published on March 30, 2008, in the *New York Times*. The authors wished to decide how unlikely Joe DiMaggio's streak was. They used a comprehensive collection of baseball statistics stretching from 1876 to 2005.

The specific question they asked was, how likely was it that any player in baseball history achieved a streak of at least 56 games? They assigned, to every baseball player from 1871 to 2005, a fixed probability of getting at least one hit in a game for that season. The technique used to calculate this probability resembles what we used to construct *U*, used in the CB Formula and the MF Formula. The calculation involved the number of games played, plate appearances, and hits for a season. It assumed the same number of plate appearances for each game, that the probability of getting a hit in a plate appearance is always the same, and that the plate appearances are independent of each other.

The simulation was done for all players in the history of baseball from 1876 to 2005. The simulation used the value of *U* assigned to each player and the random number generator (basically the RANDBETWEEN function) to find a result for each plate appearance (hit or no hit). A computer program performed the simulation. The history of baseball was replicated 10000 times. For each run of the simulation for all players in the history of baseball, they found the player who held the longest hitting streak, when the player achieved the streak, and how many games were involved in their longest hitting streak. The results of the 10,000 simulations are displayed in the graphs in Figures 16.1 and 16.2.

Some of the observations made by the authors concerning the simulation were as follows:

1. The record streaks ranged in length from 39 to 109.
2. The median was 53.
3. The mode for the simulated record streaks was 51.
4. The frequency distribution is slightly right skewed.
5. The likeliest time for the longest streak to have occurred was in the nineteenth century.
6. The year 1941 was one of the least likely years for the longest streak, while 1894 was the most likely year. The longest streak occurred only 19 times in 1941; the longest streak occurred 1290 times in 1894.
7. In those 10000 universes, many other players held the record more often than DiMaggio. Joe DiMaggio was number 56 on the list of the most likely players to have the longest streak (what a coincidence). Joe had the longest streak 29 times, with one time occurring in 1941. Remember,

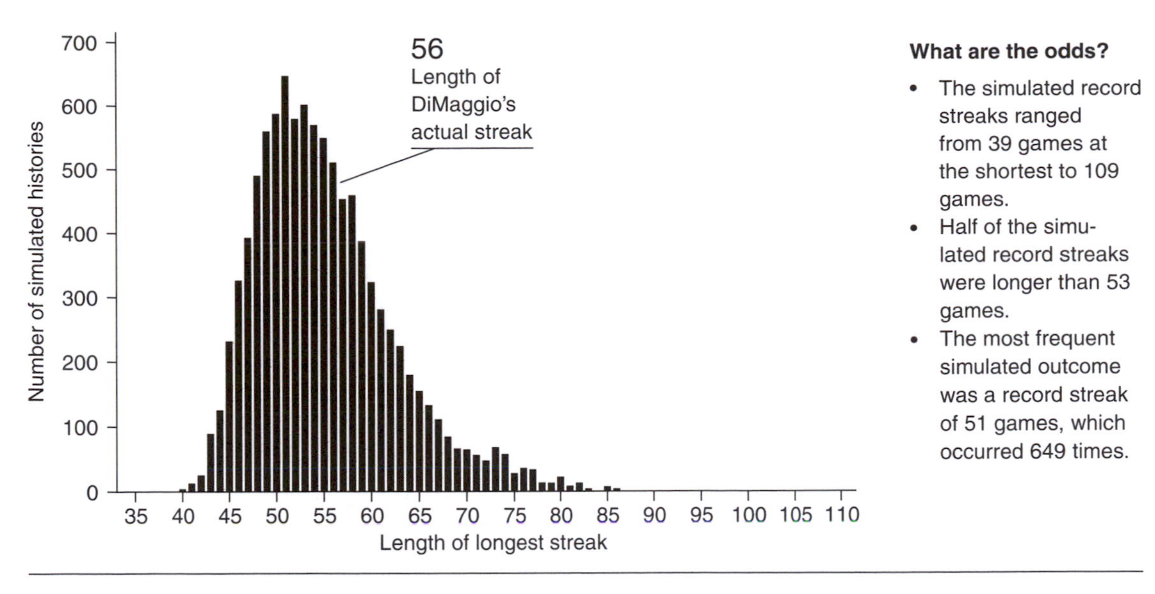

56
Length of
DiMaggio's
actual streak
Number of simulated histories
Length of longest streak
What are the odds?
The simulated record streaks ranged from 39 games at the shortest to 109 games.
Half of the simulated record streaks were longer than 53 games.
The most frequent simulated outcome was a record streak of 51 games, which occurred 649 times.

Figure 16.1 Frequency distribution of the longest hitting streak for each of the 10000 simulations by the length of the longest streak. Adapted by permission of Steve Strogatz.

1/10000 = .0001 was the probability of DiMaggio achieving his 56-game hitting streak from the MF Formula.

8. The two players most likely to have the longest streak were Hugh Duffy and Willie Keeler (both nineteenth-century players).
9. The longest streak was between 50 and 64 games 66% of the time.
10. At least 42% of the time there was at least one streak as long as 56 games. DiMaggio's streak was broken just 2.5% of the time since 1941.

Two observations of my own are as follows:

11. Figure 16.2 shows that, after 1900, the period between 1920 and 1930 was the most likely to have the longest hitting streaks. In the next chapter, the period from 1920 to 1930 will be shown to be the era of high batting averages and low base-on-balls percentages. Both of these properties contribute to a player having a long hitting streak.
12. As we shall see later in this chapter, Joe DiMaggio had a 61-game hitting streak in the minors.

What observations can you make about the graphs in Figure 16.1 and 16.2?

Academic studies question whether DiMaggio's streak is evidence of a spike in his ability that exceeded his everyday talent, rather than just a chance occurrence to be expected from a talented player. In their article, Arbesman and Strogatz stated, "More than half the time, or in 5,295 baseball universes, the record for the longest hitting streak exceeded 53 games. Two-thirds of the time, the best streak was between 50 and 64 games. In other words, a streak of 56 games or longer is not at all an unusual occurrence. Forty-two percent of the simulated baseball histories have a streak of DiMaggio's length or

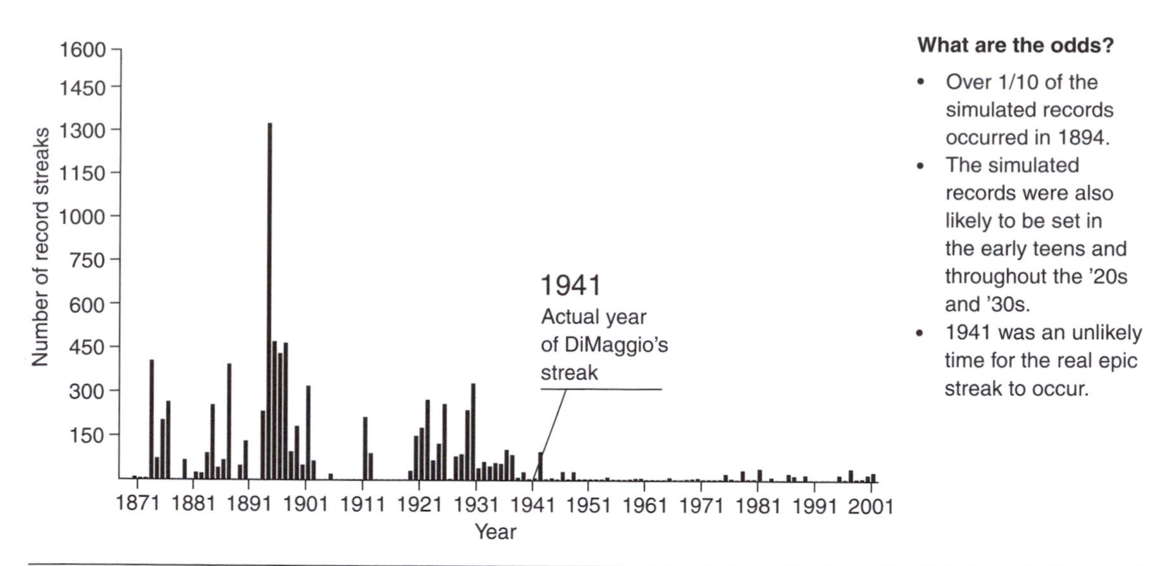

1600
1450
1300
1150
1000
750
600
450
300
150
Number of record streaks
1871 1881 1891 1901 1911 1921 1931 1941 1951 1961 1971 1981 1991 2001
Year
1941
Actual year of DiMaggio's streak
What are the odds?
Over 1/10 of the simulated records occurred in 1894.
The simulated records were also likely to be set in the early teens and throughout the '20s and '30s.
1941 was an unlikely time for the real epic streak to occur.

Figure 16.2 Frequency distribution of the longest hitting streak for each of the 10000 simulations by the year of the longest streak. Adapted by permission of Steve Strogatz.

longer. You shouldn't be too surprised that someone, at some time in the history of the game, accomplished what DiMaggio did."

These statements, diminishing the fabulous feat of Joe DiMaggio, brought a response from Carl Bialik in an article entitled "In Defense of Joe DiMaggio," written on April 3, 2008. Bialik questioned the validity of the following assumptions used by Arbesman and Strogatz:

1. They assigned a probability to each player, based on the player's actual plate appearances, to represent his chance of getting a hit in a plate appearance. This same probability was used for each plate appearance for the season. Looking at Table 1.1, we observe that DiMaggio faced four Hall of Fame pitchers during his streak. They were Hal Newhouser (twice), Bob Feller (twice), Lefty Grove, and Ted Lyons. Also, in three of the games he wound up with only two actual at bats. One of those games was against the great Bob Feller (who threw with a velocity of 100 mph). The plate appearances against a Hall of Fame pitcher have to be assigned a different probability than those against an ordinary pitcher.
2. The authors assumed that the player had the same number of plate appearances in each game for the season. Even if we can assume that the number of plate appearances in each game is off by at most one, the plate appearances that result in an at bat are the only productive ones. If a player has four plate appearances and is walked three times, he must get a hit in one plate appearance or the streak will be broken.
3. The authors assumed that the probability of getting at least one hit in a game is the same for each game in the season. Other factors such as a rain-shortened or lopsided game can drastically change the number of plate

appearances. A player who is rested for a game, but used as a pinch hitter, will have at most one chance of getting a hit. A player who is injured during a game may have fewer plate appearances. I am sure you can think of many other factors that would alter a player's probability of getting a hit in a game.

Bialik presented two other reasons that show the specialness of DiMaggio's streak. The second-longest streak of 44 games, held jointly by Pete Rose (1978) and Willie Keeler (1897), exhibits a gap of 12 games (very significant). Finally, Bialik points out that even the simulations show that DiMaggio's streak was broken just 2.5% of the time since 1941.

16.3.4. Hitting Streaks of at Least 20 Games

Our data come from the article "Hitting Streaks: Complete List of Major League Players with Batting Skeins of 20 or More Games in One Season," by Bob Kuenster, which appeared in *Baseball Digest*.

Bob's data consist of 502 documented hitting streaks of 20 or more games, recorded through the 2004 season. Of those, 25 occurred prior to 1900. A total of 37 players had a streak of 30 or more games. Willie Keeler's 44-game streak is tops among left-handed batters, and Pete Rose's 44-game streak heads the list of switch-hitters. Rose also holds the Major League record for most hitting streaks of more than 20 with seven. The most in one season, three, is held by Tris Speaker in 1912.

Table 16.6 shows the frequency distribution of all recorded hitting streaks of 20 or more games through 2004. Table 16.7 shows the frequency distribution of all recorded hitting streaks of 20 or more games through 1941. Table 16.8 shows the frequency distribution of all recorded hitting streaks of 20 or more games from 1942 through 2004. Table 16.9 provides some of the descriptive statistics for the data in Tables 16.6–16.8.

You will be asked in the exercises that follow to use what you have learned about descriptive statistics and probability from the earlier chapters to analyze the data in the tables.

Exercise 16.15. Construct a frequency histogram for each of the frequency distributions in Tables 16.6, 16.7, and 16.8.

Exercise 16.16. Decide on the shape of each of the distributions in Tables 16.6, 16.7, and 16.8 (left skewed, right skewed, or symmetric).

Exercise 16.17. Let the *X*-variable be the midpoint of each class interval in Table 16.6. Construct the discrete probability distribution for the variable *X*.

Exercise 16.18. For the discrete probability distribution *X* in Exercise 16.17, find the mean and variance.

Table 16.6 **Frequency distribution of all recorded hitting streaks of 20 or more games through 2004**

Class interval	Frequency
20–21	198
22–23	124
24–25	65
26–27	49
28–29	30
30–31	19
32–33	3
34–35	6
36–37	1
38–39	1
40–41	2
42–43	1
44–45	2
46–47	0
48–49	0
50–51	0
52–53	0
54–55	0
56–57	1
Total	502

Table 16.7 Frequency distribution of all recorded hitting streaks of 20 or more games through 1941

Class interval	Frequency
Before 1900	25
1900–1909	20
1910–1919	19
1920–1929	50
1930–1941	72
Total	186

Table 16.8 Frequency distribution of all recorded hitting streaks of 20 or more games from 1942 through 2004

Class interval	Frequency
1942–1949	30
1950–1959	43
1960–1969	34
1970–1979	44
1980–1989	51
1990–1999	69
2000–2004	45
Total	316

16.4. Ted Williams's 84-Game On-Base Streak

Many baseball people feel that on-base percentage for a team is a better predictor of runs scored than batting average. We saw in Chapter 5 that, for both Bonds and Aaron, their on-base percentage was a better predictor of R/27 (runs per 27 outs) than their batting averages.

Officially, there are three ways to get credit for getting on base safely. These are getting a hit (H), getting a base on balls (BB), or being hit by a pitch (HBP). Reaching base by virtue of a fielder's choice, an error, a dropped third strike, or an interference play does not count.

As mentioned in Chapter 3, the concept of on-base average or percentage (OBA or OBP) was first introduced in the mid-1950s by Allan Roth and Branch Rickey. Allan was the first full-time baseball statistician, while Branch was the manager who broke the color barrier, signing Jackie Robinson to the Brooklyn Dodgers. These two men showed the strong correlation between OBP and runs scored, but many years passed before the baseball world accepted OBP as an important statistic.

In his article "Ted Williams' On-Base Performances in Consecutive Games," published in the *Baseball Research Journal*, Herm Krabbenhoft outlines his procedure for finding the player with the longest consecutive-game on-base streak. Since OBP did not exist for most of baseball's history, he could not just go to the record books to find the player.

Krabbenhoft started his research journey by looking at Joe DiMaggio's 56-game hitting streak. By inspecting DiMaggio's games in the 1941 sea-

Table 16.9 Descriptive statistics for the data in Tables 16.6–16.8

Statistics	Table 16.6	Table 16.7	Table 16.8
Minimum	20	20	20
Maximum	56	56	44
Mean	23.48	24.38	22.94
Median	22	23	22
Mode	20	20	20
Range	36	36	20
Sample standard deviation	4.04	4.79	3.42

son, he discovered that Joe had a 74-game on-base streak wrapped around his 56-game hitting streak.

In considering players who might be good candidates for the record, Krabbenhoft said,

> Ted Williams appeared to be a particularly good candidate. That's because the Splendid Splinter accomplished an extraordinary on-base performance record, as indicated by the following:
>
> - He had the highest single season OBP in ML history of .553 in 1941. (Barry Bonds surpassed it with a .582 mark in 2002.)
> - Ted fashioned the highest career OBP in ML history (.482).
> - Ted holds the mark for most seasons leading the league in OBP (12).
> - The *Sporting News Baseball Record Book* lists Williams with the Major League record for most consecutive plate appearances getting on-base safely–16 (in 1957).

In addition to Ted Williams, Krabbenhoft studied a total of approximately 12,000 players who satisfied certain criteria. These criteria included the following:

- Those players who were in the top five in OBP for a year.
- Those players who were in the top 25 players in (hits plus walks) for a year.
- Those players with consecutive-game hitting streaks of at least 20 games.

The top 10 players with respect to longest consecutive-game on-base streaks, according to the *SABR Record Book* and Krabbenhoft's research, are given in Table 16.10.

Table 16.10 **Longest consecutive-game on-base streaks**

Length of streak	Name	League	Team	Year
84	Ted Williams	AL	Boston	1949
74	Joe DiMaggio	AL	New York	1941
69	Ted Williams	AL	Boston	1941
65	Ted Williams	AL	Boston	1948
64	Bill Joyce	AA	Boston	1891
63	Orlando Cabrera	AL	Los Angeles	2006
60	George Van Haltren	NL	Pittsburgh	1893
58	Duke Snider	NL	Brooklyn	1954
58	Barry Bonds	NL	San Francisco	2003
57	Wade Boggs	AL	Boston	1985

Table 16.11 *U*-value and probability from the MF Formula of the players in Table 16.10 achieving an 84-game on-base streak

Year	Streak length	Player	#G	#BB	#HBP	#SF + #SH	#AB	#H	#PA	p (#OB/ #PA)	U	MF Formula probability
1949	84	Ted Williams	155	162	2	0	566	194	730	0.490	0.95604	0.09351
1941	74	JoeDimaggio	139	76	3	0	541	193	620	0.439	0.92079	0.00523
1941	69	Ted Williams	143	147	3	0	456	185	606	0.553	0.96526	0.15463
1948	65	Ted Williams	137	126	3	0	509	188	638	0.497	0.95683	0.08037
1891	64	Bill Joyce	65	63	5	0	243	75	311	0.460	0.94557	0.00000
2006	63	Orlan. Cabrera	153	51	3	14	607	171	675	0.333	0.82958	0.00000
1893	60	G. Van Haltren	124	75	2	0	529	179	606	0.422	0.93043	0.00885
1954	58	Duke Snider	149	84	4	7	584	199	679	0.423	0.91507	0.00377
2003	58	Barry Bonds	130	148	10	2	390	133	550	0.529	0.95683	0.07317
1985	57	Wade Boggs	161	96	4	5	653	240	758	0.449	0.93690	0.02448

Table 16.11 displays the value *U*, the probability of a player getting on base at least once in a typical game, and the probability generated from the MF Formula for each of the players listed in Table 16.10 achieving an 84-game on-base streak. The MF Formula for the probability of a player having an 84-game on-base streak is obtained by using $s=84$ and $p=\#OB/\#PA$ in the Generalized MF Formula.

The MF Formula for an 84-game on-base streak is

$D(n)=D(n-1)+[1-D(n-85)]*(1-U)*U^{84}$, where $n>84$,
Define $D(0)=D(1)=D(2)=...=D(83)=0$, $D(84)=U^{84}$,
$U=[1-(1-p)^j]*(1-.kl)+[1-(1-p)^{(j+1)}]*.kl$, $p=\#OB/\#PA$,
$j.kl=\#PA/\#G$.

An often-asked question is, which of the two streaks is harder to duplicate? In article 5, Krabbenhoft gives the answer in terms of approachability. He states,

> Since DiMaggio achieved his streak in 1941, the closest any major league player has come to it was the 44-game hitting streak by Pete Rose in 1978. Forty-four is 78.6% of the way to 56. Since Williams achieved his 84-game streak in 1949, the closest any major league player has come to it were the 58 consecutive game on-base streak by Duke Snider in 1954 and Barry Bonds in 2003. Fifty-eight is 69% of the way to 84. So, with the above approachability considerations in mind, it can be argued that Teddy Ballgame's 84 game on-base safely streak may be the greatest batting achievement of all.

Krabbenhoft also quotes Ted Williams himself saying, "I believe there isn't a record on the books that will be tougher to break than Joe DiMaggio's 56-game hitting streak." Since Krabbenhoft's article was published in 2004, Orlando Cabrera recorded a consecutive-game on-base streak of 63 games in 2006. Sixty-three is 75% of the way to 84. This blows a hole in the approachability argument.

I look at the question differently. Since in the 1941 season Ted Williams batted .406 and Joe DiMaggio had the 56-game hitting streak, it seemed reasonable to calculate their probabilities of achieving these two streaks for that year. Applying the (MF or SR) Formula, we obtain, for the year 1941, the following:

- From Table 16.2, the probability of Joe DiMaggio achieving a 56-game hitting streak was .0001, whereas the probability of Ted Williams achieving a 56-game hitting streak was .00002.
- From Table 16.11, the probability of Joe DiMaggio achieving an 84-game on-base streak was .0052, whereas the probability of Ted Williams achieving an 84-game on-base streak was .1546.

Based on the probabilities calculated above, I agree with Ted Williams that the 56-game hitting streak would be the more difficult to duplicate.

Exercise 16.19. Write the SR Formula for the probability of achieving an 84-game on-base streak. Discuss the differences between the (MF and SR) Formula for a 56-game hitting streak and that for an 84-game on-base streak.

Exercise 16.20. Write the MF Formula that represents the probability of a player achieving a 22-game base-on-balls streak. For p you must use #BB/#PA. The SR Formula can only be used if the number of games is less than or equal to $[3 * (\text{length of streak}) + 1] = 3 * 22 + 1 = 67$.

The following exercises refer to Table 16.11.

Exercise 16.21. What is the meaning of U in the table?

Exercise 16.22. Compare the probabilities in Table 16.2 to the probabilities in Table 16.11. What does this comparison say about the difficulty in achieving these two streaks?

Exercise 16.23. Choose one player in Table 16.2 and one player in Table 16.11. Show all the necessary work to reproduce his row statistics. You can use either the MF Formula or the SR Formula to calculate the probabilities.

Exercise 16.24. Use the MF probabilities from Tables 16.2 and 16.11 to compare the 56-game hitting streak of DiMaggio with the 84-game on-base streak of Williams.

16.5. Applying Statistical Inference Techniques to Examine Simulations

In this section, we use the technique of confidence intervals to draw conclusions about certain populations.

From article 4, the frequency distribution in Table 16.4, representing a simulation for 100000 seasons of Joe DiMaggio's 1941 season, has the following descriptive statistics for the variable "length of streak":

1. The shortest was seven games.
2. The longest was 70 games.
3. The mean length of the 100000 streaks was 18.70 games.
4. The sample standard deviation for the length of the 100000 streaks was 4.77 games.
5. The sample proportion of seasons with hitting streaks of 56 or more games is equal to 23/100000 = .00023.

The theoretical population, for the confidence intervals that follow, is the length of all hitting streaks for any player whose season consisted of 139 games in which his probability of getting at least one hit in a game was $U = .817$. The sample selected is the simulation of 100000 seasons. The results appear in Table 16.4. Using this sample, we find the following confidence intervals.

16.5.1. A 95% Confidence Interval for the Population Mean Length of a Hitting Streak

Since the sample size is greater than 30, we can use Formula (14.3) in Section 14.5. The formula is

$$\bar{x} - z_L * (s/\sqrt{n}) < \mu < \bar{x} + z_L * (s/\sqrt{n}).$$

For the formula, $L = 0.95$, $z_L = 1.96$, $\bar{x} = 18.70$, $s = 4.77$, $n = 100000$.

The margin of error is $1.96 * (4.77/\sqrt{100000}) = .030$. Note that both $\bar{x}$ and s are calculated from the results in Table 16.4.

Applying Formula (14.3),

$$18.70 - .030 < \mu < 18.70 + .030,$$
$$18.67 < \mu < 18.73.$$

With a level of confidence of 95%, we can conclude that the *true* population mean length of a hitting streak for a season of 139 games with $U = 8.17$ lies between 18.67 and 18.73. This means that we would expect a player with Joe DiMaggio's batting statistics for 1941 to have a *true* mean longest hitting streak of between 18.67 and 18.73 games. This provides more evidence favoring the spike in Joe's ability during his 56-game hitting streak.

Exercise 16.25. What is the margin of error for the above confidence interval?

16.5.2. A 95% Confidence Interval for the Population Proportion for the Occurrence of a 56-Game Hitting Streak

We use Formula (14.1) in Section 14.3:

$$\hat{p}-z_L*\sqrt{(\hat{p}(1-\hat{p})/n}<p<\hat{p}+z_L*\sqrt{(\hat{p}(1-\hat{p})/n}.$$

For the formula, $L=0.95, z_L=1.96, \hat{p}=.00023$ (see Table 16.5), $n=100000$. The margin of error is $1.96*\sqrt{.00023*.99977/100000}=.000094$

Applying the formula gives

$$.00023-.000094<p<.00023+.000094,$$
$$.00014<p<.00032.$$

With a level of confidence of 95%, we conclude that the population proportion for the occurrence of a 56-game hitting streak lies between 14/100000 and 32/100000. This means that for every 100000 seasons we would expect a player with Joe DiMaggio's 1941 batting statistics to have a frequency of between 14 and 32 seasons with a 56-game hitting streak. Note that since both $n*\hat{p}$ and $n*(1-\hat{p})$ are greater than 5, Formula (14.1) above can be used.

What do the above two confidence intervals say about the accomplishment of Joe DiMaggio in achieving the 56-game hitting streak?

Exercise 16.26. Find a 95% confidence interval estimate for the population proportion of hitting streaks of at least 41 games for all players whose season consisted of 139 games in which his probability of getting at least one hit in a game was $U=.817$. Use for your sample the 100000 simulations given in Table 16.4.

We conclude this chapter by presenting other notable batting streaks in baseball.

16.6. Other Notable Streaks (since 1900)

Consecutive-Game Batting Streaks with *at Least One Success* in Each Game

- Most consecutive games with at least one home run (8): Dale Long (1956), Don Mattingly (1987), and Ken Griffey Jr. (1993).
- Most consecutive games with at least one extra-base hit (14): Paul Waner (1927) and Chipper Jones (2006).

- Most consecutive games scoring at least one run (18): Red Rolfe (1939) and Kenny Lofton (2000).
- Most consecutive games with at least one RBI (17): Ray Grimes (1922).
- Most consecutive games with at least one base on balls (22): Roy Cullenbine (1947).
- Most consecutive games with at least one triple (5): John "Chief" Wilson (1912).

Consecutive-Game Batting Streaks with *More Than One Success* in Each Game

- Most consecutive games with two or more hits in each game (13): Rogers Hornsby (1923).
- Most consecutive games with three or more hits in each game (6): Jimmy Johnston (1923) and George Brett (1976).

Consecutive-Game Batting Streak with *Zero Successes* in Each Game

- Most consecutive games without striking out (115): Joe Sewell (1929). In this application success is striking out. Zero successes means not striking out in a game.

Consecutive Plate Appearance Batting Streaks

- Most consecutive plate appearances with a hit (12): Walt Dropo (1952). Pinky Higgins (1938) had 12 consecutive hits in 12 consecutive at bats (it took him 14 plate appearances to get the 12 at bats). Which of these two streaks do you think would be harder to duplicate? Why? These streaks will be revisited in the chapter problems.
- Most consecutive plate appearances getting on base (16): Ted Williams (1957).
- Most consecutive plate appearances with a base on balls (7): This record is held by five players. The most recent player to achieve this was Barry Bonds (2004).

How can we use the MF Formula for the other consecutive-game batting streaks? We rewrite the Generalized MF Formula by evaluating p as the #successes/#PA.

New Form for the Generalized MF Formula

$$D(n) = D(n-1) + \{1 - D[n-(s+1)]\} * (1-U) * U^s; \; n > s,$$
$$D(0) = D(1) = D(2) = \ldots = D(s-1) = 0, \; D(s) = U^s,$$
$$U = [1-(1-p)^j] * (1-.kl) + [1-(1-p)^{(j+1)}] * .kl,$$
p = #successes/#PA, $j.kl$ = #PA/#G, s is streak length.

As an example, we choose to look at the streak for consecutive games scoring at least one run. For this streak $s = 18$ and success is scoring a run. To use the Generalized MF Formula, we need the following information for a player: the season, the number of games played by the player in that season, the number of plate appearances in that season, and the number of runs scored in that season. In this application, $p = \#R/\#PA$, $j.kl = \#PA/\#G$, and $s = 18$ (the current record).

In the case of zero successes, we make an adjustment for the calculation of U. Success is defined as striking out in a plate appearance. As an example, we look at Joe Sewell's 1929 record of 115 consecutive games without striking out. For that season, p is the probability of striking out in a plate appearance and is equal to #SO/#PA. In 1929, Sewell struck out four times in 626 plate appearances. Therefore, the probability of striking out in a plate appearance is $p = 4/626 = .0063$. Since Sewell played in 152 games, his $\#PA/\#G = 4.12$. Hence, $j = 4$ and $.kl = .12$. Since p is the probability of striking out in a plate appearance, $(1 - p)$ is the probability of *not* striking out in a plate appearance and $(1 - p)^j$ is the probability of *not* striking out in j plate appearances. Therefore, the calculation of U is

$$\begin{aligned} U &= [(1-p)^j] * (1 - .kl) + [(1-p)^{(j+1)}] * .kl \\ &= [(1 - .0063)^4] * .88 + [(1 - .0063)^5] * .12. \end{aligned}$$

U is the probability of Joe Sewell not striking out in a typical game in the 1929 season. Try finishing the calculation for U. Are you surprised with the result?

For the exercises that follow you will have to retrieve a player's yearly statistics. You can use either the website mlb.com or baseball-reference.com to retrieve the data. The players who achieved the streaks are mentioned in Section 16.6 (Other Notable Streaks).

Exercise 16.27. Pick the streak of consecutive games with at least one home run. For that streak use $s = 8$.

a. Write the formula for U (the probability of having at least one home run in a typical game).
b. Write the Generalized CB Formula with the U from (a).
c. Write the Generalized MF Formula with the U from (a).
d. Use both of these formulas to calculate the probability that the players who actually achieved the streak duplicate the streak. Use their statistics from the year they accomplished the streak.
e. Why can't we use the Generalized SR Formula?

Exercise 16.28. Repeat Exercise 16.27 for the streak of consecutive games with at least one triple. For that streak use $s = 5$.

Exercise 16.29. Repeat Exercise 16.27 for the streak of consecutive games with at least one extra-base hit. For that streak use $s = 14$.

Exercise 16.30. For your two players, pick out their best year and use the Generalized MF Formula to calculate the following:

a. The probability that they have a consecutive-game streak of at least one run scored. Use $s = 18$.
b. The probability that they have a consecutive-game streak of at least one home run. Use $s = 8$.
c. The probability that they have a consecutive-game streak of at least one extra-base hit. Use $s = 14$.
d. The probability that they have a consecutive-plate-appearance streak of getting on base. Use $s = 16$.

16.6.1. Relating the Binomial Model to the CB, MF, and SR Formulas

Each plate appearance represents a trial for a binomial experiment. The assumptions needed to apply the (CB, MF, SR) Formulas are the same assumptions as used in the binomial model. Why is this so? The following information is needed to calculate the value for U for a player:

- For each outcome of a plate appearance, what constitutes the event success and what constitutes the event failure.
- The number of plate appearances (#PA) he had for the season.
- The number of games (#G) he played in for the season.
- The total number of successes for the season (#successes).
- Define $p = \#successes/\#PA$.
- Define $j.kl = \#PA/\#G$ as the fixed number of plate appearances for each game.
- The minimum number of successes x required in a game for the streak to continue.
- $\Pr(X \geq x$, for n trials$) = 1 - \Pr[X \leq (x - 1)]$; Pr(of at least x successes in n trials), $x \geq 1$.
- $U = [\Pr(X \geq x)$, for j trials$] * (1 - .kl) + [\Pr(X \geq x)$, for $(j+1)$ trials$] * .kl$; $p = \#successes/\#PA$ and $j.kl = \#PA/\#G$.

We now look at two cases for the minimum number of successes x. The calculations are done using the formula for binomial probabilities, studied in Chapter 11.

Case 1: If $x = 1$,

$$\begin{aligned}\Pr(X \geq 1, \text{ for } j \text{ trials}) &= 1 - \Pr[X \leq (x - 1)] \\ &= 1 - \Pr[X \leq (1 - 1)] = 1 - \Pr(X = 0) \\ &= 1 - [{}_jC_0 * p^0(1 - p)^j] \\ &= 1 - [1 * 1 * (1 - p)^j] = 1 - (1 - p)^j,\end{aligned}$$

$$\begin{aligned}\Pr[X \geq 1, \text{ for } (j+1) \text{ trials}] &= 1 - \Pr[X \leq (x-1)] \\ &= 1 - \Pr[X \leq (1-1)] = 1 - \Pr(X=0) \\ &= 1 - [{}_{(j+1)}C_0 * p^0(1-p)^{(j+1)}] \\ &= 1 - [1 * 1 * (1-p)^{(j+1)}] = 1 - (1-p)^{(j+1)}.\end{aligned}$$

Observe that these are the probabilities used in the formula for U:

$$U = [1-(1-p)^j] * (1-.kl) + [1-(1-p)^{(j+1)}] * .kl.$$

Case 2: If $x > 1$,

$$\Pr(X \geq x, \text{ for } j \text{ trials}) = 1 - \Pr[X \leq (x-1)],$$
$$\Pr[X \geq x, \text{ for } (j+1) \text{ trials}] = 1 - \Pr[X \leq (x-1)].$$

Use one of the methods provided in Chapter 11 to calculate $\Pr[X \leq (x-1)]$. Finally, we look at the special case of zero successes for j trials and $(j+1)$ trials:

$$\Pr(X=0, \text{ for } j \text{ trials}) = {}_jC_0 * p^0(1-p)^j = (1-p)^j,$$
$$\Pr[X=0, \text{ for } (j+1) \text{ trials}] = (1-p)^{(j+1)},$$
$$U = [(1-p)^j] * (1-.kl) + [(1-p)^{(j+1)}] * .kl.$$

Observe that this is the formula we used for Joe Sewell's streak of games not striking out. For this formula success in a plate appearance was striking out.

Using the above information, we are able to look at consecutive-game streaks with more than one success per game or zero successes per game.

Exercise 16.31. Pick the streak of consecutive games with at least two hits. For that streak use $s = 13$.

a. Write the formula for U (the probability of having at least two hits in a typical game).
b. With the U from (a) express the CB Formula.
c. With the U from (a) express the MF Formula.
d. Use both the Generalized CB Formula and the Generalized MF Formula to calculate the probability that the player who actually achieved the streak duplicates the streak. Use his statistics for the year he accomplished the streak.
e. Why can't we use the Generalized SR Formula?

Exercise 16.32. Pick the streak of consecutive games with at least three hits. For that streak use $s = 6$.

a. Write the formula for U.
b. With the U from (a) express the CB Formula.
c. With the U from (a) express the MF Formula.

d. Use both the Generalized CB Formula and the Generalized MF Formula to calculate the probability that the players who actually achieved the streak duplicate the streak. Use their statistics from the year they accomplished the streak.
e. Why can't we use the Generalized SR Formula?

Exercise 16.33. Pick the streak of consecutive games with zero strikeouts. For that streak use $s = 115$.

a. Write the formula for *U*.
b. With the *U* from (a) express the Generalized CB Formula.
c. With the *U* from (a) express the Generalized MF Formula.
d. Use both the Generalized CB Formula and the Generalized MF Formula to calculate the probability that the player who actually achieved the streak duplicates the streak.
e. Why *can* we use the Generalized SR Formula?
f. With the *U* from (a) express the Generalized SR Formula and calculate the probability.
g. Does your result for (f) agree with your result for (c)?

16.7. Minor League Consecutive-Game Hitting Streaks

Having examined the 56-game hitting streak of Joe DiMaggio, it is natural to ask the following question: What is the minor league record for a consecutive-game hitting streak?

16.7.1. The Longest Consecutive-Game Hitting Streaks in Minor League History (as of 2007)

Table 16.12 provides a list of the Minor League players with the longest consecutive-game hitting streaks.

When Joe DiMaggio played for the San Francisco Seals of the Pacific Coast League (PCL) in 1933, he had a 61-game hitting streak with a total of 104 hits–five games longer than his Major League record!

Getting back to the argument whether Joe DiMaggio's 56-game hitting streak was due to a spike in his talent or just chance, the fact that he achieved the 56-game hitting streak in the Major Leagues eight years after his Minor League streak seems to substantiate the argument of how special Joe DiMaggio was as a hitter. What do you think?

16.7.2. Last Player in Professional Baseball with a Hitting Streak of More than 44 Games (as of 2009)

James McOwen, a 23-year-old outfielder for the Seattle Mariner's Class A affiliate High Desert, established the sixth-longest hitting streak in Minor

Table 16.12 **Longest consecutive-game hitting streaks in Minor League history (as of 2007)**

Number	Player	Team	League	Year
69	Joe Wilhoit	Wichita	WL	1919
61	Joe DiMaggio	San Francisco	PCL	1933
55	Roman Mejias	Waco	BSL	1954
50	Otto Pahlman	Danville	III	1922
49	Jack Ness	Oakland	PCL	1915
49	Harry Chosen	Mobile	SA	1945
46	Johnny Bates	Nashville	SA	1925
43	Brandon Watson	Columbus	IL	2007
43	Eddie Marshall	Milwaukee	AA	1935
43	Howie Bedell	Louisville	AA	1961
43	Orlando Moreno	Big Spring	LHL	1947
42	Jack Lelivelt	Rochester	IL	1912
42	Herbert Chapman	Gadsden	SEL	1950
40	Frosty Kennedy	Plainville	WT	1953
38	Mitch Hilligoss	Charleston	SAL	2007
38	Hubert Mason	Binghamton	EL	1925
38	Paul Owens	Orleans	PONY	1951

League history. He is a left-handed batter with good speed and was a sixth-round draft choice in 2007. His 45-game hitting streak began on May 9, 2009, and ended on July 10, 2009. It is the longest streak in 55 years.

James, like most players with a long hitting streak, had many close calls. With the streak at 20, his game was suspended in the eighth inning with him being hitless. The next day he came up in the ninth inning and fell behind 0-2 in the count before lining a single to center. In seven of the games, he got his only hit in his last at bat. In four of these at bats he got his only hit with two strikes–one of these was a broken-bat blooper.

During his 45-game hitting streak, McOwen compiled a .398 batting average with five home runs and seven triples. Last year for the same team, McOwen batted .263 with 103 strikeouts and 44 walks in 126 games. Before starting his streak on May 10, he was batting only .270.

In his final year in college ball at Florida International University, he hit .414.

What do you think happened to account for McOwen going from a mediocre Minor League hitter to a .398 hitter during his streak? Was this just chance in his case?

Exercise 16.34. Let $\#G = 126$, $p = .263$, $j.kl = 4.00$, and $s = 45$. Calculate the U value. Using the Generalized SR Formula, find the probability that McOwen had a 45-game hitting streak. Why can we use the Generalized SR Formula?

16.7.3. The Consecutive-Game Hitting Streak of Joe Wilhoit

Joe Wilhoit had the longest hitting streak in the history of professional baseball at 69 games for the Wichita Witches of the Class A Western League. Before discussing his miraculous streak in the Minor Leagues, a brief summary of his Major League statistics is given in Table 16.13.

After spending the three previous seasons as a part-time player for the Pirates, Giants, and Braves (he played for New York in the 1917 World Series), Joe found himself back in the minors. In 1919, he signed with Seattle in the PCL. In 17 games he batted just .165. The owner of Wichita needed an outfielder, so he traded his best pitcher for Wilhoit. What a fabulous trade!

His streak began on June 14, 1919, with an infield hit. Joe was a left-handed hitter with some power. For the first 12 games of the streak he batted .510 (25 for 49). When his streak reached 29 games, a few Major League teams offered to buy his contract. If the owner had sold his contract, the streak would have ended. The interest in the streak led to an increase in attendance. The owner of the Wichita team decided that Joe was worth more by keeping him.

After the streak hit 50 games, it was getting national attention. *Sporting News* reported in early August, "Never in the annals of Western League baseball has so much interest and enthusiasm been aroused among fans as has been caused by the feat of this hard-hitting outfielder. . . . The fans talked of nothing else."

Throughout the streak there were close calls. In a few games, he got his only hit in his last at bat. On August 14, in a doubleheader at Omaha, his

Table 16.13 Major League hitting statistics for Joe Wilhoit

Season	Team	G	AB	R	H	2B	3B	HR	RBI	TB	BB	SO	SB	CS	OBP	SLG	AVG
1916	Boston Braves	116	383	44	88	13	4	2	38	115	27	45	18	—	.272	.300	.230
1917	Boston Braves	54	186	20	51	5	0	1	10	59	17	15	5	—	.321	.317	.274
1917	New York Giants	34	50	9	17	2	2	0	8	23	8	5	0	—	.424	.460	.340
1917	Pittsburgh Pirates	9	10	0	2	0	0	0	0	2	1	1	0	—	.273	.200	.200
1918	New York Giants	64	135	13	37	3	3	0	15	46	17	14	4	—	.346	.341	.274
1919	Boston Red Sox	6	18	7	6	0	0	0	2	6	5	2	1	—	.478	.333	.333
Career totals		283	782	93	201	23	9	3	73	251	75	82	28	—	.312	.321	.257

streak almost ended at 61. Joe was hitless going into the bottom of the ninth inning in a tied game. Wichita had the potential winning run thrown out at home plate. This sent the game into extra innings. Joe promptly extended his streak by hitting a walk-off home run. In the second game of the doubleheader, Joe was hitless in his first three at bats. He put down a bunt and, with Wichita trailing 9 to 2, fans say the third baseman just held the ball.

The streak ended at 69 games, but while it lasted, Wilhoit hit .512 (153 for 299), including 24 doubles, nine triples, and five home runs. He walked 34 times and had two or more hits in 50 games. Therefore, in 19 games he had just one hit. How does this compare with DiMaggio and his streak?

Before being sold to the Boston Red Sox, Wilhoit batted .422 in 128 games for Wichita. He led the league with 222 hits. After joining the Red Sox, he batted .333 (6 for 18). He played in the outfield with a 24-year-old named Babe Ruth. His time with the Red Sox ended when Boston released him in 1920. He spent the next three years in the minors, but he would never return to the majors. In 1923, at the age of 32, he batted .360 in 172 games with Salt Lake City in the PCL. "The Wichita Wonder" realized that because of his age he was no longer a Major League prospect and so retired from baseball. He tragically died at the age of 44 from a brief illness.

The question asked was, what did he do to raise himself from a below average Major League hitter on a major downswing to achieving the most celebrated record-setting hitting streak? It was reported that the owner of the Wichita team suggested that he use a much lighter bat. He followed the owner's suggestion and used this new bat during the streak and for the rest of the year. Could the bat have made such a difference? Finally, we ask the same question as we did for Joe DiMaggio. Did Joe Wilhoit achieve his 69-game hitting streak as a result of a sudden increase in ability, or was it just a random chance occurrence?

Exercise 16.35. Compare the statistics of Wilhoit during his streak with the statistics of DiMaggio during his streak. What are the differences and similarities?

Exercise 16.36. The hitting statistics for Joe Wilhoit for his 1919 season with the Wichita Witches are given as follows:

#G = 128, #AB = 526, #BB = 65, #H = 222, #HBP = 0.

Use the SR Formula to calculate the probability of Wilhoit having a 56-game hitting streak at some point in the 1919 season. (Hint: You must first calculate U using $s = 56$.)

16.8. What Will It Take for a Player to Equal or Exceed Joe DiMaggio's Streak?

Since Joe DiMaggio achieved the 56-game streak twice, once in the majors and once in the minors, the batting characteristics of Joe DiMaggio might help answer the question.

Some observations about Joe DiMaggio's 56-game hitting streak, from Table 1.1 and Table 3.6, are as follows:

- In 75% of the 56 games, Joe faced the same starting pitcher in all his plate appearances. This would never happen in today's game.
- In over 60% of the 56 games, Joe got just one hit.
- Joe had 193 hits in 541 at bats for that year, giving him a batting average of .357.
- During his streak, Joe had 91 hits in 223 at bats, giving him an average of .408.
- For 1941, Joe had 76 walks and struck out only 13 times.
- He faced five Hall of Fame pitchers during the streak.
- He averaged 3.98 at bats during the streak. Of those games, there were 11 games with three at bats and three games with two at bats.

From DiMaggio's streak results, the next player to equal Joe's streak should satisfy as many of the following properties as possible:

- Average close to four at bats per game. The more at bats per game, the better the chance of a hit. This means that the player should have a low BBA (base-on-balls average).
- Have a very low strikeout average. By making contact with the ball, at least there is a chance of a hit.
- Be able to bat over .400 for a streak of 56 games.
- Play in a home park that favors hitters.
- Possess enough speed to beat out infield hits.
- Be able to hit with power, creating hard-hit line drives.
- Bat left-handed or be a switch-hitter.

Since 1913, the two players closest to Joe DiMaggio's streak were Pete Rose in 1978 (44) and George Sisler in 1922 (41). George Sisler also had a 34-game streak in 1925. Since 2000, the two players that meet several of the above properties are Ichiro Suzuki and Albert Pujols, although Albert is a right-handed hitter. Suzuki had five seasons where he had at least one 20-game hitting streak. The spreadsheet in Table 16.14 compares DiMaggio (1941) with Sisler (1922, 1925), Rose (1978), Pujols (2003), and Suzuki (2001, 2004, 2006, and 2007).

Table 16.14 **Comparison of DiMaggio (1941) with Sisler (1922, 1925), Rose (1978), Pujols (2003), and Suzuki (2001, 2004, 2006, and 2007)**

Player	Year	#G	#AB	#H	#BB	#SO	BBA	SOA	#AB/#G	AVG	Length
Joe DiMaggio	1941	139	541	193	76	13	0.123	0.024	3.89	0.357	56
Pete Rose	1978	159	655	198	62	30	0.086	0.046	4.12	0.302	44
George Sisler	1922	142	586	246	49	14	0.077	0.024	4.13	0.420	41
George Sisler	1925	150	649	224	27	24	0.040	0.037	4.33	0.345	34
Ichiro Suzuki	2001	157	692	242	30	53	0.042	0.077	4.41	0.350	21
Ichiro Suzuki	2001	157	692	242	30	53	0.042	0.077	4.41	0.350	23
Ichiro Suzuki	2004	161	704	262	49	63	0.065	0.089	4.37	0.372	21
Ichiro Suzuki	2006	161	695	224	49	71	0.066	0.102	4.32	0.322	20
Ichiro Suzuki	2007	161	678	238	49	77	0.067	0.114	4.21	0.351	25
Albert Pujols	2003	157	591	212	79	65	0.118	0.110	3.76	0.359	30

Exercise 16.37. Make a table similar to Table 16.14 for the following modern-day players: Paul Molitor (1987), Luis Castillo (2002), Jimmy Rollins (2005), Chase Utley (2006), Chipper Jones (2008), and Ryan Zimmerman (2009).

Exercise 16.38. Compare the statistics of the players in Exercise 16.37 with those in Table 16.14.

16.9. Joe DiMaggio's Streak Halted at 56 Games

Joe DiMaggio's streak ended in a night game at Cleveland's Municipal Stadium on July 17, 1941, in front of a record 67468 roaring fans. In his four plate appearances, Joe walked once and hit three ground balls. The first ground ball was a rocket hit down the third base line, and it took a great backhand play by Cleveland's Ken Keltner to deny Joe a hit. When Joe was asked what upset him the most about the streak being broken, Joe quipped, "If I just hit in that 57th game I could have signed a promotional deal with the Heinz 57 Varieties company for 10,000 dollars."

Joe walked on a 3-2 pitch in the fourth inning. In the seventh, he sent another rocket to Keltner and was thrown out at first by a couple of steps. In his final plate appearance, he slapped a routine grounder to shortstop Lou Boudreau, who barehanded it and turned the double play.

Joe's streak began more than two months before, on May 15, when he blooped a single to right field off of the White Sox's Ed Smith. During the streak, he faced every American League team at least once.

Joe's greatness showed when he promptly started a new hitting streak, which lasted 16 games. This supports the argument of a spike in Joe's talent.

Without Keltner's great play, though, the streak might have reached 73 games. As mentioned earlier, Joe accomplished a 61-game hitting streak while playing for the Seals in the PCL.

16.10. Further Reading

Those students who are interested in reading a more in-depth study of batting streaks are encouraged to get the book *Mathematics and Sports* published by the Mathematics Association of America in 2010. The author, with the help of a math major, has written Chapter 4 in that book, entitled "Streaking: Finding the Probability for a Batting Streak."

In particular, if you are interested in a book just on the 56-game hitting streak, I recommend *56: Joe Dimaggio and the Last Magic Number in Sports* by Kostya Kennedy. This book follows the streak in 1941 and its relationship to the buildup for World War II.

Chapter Summary

Three functions are provided to calculate the probability of several players duplicating various batting streaks in baseball. The three functions are the CB Formula, the MF Formula, and the SR Formula. The CB Formula was shown to be mathematically incorrect, the MF Formula is a recursive formula, and the SR Formula is a piecewise function composed of polynomials.

All three functions have for their domains the first n games of a player's season. The CB Formula uses the probability product rule for independent events even though the events are not independent. The CB Formula should not be used. The MF Formula requires several steps to find the probability. It is cumbersome to use. The SR Formula is much easier to use. However, the SR Formula is only correct when n is less than or equal to $3s+1$, where s is the length of the streak. In order to use the SR Formula for small streaks where $n>3s+1$, additional polynomial pieces would have to be added. Since a player's season consists of 162 games or less, for such streaks as Joe DiMaggio's 56-game hitting streak and Ted Williams's 84-game on-base streak, the SR Formula is equivalent to the MR Formula.

Two categories of streaks are discussed. The first category is streaks based on games; the second category is streaks based on plate appearances. Ted Williams had record streaks of both types. He holds the record for consecutive games getting on base at least one time (84 games in 1949). He also holds the modern-day record of consecutive plate appearances getting on base (16 times in 1957).

Much of the chapter compares Joe DiMaggio's 56-game hitting streak with Ted Williams's 84-game on-base streak. The probability of achieving each of these two streaks is calculated for several great hitters of the past and present. The results indicated that DiMaggio's 56-game hitting streak would be the harder one to duplicate.

The chapter also looks at batting streaks involving more than one success in each game or having zero successes in each game. The probability of achieving such streaks as Rogers Hornsby's 13-game streak of getting at least two hits in each game and Joe Sewell's 115-game streak of not striking out are calculated.

Various simulations of Joe DiMaggio's 56-game hitting streak, done by other researchers, are presented. The results of the simulations are studied using frequency distributions and various descriptive measures. The inferential technique of confidence intervals is applied to the sample results. Based on DiMaggio's 1941 batting statistics, both a 95% confidence interval for the population mean length of a hitting streak and a 95% confidence interval estimate for the population proportion for the occurrence of a 56-game hitting streak are calculated.

The Binomial Formula can be used in the calculations for the MF and SR Formulas. For example, in a batting streak each game is a trial, the total number of games a player plays is the number of trials, and success is achieving the desired outcome or outcomes in each game of the streak. Since the MF and SR Formulas require independence between games and the same probability of success for each game, the Binomial Formula can be used to calculate the value U used in both formulas.

Another player named Joe is discussed in the chapter. That is Joe Wilhoit, who holds the professional baseball record of a 69-game hitting streak. He accomplished this while playing for the Wichita Witches in the Class A Western League.

The chapter ends with a discussion of the day Joe DiMaggio's 56-game streak ended.

CHAPTER PROBLEMS

1. Set up a spreadsheet formula in Excel to compute the MF Formula, the CB Formula, and the SR Formula.
2. Using these three formulas, compute the probabilities of your two players achieving a 56-game hitting streak and an 84-game on-base streak. For each player, choose what you consider their best year to achieve each of these feats. Compare your results for the two players.
3. Repeat problem 2, but for achieving the streak of at least one RBI in 17 consecutive games. You cannot use the Generalized SR Formula.

4. Which of the two streaks, Joe's streak or Ted's streak, is harder to duplicate? Do not use the approachability argument given in this chapter. You can use any other information in this chapter to answer this question.
5. Choose one of the streaks in Section 16.6 that you have not done before. Establish the three generalized formulas for your chosen streak. Discuss whether the Generalized SR Formula can be used.
6. Choose an active player whom you believe would be the best candidate to duplicate DiMaggio's 56-game hitting streak, and choose a year for that player. Use the SR Formula to calculate the probability of that player duplicating Joe's streak.
7. Repeat problem 6, but this time use the 84-game on-base streak of Williams.

How would you adjust the Generalized MF Formula for consecutive-plate-appearance or consecutive-at-bat streaks? Hint: Each plate appearance or at bat results in either a success or a failure, with p being the probability of success. You can think of a plate appearance or an at bat as a game with only one plate appearance. What is $j.kl$ equal to? What does U become?

The next three questions deal with streaks involving plate appearances and at bats. For these streaks each individual plate appearance or at bat is a treated as a game. In particular, we look at Walt Dropo's streak of 12 consecutive hits in 12 consecutive plate appearances and his streak of 12 consecutive hits in 12 consecutive at bats. Clearly, the plate appearance streak implies the at bat streak since each plate appearance that results in a hit is an at bat. Can you give an example of a streak with 12 consecutive hits in 12 consecutive at bats that would not satisfy 12 consecutive hits in 12 consecutive plate appearances? Both these streaks occurred in 1952 with the same 12 consecutive plate appearances.

For 1952, Walt Dropo had 633 plate appearances and 591 at bats with 163 hits. Success is getting a hit. Find p for both of Walt Dropo's streaks in 1952. Show $U = p$ in each case.

8. Use both the Generalized MF Formula and the Generalized SR Formula to find the probability of Walt Dropo achieving his at-bat streak in 1952. Take both calculations to 10 decimal places. Why do these two formulas give a different result? Calculate the difference between the two formulas. Which formula is correct?
9. Redo problem 8 for Walt Dropo's streak of 12 consecutive hits in 12 consecutive plate appearances.
10. From your results in the above two problems, which streak was more difficult for Walt Dropo to achieve? Do the probability results make sense? Why or why not?

Chapter 17

Mission Impossible: Batting .400 for a Season

17.1. Introduction

The National League was created in 1876, the American League in 1901. From 1876 to 1901, the rules concerning the number of balls for a walk, the distance of the pitching mound from home, and whether a foul ball was considered a strike were in flux. Before 1913, a player's yearly number of strikeouts was not accurately recorded. For these reasons, we will only look at batting averages for players from 1913 to 2007.

Prior to 1876, a player's batting performance was evaluated by dividing his total number of hits for a season by the total number of games he played in. The unfairness of this method was clear. With the creation of the National League, the batting average was invented. Instead of dividing the number of hits by the number of games, the number of hits for the season was divided by the number of at bats for that season.

From 1876 until today, the batting average has been considered the "gold standard" for judging a player's performance. The player with the highest batting average, at the end of a season in each league, is crowned the "Batting Champion." If a player, at the end of a season, leads his league in batting average, home runs, and runs batted in, he is said to have won the "Triple Crown."

Over the years, many other batting statistics have been established. Many baseball people judge the importance of a batting statistic by the number of runs that statistic creates for their team. A batting average has two major weaknesses: it gives no credit for a player reaching base by a walk, and it treats all types of hits (singles, doubles, triples, and home runs) as equally important. Today, on-base percentage, on-base plus slugging, and runs created are considered more important in terms of run creation.

Regardless of its shortcomings, the batting average is still the first statistic mentioned when describing batting performance. Attaining a batting average of at least .400 has the same importance as hitting at least 60 home runs (before 1998) or having a hitting streak of at least 56 games.

At the end of this chapter, you will be asked to answer the following question: Based on a minimum of 512 plate appearances in a year, will a player ever have a batting average greater than or equal to .400 again? You can answer yes or no, but your answer must be justified by using the statistical techniques presented in this book.

In Major League Baseball, since 1913 (based on a minimum of 3.1 plate appearances per scheduled game in a season) only six different players have batted at least .400 in a single season, for a total of nine times. These players are listed in Table 17.1.

Rogers Hornsby batted over .400 three times and George Sisler twice. Harry Heilmann, Ty Cobb, Bill Terry, and Ted Williams each did it once.

Of the six players, Cobb, Williams, Sisler, and Terry batted left-handed, while Heilmann and Hornsby batted right-handed. As there are many more right-handed pitchers than left-handed pitchers and a left-handed batter is a step closer to first base, baseball experts feel that it is an advantage to bat left-handed.

Except for Ted Williams, all the .400 seasons were between 1920 and 1930. Because of the many high batting averages, the era from 1920 to 1930 is often referred to as "the lively ball era."

In this chapter, batting .400 will be analyzed by studying the following:

- Two formulas that relate batting average (AVG) to in-play batting average (IPBA) and strikeout average (SOA).
- The variables AVG, IPBA, SOA, and the in-play home run average (IPHR) during the years from 1920 to 1930 (the lively ball era). IPHR can also be symbolized as IPHRA.

Table 17.1 **Major League players who batted at least .400 for a season (since 1913)**

Year	Player	League	Team	Average
1924	Rogers Hornsby	NL	St Louis	.424
1922	George Sisler	AL	St Louis	.420
1920	George Sisler	AL	St Louis	.407
1941	Ted Williams	AL	Boston	.406
1923	Harry Heilmann	AL	Detroit	.403
1925	Rogers Hornsby	NL	St Louis	.403
1922	Rogers Hornsby	NL	St Louis	.401
1922	Ty Cobb	AL	Detroit	.401
1930	Bill Terry	NL	New York	.401

- The variables AVG, IPBA, SOA, and the IPHR for the years 1994–2007 (the home run era).
- The batting characteristics of the players who either batted .400 since 1913 or batted close to .400 after 1941.
- The batting characteristics of Ted Williams (the last player to bat .400).
- The 36 players who for the years 2000 to 2007 had a high batting average and an IPBA of at least .400, for a total of 56 times.
- Two methods for comparing baseball statistics between two or more players.

Ted Williams was the last Major League player to bat over .400. The last *player* in a professional baseball league to do so was Joanne Weaver, in 1954, in the Women's Professional Baseball League. The last man to accomplish this feat was Artie Wilson, in 1948, in the Negro Leagues.

The tools used for the analysis come from descriptive and inferential statistics, which were covered in Chapters 1–15 in this book.

We discuss what it will take for a player, in the future, to have a batting average of at least .400 for a season.

17.2. The Statistics AVG, IPBA, SOA, IPHR, and BBA

We review a few definitions from Tables 3.2 and 3.3.

AB: An official at bat is credited if a batter either puts a ball in play (an in-play at bat or IPAB) or strikes out (SO). Recall that BB, HBP, SF, and SH are *not* considered an at bat. IPAB = AB − {SO}.
PA: An official plate appearance is credited for an AB, a BB, a HBP, or a SF.
H: A hit consists of the following outcomes of a plate appearance: a single (1B), double (2B), triple (3B), or home run (HR).
#AB: Denotes the number of at bats.
#IPAB: #AB − #SO, denoting the number of at bats resulting in a ball put in play. By definition, home runs are included; a SF is excluded.
#SO: Denotes the number of at bats that result in a strikeout.
AVG or BA: #H/#AB.
IPBA (in-play batting average): #H/#IPAB = #H/(#AB − #SO).
SOA (strikeout average): #SO/#AB.
IPHR (in-play home run average): #HR/#IPAB = #HR/(#AB − #SO).
BBA (base-on-balls average): #BB/#PA = #BB/(#AB + #BB + #HBP + #SF).

Throughout this chapter, AVG, IPBA, SOA, and IPHR are referred to as the "*four statistics.*"

Using algebra, a player's batting average can be broken down into the product of the IPBA and (1 − SOA):

$$\begin{aligned}\text{AVG} &= \#\text{H}/\#\text{AB} = (\#\text{H}/\#\text{AB}) * (\#\text{IPAB}/\#\text{IPAB}) \\ &= (\#\text{H}/\#\text{IPAB}) * (\#\text{IPAB}/\#\text{AB}) \\ &= \text{IPBA} * (\#\text{IPAB}/\#\text{AB}) = \text{IPBA} * [(\#\text{AB} - \#\text{SO})/\#\text{AB}] \\ &= \text{IPBA} * [(\#\text{AB}/\#\text{AB} - \#\text{SO}/\#\text{AB})] \\ &= \text{IPBA} * (1 - \text{SOA}). \end{aligned} \tag{17.1}$$

Example 17.1. In 2009, Seattle's Ichiro Suzuki had the following batting line:

PA	AB	H	SO	HR	BB	HBP	SF	SH
678	639	225	71	11	32	4	3	0

AVG = 225/639 = .352,

IPBA = 225/(639 − 71) = .396,

SOA = 71/639 = .111,

AVG = IPBA * (1 − SOA) = .396 * (1 − .111) = .396 * .888 = .352 (rounded),

IPHR = 11/(639 − 71) = .019,

BBA = 32/(639 + 32 + 4 + 3) = .047,

IPBA − AVG = [225/(639 − 71)] − (225/639) = .396 − .352 = .044 (rounded),

IPBA * SOA = 396 * .111 = .044.

Observe that

$$\begin{aligned}(\text{IPBA} - \text{AVG}) &= \text{IPBA} - [\text{IPBA} * (1 - \text{SOA})] = \text{IPBA} - (\text{IPBA} - \text{IPBA} * \text{SOA}) \\ &= \text{IPBA} - \text{IPBA} + \text{IPBA} * \text{SOA} \\ &= \text{IPBA} * \text{SOA}. \end{aligned} \tag{17.2}$$

The outcome of an at bat is either putting the ball in play or striking out. The IPBA of a player measures a player's batting average restricted to those at bats where the ball is put in play (a home run is considered putting a ball in play). The IPBA and the SOA are both determined by a combination of skill and chance. Baseball aficionados feel that chance is more important in determining the IPBA than the SOA. The skill component of the IPBA is a player's ability to direct a batted ball to a place where there are no fielders. The skill component of the SOA is the ability of a player to put the ball in play before striking out.

If a player has no strikeouts, his AVG will equal his IPBA. Otherwise, his AVG is strictly less than his IPBA. Of course, all players strike out. As a consequence of Formula (17.1), the two statistics IPBA and SOA determine a player's AVG. If a player can find a way to increase his IPBA and decrease his SOA at the same time, he will increase his AVG. The following questions, concerning various samples, are explored throughout this chapter:

- How well does IPBA predict AVG?
- How well does SOA predict AVG?
- How well does IPHR predict IPBA?

- How well does IPHR predict SOA?
- How well does each of the statistics IPBA, SOA, and IPHR predict (IPBA – AVG)?

Substituting .400 for AVG in Formula (17.1) gives the following equation:

$.400 = IPBA * (1 - SOA).$

There are infinite choices for pairs of numbers for a player's IPBA and SOA that satisfy the above equation. Algebraically, we could choose .800 for the IPBA and .500 for the SOA. The above equation would be satisfied. However, no player in the history of baseball has come close to these numbers. The pairs of numbers selected are restricted to what is possible with respect to the physical experiment of batting in the Major Leagues.

By looking at players who *have* achieved a .400 AVG or batted *close* to .400, we will be able to find realistic numbers for the IPBA and SOA. The question is, because of the way baseball is played today, how likely is it for a current player to achieve these numbers for their IPBA and SOA?

Exercise 17.1. Based on your opinion, starting with 100% for each of the statistics IPBA, SOA, and IPHR, assign a percent to ability and a percent to chance. Give reasons for your decision.

17.3. The Two Leagues and the Four Statistics AVG, IPBA, SOA, and IPHR for the Years 1913–2007

Table 17.2 provides these four statistics for each league for the years from 1913 to 2007. Each statistic is calculated for each league by considering the outcomes of all at bats for all players in that league for that year.

Table 17.3 gives, for the National League, the top 15 yearly AVGs sorted in descending order and the bottom 15 yearly AVGs sorted in ascending order.

The following are observations from Table 17.3:

- Except for 1920, all the years from 1920 to 1930 belong to the list of the highest batting averages.
- Except for 1913, 1918, and 1919, all the years from 1913 to 1919 belong to the list of the lowest batting averages.

Exercise 17.2. Calculate the mean for each column in Table 17.3. Determine which of IPBA, SOA, or IPHR has the greatest effect on AVG. Determine which has the least effect on AVG.

Figure 17.1 is a time-series graph for the four statistics for the National League (1913–2007). Figure 17.2 is a time-series graph for the four statistics for the American League (1913–2007).

Table 17.2 AVG, IPBA, SOA, and IPHR for the years 1913–2007 for the two leagues

	NL				AL			
Year	AVG	IPBA	SOA	IPHR	AVG	IPBA	SOA	IPHR
1913	0.262	0.295	0.111	0.008	0.256	0.290	0.117	0.005
1914	0.251	0.283	0.114	0.007	0.248	0.283	0.125	0.004
1915	0.248	0.280	0.116	0.006	0.248	0.282	0.121	0.005
1916	0.247	0.280	0.118	0.007	0.248	0.281	0.115	0.004
1917	0.249	0.280	0.108	0.005	0.248	0.276	0.103	0.004
1918	0.254	0.278	0.086	0.005	0.254	0.278	0.090	0.003
1919	0.258	0.282	0.088	0.006	0.268	0.296	0.095	0.007
1920	0.270	0.295	0.086	0.007	0.283	0.310	0.087	0.010
1921	0.289	0.315	0.080	0.012	0.292	0.319	0.084	0.012
1922	0.292	0.317	0.079	0.013	0.285	0.311	0.085	0.014
1923	0.286	0.310	0.079	0.014	0.283	0.309	0.086	0.011
1924	0.283	0.308	0.080	0.013	0.290	0.314	0.077	0.010
1925	0.292	0.316	0.079	0.016	0.292	0.316	0.078	0.014
1926	0.280	0.304	0.080	0.011	0.281	0.307	0.083	0.011
1927	0.282	0.307	0.082	0.012	0.286	0.310	0.079	0.011
1928	0.281	0.306	0.081	0.016	0.281	0.308	0.088	0.013
1929	0.294	0.320	0.081	0.019	0.284	0.310	0.085	0.015
1930	0.303	0.333	0.088	0.022	0.288	0.318	0.095	0.017
1931	0.277	0.304	0.090	0.013	0.278	0.307	0.093	0.015
1932	0.276	0.303	0.088	0.016	0.277	0.305	0.093	0.018
1933	0.266	0.290	0.083	0.012	0.273	0.300	0.091	0.016
1934	0.279	0.309	0.097	0.017	0.279	0.310	0.100	0.018
1935	0.277	0.306	0.094	0.017	0.280	0.308	0.092	0.017
1936	0.278	0.308	0.096	0.015	0.289	0.319	0.092	0.019
1937	0.272	0.304	0.107	0.016	0.281	0.313	0.103	0.021
1938	0.267	0.296	0.096	0.016	0.281	0.312	0.100	0.023
1939	0.272	0.302	0.099	0.017	0.279	0.310	0.101	0.021
1940	0.264	0.293	0.101	0.018	0.271	0.305	0.110	0.023
1941	0.258	0.288	0.103	0.016	0.266	0.297	0.103	0.019
1942	0.249	0.277	0.100	0.014	0.257	0.285	0.099	0.014
1943	0.258	0.285	0.096	0.011	0.249	0.279	0.107	0.013
1944	0.261	0.287	0.092	0.015	0.260	0.289	0.099	0.012
1945	0.265	0.291	0.090	0.015	0.255	0.284	0.101	0.011
1946	0.256	0.286	0.106	0.015	0.256	0.292	0.124	0.018
1947	0.265	0.297	0.107	0.023	0.256	0.287	0.110	0.018
1948	0.261	0.294	0.112	0.023	0.266	0.296	0.102	0.019
1949	0.262	0.294	0.107	0.025	0.263	0.294	0.105	0.021
1950	0.261	0.296	0.118	0.029	0.271	0.303	0.107	0.026
1951	0.260	0.292	0.111	0.027	0.262	0.294	0.108	0.022
1952	0.253	0.289	0.125	0.025	0.253	0.288	0.122	0.021
1953	0.266	0.304	0.124	0.032	0.262	0.297	0.116	0.023

Table 17.2 **AVG, IPBA, SOA, and IPHR for the years 1913–2007 for the two leagues,** ***continued***

	NL				AL			
Year	AVG	IPBA	SOA	IPHR	AVG	IPBA	SOA	IPHR
1954	0.265	0.302	0.121	0.030	0.257	0.293	0.122	0.022
1955	0.259	0.297	0.130	0.035	0.258	0.297	0.129	0.026
1956	0.256	0.296	0.136	0.034	0.260	0.302	0.138	0.030
1957	0.260	0.304	0.143	0.032	0.255	0.296	0.138	0.028
1958	0.262	0.307	0.147	0.033	0.254	0.297	0.145	0.030
1959	0.260	0.308	0.154	0.032	0.253	0.296	0.145	0.030
1960	0.255	0.304	0.162	0.029	0.255	0.298	0.143	0.030
1961	0.262	0.311	0.157	0.034	0.256	0.301	0.152	0.033
1962	0.261	0.311	0.163	0.031	0.255	0.301	0.155	0.033
1963	0.245	0.297	0.174	0.027	0.247	0.297	0.168	0.033
1964	0.254	0.305	0.167	0.026	0.247	0.302	0.180	0.034
1965	0.249	0.302	0.174	0.029	0.242	0.294	0.177	0.031
1966	0.256	0.308	0.168	0.030	0.240	0.292	0.176	0.031
1967	0.249	0.301	0.172	0.024	0.236	0.289	0.184	0.027
1968	0.243	0.294	0.173	0.020	0.230	0.280	0.180	0.025
1969	0.250	0.304	0.177	0.027	0.246	0.295	0.165	0.030
1970	0.258	0.312	0.172	0.031	0.250	0.300	0.167	0.032
1971	0.252	0.300	0.160	0.025	0.247	0.294	0.161	0.027
1972	0.248	0.298	0.167	0.026	0.239	0.286	0.165	0.023
1973	0.254	0.303	0.159	0.028	0.259	0.305	0.149	0.028
1974	0.255	0.301	0.151	0.023	0.258	0.302	0.144	0.024
1975	0.257	0.302	0.148	0.022	0.258	0.302	0.145	0.026
1976	0.255	0.298	0.146	0.020	0.256	0.297	0.139	0.020
1977	0.262	0.311	0.157	0.029	0.266	0.312	0.145	0.030
1978	0.254	0.300	0.152	0.023	0.261	0.301	0.133	0.025
1979	0.261	0.307	0.150	0.025	0.270	0.311	0.132	0.030
1980	0.259	0.305	0.149	0.022	0.269	0.310	0.133	0.027
1981	0.255	0.299	0.145	0.019	0.256	0.296	0.136	0.024
1982	0.258	0.305	0.155	0.023	0.264	0.307	0.140	0.031
1983	0.255	0.305	0.164	0.025	0.266	0.309	0.141	0.028
1984	0.255	0.306	0.166	0.023	0.264	0.310	0.149	0.030
1985	0.252	0.301	0.162	0.026	0.261	0.308	0.152	0.033
1986	0.253	0.308	0.177	0.028	0.262	0.315	0.169	0.036
1987	0.261	0.316	0.176	0.033	0.265	0.320	0.173	0.041
1988	0.248	0.298	0.168	0.023	0.259	0.309	0.160	0.029
1989	0.246	0.298	0.173	0.025	0.261	0.310	0.160	0.027
1990	0.256	0.309	0.169	0.028	0.259	0.310	0.165	0.028
1991	0.250	0.303	0.175	0.027	0.260	0.312	0.167	0.030
1992	0.252	0.304	0.173	0.023	0.259	0.308	0.158	0.027
1993	0.264	0.319	0.172	0.031	0.267	0.320	0.167	0.032

(continued)

Table 17.2 AVG, IPBA, SOA, and IPHR for the years 1913–2007 for the two leagues, *continued*

	NL				AL			
Year	AVG	IPBA	SOA	IPHR	AVG	IPBA	SOA	IPHR
1994	0.267	0.327	0.184	0.034	0.273	0.330	0.174	0.039
1995	0.263	0.326	0.193	0.034	0.270	0.327	0.174	0.038
1996	0.262	0.327	0.196	0.036	0.277	0.337	0.178	0.042
1997	0.263	0.328	0.198	0.035	0.271	0.333	0.187	0.039
1998	0.262	0.326	0.197	0.036	0.271	0.333	0.184	0.039
1999	0.268	0.332	0.193	0.040	0.275	0.334	0.179	0.041
2000	0.266	0.330	0.195	0.042	0.276	0.336	0.178	0.042
2001	0.261	0.328	0.203	0.042	0.267	0.328	0.186	0.039
2002	0.259	0.322	0.195	0.037	0.264	0.323	0.183	0.039
2003	0.262	0.324	0.192	0.038	0.267	0.324	0.176	0.039
2004	0.263	0.326	0.195	0.040	0.270	0.331	0.185	0.041
2005	0.262	0.324	0.192	0.036	0.268	0.325	0.176	0.038
2006	0.265	0.329	0.196	0.040	0.275	0.336	0.181	0.040
2007	0.266	0.330	0.195	0.038	0.271	0.333	0.188	0.035

Table 17.3 The 15 highest and 15 lowest National League yearly AVGs for the years 1913–2007

	Fifteen highest NL AVGs					Fifteen lowest NL AVGs			
Year	AVG	IPBA	SOA	IPHR	Year	AVG	IPBA	SOA	IPHR
1930	0.303	0.333	0.088	0.022	1968	0.243	0.294	0.173	0.020
1929	0.294	0.320	0.081	0.019	1963	0.245	0.297	0.174	0.027
1922	0.292	0.317	0.079	0.013	1989	0.246	0.298	0.173	0.025
1925	0.292	0.316	0.079	0.016	**1916**	0.247	0.280	0.118	0.007
1921	0.289	0.315	0.080	0.012	**1915**	0.248	0.280	0.116	0.006
1923	0.286	0.310	0.079	0.014	1988	0.248	0.298	0.168	0.023
1924	0.283	0.308	0.080	0.013	1972	0.248	0.298	0.167	0.026
1927	0.282	0.307	0.082	0.012	1942	0.249	0.277	0.100	0.014
1928	0.281	0.306	0.081	0.016	1967	0.249	0.301	0.172	0.024
1926	0.280	0.304	0.080	0.011	1965	0.249	0.302	0.174	0.029
1934	0.279	0.309	0.097	0.017	**1917**	0.249	0.280	0.108	0.005
1936	0.278	0.308	0.096	0.015	1991	0.250	0.303	0.175	0.027
1935	0.277	0.306	0.094	0.017	1969	0.250	0.304	0.177	0.027
1931	0.277	0.304	0.090	0.013	**1914**	0.251	0.283	0.114	0.007
1932	0.276	0.303	0.088	0.016	1992	0.252	0.304	0.173	0.023

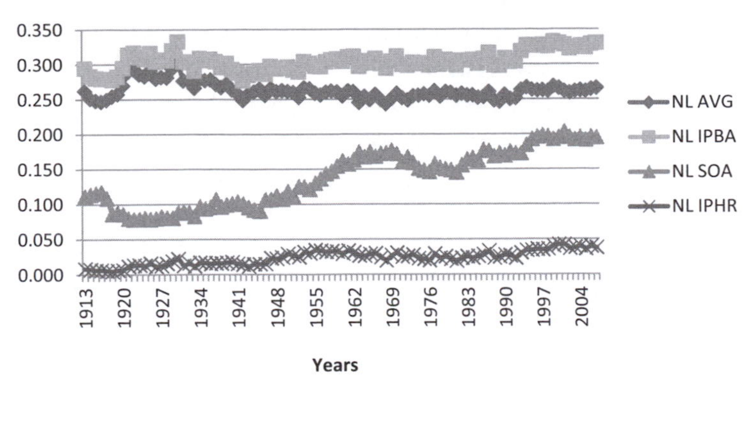

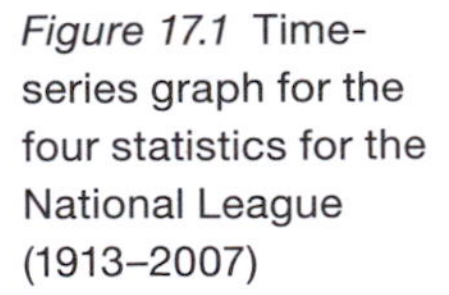

Figure 17.1 Time-series graph for the four statistics for the National League (1913–2007)

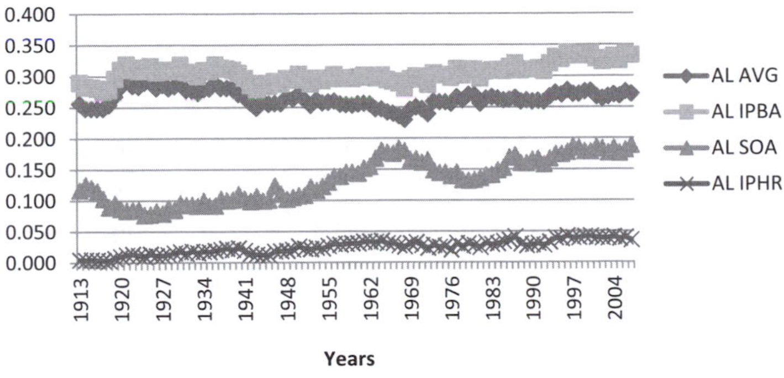

Figure 17.2 Time-series graph for the four statistics for the American League (1913–2007)

Table 17.4 Mean and standard deviation for AVG, IPBA, SOA, and IPHR for each league from 1913 to 2007

	Mean				Standard deviation			
League	AVG	IPBA	SOA	IPHR	AVG	IPBA	SOA	IPHR
National	.262	.304	.138	.024	.012	.013	.040	.010
American	.264	.305	.134	.024	.013	.015	.035	.010

Table 17.4 provides the mean and standard deviation for AVG, IPBA, SOA, and IPHR for each league for the 95 years from 1913 to 2007. The calculation of the mean and standard deviation will be done by using the formulas presented in Chapter 2, with the data coming from each of the eight columns in Table 17.2. For example, the mean for the NL AVGs is found by summing the NL AVGs column in Table 17.2 and then dividing that total by 95. The formula used for the standard deviation is the one for sample standard deviation.

17.3.1. The Four Statistics for the Years 1913–2007

For all four statistics, the time-series graphs for the NL (Fig. 17.1) and AL (Fig. 17.2) are remarkably similar. Table 17.4 shows that the means and standard deviations are similar for both leagues. This suggests that hitting and pitching talent was evenly distributed between the two leagues from 1913 to 2007.

Since AVG = IPBA * (1 − SOA), a study of AVG should involve the two variables IPBA and SOA. We select the National League (NL), but note that what follows also applies to the American League.

We begin by looking at the IPBA for the NL as shown in Table 17.2. For 1920–1930, most of the IPBAs were between .305 and .320. For 1931–1951, most of the IPBAs were between .290 and .310. For 1952–1959, most of the IPBAs were between .289 and .308. For 1960–1993, most of the IPBAs were between .300 and .315. Starting in 1994, the IPBAs were consistently between .324 and .332. Figure 17.1 illustrates that until 1994 the IPBAs stayed about the same, with a slight increase for 1994–2007.

We now turn our attention to the SOA for the NL as shown in Table 17.2. For 1920–1930, the SOAs ranged between .079 and .086. For 1931–1951, the SOAs ranged between .083 and .118. For 1952–1959, the SOAs ranged between .121 and .154. For 1960–1993, the SOAs ranged between .145 and .177. For 1994–2007, the SOAs ranged between .184 and .203.

Figure 17.1 shows a widening gap between the AVG and the IPBA as the years approached 2007. This can be attributed to the sharp increase in SOA as the years approached 2007.

But what effect did these two variables have on the AVG for the years 1913–2007? For 1920–1930, most of the NL AVGs were between .285 and .303. In fact, 1930 is the only year in Major League history where one of the leagues had an AVG that exceeded .300. For 1931–1969, the NL AVGs declined from the high .270s to around .250. For 1970–1992, they stayed in the mid-.250s. For 1994–2007, they rose to the mid-.260s. Even though the years from 1994 to 2007 had the highest IPBAs, their high SOAs kept the AVGs down. Figure 17.1 shows that even though the IPBAs were considerably higher in 1994–2007, compared with the years 1920–1930, the AVGs for 1920–1930 were higher.

The variation of the SOA is shown through the standard deviation. Looking at Table 17.4, the standard deviation for the SOA, for both leagues, is over 3 times as high as the standard deviation for AVG and for IPBA. This indicates a greater variability for SOA. Can you see this by looking at the graphs in Figures 17.1 and 17.2?

The real difference in the behavior of the four statistics is found when comparing different eras (time periods). The beginning and ending years for each era are open to debate.

The period 1900–1919 was called the "dead ball era." The ball had a hard rubber core with a soft cover. As the game continued, it got softer and dirtier, and at most two balls were used in a game.

The period 1920–1930 was called the "lively ball era." It is alleged that a new ball was introduced in 1920, which the players insisted was livelier than its predecessor. The new ball was stitched more tightly, making the ball harder and livelier. The spitball and other trick pitches were abolished. Because of the death of Ray Chapman in 1920, after being struck by a pitched ball, more shiny white balls were put into play in a game.

The next era begins in 1931. For most of the 1930s averages stayed high, but lower than the 1920s. They started to decline through the 1940s. This era finishes with the end of World War II in 1945, which marked the return of professional baseball players from the war.

During 1946–1968, averages continued to decline. This era saw the advent of the relief pitcher, so starting pitchers were not expected to complete every game. New pitches were developed, like the slider, and night baseball became more prevalent. The expansion of the two leagues began in the 1960s, which caused an increase in travel.

In 1969, batters were given an advantage when the pitcher's mound was lowered and the strike zone was tightened. In the 1970s, the split-fingered fastball was introduced, which reduced the batting averages. The American League had slightly higher averages than the National League, because of the "designated hitter rule" of 1973. This era ends in 1993.

The period 1994–2007 marks the "home run era." Smaller new ballparks were introduced. The strike zone was further tightened. Personal trainers worked with batters to increase their strength. A new emphasis on hitting home runs emerged. The baseball was wound tighter, causing it to travel faster and farther. Also, new and controversial nutritional programs were introduced.

Some other observations about AVGs from Table 17.4 and the two time-series graphs in Figures 17.1 and 17.2 are as follows:

- For part of the dead ball era (1913–1919), both leagues have yearly AVGs of around .250.
- From 1920 to 1930, the league AVG increased to close to .300 in both leagues.
- Between 1931 and 1945, the league AVG gradually dropped to around .260 in both leagues.
- The league AVG stayed around .260, gradually dropping to below .250, for both leagues between 1946 and 1968.
- From 1969 to 1993, the National League AVG rose to the upper .250s, while that of the American League increased to the low .260s. (The slightly higher AL AVG was due to the designated hitter in the AL.)
- AVGs increased slightly from 1994 to 2007 for both leagues.

- The mean for the AVGs from 1913 to 2007 was .262 for the NL and .264 for the AL.
- The standard deviation for the AVGs for the years 1913–2007 was .012 for the NL and .013 for the AL.
- The closeness of the mean and standard deviation for AVGs between the leagues seems to indicate that the talent, on a year-by-year basis, was similar in both leagues.
- In general, except for the higher batting averages in the1920s and 1930s, the AVGs for both leagues stayed at about .260 until 1994. From 1994 to 2007, the AL AVGs rose to about .270 and the NL AVGs rose to about .265.

Exercise 17.3. Using Table 17.4 and Figures 17.1 and 17.2, make your own observations for both leagues about the statistics AVG, IPBA, SOA, and IPHR. What is different about the standard deviation of the SOA in comparison with the other three statistics?

Exercise 17.4. Construct time-series graphs for the two leagues, similar to Figures 17.1 and 17.2, for the time periods 1931–1946 and 1969–1993.

Exercise 17.5. Create a table similar to Table 17.4 for the time periods 1931–1946 and 1969–1993.

Exercise 17.6. Make some observations about AVGs for the two leagues for the time periods 1931–1946 and 1969–1993.

Exercise 17.7. Make some observations about IPHRs for the two leagues for the time periods 1931–1946 and 1969–1993.

17.3.2. Relationships between the Four Statistics for the National League, 1913–2007

The statistical techniques used to find relationships between two quantitative variables are regression and correlation analysis. We review these topics (covered in Chapter 5) now. Given a sample of n objects with an X-variable and Y-variable assigned to each object, the sample regression line is the one that comes closest to the n ordered pairs, (x, y). The equation of the sample regression line is $y' = m * x + b$. As discussed in Chapter 5, the easiest way to find the equation for a regression line and the corresponding correlation coefficient is by using Excel. The slope of the line is m. The y-intercept is b. The symbol y' represents the expected y'-value for a given x–value, whereas the symbol y represents the observed y-value for a given x-value. The regression line is used to make a prediction about the variable Y given a value for the variable X. The closer $|r|$ is to 1, the more reliable the prediction.

The statistic r^2 is the *sample coefficient of determination*, and it represents the ratio of the variation of observed y-values around $\bar{y}$ to the variation of the expected y'-values around $\bar{y}$. The statistic r is called the *sample correlation coefficient* and is defined as

$$r = +\sqrt{r^2}, \text{ if } m > 0,$$
$$r = -\sqrt{r^2}, \text{ if } m < 0.$$

The following guidelines for the interpretation of r *and* r^2 were established in Chapter 5:

Strong positive correlation	$.70 \le r \le 1.00$, $.49 \le r^2 \le 1.00$, and $m > 0$
Moderate positive correlation	$.40 \le r < .70$, $.16 \le r^2 < .49$, and $m > 0$
Weak positive correlation	$.20 < r < .40$, $.04 < r^2 < .16$, and $m > 0$
No or poor relationship	$-.20 \le r \le .20$, $0 \le r^2 \le .04$, and any m
Weak negative correlation	$-.40 < r < -.20$, $.04 < r^2 < .16$, and $m < 0$
Moderate negative correlation	$-.70 < r \le -.40$, $.16 \le r^2 < .49$, and $m < 0$
Strong negative correlation	$-1.00 \le r \le -.70$, $.49 \le r^2 \le 1.00$, and $m < 0$

A positive correlation means that as the x-values increase, the y-values increase. A negative correlation means that as the x-values increase, the y-values decrease. As $|r|$ gets closer to 1, the regression line comes closer to the ordered pairs. If $|r| = 1$, then every ordered pair lies on the regression line. If $r = 0$, the regression line is horizontal and we say that there is no correlation between the two variables. The closer $|r|$ is to zero, the weaker the correlation.

The analysis that follows is based on the sample of 95 National League years from 1913 to 2007. For each year, from the five variables AVG, IPBA, SOA, IPHR, and (IPBA – AVG) one is chosen as the independent X-variable and one is chosen as the dependent Y-variable. The results that follow would be similar if we had chosen the 95 American League years. (Why?)

For the years 1913–2007, Figure 17.3 shows the regression line with the equation $y' = -0.158x + 0.284$ and $r^2 = .285$. Since $r^2 = .285$ and the slope of the regression line is negative, $r = -.534$. The independent variable is X=SOA, and the dependent variable is Y = AVG. This correlation coefficient shows a moderate negative correlation between SOA and AVG (as the yearly SOA increases, the yearly AVG decreases). The interpretation of the regression line equation is that if $X = .000$ (which means that the SOA = .000), we get the starting point of .284 for AVG.

Each point change in X means either an increase or decrease by .001. For each 10 points (.010) added to the SOA, the AVG decreases by $(-0.1587 * .010) = -.0016$ (approximately 2 points).

If SOA = .100, then we would predict that the AVG = $.284 + (-0.1587 * .100) = .284 - .016 = .268$.

An increase of 100 points in SOA produces a decrease of 16 points in AVG. How do these empirical data relate to Formula (17.1)?

For the years 1913–2007, Figure 17.4 shows $r^2 = .197$ and a sample regression line with a positive slope. Since $r = .44$, there is a moderate positive correlation between IPBA and AVG (as the yearly IPBA increases, the yearly AVG increases).

For the years 1913–2007, we see that the SOA has a moderate negative correlation with the AVG and the IPBA has a moderate positive correlation with the AVG. The question to be asked now is, what is the relationship between the SOA and the IPBA?

For the next exercise, you are asked to see what a decrease or increase in SOA does to the IPBA.

Exercise 17.8. Let $X = \text{SOA}$ and $Y = \text{IPBA}$. Use Excel to construct the regression line for the years 1913–2007 for the National League. Find the coefficient of determination r^2 and the sample correlation coefficient r. Compare these results with the results in Figures 17.3 and 17.4. In your opinion, does a decrease in IPBA cause a decrease in SOA, does a decrease in SOA cause a decrease in IPBA, or are the results just a coincidence? How do all these empirical results relate to Formula (17.1)?

The next two figures show a strong positive correlation between the IPHR and both the IPBA and the SOA for the years 1913–2007.

Figure 17.5 shows $r^2 = .474$ and a sample regression line with a positive slope. Since $r = .69$ (very close to .70), there is a strong positive correlation between IPHR and IPBA. One reason for this strong positive correlation is that a home run hitter, being a power hitter, tends to hit harder and deeper balls.

Figure 17.6 shows $r^2 = .670$ and a sample regression line with a positive slope. Since $r = .82$, there is a strong positive correlation between IPHR and SOA. Most baseball people agree that one of the reasons for an increase in yearly strikeouts is that, in recent years, players tend to swing harder, attempting to hit home runs.

Figures 17.7, 17.8, and 17.9 investigate the variable (IPBA – AVG). Two requirements are needed to average at least .400 for a season. First, a player must have an IPBA of at least .400. Then the difference of (IPBA – AVG)

Table 17.5 **Correlations for the independent variables $X = $ (IPBA, IPHR, and SOA) and the dependent variable $Y = $ (IPBA – AVG) for the National League**

X	Y	r^2	r	Conclusion	Figure
IPBA	(IPBA – AVG)	0.38	0.61	Moderate positive correlation	17.8
IPHR	(IPBA – AVG)	0.72	0.85	Strong positive correlation	17.9
SOA	(IPBA – AVG)	0.98	0.99	Strong positive correlation	17.7

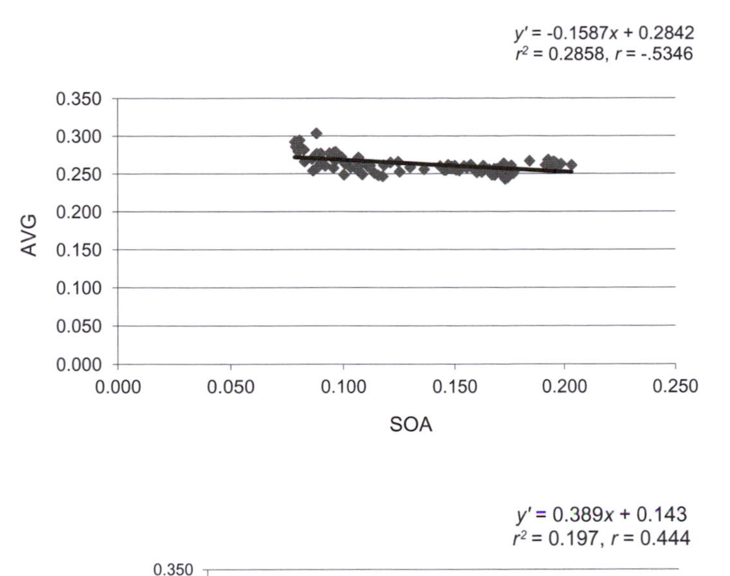

Figure 17.3 Regression line, equation, and correlation coefficient for X = SOA, Y = AVG (1913–2007, NL)

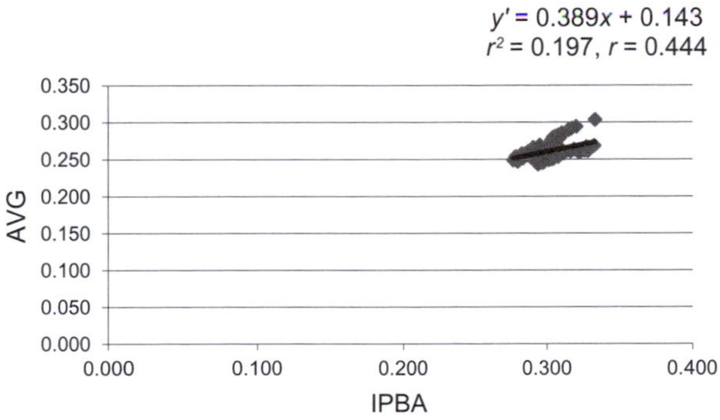

$y' = 0.389x + 0.143$
$r^2 = 0.197, r = 0.444$

Figure 17.4 Regression line, equation, and correlation coefficient for X = IPBA, Y = AVG (1913–2007, NL)

must be small enough that when it is subtracted from IPBA the result is at least .400. Since 1941, many players have achieved an IPBA of at least .400. However, no player has had an AVG of at least .400. Table 17.5 gives the correlations for the independent variables X = (IPBA, IPHR, and SOA) and the dependent variable Y = (IPBA – AVG) for the National League.

Since X has a positive correlation with (IPBA – AVG), in all three cases, as X gets smaller, so does (IPBA – AVG). Considering the year-by-year performance of the National League, the strongest predictor of (IPBA – AVG) for a year is SOA. IPHR is also a strong predictor of (IPBA – AVG). Reducing the SOA appears to be the best way of reducing the gap between the statistics IPBA and AVG for a typical year in the National League. From Figure 17.6, a decrease in IPHR results in a decrease in SOA.

The next section performs the same analysis as done in this section, except that we restrict the analysis to the two eras 1994–2007 and 1920–1930.

The reason for looking at certain subgroups of a population is that important information can be lost by looking at only the entire group of objects. The era from 1920 to 1930 (the lively ball era) produced all but one of

Figure 17.5 Regression line, equation, and correlation coefficient for X = IPHR, Y = IPBA (1913–2007, NL)

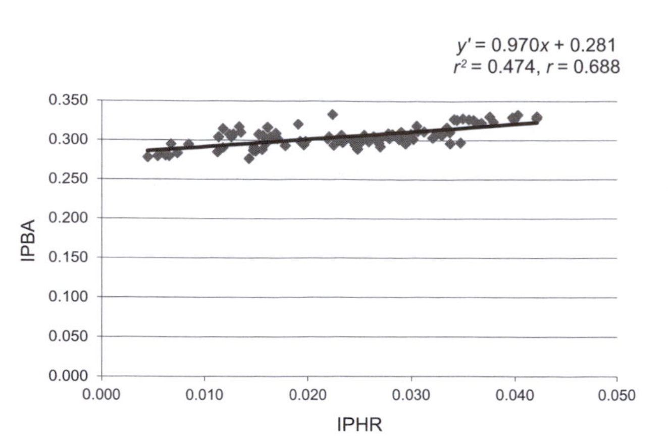

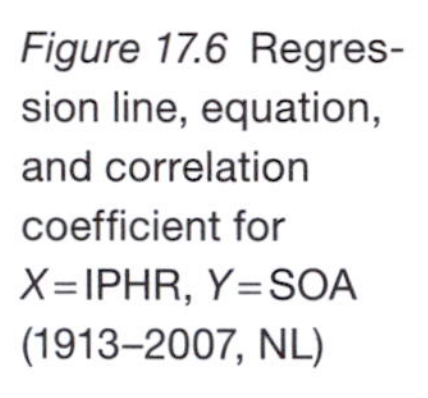

Figure 17.6 Regression line, equation, and correlation coefficient for X = IPHR, Y = SOA (1913–2007, NL)

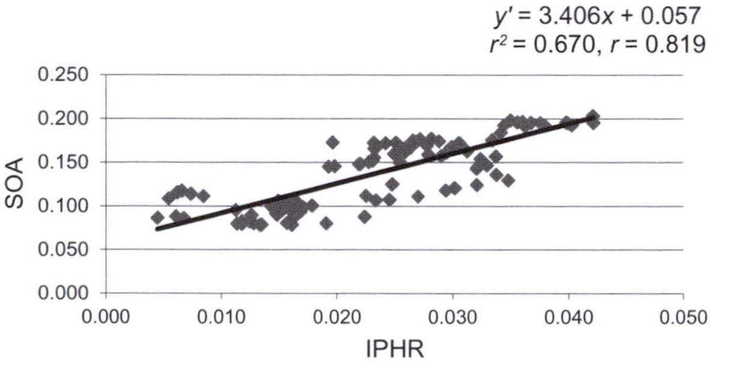

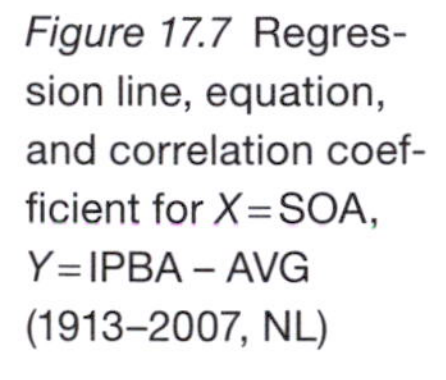

Figure 17.7 Regression line, equation, and correlation coefficient for X = SOA, Y = IPBA – AVG (1913–2007, NL)

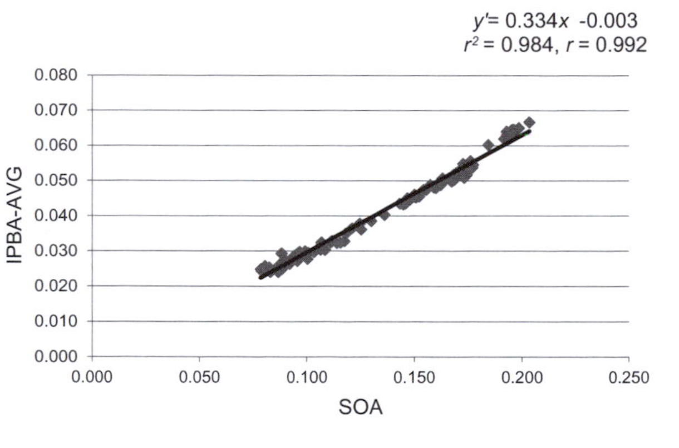

Figure 17.8 Regression line, equation, and correlation coefficient for X = IPBA, Y = IPBA – AVG (1913–2007, NL)

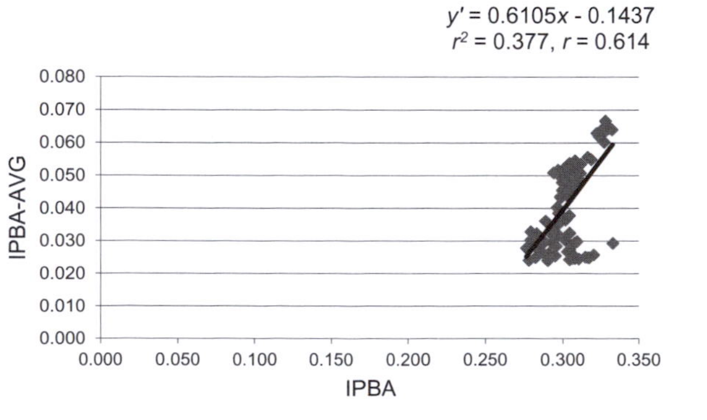

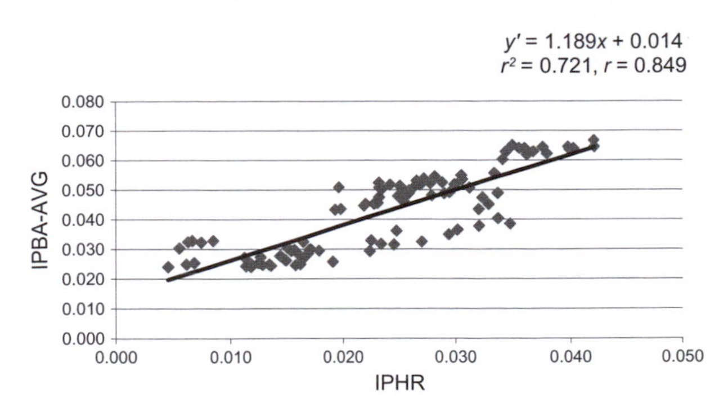

Figure 17.9 Regression line, equation, and correlation coefficient for X = IPHR, Y = IPBA − AVG (1913–2007, NL)

the .400 hitters since 1913, the exception being Ted Williams. The era beginning in 1994 represents baseball today and in the future. Remember, the question was, can we conclude that, in the future, a player will have a batting average of at least .400?

17.4. The Two Leagues and the Four Statistics for 1920–1930 and 1994–2007

Figures 17.10 and 17.11 contain time-series graphs for AVG, IPBA, SOA, and IPHR for both leagues for the years 1994–2007. Figures 17.12 and 17.13 contain the same time series graphs for 1920–1930. Table 17.6 supplies the descriptive measures of mean and standard deviation for the four statistics AVG, IPBA, SOA, and IPHR for the years 1994–2007, while Table 17.7 does so for 1920–1930.

17.4.1. Comparison of AVG, IPBA, SOA, and IPHR between the Two Leagues and between the Two Eras

There are two comparisons to be made. The first is between the two leagues in the same era; the second is between the two eras. The observations below make use of Figures 17.10–17.13 and Tables 17.6 and 17.7.

Comparison between the Two Leagues

- For 1920–1930, the two leagues are similar in terms of the four statistics. The only difference is that the National League shows slightly *more* variation (as measured by the standard deviation) for AVG, IPBA, and IPHR and *less* variation for SOA.
- Between 1994 and 2007, the mean of the statistics AVG, IPBA, and IPHR is slightly higher for the American League. The SOA is slightly higher for the National League. These differences can be attributed to the designated hitter rule.

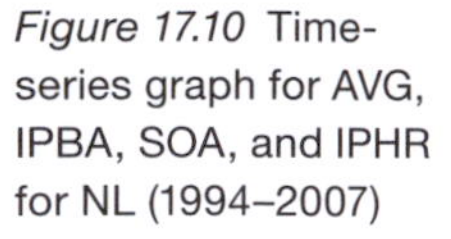

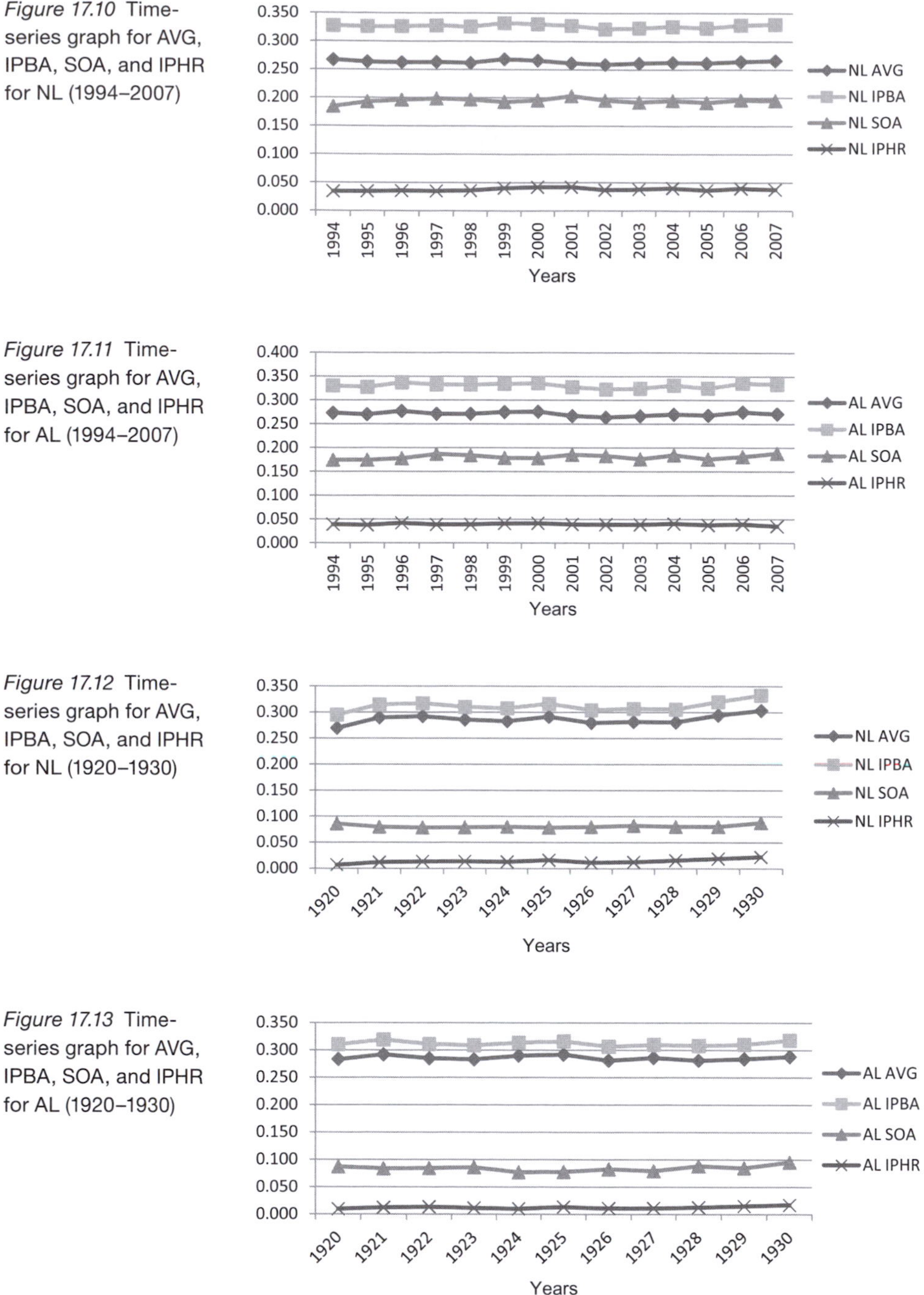

Figure 17.10 Time-series graph for AVG, IPBA, SOA, and IPHR for NL (1994–2007)

Figure 17.11 Time-series graph for AVG, IPBA, SOA, and IPHR for AL (1994–2007)

Figure 17.12 Time-series graph for AVG, IPBA, SOA, and IPHR for NL (1920–1930)

Figure 17.13 Time-series graph for AVG, IPBA, SOA, and IPHR for AL (1920–1930)

Comparison between the Two Eras

- The AVG is about 25 points higher from 1920 to 1930. One explanation is that the SOA from 1994 to 2007 is more than twice that from 1920 to 1930.

Table 17.6 **Mean and standard deviation for the four statistics AVG, IPBA, SOA, and IPHR for the years 1994–2007**

	Mean				Standard deviation			
League	AVG	IPBA	SOA	IPHR	AVG	IPBA	SOA	IPHR
National	.263	.327	.195	.038	.003	.003	.004	.003
American	.271	.330	.180	.039	.004	.004	.005	.002

Table 17.7 **Mean and standard deviation for the four statistics AVG, IPBA, SOA, and IPHR for the years 1920–1930**

	Mean				Standard deviation			
League	AVG	IPBA	SOA	IPHR	AVG	IPBA	SOA	IPHR
National	.287	.312	.081	.014	.009	.010	.003	.004
American	.286	.312	.084	.013	.004	.004	.005	.002

- The IPBA is about 18 points lower from 1920 to 1930. One explanation could be that the IPHR for 1994–2007 is close to 3 times that of 1920–1930. Another possibility is that a third variable could be causing both the IPBA and the IPHR to be higher in the period 1994–2007. One possible third variable could be that the bats of today are better constructed for hitting. Other possible explanations could be the size of the players and the smaller dimensions of the modern stadiums.
- The difference between the AVG and the IPBA is almost constant during both eras. One explanation is that the SOA is almost constant in both eras.
- The difference between the AVG and the IPBA for the years 1920–1930 is approximately half that for the years 1994–2007. This can be explained by the smaller SOA for the years 1920–1930.

The trend of a higher SOA for Major League Baseball for 1994–2007 will make it difficult for a batter to hit .400. Since 1913, except for Ted Williams, all the players who batted at least .400 achieved this feat between 1920 and 1930.

17.4.2. Relationships between AVG, IPBA, SOA, and IPHR for the National League for 1920–1930 and 1994–2007

During the "lively ball era," 1920–1930, batting averages jumped from .250 to close to .300. The period from 1994 to 2007, the home run era, saw the 60-home-run number eclipsed several times and the IPHR jump from about .030 to .040.

Figures 17.14–17.20 compare the relationships between the four statistics for the two eras. These relationships are shown through the use of regression and correlation analysis.

Figure 17.14b shows that in the home run era (1994–2007), there is a moderate negative correlation between SOA and AVG. Figure 17.14a shows that SOA had virtually no effect on AVG during the lively ball era (1920–1930). Figures 17.15a and 17.15b show a strong positive correlation between IPBA and AVG for both eras. The correlation is stronger for 1920–1930 (r is almost equal to 1). This follows from Formula (17.1) and Figure 17.14a. Substituting .400 for the y-variable in each of the two regression equations and solving for x gives .510 for the home run era (1994–2007) and .437 for the lively ball era (1920–1930). This means that to have a year with an AVG of at least .400 for the era 1994–2007 you would need an IPBA of .510; for the era 1920–1930 you would need an IPBA of only .437.

Exercise 17.9. Use Figures 17.16a and 17.16b to compare the relationship between IPHR and IPBA for the years in the two eras. Using the two regression equations, find the minimum IPHR needed to have an IPBA of at least .400 for each era.

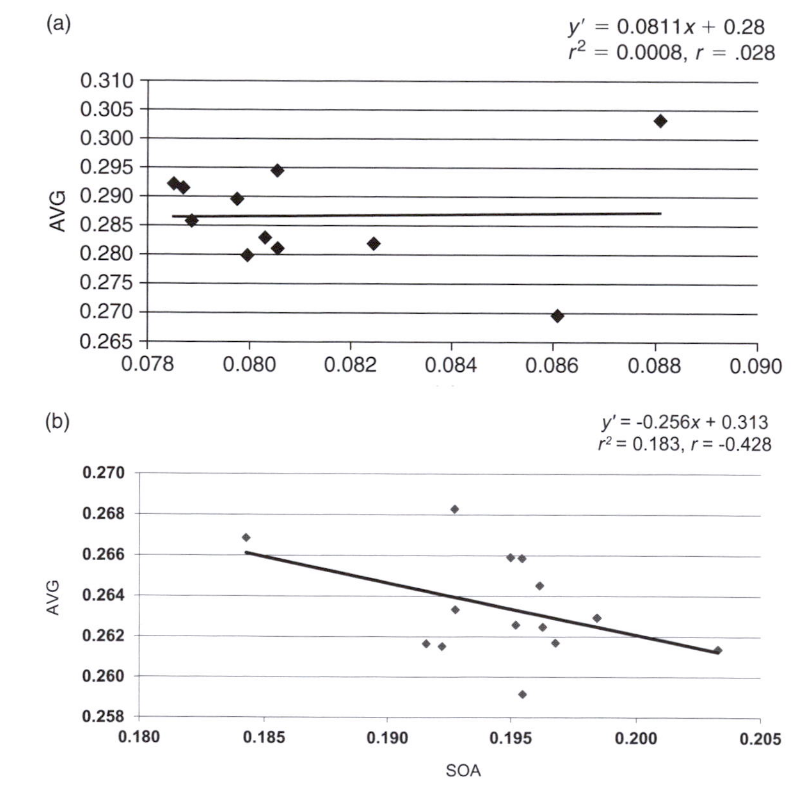

Figure 17.14 (a) Regression line, equation, and correlation coefficient for X=SOA, Y=AVG (1920–1930, NL). (b) Regression line, equation, and correlation coefficient for X=SOA, Y=AVG (1994–2007, NL).

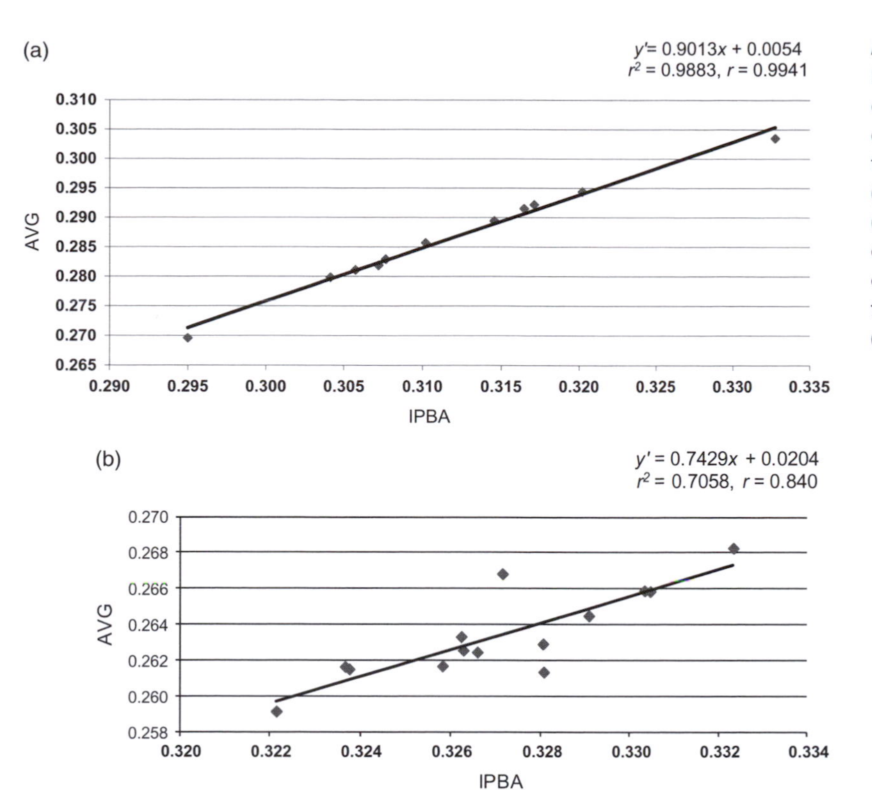

Figure 17.15 (a) Regression line, equation, and correlation coefficient for $X = \text{IPBA}$, $Y = \text{AVG}$ (1920–1930, NL). (b) Regression line, equation, and correlation coefficient for $X = \text{IPBA}$, $Y = \text{AVG}$ (1994–2007, NL).

(a)

y' = 2.1165x + 0.282
r² = 0.7791, r = 0.8827

IPBA

0.335
0.330
0.325
0.320
0.315
0.310
0.305
0.300
0.295
0.290

0.000
0.005
0.010
0.015
0.020
0.025

IPHR

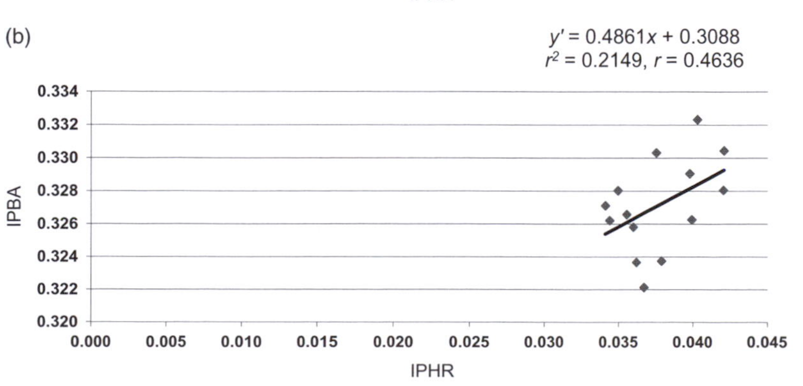

Figure 17.16 (a) Regression line, equation, and correlation coefficient for $X = \text{IPHR}$, $Y = \text{IPBA}$ (1920–1930, NL). (b) Regression line, equation, and correlation coefficient for $X = \text{IPHR}$, $Y = \text{IPBA}$ (1994–2007, NL).

Figure 17.17 (a) Regression line, equation, and correlation coefficient for X = IPHR, Y = SOA (1920–1930, NL). (b) Regression line, equation, and correlation coefficient for X = IPHR, Y = SOA (1994–2007, NL).

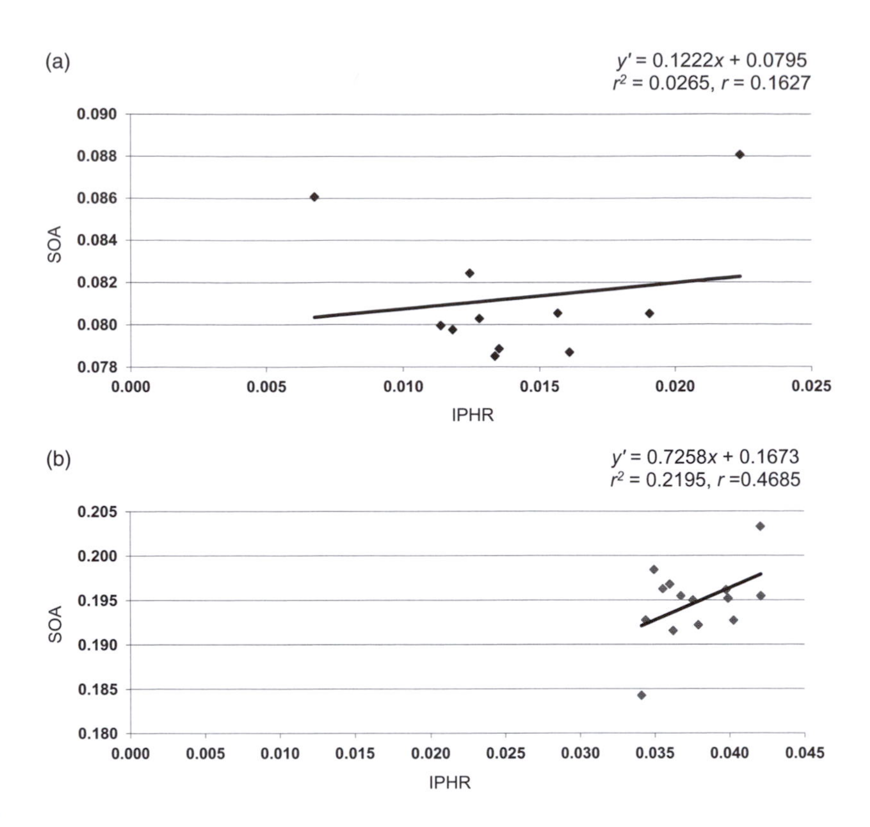

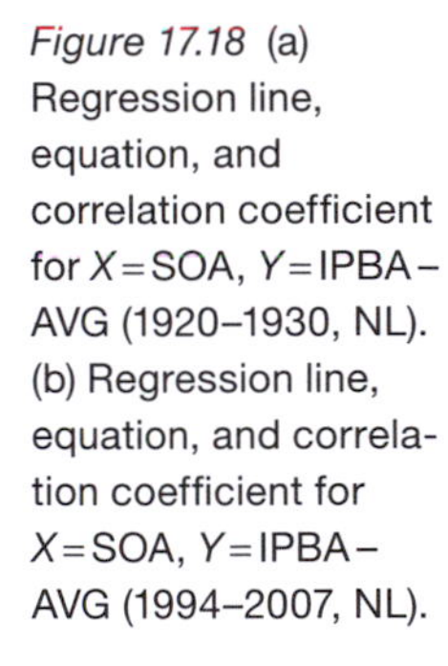

Figure 17.18 (a) Regression line, equation, and correlation coefficient for X = SOA, Y = IPBA – AVG (1920–1930, NL). (b) Regression line, equation, and correlation coefficient for X = SOA, Y = IPBA – AVG (1994–2007, NL).

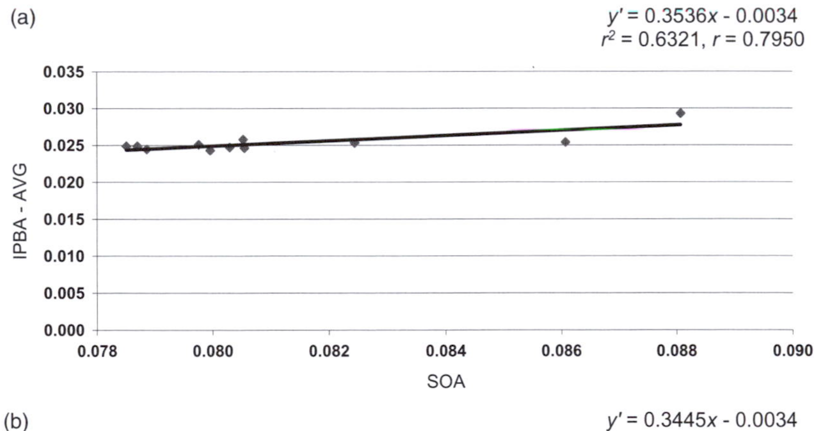

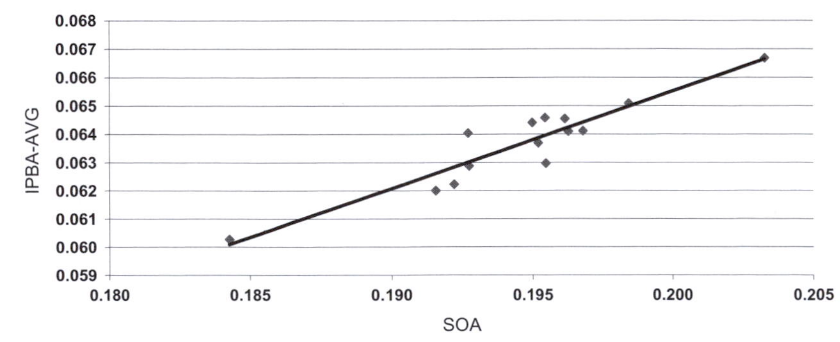

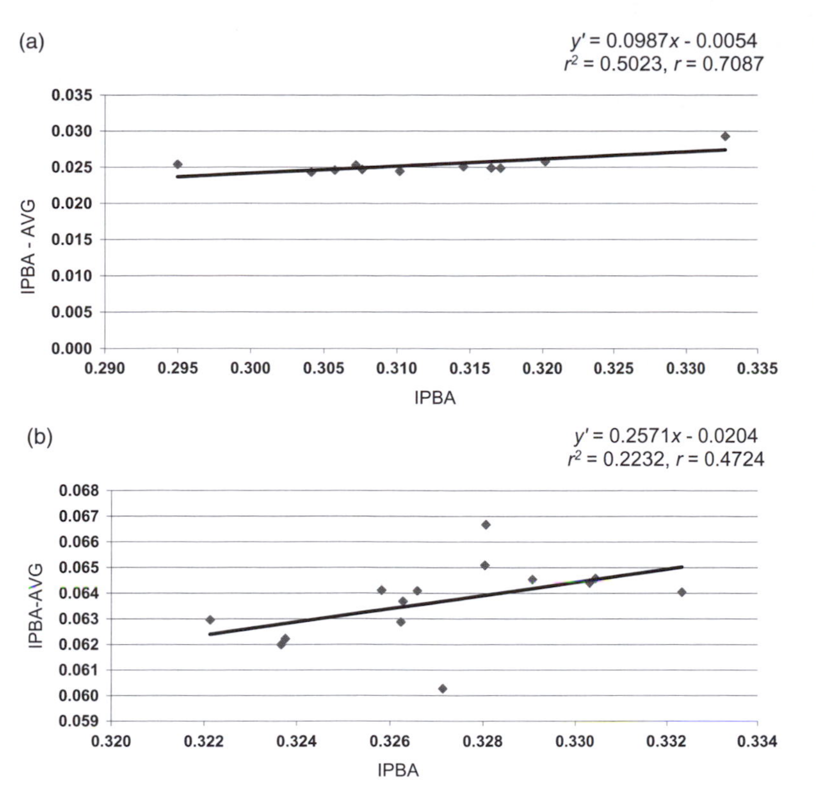

Figure 17.19 (a) Regression line, equation, and correlation coefficient for $X = \text{IPBA}$, $Y = \text{IPBA} - \text{AVG}$ (1920–1930, NL). (b) Regression line, equation, and correlation coefficient for $X = \text{IPBA}$, $Y = \text{IPBA} - \text{AVG}$ (1994–2007, NL).

(a)
y' = 0.2201x + 0.0222
r2 = 0.435, r = 0.6595
IPBA - AVG
IPHR

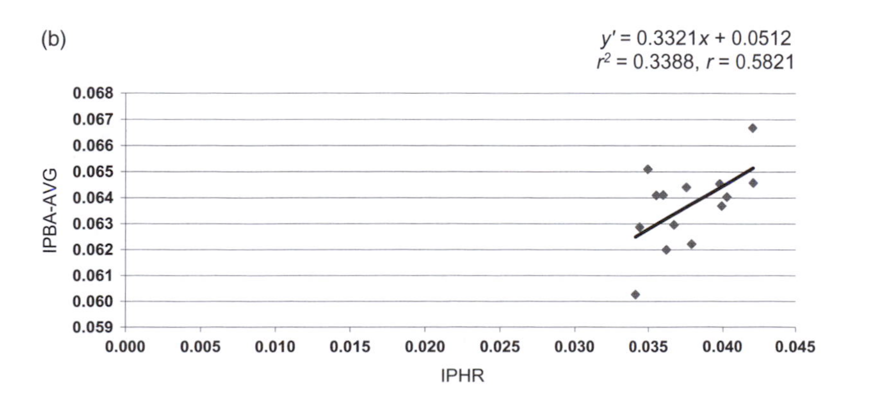

Figure 17.20 (a) Regression line, equation, and correlation coefficient for $X = \text{IPHR}$, $Y = \text{IPBA} - \text{AVG}$ (1920–1930, NL). (b) Regression line, equation, and correlation coefficient for $X = \text{IPHR}$, $Y = \text{IPBA} - \text{AVG}$ (1994–2007, NL).

Figure 17.17b shows a moderate positive correlation between IPHR and SOA for the years 1994–2007. Figure 17.17a shows that there is a poor correlation between IPHR and SOA for the years 1920–1930.

Exercise 17.10. How would you explain the stronger positive correlation for the period 1994–2007 in Figure 17.17b?

Figures 17.18a and 17.18b show a strong positive correlation between SOA and (IPBA – AVG) for both eras. This relationship is probably explained by Formula (17.2).

Figure 17.19b shows a moderate positive correlation between IPBA and (IPBA – AVG) for the period 1994–2007. Figure 17.19a shows that, for the period 1920–1930, there is a strong positive correlation.

Exercise 17.11. How would you explain the stronger positive correlation in Figure 17.19a for the period 1920–1930?

Figure 17.20a shows a moderate positive correlation between IPHR and (IPBA – AVG) for the period 1920–1930. Figure 17.20b, for the period 1994–2007, shows a moderate positive correlation.

Exercise 17.12. How would you explain the similar moderate positive correlation between IPHR and (IPBA – AVG) for the two periods?

To summarize:

- For the era 1920–1930, AVG is almost completely determined by the IPBA.
- For the era 1994–2007, AVG is determined by a combination of SOA and IPBA.

In the next section, our attention shifts to the last player to hit .400, Ted Williams.

17.5. Ted Williams: The Last Player to Hit .400 for a Season

Many people consider Ted Williams the greatest hitter of all time. In the concluding chapter, we will attempt to choose the 10 best hitters of all time. Ted possessed the rare combination of being a power hitter who hit for a high average. He was difficult to strike out and, as a result of his exceptional eyesight, drew a great many walks each year. These skills helped Ted hit over .300 for 18 of his 19 years in the Major Leagues. His career batting average was .344. Ted led the league in base on balls six times and had the highest career ratio of base on balls to strikeouts. He won six batting titles and led the league at the age of 39 with a .388 average. Williams made the All-Star Team every year he played and holds the record of getting on base in 84 consecutive games. Ted and Rogers Hornsby are the only two players since 1900 to win the Triple Crown twice. He would have been the only three-time winner except that in 1949 he lost the batting title to George Kell by .00015 (.34291 to .34276). He is one of only six players, since 1913, to have hit over .400 in a season, and at the age of 23 he was the youngest player to have done so. It should be noted that Joe Jackson was the youngest player to hit .400 for a season at the age of 21 in 1911.

To illustrate what a great hitter Ted Williams was, we present Table 17.8, which contains the year-by-year data for the four statistics for Williams compared with that of the American League.

For this comparison, we have chosen his active years from 1939 through 1959. The years 1943–1945 are omitted since during those years Ted served in World War II. The years 1952 and 1953 are also not shown because Ted had only 10 and 91 at bats, respectively. He served in the Korean War for those two years.

The next four figures present time-series graphs comparing Ted's year-by-year performance with that of the American League (AL).

What do Table 17.8 and Figures 17.21–17.24 tell us about Ted Williams as a hitter in his era? Except for 1959, when he was 41, his AVG, IPBA, and IPHR were considerably higher than the American League's AVG, IPBA, and IPHR. At the same time, his SOA was substantially lower than the American League's SOA.

Table 17.8 Year-by-year data for the *four statistics* for Williams and for the American League

	AVG		IPBA		SOA		IPHR	
Year	AL	Williams	AL	Williams	AL	Williams	AL	Williams
1939	0.279	0.327	0.310	0.369	0.101	0.113	0.021	0.062
1940	0.271	0.344	0.305	0.381	0.110	0.096	0.023	0.045
1941	0.266	0.406	0.297	0.431	0.103	0.059	0.019	0.086
1942	0.257	0.356	0.285	0.395	0.099	0.098	0.014	0.076
1946	0.256	0.342	0.292	0.374	0.124	0.086	0.018	0.081
1947	0.256	0.343	0.287	0.376	0.110	0.089	0.018	0.067
1948	0.266	0.369	0.296	0.402	0.102	0.081	0.019	0.053
1949	0.263	0.343	0.294	0.375	0.105	0.085	0.021	0.083
1950	0.271	0.317	0.303	0.339	0.107	0.063	0.026	0.089
1951	0.262	0.318	0.294	0.348	0.108	0.085	0.022	0.062
1954	0.257	0.345	0.293	0.376	0.122	0.083	0.022	0.082
1955	0.258	0.356	0.297	0.385	0.129	0.075	0.026	0.095
1956	0.260	0.345	0.302	0.382	0.138	0.098	0.030	0.066
1957	0.255	0.388	0.296	0.432	0.138	0.102	0.028	0.101
1958	0.254	0.328	0.297	0.373	0.145	0.119	0.030	0.072
1959	0.253	0.254	0.296	0.282	0.145	0.099	0.030	0.041

In comparing two players, the time period in which they played is important.

The next section adjusts a player's batting statistics for the years in which he played. The adjusted statistics are then used to compare the players.

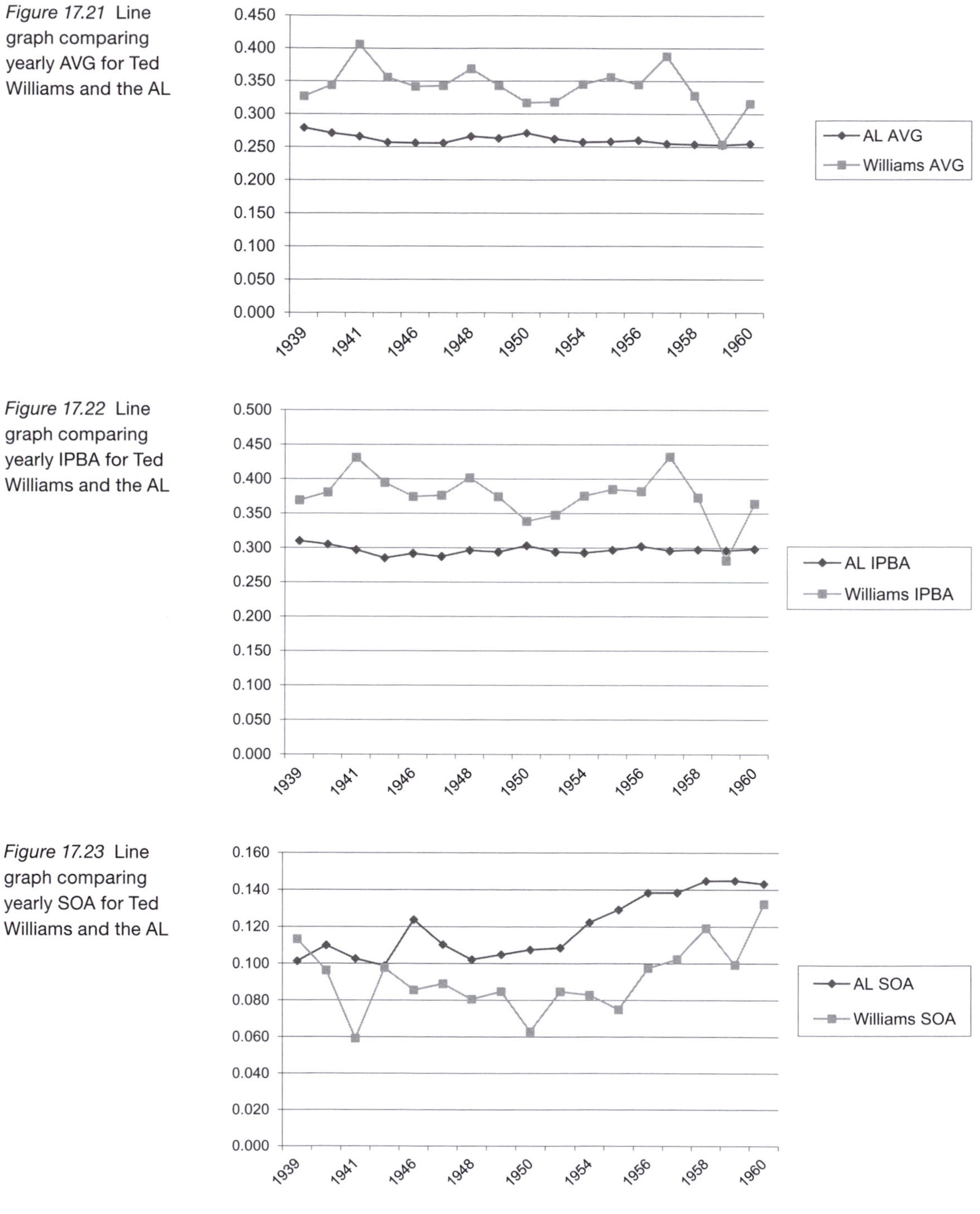

Figure 17.21 Line graph comparing yearly AVG for Ted Williams and the AL

Figure 17.22 Line graph comparing yearly IPBA for Ted Williams and the AL

Figure 17.23 Line graph comparing yearly SOA for Ted Williams and the AL

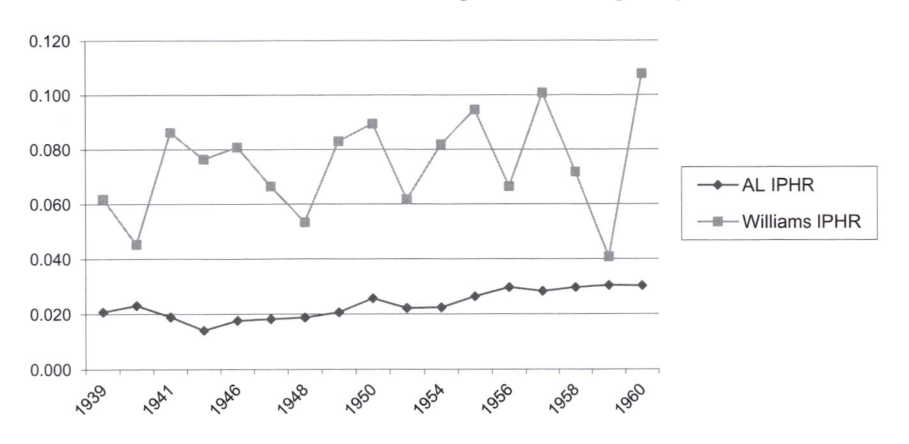

Figure 17.24 Line graph comparing yearly IPHR for Ted Williams and the AL

17.6. Two Methods to Compare Players

Both methods compare a *player's* statistic with the *league's* statistic for that year. We now outline both methods. We use the four statistics (AVG, IPBA, SOA, and IPHR) in this chapter, but these methods can be used with individual statistics or any basket of statistics.

17.6.1. Method 1: The Sum Method

This method produces one sum for the four statistics. The sum is generated by following these rules:

Rule 1: For AVG, IPBA, and IPHR, subtract the league statistics from the player's statistic. A positive difference results if the player's statistic is greater than the league's. The larger the positive number, the more superior the player is to the league.

Rule 2: For SOA, we subtract the player's statistic from the league's statistic. A positive difference means that the player did better than the league.

Rule 3: The player with the larger sum is considered superior with respect to the four statistics. (If other statistics are to be used, rule 1 is used when a larger number is better and rule 2 is used when a smaller number is better.)

Table 17.9 demonstrates the Sum Method to compare these four statistics between Ted Williams (1941) and Rogers Hornsby (1924). The statistics for each player are taken from Table 17.10 below. The reason for choosing Hornsby and 1924 is that Hornsby's AVG of .424 is the highest since 1913.

The sum of the differences for Williams is .374, and the sum of the differences for Hornsby is .340. We conclude, from the Sum Method, that the performance for Williams (1941) was superior to the performance for Hornsby (1924) with respect to the four statistics.

Table 17.9 **Using the Sum Method to compare the *four statistics* between Ted Williams (1941) and Rogers Hornsby (1924)**

Statistic	Williams (1941)	League (1941)	Sum of differences	Hornsby (1924)	League (1924)	Sum of differences
AVG	.406	.266	+.140	.424	.283	+.141
IPBA	.431	.297	+.123	.450	.308	+.142
SOA	.059	.103	+.044	.060	.080	+.020
IPHR	.086	.019	+.067	.050	.013	+.037
Total			+.374			+.340

As noted earlier, this method can be used to compare two or more players, using any basket of statistics. It would be interesting to use the Sum Method with the basket of statistics AVG, OBP, and SLG.

Exercise 17.13. Use the Sum Method to compare Williams and DiMaggio with respect to the combination consisting of AVG, OBP, and SLG for the 1941 season.

This method can also be used to compare two players for any group of years. For example, we could compare two players for the years that correspond to when both players were 27–32 years old. We would perform the Sum Method for each year and then add the corresponding totals. The player with the highest total for the 6 years would be considered better.

If multiple years are used, it is necessary to use the same number of years for both players. Also, a certain minimum number of plate appearances for each year should be required.

17.6.2. Method 2: The Adjustment Formula Introduced in Chapter 3

This method can be used for any statistic. We demonstrate the Adjustment Method by comparing AVG between Williams (1941) and Hornsby (1924).

In Chapter 3, we adjusted a player's yearly statistic based on that statistic for the league for that year. The time-adjusted statistic was symbolized by AdjT_(statistic) for (year). We use this formula to compare the AVG for Williams (1941) with the AVG for Hornsby (1924). Remember that BA is another abbreviation for batting average. The formula for adjusted batting average is

AdjT_BA for *xyzw* = [(*xyzw*BA)/(league *xyzw*BA)] * (Standard BA),

where *xyzw* is the year and the Standard BA is equal to .257. For Hornsby,

AdjT_BA for 1924 = [1924BA/(league 1924BA)] * (Standard BA)
= (.424/.283) * .257 = .385.

For Williams,

AdjT_BA for 1941 = [1941BA/(league 1941BA)] * (Standard BA)
= (.406/.266) * .257 = .392.

We can conclude that, relative to their time period, the .406 AVG of Williams can be considered higher than the .424 AVG of Hornsby.

The Adjustment Method can be used to compare two or more players with respect to any statistic. It would be interesting to use the Adjustment Method with such statistics as OBP and SLG.

Use Tables 17.2 and 17.10 for the next two exercises. Table 17.10 shows the batting statistics for all players who batted .400 for a season since 1913 combined with the players who came closest to batting .400 for a season since 1941.

Exercise 17.14. Choose another .400 hitter to compare with Williams (1941). Perform the Sum Method applied to the four statistics and the Adjustment Method applied to AVG. Which player is superior and why?

Exercise 17.15. Use the Sum Method to compare Williams (1941) and Williams (1957) with respect to the basket of statistics consisting of AVG, OBP, and SLG. What conclusion can you draw?

Table 17.10 **All players who batted .400 for a season (since 1913) combined with Carew (1977), Brett (1980), Gwynn (1994), Walker (1999), and Williams (1957)**

Year	Player	G	AB	H	HR	BB	SO	AVG	IPBA	SOA	IPHR	BBA	OBP	SLG
1977	Carew	155	616	239	14	69	55	0.388	0.426	0.089	0.025	0.101	0.449	0.570
1980	Brett	117	449	175	24	58	22	0.390	0.410	0.049	0.056	0.114	0.454	0.664
1994	Gwynn	110	419	165	12	48	19	0.394	0.413	0.045	0.030	0.103	0.454	0.568
1999	Walker	127	438	166	37	57	52	0.379	0.430	0.119	0.096	0.115	0.458	0.710
1957	Williams	132	420	163	38	119	43	0.388	0.432	0.102	0.101	0.221	0.526	0.731
1923	Heilmann	144	524	211	18	74	40	0.403	0.436	0.076	0.037	0.124	0.481	0.632
1922	Hornsby	154	623	250	42	65	50	0.401	0.436	0.080	0.073	0.094	0.459	0.722
1924	Hornsby	143	536	227	25	89	32	0.424	0.450	0.060	0.050	0.142	0.507	0.696
1925	Hornsby	138	504	203	39	83	39	0.403	0.437	0.077	0.084	0.141	0.489	0.756
1920	Sisler	154	631	257	19	46	19	0.407	0.420	0.030	0.031	0.068	0.449	0.632
1922	Sisler	142	586	246	8	49	14	0.420	0.430	0.024	0.014	0.077	0.467	0.594
1930	Terry	154	633	254	23	57	33	0.401	0.423	0.052	0.038	0.083	0.452	0.619
1922	Cobb	137	526	211	4	55	24	0.401	0.420	0.046	0.008	0.095	0.462	0.565
1941	Williams	143	456	185	37	147	27	0.406	0.431	0.059	0.086	0.244	0.553	0.735

Exercise 17.16. Try to find another player who for a year is superior to Williams (1941) with respect to the Sum Method applied to the four statistics and the Adjustment Method applied to AVG. Two players worth trying are Stan Musial (1948) and Barry Bonds (2003).

17.7. Players Closest to Batting .400 for a Season since 1941

We now look at the four statistics BA, IPBA, SOA, and IPHR for the players who batted close to .400 for a season since 1941.

The five players who came closest to hitting .400 since 1941 were Ted Williams (.388 in 1957), Rod Carew (.388 in 1977), George Brett (.390 in 1980), Tony Gwynn (.394 in 1994), and Larry Walker (.379 in 1999). We look at the four statistics for these players applied to the season in which they almost hit .400 along with the 1941 season for Ted Williams. All these players hit from the left side, and the IPBA was above .400 for each one. The highest SOA belongs to Walker, the lowest to Gwynn. As the SOA increases, the spread between BA and IPBA increases. The largest difference between IPBA and AVG belongs to Walker, the smallest to Gwynn. Williams's IPBA was one point higher in 1957 than in 1941, but his SOA was also significantly higher in 1957. This may explain why his AVG fell below .400 in 1957.

Table 17.11 provides the statistics BA, IPBA, and SOA for these players. The IPHR statistic can be found in Table 17.10.

Figures 17.25 and 17.26 provide a line graph and bar graph comparison of the four statistics BA, IPBA, SOA, and IPHR for the players in Table 17.11.

What observations about the players in Table 17.11 can you make from these two figures?

Exercise 17.17. Table 17.12 presents the 11 players who came closest to batting .400 for a season since 1941.

Table 17.11 **Statistics for the five players who came closest to batting .400 (since 1941)**

Year	Batting side	Player	BA	IPBA	SOA
1941	Left	Williams	0.406	0.431	0.059
1957	Left	Williams	0.388	0.432	0.102
1977	Left	Carew	0.388	0.426	0.089
1980	Left	Brett	0.390	0.410	0.049
1994	Left	Gwynn	0.394	0.413	0.045
1999	Left	Walker	0.379	0.430	0.119

a. Complete Table 17.12.
b. Construct a comparative line graph similar to Figure 17.25.
c. Construct a comparative side-by-side bar graph similar to Figure 17.26.
d. Use (a), (b), and (c) to make observations about AVG, IPBA, SOA, IPHR, and BBA for these players.

We now ask the following two questions:

1. Leaving the number of at bats the same for each of these players, how many more hits were needed for them to become a .400 hitter?
2. Leaving the number of at bats and the IPBA the same for each of these players, how many fewer strikeouts were needed for them to become a .400 hitter?

We answer these two questions for Carew and leave the other players as an exercise.

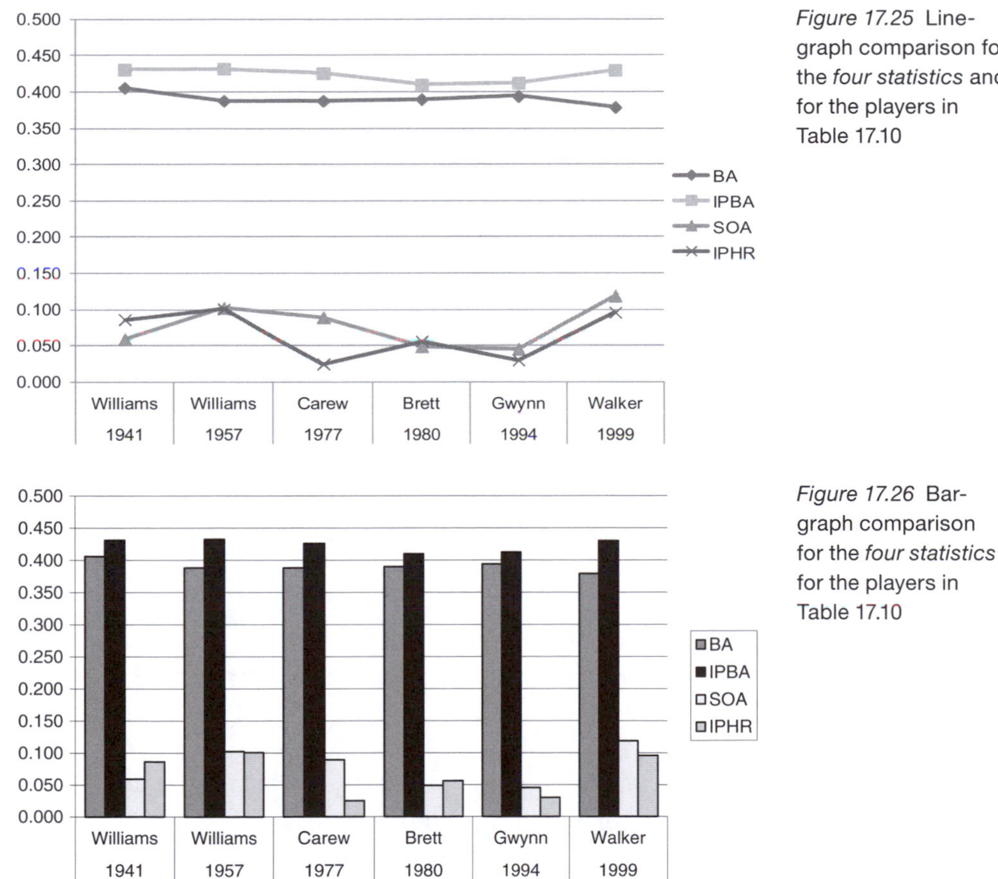

Figure 17.25 Line-graph comparison for the *four statistics* and for the players in Table 17.10

Figure 17.26 Bar-graph comparison for the *four statistics* for the players in Table 17.10

Table 17.12 **Batting statistics for the 11 players who came closest to batting .400 for a season (since 1941)**

Player	Year	AB	H	SO	BB	SF	HBP	HR	AVG	IPBA	SOA	IPHR	BBA
Tony Gwynn	1994	419	165	19	48	5	3	12					
George Brett	1980	449	175	22	58	7	1	24					
Ted Williams	1957	420	163	43	119	2	5	38					
Rod Carew	1977	616	239	55	69	5	3	14					
Larry Walker	1999	438	166	52	57	6	12	37					
Stan Musial	1948	611	230	34	79	0	3	39					
Todd Helton	2000	580	216	61	103	10	4	42					
Nomar Garciaparra	2000	529	197	50	61	7	2	21					
Ichiro Suzuki	2004	704	262	63	49	3	4	8					
Tony Gwynn	1997	592	220	28	43	12	3	17					
Andres Galarraga	1993	470	174	73	24	6	6	22					

Tables 17.13 and 17.14 were constructed using Excel. Table 17.13 simply increases the number of hits, leaving the number of at bats the same. Table 17.14 keeps the number of at bats and the IPBA the same while reducing the number of strikeouts. This results in an increase in the number of hits and therefore an increase in AVG.

According to Table 17.13, given his 616 at bats, Carew would have hit .400 with just eight more hits. While maintaining the same number of at bats and IPBA, just by reducing his number of strikeouts by seventeen, Carew would have batted .400.

Table 17.13 **Number of additional hits needed by Carew in 1977 to bat .400, keeping the number of at bats the same**

AB	H	AVG
616	239	0.387987
616	240	0.389610
616	241	0.391234
616	242	0.392857
616	243	0.394481
616	244	0.396104
616	245	0.397727
616	246	0.399351
616	247	0.400974

Exercise 17.18. Keeping the at bats the same for their year, determine the number of additional hits needed for Williams (1957), Brett (1980), Walker (1999), and Gwynn (1994) to bat .400 for their year.

Exercise 17.19. Keeping the at bats and IPBA the same for their year, determine the reduction in number of strikeouts needed for these players to bat .400 for their year.

After completing these two exercises, which player do you think had the best chance of hitting .400 for their year?

We now look at the players who batted .400 for a season, from 1913 to 1941, combined with those players who came closest to batting .400 for a season after 1941.

Table 17.14 **Number of fewer strikeouts needed by Carew in 1977 to bat .400, keeping the number of at bats and the IPBA the same**

AB	H	SO	IPBA	AVG
616	239	55	0.426025	0.387987
616	239	54	0.426025	0.388679
616	240	53	0.426025	0.389370
616	240	52	0.426025	0.390062
616	241	51	0.426025	0.390753
616	241	50	0.426025	0.391445
616	242	49	0.426025	0.392137
616	242	48	0.426025	0.392828
616	242	47	0.426025	0.393520
616	243	46	0.426025	0.394211
616	243	45	0.426025	0.394903
616	244	44	0.426025	0.395595
616	244	43	0.426025	0.396286
616	245	42	0.426025	0.396978
616	245	41	0.426025	0.397669
616	245	40	0.426025	0.398361
616	246	39	0.426025	0.399053
616	246	38	0.426025	0.399744

17.8. Players Listed in Table 17.10

17.8.1. Comparison of AVG, IPBA, SOA, and IPHR for Players in Table 17.10

Figure 17.27 is a line graph comparing the players in Table 17.10 with respect to IPBA, SOA, BBA, AVG, and IPHR. Figure 17.28 depicts the same information as Figure 17.27 except that the graph is a side-by-side bar graph.

Up to now in this chapter, the BBA (base-on-balls average) has not been discussed. Figure 17.27 shows the superiority of Ted Williams, with respect to BBA, as compared with all the other great hitters listed in Table 17.10. In 1941 and 1957, Ted's BBA was above .200. All the other players in the table had a BBA of less than .150. Even though this chapter is not involved with OBP, because of Williams's very high BBA we see, from Table 17.10, that Williams had the two highest yearly OBPs. From our work in Chapter 16, we learned that Williams owns the longest on-base streak in the history of baseball.

A few more remarks are needed about the importance of a player's ability to draw walks. We will establish shortly that there has not been a player in the history of baseball who we would consider a *true* .400 hitter for a season. From the *Law of Large Numbers*, the more at bats a player has, the more his

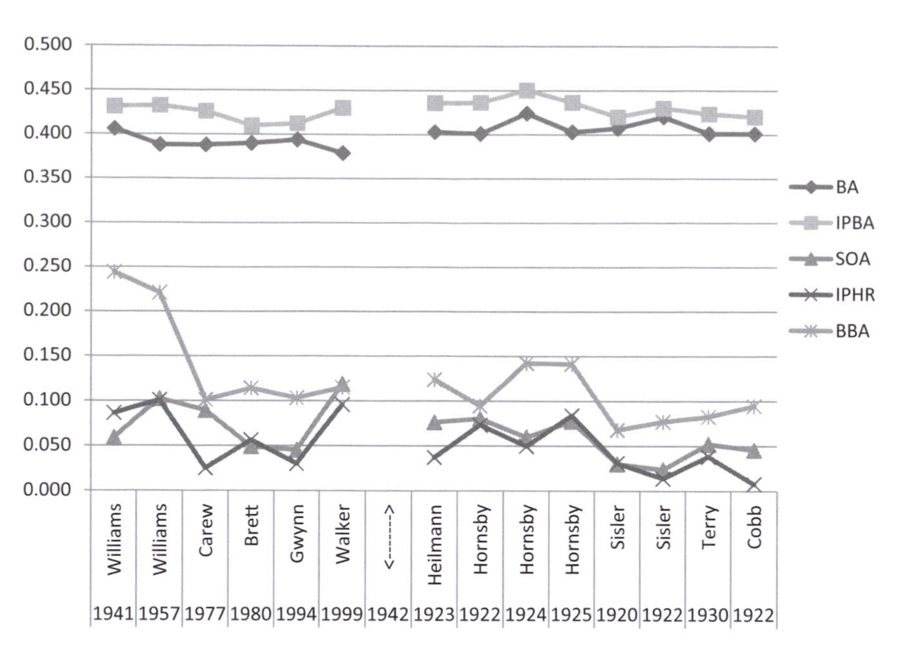

Figure 17.27 Line-graph comparison for the *four statistics* and BBA for the players in Table 17.10

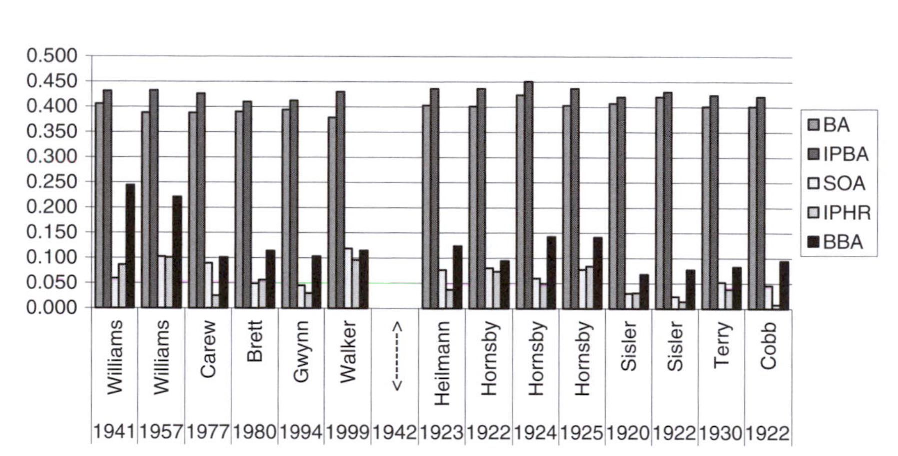

Figure 17.28 Bar-graph comparison for the *four statistics* and BBA for the players in Table 17.10

batting average approaches his *true* batting average. Since a walk is counted as a plate appearance but not as an at bat, getting a walk benefits a player in two ways. An increase in the number of plate appearances helps qualify a player for the batting title. At the same time, not increasing the number of at bats helps a player maintain a higher batting average. Tony Gwynn was hitting .394 in 1994 before the players' strike halted baseball. He only had a total of 419 at bats for that year. If he had completed the season, most baseball experts feel that it is more likely that his batting average would have been lower than .394. Of course, we will never know. Drawing many walks in a season will have a positive effect on a player's batting average for these additional reasons:

- A player tends to swing at pitches that are in the strike zone. These pitches are easier to hit hard.
- A player will see more of a pitcher's pitches before hitting. Seeing more pitches definitely helps the hitter.
- A player is more likely to have a favorable count in the at bat. A favorable count forces the pitcher to throw a better pitch to hit.

We leave the discussion of the role of walks and the BBA statistic to the interested reader.

Many baseball people would pick Albert Pujols and Ichiro Suzuki as the players most likely to hit .400 since 2000. Pujols and Suzuki are definitely different types of hitters. Albert Pujols is a right-handed power hitter. Ichiro Suzuki is a left-handed slap hitter who occasionally demonstrates power. Suzuki is a leadoff batter but does not draw many walks. For this reason, his yearly at bats are very high compared with Pujols.

Exercise 17.20. Table 17.15 shows the statistics for Ichiro Suzuki (2004), Albert Pujols (2003), and Ted Williams (1941).

a. Compare these three players for their given years by constructing a line-graph comparison of AVG, IPBA, SOA, BBA, and IPHR.
b. Compare these three players for their given years by constructing a bar-graph comparison of AVG, IPBA, SOA, BBA, and IPHR.
c. Assuming that their at bats stayed the same, how many additional hits did Suzuki and Pujols need to bat .400 for their given years?
d. Assuming that their at bats and IPBA stayed the same for that year, determine the reduction in number of strikeouts needed for Suzuki and Pujols to bat .400 for that year.
e. Compare these three players for their given years by using the Sum Method from Section 17.6, with a basket of statistics consisting of AVG, IPBA, SOA, IPHR, and BBA.
f. Compare these three players by using the Adjustment Method from Section 17.6 for the statistic AVG.
g. Which player, Suzuki or Pujols, do you think had a better chance of batting .400 for the years listed in the table? Give your reasons.

Table 17.15 Statistics for Ichiro Suzuki (2004), Albert Pujols (2003), and Ted Williams (1941)

Year	Player	G	AB	H	HR	BB	SO	AVG	IPBA	SOA	IPHR	BBA	OBP	SLG
1941	Williams	143	456	185	37	147	27	0.406	0.431	0.059	0.086	0.244	0.553	0.735
2004	Suzuki	161	704	262	8	49	63	0.372	0.409	0.089	0.012	0.065	0.414	0.455
2003	Pujols	157	591	212	43	79	65	0.359	0.403	0.110	0.082	0.118	0.439	0.667

17.8.2. Relationships between AVG, IPBA, SOA, and IPHR for the Players in Table 17.10

Figures 17.29–17.35 use the statistics for the 14 players listed in Table 17.10. These 14 players consist of those nine players who batted at least .400 since 1913 along with the five players who came closest to batting .400 since 1941.

Figure 17.29 contains the regression line that uses IPBA to predict AVG. Figure 17.30 contains the regression line that uses SOA to predict AVG. For both graphs, we locate the x-value corresponding to a y-value of .400.

We rewrite the two linear regression equations as

$$\text{AVG} = 0.539 * \text{IPBA} + 0.1696, \quad \text{(a)}$$
$$\text{AVG} = -0.2898 * \text{SOA} + 0.4190. \quad \text{(b)}$$

Formula (a) says that each addition of 10 points (.010) to the IPBA will increase a player's AVG by 0.539 * 0.010 = .0054 (approximately 5.5 points).

Formula (b) says that each addition of 10 points to the SOA will decrease a player's AVG by −0.2898 * 0.010 = −.0029 (approximately 3 points).

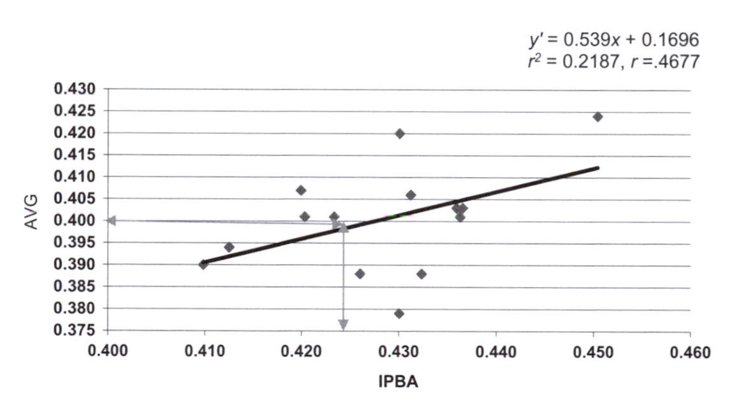

Figure 17.29 Regression line, equation, and correlation coefficient for X = IPBA, Y = AVG (players in Table 17.10)

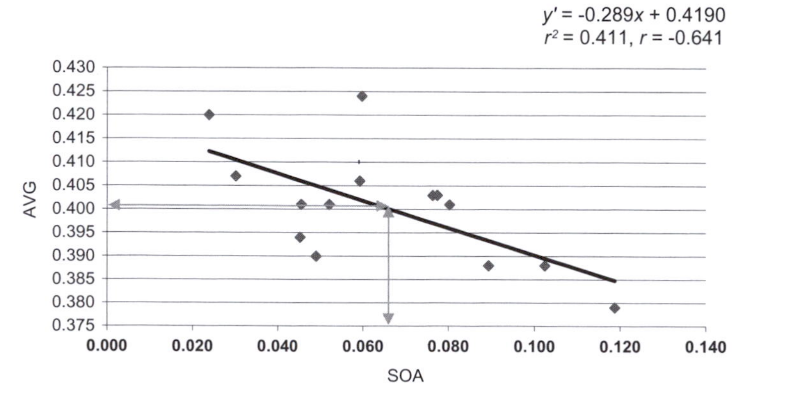

Figure 17.30 Regression line, equation, and correlation coefficient for X = SOA, Y = AVG (players in Table 17.10)

We let AVG = .400 in each of these equations and solve for the other variable. We solve for IPBA in Formula (a):

.400 = .539 * IPBA + .1696,
IPBA = (.400 − .1696)/.539 = .427.

Exercise 17.21. Letting AVG = .400, solve for SOA in Formula (b) above.

After completing Exercise 17.21, you should have gotten an SOA equal to .066. In Figures 17.29 and 17.30, we locate the ordered pairs (.427, .400) and (.066, .400) on their corresponding regression lines.

We use these results to claim that the best chance a player has to hit .400 for a season is to have an IPBA of at least .427 and an SOA of less than .066.

Exercise 17.22. Using the regression equation from Figure 17.31, find the x-value for the variable X = IPHR that generates y = .400 for Y = AVG. Construct a horizontal line through y = .400 and then draw the vertical line through the intersection of the horizontal line and the regression line. Does the intersection of the vertical line with the x-axis correspond with your algebraic result?

We now interpret the results in Figures 17.29–17.35. Figure 17.29 shows a moderate positive correlation between IPBA and AVG. Figure 17.30 shows a negative correlation (close to strong) between SOA and AVG.

Figure 17.31 shows a weak negative correlation between IPHR and AVG. The addition of .010 (10 points) in the IPHR will result in a decrease in a player's AVG of −0.148 * .010 = −.0015 (approximately 1.5 points).

We now turn our attention to Figures 17.32a and 17.32b. Figure 17.32a shows a strong positive correlation between IPHR and SOA, and Figure 17.32b shows a weak positive correlation between IPHR and IPBA. Setting SOA = .066, we solve for x to obtain an IPHR of x = .054. This result is shown in Figure 17.32a. Therefore, an IPHR of less than .054 would give an

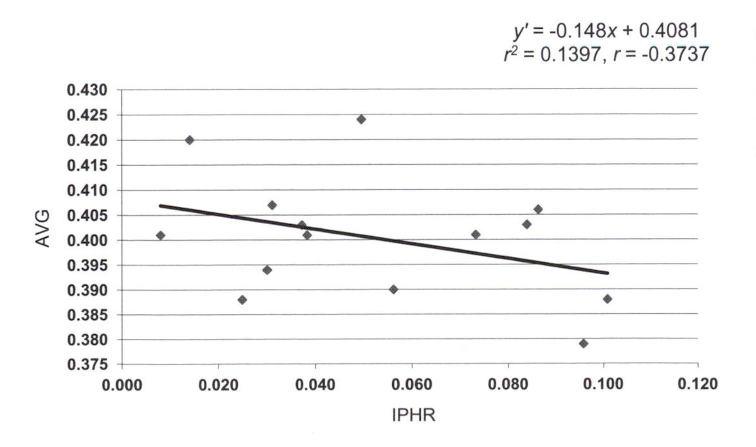

Figure 17.31 Regression line, equation, and correlation coefficient for X = IPHR, Y = AVG (players in Table 17.10)

SOA of less than .066. From our prior work, an SOA of less than .066 yields an AVG of at least .400.

Figure 17.32a shows that an increase in IPHR results in an increase in SOA. Figure 17.30 shows that an increase in SOA decreases the AVG. These two relationships help to explain why an increase in IPHR results in a decrease in AVG. We demonstrate this by finding the AVG as a result of an increase in IPHR in two ways: first using both Figures 17.30 and 17.32a, and then just using Figure 17.31. An increase of .010 in IPHR in Figure 17.32a results in an increase of .006 in SOA. An increase of .006 in SOA in Figure 17.30 results in a decrease of .0017 in AVG. Substituting .010 for IPHR in Figure 17.31 decreases the AVG by .0015.

To have an AVG of at least .400, one must first have an IPBA of at least .400. To stay above .400 for your AVG, the difference (IPBA – AVG) must be small. Figure 17.33 shows a strong positive correlation between SOA and (IPBA – AVG). Letting SOA = .066, we get from the regression equation (IPBA – AVG) = .028. This says that for a player to have an AVG of .400, he needs an IPBA of at least .428. As SOA gets smaller, so does (IPBA – AVG).

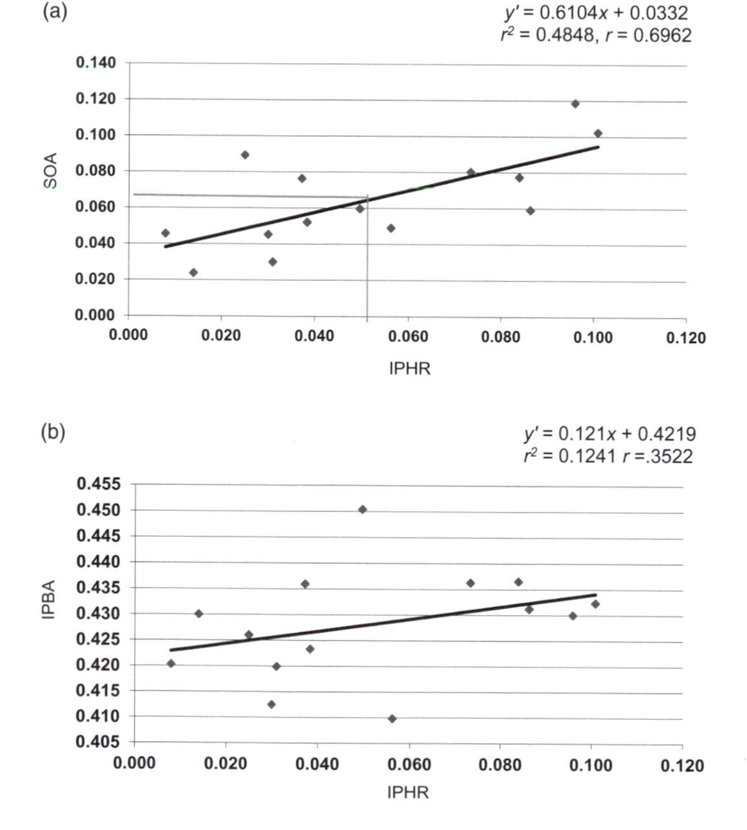

Figure 17.32 (a) Regression line, equation, and correlation coefficient for X = IPHR, Y = SOA (players in Table 17.10). (b) Regression line, equation, and correlation coefficient for X = IPHR, Y = IPBA (players in Table 17.10).

This says that if we can lower the SOA, then a lower IPBA will be needed to have an AVG of at least .400.

Exercise 17.23. How does the above result compare with solving the two regression equations for $y = .400$ in Figures 17.29 and 17.30?

The regression equations in Figures 17.33, 17.34, and 17.35 use the statistics SOA, IPBA, and IPHR to predict (IPBA – AVG). The results show that

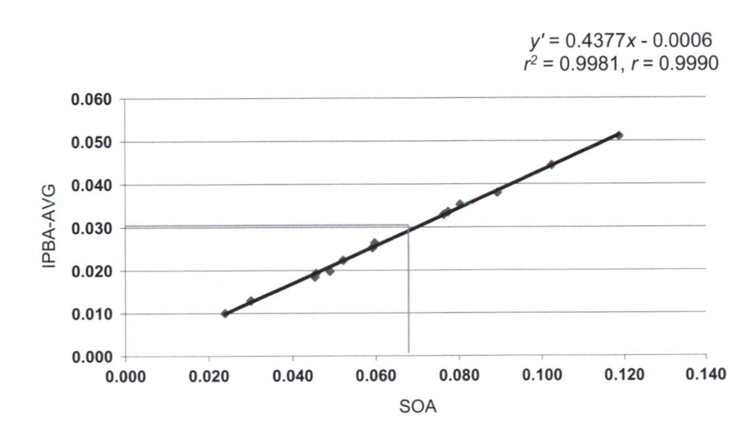

Figure 17.33 Regression line, equation, and correlation coefficient for $X = \text{SOA}$, $Y = \text{IPBA} - \text{AVG}$ (players in Table 17.10)

$y' = 0.461x - 0.1696$
$r^2 = 0.17$, $r = 0.4123$

IPBA-AVG: 0.060, 0.050, 0.040, 0.030, 0.020, 0.010, 0.000

IPBA: 0.400, 0.410, 0.420, 0.430, 0.440, 0.450, 0.460

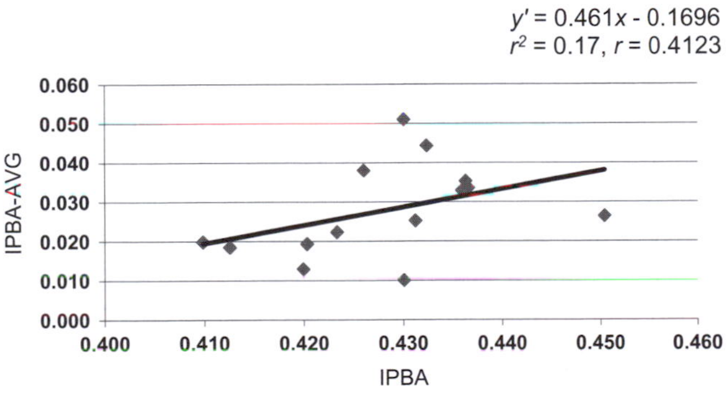

Figure 17.34 Regression line, equation, and correlation coefficient for $X = \text{IPBA}$, $Y = \text{IPBA} - \text{AVG}$ (players in Table 17.10)

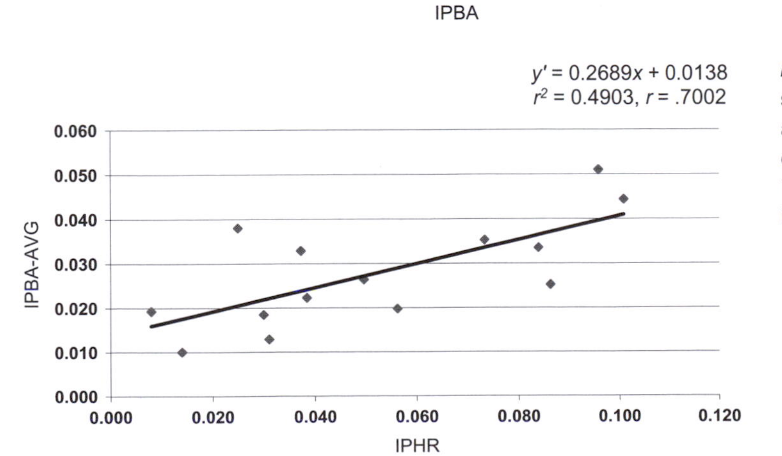

Figure 17.35 Regression line, equation, and correlation coefficient for $X = \text{IPHR}$, $Y = \text{IPBA} - \text{AVG}$ (players in Table 17.10)

the strongest positive correlation exists for SOA. Also, IPHR showed a strong positive correlation. IPBA showed a moderate positive correlation. These regression equations show that the best way to reduce the difference between IPBA and AVG is to reduce your SOA.

17.9. A *True* .400 Batting Average for a Year

We take the time to review the meaning of a player having a *true* AVG of .400 for a year. A *true* AVG for a year represents the population proportion of hits to at bats, based on a theoretical infinite population of at bats for that player for that year. The population proportion is different from the sample proportion (actual yearly AVG). The sample proportion is calculated from a player's actual at bats for a year. To understand the concept of the population proportion for an infinite population, we give the following example. Consider flipping a coin an infinite number of times. Physically, one cannot perform this task an infinite number of times. However, we know theoretically that the population proportion for the number of heads to the number of tosses would be 1/2. The reason we can assign a population proportion to the experiment of flipping a coin an infinite number of times is that each flip of the coin results in two equally likely outcomes determined strictly by chance. For an at bat, the two results of getting a hit or making an out are not equally likely and are determined by a combination of skill and chance. Therefore, the *true* AVG (population proportion) will never be known. The *Law of Large Numbers* states that the more times an experiment is repeated, the closer the sample proportion will get to the population proportion. As a result, we can conclude, for experiments (repeatedly tossing a coin or repeatedly batting), that the sample proportion will approach the population proportion. Since a player's *true* AVG will never be known, we use the technique of hypothesis testing for one population proportion to define the statement that a player is a *true* .400 hitter. A more general definition for a *true .xyz* hitter is given next.

Definition 17.1. A player will be called a *true .xyz* hitter for a given year if we can conclude, with a level of significance of $\alpha=0.05$, that his *true* batting average (*true* AVG) for that year is greater than (*.xyz* – .001). Alternatively, the *p*-value is less than 0.05.

Applying the above definition to a *true* .400 hitter, we let $.xyz=.400$ and $(.xyz-.001)=.399$. The null hypothesis (H_0), which is assumed true at the beginning, is the statement that $p=$ the *true* AVG$\leq .399$. That is, we are claiming at the beginning that there has never been a *true* .400 hitter in Major League history. The statement that the *true* AVG$>.399$ will be the alternate hypothesis (H_A). The null hypothesis will be rejected only if, under the assumption that the null hypothesis is true, the test statistic, calculated from the sample yearly AVG, has a probability of less than 5% of occurring by chance. From our definition, the rejection of the null hypothesis means that the player is a *true* .400 hitter.

With the substitution of OBP or SOA for AVG, the above definition can be used to show that a player's *true* OBP or *true* SOA for a given year is less than or more than a specific value. Why can't we use the above definition for the statistic SLG?

In the previous section, we looked at the five players who came closest to batting .400 for one season since 1941. For each of these players, their yearly AVG is a sample proportion. Using these sample proportions, we can construct 95% confidence intervals to estimate their *true* AVG for their seasons.

We begin with Rod Carew's sample proportion of .388 in 1977. For that year, he had $n=616$ at bats and $x=239$ hits. His sample proportion $\hat{p}=x/n=239/616=.388$ (rounded to three decimal places).

The margin of error $=1.96*\sqrt{(\hat{p}(1-\hat{p})/n}=1.96*\sqrt{.388*.612/616}=.0385$. The 95% confidence interval for Carew's *true* AVG is constructed as follows:

$$
\begin{aligned}
.388-.0385<p<.388+.0385,\\
.3495<p<.4265,\\
.350<p<.427.
\end{aligned}
$$

We can say with 95% confidence that Carew's *true* AVG for the 1977 season was between .350 and .427. What does this mean? Since .400 lies in the confidence interval, there is some evidence that for 1977 Carew's *true* AVG could have been .400. Since the number of values in an interval is infinite, Carew could have been a *true* .360 hitter or a *true* .365 hitter just as easily as a *true* .400 hitter. In other words, just because .400 lies in the 95% confidence does not mean that we can conclude that he was a *true* .400 hitter for that year.

We now look at Williams for 1941. In that year, Williams batted .406 by getting 185 hits in 456 at bats. The 95% confidence interval is

$$
\begin{aligned}
.406-1.96*\sqrt{.406*.594/456}<p<.406+1.96*\sqrt{.406*.594/456},\\
.406-.0451<p<.406+.0451,\\
.361<p<.451.
\end{aligned}
$$

Since .400 lies in the confidence interval, there is some evidence that Ted could have been a *true* .400 hitter for 1941.

Can we conclude from the above confidence interval that Williams was a *true* .400 hitter for 1941? The answer is no. Notice that the conclusion for

Carew is the same as for Williams. Even though Williams did bat over .400 while Carew batted less than .400, the difference between their performances could be due more to chance than to ability. In fact, going into the last day of the 1941 season, Williams had a batting average of .400. The last day consisted of a doubleheader. Williams was given the option of playing or not playing. He chose to play and went 6 for 8 in the doubleheader, raising his batting average to .406.

Exercise 17.24. Assuming that Williams had eight at bats in the last doubleheader, how many hits did he need to maintain his .400 AVG?

Since 1913, Rogers Hornsby (1924) had the highest yearly batting average. For that year, Hornsby batted .424 for the Cardinals with 227 hits in 536 at bats. His 95% confidence interval for AVG for 1924 was

$$.424-1.96*\sqrt{.424*.576/536}<p<.424+1.96*\sqrt{.424*.576/536},$$
$$.424-.0418<p<.424+.0418,$$
$$.382<p<.466.$$

Exercise 17.25. What can we conclude about Hornsby being a *true* .400 hitter for 1924?

Exercise 17.26. Construct a 95% confidence interval estimate for the *true* batting average of Brett (1980), Gwynn (1994), Walker (1999), and Williams (1957).

How do we show that a player is a *true*.400 hitter? Our definition of a *true* .400 hitter is set up to use hypothesis testing. Remember, our claim is that there never has been a *true* .400 hitter. As an example of this process, we consider Rogers Hornsby's 1924 season. We now provide the eight steps in hypothesis testing.

Step 1: Can we conclude, with a level of significance of 0.05, that Rogers Hornsby's *true* batting average for 1924 was greater than .399?
Step 2: H_0: $p \leq .399$ (claim); H_A: $p > .399$.
Step 3: Level of significance: $\alpha = 0.05$.
Step 4: Sample data: $n = 536$, $\hat{p} = 227/536 = .424$. There are no assumptions for the X-population. Since np and $n(1-p)$ are both greater than 5, the sampling distribution can be assumed normal.
Step 5: Convert the test statistic $\hat{p}$ to the standard test statistic z:

$$\begin{aligned} z &= (\hat{p}-p)/\sqrt{p(1-p)/n} \\ &= (.424-.399)/\sqrt{.399*.601/536} \\ &= 1.18. \end{aligned}$$

Step 6: The acceptance and rejection regions are set up in Figure 17.36.

Step 7: Statistical decision: Since the test statistic lies under the acceptance region, we accept H_0.

Step 8: We *cannot* conclude that Hornsby was a true .400 hitter for 1924.

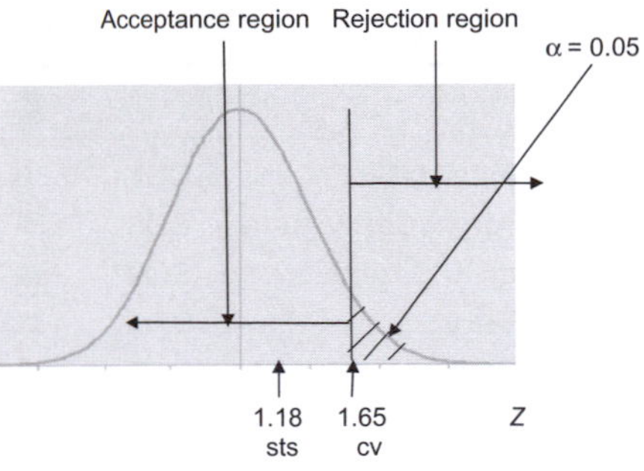

Figure 17.36 Acceptance and rejection regions for the hypothesis test for Hornsby being a true .400 hitter for the year 1924

It should be clear that the technique of hypothesis testing can be used to show that any player's *true* baseball statistic is above or below a given number. However, the *true* baseball statistic must be either a mean or a proportion.

Exercise 17.27. Find the *p*-value for the above hypothesis test.

Exercise 17.28. What is the minimum yearly AVG needed by Rogers Hornsby in 1924 to be considered a *true* .400 hitter for that year? Use hypothesis testing to verify that Hornsby would be a true .400 hitter with your answer. (Use for the sample size the number of at bats for Hornsby in 1924.)

Exercise 17.29. Can we conclude that Williams for the 1941 season was a *true* .400 hitter?

17.10. How Does a Player, Today and in the Future, Achieve a Yearly AVG of at Least .400?

In this section, we search for those players whom, for the seasons 2000–2007, we believe had the best chance of hitting .400 for a season. The players chosen met the following three requirements for their respective seasons:

1. Rank in the top 40 in AVG for that year.
2. Have an IPBA of at least .400 for that year.
3. Qualify for the batting title (at least 502 plate appearances) for that year.

These three requirements were met by 36 players, a total of 56 times.

Several tables will be used in the analysis. Tables 17.16–17.19 show the 36 players who met the above three requirements. Since the 36 players met the requirements a total of 56 times, we refer to the players in each of the tables as the *56 players.*

Table 17.16 shows the 56 players sorted in descending order based on AVG.

Of the top 28 AVGs, 7 belonged to players on the Colorado Rockies. One might conclude that their favorable home ballpark contributed to this statistic.

Table 17.17 shows the 56 players sorted in descending order based on IPBA.

Table 17.16 **The 56 players sorted in descending order based on AVG**

Rank	Player	Team	AB	H	HR	BB	SO	AVG	IPBA	SOA	IPHR	Year
1	Todd Helton	COL	580	216	42	103	61	0.372	0.416	0.105	0.072	2000
2	Nomar Garciaparra	BOS	529	197	21	61	50	0.372	0.411	0.095	0.040	2000
3	Ichiro Suzuki	SEA	704	262	8	49	63	0.372	0.409	0.089	0.011	2004
4	Barry Bonds	SFO	403	149	46	198	47	0.370	0.419	0.117	0.114	2002
5	Magglio Ordonez	DET	595	216	28	76	79	0.363	0.419	0.133	0.047	2007
6	Barry Bonds	SFO	373	135	45	232	41	0.362	0.407	0.110	0.121	2004
7	Albert Pujols	STL	591	212	43	79	65	0.359	0.403	0.110	0.073	2003
8	Todd Helton	COL	583	209	33	111	72	0.358	0.409	0.123	0.057	2003
9	Darin Erstad	ANA	676	240	25	64	82	0.355	0.404	0.121	0.037	2000
10	Manny Ramirez	CLE	439	154	38	86	117	0.351	0.478	0.267	0.087	2000
11	Larry Walker	COL	497	174	38	82	103	0.350	0.442	0.207	0.076	2001
12	Manny Ramirez	BOS	436	152	33	73	85	0.349	0.433	0.195	0.076	2002
13	Todd Helton	COL	547	190	32	127	72	0.347	0.400	0.132	0.059	2004
14	Carlos Delgado	TOR	569	196	41	123	104	0.344	0.422	0.183	0.072	2000
15	Derek Jeter	NYY	623	214	14	69	102	0.343	0.411	0.164	0.022	2006
16	Jason Giambi	OAK	520	178	38	129	83	0.342	0.407	0.160	0.073	2001
17	Barry Bonds	SFO	390	133	45	148	58	0.341	0.401	0.149	0.115	2003
18	Matt Holliday	COL	636	216	36	63	126	0.340	0.424	0.198	0.057	2007
19	Melvin Mora	BAL	550	187	27	66	95	0.340	0.411	0.173	0.049	2004
20	Miguel Cabrera	FLA	576	195	26	86	108	0.339	0.417	0.188	0.045	2006
21	Derek Jeter	NYY	593	201	15	68	99	0.339	0.407	0.167	0.025	2000
22	Jorge Posada	NYY	506	171	20	74	98	0.338	0.419	0.194	0.040	2007
23	Todd Helton	COL	587	197	49	98	104	0.336	0.408	0.177	0.083	2001
24	Derrek Lee	CHC	594	199	46	85	109	0.335	0.410	0.184	0.077	2005
25	Jeffrey Hammonds	COL	454	152	20	44	83	0.335	0.410	0.183	0.044	2000
26	Jeff Kent	SFO	587	196	33	90	107	0.334	0.408	0.182	0.056	2000
27	Ivan Rodriguez	DET	527	176	19	41	91	0.334	0.404	0.173	0.036	2004
28	Jason Giambi	OAK	510	170	43	137	96	0.333	0.411	0.188	0.084	2000
29	David Ortiz	BOS	549	182	35	111	103	0.332	0.408	0.188	0.064	2007
30	Lance Berkman	HOU	577	191	34	92	121	0.331	0.419	0.210	0.059	2001
31	Bret Boone	SEA	623	206	37	40	110	0.331	0.402	0.177	0.059	2001
32	C.Figgins	LAA	442	146	3	51	81	0.330	0.404	0.183	0.007	2007
33	Sammy Sosa	CHC	577	189	64	116	153	0.328	0.446	0.265	0.111	2001
34	Barry Bonds	SFO	476	156	73	177	93	0.328	0.407	0.195	0.153	2001
35	David Wright	NYM	604	196	30	94	115	0.325	0.401	0.190	0.050	2007
36	Miguel Cabrera	FLA	613	198	33	64	125	0.323	0.406	0.204	0.054	2005
37	Alex Rodriguez	NYY	605	194	48	91	139	0.321	0.416	0.230	0.079	2005
38	Manny Ramirez	BOS	449	144	35	100	102	0.321	0.415	0.227	0.078	2006
39	Erubiel Durazo	OAK	511	164	22	56	104	0.321	0.403	0.204	0.043	2004
40	Sammy Sosa	CHC	604	193	50	91	168	0.320	0.443	0.278	0.083	2000
41	Miguel Cabrera	FLA	588	188	34	79	127	0.320	0.408	0.216	0.058	2007
42	Alex Rodriguez	TEX	632	201	52	75	131	0.318	0.401	0.207	0.082	2001
43	Alex Rodriguez	SEA	554	175	41	100	121	0.316	0.404	0.218	0.074	2000

Table 17.16 **The 56 players sorted in descending order based on AVG, *continued***

Rank	Player	Team	AB	H	HR	BB	SO	AVG	IPBA	SOA	IPHR	Year
44	Jermaine Dye	CHW	539	170	44	59	118	0.315	0.404	0.219	0.082	2006
45	Ryan Howard	PHI	581	182	58	108	181	0.313	0.455	0.312	0.100	2006
46	Jim Edmonds	STL	476	148	28	86	134	0.311	0.433	0.282	0.059	2002
47	Travis Hafner	CLE	482	150	28	68	111	0.311	0.404	0.230	0.058	2004
48	Travis Hafner	CLE	454	140	42	100	111	0.308	0.408	0.244	0.093	2006
49	Manny Ramirez	BOS	529	162	41	81	147	0.306	0.424	0.278	0.078	2001
50	Phil Nevin	SDG	546	167	41	71	147	0.306	0.419	0.269	0.075	2001
51	Jason Bay	PIT	599	183	32	95	142	0.306	0.400	0.237	0.053	2005
52	Travis Hafner	CLE	486	148	33	79	123	0.305	0.408	0.253	0.068	2005
53	Jim Thome	CLE	480	146	52	122	139	0.304	0.428	0.290	0.108	2002
54	Jim Edmonds	STL	500	152	30	93	136	0.304	0.418	0.272	0.060	2001
55	Jim Edmonds	STL	498	150	42	101	150	0.301	0.431	0.301	0.084	2004
56	B.J. Upton	TAM	474	142	24	65	154	0.300	0.444	0.325	0.051	2007

Table 17.18 shows the 56 players sorted in ascending order based on SOA. Observe that in this case having a smaller SOA is better.

Table 17.19 shows the 56 players sorted in descending order based on IPHR.

Table 17.20 shows a list of the 10 players who appear more than once in Table 17.16.

Table 17.21 shows, by rank, the top 10 players in AVG. For each of these players, their rank for SOA, IPBA, and IPHR is also provided.

Here are some observations from Table 17.21:

- Each player in the top nine for AVG is also in the top 10 for SOA.
- The top three players for AVG are also the top three in SOA.
- Manny Ramirez ranked 10th in AVG. However, his SOA of .267 ranked 47th. How was he able to have such a high rank by AVG? The answer is that he ranked first in IPBA. His IPBA was an astronomical .478.
- Ramirez is an outlier when compared with the top nine players in rank in AVG. The top nine players in rank in AVG had ranks from 16 to 49 in IPBA.

Table 17.21 shows how strong the role of SOA is in determining a player's AVG.

Figures 17.37–17.41 explore relationships between the four statistics applied to the sample of the 56 players in Table 17.16.

The regression equation in Figure 17.37 shows that SOA has a strong negative correlation with AVG. The regression equation in Figure 17.38 shows a poor correlation between IPBA and AVG. Putting these two results

Table 17.17 **The 56 players sorted in descending order based on IPBA**

Rank	Player	Team	AB	H	HR	BB	SO	AVG	IPBA	SOA	IPHR	Year
1	Manny Ramirez	CLE	439	154	38	86	117	0.351	0.478	0.267	0.087	2000
2	Ryan Howard	PHI	581	182	58	108	181	0.313	0.455	0.312	0.100	2006
3	Sammy Sosa	CHC	577	189	64	116	153	0.328	0.446	0.265	0.111	2001
4	B.J. Upton	TAM	474	142	24	65	154	0.300	0.444	0.325	0.051	2007
5	Sammy Sosa	CHC	604	193	50	91	168	0.320	0.443	0.278	0.083	2000
6	Larry Walker	COL	497	174	38	82	103	0.350	0.442	0.207	0.076	2001
7	Manny Ramirez	BOS	436	152	33	73	85	0.349	0.433	0.195	0.076	2002
8	Jim Edmonds	STL	476	148	28	86	134	0.311	0.433	0.282	0.059	2002
9	Jim Edmonds	STL	498	150	42	101	150	0.301	0.431	0.301	0.084	2004
10	Jim Thome	CLE	480	146	52	122	139	0.304	0.428	0.290	0.108	2002
11	Manny Ramirez	BOS	529	162	41	81	147	0.306	0.424	0.278	0.078	2001
12	Matt Holliday	COL	636	216	36	63	126	0.340	0.424	0.198	0.057	2007
13	Carlos Delgado	TOR	569	196	41	123	104	0.344	0.422	0.183	0.072	2000
14	Jorge Posada	NYY	506	171	20	74	98	0.338	0.419	0.194	0.040	2007
15	Lance Berkman	HOU	577	191	34	92	121	0.331	0.419	0.210	0.059	2001
16	Magglio Ordonez	DET	595	216	28	76	79	0.363	0.419	0.133	0.047	2007
17	Phil Nevin	SDG	546	167	41	71	147	0.306	0.419	0.269	0.075	2001
18	Barry Bonds	SFO	403	149	46	198	47	0.370	0.419	0.117	0.114	2002
19	Jim Edmonds	STL	500	152	30	93	136	0.304	0.418	0.272	0.060	2001
20	Miguel Cabrera	FLA	576	195	26	86	108	0.339	0.417	0.188	0.045	2006
21	Alex Rodriguez	NYY	605	194	48	91	139	0.321	0.416	0.230	0.079	2005
22	Todd Helton	COL	580	216	42	103	61	0.372	0.416	0.105	0.072	2000
23	Manny Ramirez	BOS	449	144	35	100	102	0.321	0.415	0.227	0.078	2006
24	Nomar Garciaparra	BOS	529	197	21	61	50	0.372	0.411	0.095	0.040	2000
25	Melvin Mora	BAL	550	187	27	66	95	0.340	0.411	0.173	0.049	2004
26	Derek Jeter	NYY	623	214	14	69	102	0.343	0.411	0.164	0.022	2006
27	Jason Giambi	OAK	510	170	43	137	96	0.333	0.411	0.188	0.084	2000
28	Derrek Lee	CHC	594	199	46	85	109	0.335	0.410	0.184	0.077	2005
29	Jeffrey Hammonds	COL	454	152	20	44	83	0.335	0.410	0.183	0.044	2000
30	Todd Helton	COL	583	209	33	111	72	0.358	0.409	0.123	0.057	2003
31	Ichiro Suzuki	SEA	704	262	8	49	63	0.372	0.409	0.089	0.011	2004
32	Jeff Kent	SFO	587	196	33	90	107	0.334	0.408	0.182	0.056	2000
33	Travis Hafner	CLE	454	140	42	100	111	0.308	0.408	0.244	0.093	2006
34	David Ortiz	BOS	549	182	35	111	103	0.332	0.408	0.188	0.064	2007
35	Todd Helton	COL	587	197	49	98	104	0.336	0.408	0.177	0.083	2001
36	Miguel Cabrera	FLA	588	188	34	79	127	0.320	0.408	0.216	0.058	2007
37	Travis Hafner	CLE	486	148	33	79	123	0.305	0.408	0.253	0.068	2005
38	Jason Giambi	OAK	520	178	38	129	83	0.342	0.407	0.160	0.073	2001
39	Barry Bonds	SFO	476	156	73	177	93	0.328	0.407	0.195	0.153	2001
40	Derek Jeter	NYY	593	201	15	68	99	0.339	0.407	0.167	0.025	2000
41	Barry Bonds	SFO	373	135	45	232	41	0.362	0.407	0.110	0.121	2004
42	Miguel Cabrera	FLA	613	198	33	64	125	0.323	0.406	0.204	0.054	2005

Table 17.17 **The 56 players sorted in descending order based on IPBA, *continued***

Rank	Player	Team	AB	H	HR	BB	SO	AVG	IPBA	SOA	IPHR	Year
43	C.Figgins	LAA	442	146	3	51	81	0.330	0.404	0.183	0.007	2007
44	Travis Hafner	CLE	482	150	28	68	111	0.311	0.404	0.230	0.058	2004
45	Alex Rodriguez	SEA	554	175	41	100	121	0.316	0.404	0.218	0.074	2000
46	Darin Erstad	ANA	676	240	25	64	82	0.355	0.404	0.121	0.037	2000
47	Jermaine Dye	CHW	539	170	44	59	118	0.315	0.404	0.219	0.082	2006
48	Ivan Rodriguez	DET	527	176	19	41	91	0.334	0.404	0.173	0.036	2004
49	Albert Pujols	STL	591	212	43	79	65	0.359	0.403	0.110	0.073	2003
50	Erubiel Durazo	OAK	511	164	22	56	104	0.321	0.403	0.204	0.043	2004
51	Bret Boone	SEA	623	206	37	40	110	0.331	0.402	0.177	0.059	2001
52	Alex Rodriguez	TEX	632	201	52	75	131	0.318	0.401	0.207	0.082	2001
53	David Wright	NYM	604	196	30	94	115	0.325	0.401	0.190	0.050	2007
54	Barry Bonds	SFO	390	133	45	148	58	0.341	0.401	0.149	0.115	2003
55	Jason Bay	PIT	599	183	32	95	142	0.306	0.400	0.237	0.053	2005
56	Todd Helton	COL	547	190	32	127	72	0.347	0.400	0.132	0.059	2004

Table 17.18 **The 56 players sorted in ascending order based on SOA**

Rank	Player	Team	AB	H	HR	BB	SO	AVG	IPBA	SOA	IPHR	Year
1	Ichiro Suzuki	SEA	704	262	8	49	63	0.372	0.409	0.089	0.011	2004
2	Nomar Garciaparra	BOS	529	197	21	61	50	0.372	0.411	0.095	0.040	2000
3	Todd Helton	COL	580	216	42	103	61	0.372	0.416	0.105	0.072	2000
4	Barry Bonds	SFO	373	135	45	232	41	0.362	0.407	0.110	0.121	2004
5	Albert Pujols	STL	591	212	43	79	65	0.359	0.403	0.110	0.073	2003
6	Barry Bonds	SFO	403	149	46	198	47	0.370	0.419	0.117	0.114	2002
7	Darin Erstad	ANA	676	240	25	64	82	0.355	0.404	0.121	0.037	2000
8	Todd Helton	COL	583	209	33	111	72	0.358	0.409	0.123	0.057	2003
9	Todd Helton	COL	547	190	32	127	72	0.347	0.400	0.132	0.059	2004
10	Miguel Ordonez	DET	595	216	28	76	79	0.363	0.419	0.133	0.047	2007
11	Barry Bonds	SFO	390	133	45	148	58	0.341	0.401	0.149	0.115	2003
12	Jason Giambi	OAK	520	178	38	129	83	0.342	0.407	0.160	0.073	2001
13	Derek Jeter	NYY	623	214	14	69	102	0.343	0.411	0.164	0.022	2006
14	Derek Jeter	NYY	593	201	15	68	99	0.339	0.407	0.167	0.025	2000
15	Ivan Rodriguez	DET	527	176	19	41	91	0.334	0.404	0.173	0.036	2004
16	Melvin Mora	BAL	550	187	27	66	95	0.340	0.411	0.173	0.049	2004
17	Bret Boone	SEA	623	206	37	40	110	0.331	0.402	0.177	0.059	2001
18	Todd Helton	COL	587	197	49	98	104	0.336	0.408	0.177	0.083	2001
19	Jeff Kent	SFO	587	196	33	90	107	0.334	0.408	0.182	0.056	2000
20	Carlos Delgado	TOR	569	196	41	123	104	0.344	0.422	0.183	0.072	2000
21	Jeffrey Hammonds	COL	454	152	20	44	83	0.335	0.410	0.183	0.044	2000
22	Chone Figgins	LAA	442	146	3	51	81	0.330	0.404	0.183	0.007	2007

(continued)

Table 17.18 The 56 players sorted in ascending order based on SOA, *continued*

Rank	Player	Team	AB	H	HR	BB	SO	AVG	IPBA	SOA	IPHR	Year
23	Derrek Lee	CHC	594	199	46	85	109	0.335	0.410	0.184	0.077	2005
24	Miguel Cabrera	FLA	576	195	26	86	108	0.339	0.417	0.188	0.045	2006
25	David Ortiz	BOS	549	182	35	111	103	0.332	0.408	0.188	0.064	2007
26	Jason Giambi	OAK	510	170	43	137	96	0.333	0.411	0.188	0.084	2000
27	David Wright	NYM	604	196	30	94	115	0.325	0.401	0.190	0.050	2007
28	Jorge Posada	NYY	506	171	20	74	98	0.338	0.419	0.194	0.040	2007
29	Manny Ramirez	BOS	436	152	33	73	85	0.349	0.433	0.195	0.076	2002
30	Barry Bonds	SFO	476	156	73	177	93	0.328	0.407	0.195	0.153	2001
31	Matt Holliday	COL	636	216	36	63	126	0.340	0.424	0.198	0.057	2007
32	Erubiel Durazo	OAK	511	164	22	56	104	0.321	0.403	0.204	0.043	2004
33	Miguel Cabrera	FLA	613	198	33	64	125	0.323	0.406	0.204	0.054	2005
34	Larry Walker	COL	497	174	38	82	103	0.350	0.442	0.207	0.076	2001
35	Alex Rodriguez	TEX	632	201	52	75	131	0.318	0.401	0.207	0.082	2001
36	Lance Berkman	HOU	577	191	34	92	121	0.331	0.419	0.210	0.059	2001
37	M Cabrera	FLA	588	188	34	79	127	0.320	0.408	0.216	0.058	2007
38	Alex Rodriguez	SEA	554	175	41	100	121	0.316	0.404	0.218	0.074	2000
39	Jermaine Dye	CHW	539	170	44	59	118	0.315	0.404	0.219	0.082	2006
40	Manny Ramirez	BOS	449	144	35	100	102	0.321	0.415	0.227	0.078	2006
41	Alex Rodriguez	NYY	605	194	48	91	139	0.321	0.416	0.230	0.079	2005
42	Travis Hafner	CLE	482	150	28	68	111	0.311	0.404	0.230	0.058	2004
43	Jason Bay	PIT	599	183	32	95	142	0.306	0.400	0.237	0.053	2005
44	Travis Hafner	CLE	454	140	42	100	111	0.308	0.408	0.244	0.093	2006
45	Travis Hafner	CLE	486	148	33	79	123	0.305	0.408	0.253	0.068	2005
46	Sammy Sosa	CHC	577	189	64	116	153	0.328	0.446	0.265	0.111	2001
47	Manny Ramirez	CLE	439	154	38	86	117	0.351	0.478	0.267	0.087	2000
48	Phil Nevin	SDG	546	167	41	71	147	0.306	0.419	0.269	0.075	2001
49	Jim Edmonds	STL	500	152	30	93	136	0.304	0.418	0.272	0.060	2001
50	Manny Ramirez	BOS	529	162	41	81	147	0.306	0.424	0.278	0.078	2001
51	Sammy Sosa	CHC	604	193	50	91	168	0.320	0.443	0.278	0.083	2000
52	Jim Edmonds	STL	476	148	28	86	134	0.311	0.433	0.282	0.059	2002
53	Jim Thome	CLE	480	146	52	122	139	0.304	0.428	0.290	0.108	2002
54	Jim Edmonds	STL	498	150	42	101	150	0.301	0.431	0.301	0.084	2004
55	Ryan Howard	PHI	581	182	58	108	181	0.313	0.455	0.312	0.100	2006
56	B.J. Upton	TAM	474	142	24	65	154	0.300	0.444	0.325	0.051	2007

together, we arrive at the conclusion that for these 56 players, who have the skill to have an IPBA of at least .400, their AVG is almost completely determined by their SOA. What does Figure 17.39 show?

Figure 17.40 shows a regression line with a positive slope associating X=IPHR with Y=SOA. There is a weak positive correlation of r=.24 between IPHR and SOA.

Table 17.19 **The 56 players sorted in descending order based on IPHR**

Rank	Player	Team	AB	H	HR	BB	SO	AVG	IPBA	SOA	IPHR	Year
1	Barry Bonds	SFO	476	156	73	177	93	0.328	0.407	0.195	0.153	2001
2	Barry Bonds	SFO	373	135	45	232	41	0.362	0.407	0.110	0.121	2004
3	Barry Bonds	SFO	390	133	45	148	58	0.341	0.401	0.149	0.115	2003
4	Barry Bonds	SFO	403	149	46	198	47	0.370	0.419	0.117	0.114	2002
5	Sammy Sosa	CHC	577	189	64	116	153	0.328	0.446	0.265	0.111	2001
6	Jim Thome	CLE	480	146	52	122	139	0.304	0.428	0.290	0.108	2002
7	Ryan Howard	PHI	581	182	58	108	181	0.313	0.455	0.312	0.100	2006
8	Travis Hafner	CLE	454	140	42	100	111	0.308	0.408	0.244	0.093	2006
9	Manny Ramirez	CLE	439	154	38	86	117	0.351	0.478	0.267	0.087	2000
10	Jim Edmonds	STL	498	150	42	101	150	0.301	0.431	0.301	0.084	2004
11	Jason Giambi	OAK	510	170	43	137	96	0.333	0.411	0.188	0.084	2000
12	Todd Helton	COL	587	197	49	98	104	0.336	0.408	0.177	0.083	2001
13	Sammy Sosa	CHC	604	193	50	91	168	0.320	0.443	0.278	0.083	2000
14	Alex Rodriguez	TEX	632	201	52	75	131	0.318	0.401	0.207	0.082	2001
15	Jermaine Dye	CHW	539	170	44	59	118	0.315	0.404	0.219	0.082	2006
16	Alex Rodriguez	NYY	605	194	48	91	139	0.321	0.416	0.230	0.079	2005
17	Manny Ramirez	BOS	449	144	35	100	102	0.321	0.415	0.227	0.078	2006
18	Manny Ramirez	BOS	529	162	41	81	147	0.306	0.424	0.278	0.078	2001
19	Derrek Lee	CHC	594	199	46	85	109	0.335	0.410	0.184	0.077	2005
20	Larry Walker	COL	497	174	38	82	103	0.350	0.442	0.207	0.076	2001
21	Manny Ramirez	BOS	436	152	33	73	85	0.349	0.433	0.195	0.076	2002
22	Phil Nevin	SDG	546	167	41	71	147	0.306	0.419	0.269	0.075	2001
23	Alex Rodriguez	SEA	554	175	41	100	121	0.316	0.404	0.218	0.074	2000
24	Jason Giambi	OAK	520	178	38	129	83	0.342	0.407	0.160	0.073	2001
25	Albert Pujols	STL	591	212	43	79	65	0.359	0.403	0.110	0.073	2003
26	Todd Helton	COL	580	216	42	103	61	0.372	0.416	0.105	0.072	2000
27	Carlos Delgado	TOR	569	196	41	123	104	0.344	0.422	0.183	0.072	2000
28	Travis Hafner	CLE	486	148	33	79	123	0.305	0.408	0.253	0.068	2005
29	David Ortiz	BOS	549	182	35	111	103	0.332	0.408	0.188	0.064	2007
30	Jim Edmonds	STL	500	152	30	93	136	0.304	0.418	0.272	0.060	2001
31	Bret Boone	SEA	623	206	37	40	110	0.331	0.402	0.177	0.059	2001
32	Lance Berkman	HOU	577	191	34	92	121	0.331	0.419	0.210	0.059	2001
33	Jim Edmonds	STL	476	148	28	86	134	0.311	0.433	0.282	0.059	2002
34	Todd Helton	COL	547	190	32	127	72	0.347	0.400	0.132	0.059	2004
35	Travis Hafner	CLE	482	150	28	68	111	0.311	0.404	0.230	0.058	2004
36	M Cabrera	FLA	588	188	34	79	127	0.320	0.408	0.216	0.058	2007
37	Todd Helton	COL	583	209	33	111	72	0.358	0.409	0.123	0.057	2003
38	Matt Holliday	COL	636	216	36	63	126	0.340	0.424	0.198	0.057	2007
39	Jeff Kent	SFO	587	196	33	90	107	0.334	0.408	0.182	0.056	2000
40	Miguel Cabrera	FLA	613	198	33	64	125	0.323	0.406	0.204	0.054	2005
41	Jason Bay	PIT	599	183	32	95	142	0.306	0.400	0.237	0.053	2005
42	B.J. Upton	TAM	474	142	24	65	154	0.300	0.444	0.325	0.051	2007

(continued)

Table 17.19 **The 56 players sorted in descending order based on IPHR, *continued***

Rank	Player	Team	AB	H	HR	BB	SO	AVG	IPBA	SOA	IPHR	Year
43	David Wright	NYM	604	196	30	94	115	0.325	0.401	0.190	0.050	2007
44	Melvin Mora	BAL	550	187	27	66	95	0.340	0.411	0.173	0.049	2004
45	MM, Ordonez	DET	595	216	28	76	79	0.363	0.419	0.133	0.047	2007
46	Miguel Cabrera	FLA	576	195	26	86	108	0.339	0.417	0.188	0.045	2006
47	Jeffrey Hammonds	COL	454	152	20	44	83	0.335	0.410	0.183	0.044	2000
48	Erubiel Durazo	OAK	511	164	22	56	104	0.321	0.403	0.204	0.043	2004
49	Nomar Garciaparra	BOS	529	197	21	61	50	0.372	0.411	0.095	0.040	2000
50	Jorge Posada	NYY	506	171	20	74	98	0.338	0.419	0.194	0.040	2007
51	Darin Erstad	ANA	676	240	25	64	82	0.355	0.404	0.121	0.037	2000
52	Ivan Rodriguez	DET	527	176	19	41	91	0.334	0.404	0.173	0.036	2004
53	Derek Jeter	NYY	593	201	15	68	99	0.339	0.407	0.167	0.025	2000
54	Derek Jeter	NYY	623	214	14	69	102	0.343	0.411	0.164	0.022	2006
55	Ichiro Suzuki	SEA	704	262	8	49	63	0.372	0.409	0.089	0.011	2004
56	C.Figgins	LAA	442	146	3	51	81	0.330	0.404	0.183	0.007	2007

The regression equation in Figure 17.41 shows a strong positive correlation between SOA and (IPBA – AVG).

Exercise 17.30. Compare the correlation of SOA with AVG between the players in Table 17.10 and the players in Table 17.16. Compare the correlation of IPBA with AVG between the players in Table 17.10 and the players in Table 17.16.

Exercise 17.31. Find the regression equation and correlation coefficient for *X* = IPBA and *Y* = (IPBA – AVG). Draw some conclusions about the relationship between *X* and *Y*.

Table 17.20 **The 10 players who appear more than once in Table 17.16**

Player	Frequency
Alex Rodriguez	3
Barry Bonds	4
Derek Jeter	2
Jason Giambi	2
Jim Edmonds	3
Manny Ramirez	4
Miguel Cabrera	3
Sammy Sosa	2
Todd Helton	4
Travis Hafner	3

We showed in Section 17.8.2, from our study of the 14 players in Table 17.10, that the ideal situation for a player to have a yearly AVG of at least .400 is for their IPBA to be greater than .427 and their SOA to be less than .066.

Exercise 17.32. Using Formula (17.1) (AVG = IPBA * (1 – SOA)), make a table with the following three columns: AVG, IPBA, and SOA. Let SOA range in increments of .005 (5 points) from .000 to .200. Keeping AVG fixed at .400, record all the corresponding values for IPBA.

Exercise 17.33. For the 56 players, Figure 17.37 shows a strong negative correlation between SOA and AVG, whereas Figure 17.41 shows a strong positive correlation between SOA and (IPBA – AVG). Relate these two figures to Formulas (17.1) and (17.2) at the beginning of the chapter.

Exercise 17.34. What can we conclude about the 56 players in Table 17.16 from Figure 17.39?

To gain further insight on what it would take for a current player to hit .400 for a season, we compare the nine players who had the highest yearly AVG from Table 17.16 (called *the current players*) with the nine players from Table 17.10 (called *the .400 players*) who had an AVG ≥ .400:

- The SOA for each of the current players is greater than the highest SOA for the .400 players.
- Each of the current players has an SOA > .066.
- Each IPBA for the current players is less than the lowest IPBA for the .400 players.
- Each of the current players has an IPBA < .427.

We turn our attention to those players, from 2000 to 2007, with an IPBA > .427. From Table 17.17, the top 10 players had an IPBA > .427. The problem with this group is that their SOAs were very high. The lowest SOA for these 10 players was .195. We could classify these players as sluggers who swing hard and strike out often.

From Table 17.18, only two players had an SOA less than .100. In order to have an AVG = .400 with an SOA = .100, a player would need an IPBA of at least .444. From Table 17.17, only four players had an IPBA of at least .444. Unfortunately, the lowest SOA for those four players was .265.

We now ask the following question for the theoretical population of present-day and future players meeting the three requirements set forth at the beginning of this section: Can we conclude for this population that their *true* yearly SOA is greater than .100? To answer this question, we need to define what is meant by a population of players having a *true* parameter greater than or less than a specific number. The definition below is a natural adjustment to Definition 17.1 in Section 17.9.

Definition 17.2. A population of players will have a *true* yearly parameter greater than (less than) a number c if we can conclude, with a level of significance of $\alpha = 0.05$, that the *true* yearly parameter is greater than (less than) c. Alternatively, the p-value is less than 0.05.

Definition 17.1 deals with a population consisting of all the theoretical at bats for a player for a given year. The *population proportion* is the parameter used to describe the theoretical *true* yearly batting average for a player.

Definition 17.2 deals with a population consisting of all theoretical players in the present and future who meet certain requirements for a given year. The *population mean* is the parameter used to describe the theoretical *true* yearly parameter for that population of players.

Table 17.21 **Rank for SOA, IPBA, and IPHR for the players with the top 10 AVGs**

Player	Team	AVG	Rank	SOA	Rank	IPBA	Rank	IPHR	Rank
Todd Helton	COL	0.372	1	0.105	3	0.416	22	0.072	26
Nomar Garciaparra	BOS	0.372	2	0.095	2	0.411	24	0.040	49
Ichiro Suzuki	SEA	0.372	3	0.089	1	0.409	31	0.011	55
Barry Bonds	SFO	0.370	4	0.117	6	0.419	18	0.114	4
Magglio Ordonez	DET	0.363	5	0.133	10	0.419	16	0.047	45
Barry Bonds	SFO	0.362	6	0.110	4	0.407	41	0.121	2
Albert Pujols	STL	0.359	7	0.110	5	0.403	49	0.073	25
Todd Helton	COL	0.358	8	0.123	8	0.409	30	0.057	37
Darin Erstad	ANA	0.355	9	0.121	7	0.404	49	0.037	51
Manny Ramirez	BOS	0.351	10	0.267	47	0.478	1	0.087	9
Sum of ranks			55		93		281		303

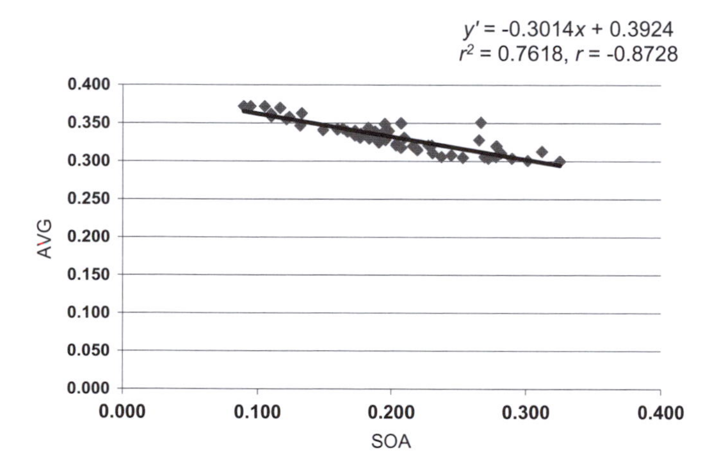

Figure 17.37 Regression line, equation, and correlation coefficient for X=SOA, Y=AVG (56 players in Table 17.16)

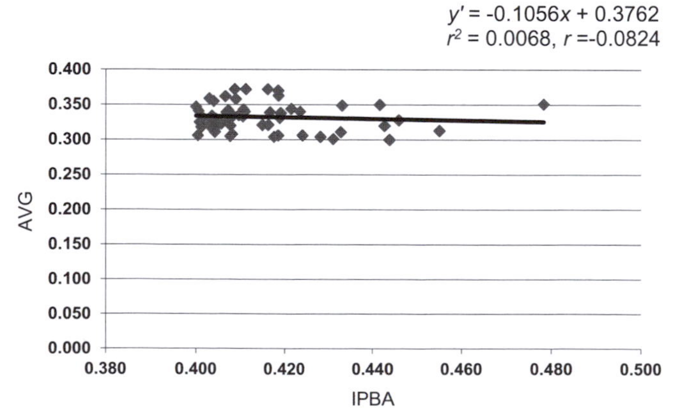

Figure 17.38 Regression line, equation, and correlation coefficient for X=IPBA, Y=AVG (56 players in Table 17.16)

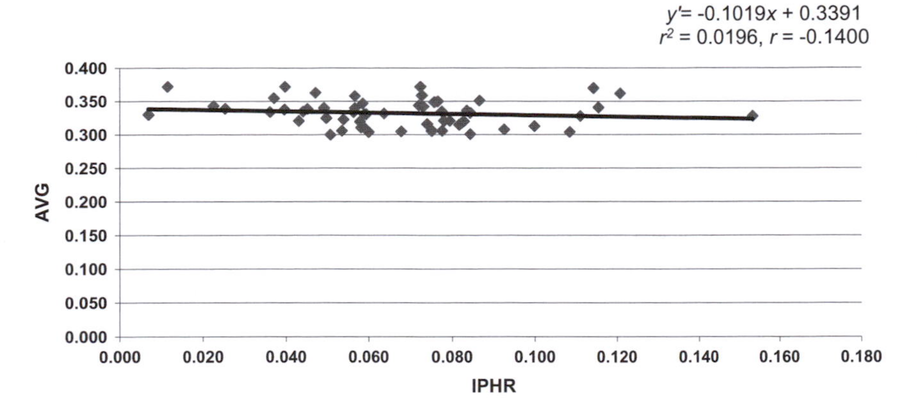

Figure 17.39 Regression line, equation, and correlation coefficient for $X = \text{IPHR}$, $Y = \text{AVG}$ (56 players in Table 17.16)

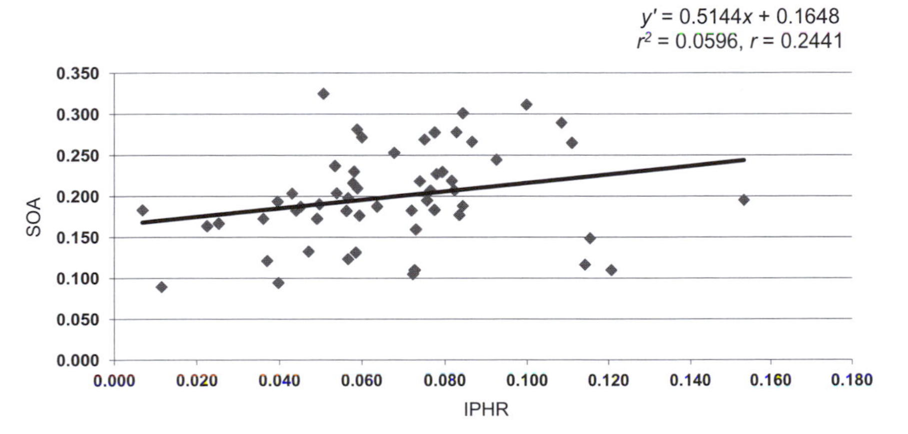

Figure 17.40 Regression line, equation, and correlation coefficient for $X = \text{IPHR}$, $Y = \text{SOA}$ (56 players in Table 17.16)

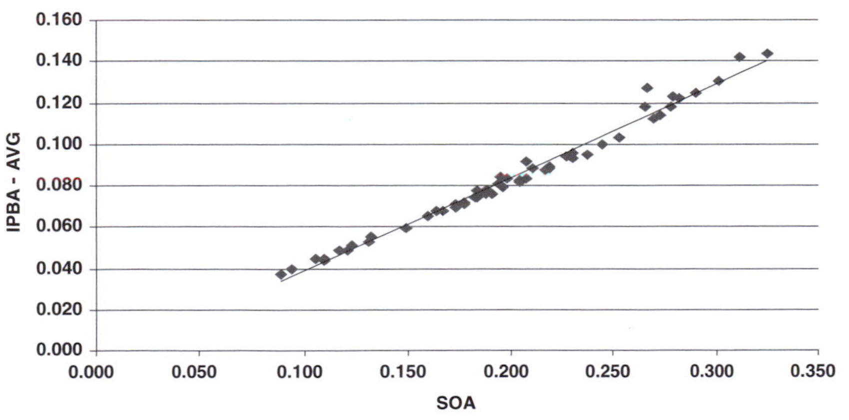

Figure 17.41 Regression line, equation, and correlation coefficient for $X = \text{SOA}$, $Y = \text{IPBA} - \text{AVG}$ (56 players in Table 17.16)

To answer the question posed above, we turn to hypothesis testing for one population mean. Our sample consists of the 56 players in Table 17.16. The sample mean ($\bar{x}$) is calculated from the sample yearly SOAs for the 56 players. The population mean (μ) is the *true* yearly SOA for the theoretical population of current and future players who meet the three requirements defined at the beginning of this section. At the onset, our claim is that this population will have an SOA greater than .100.

Step 1: Research question: Given the theoretical population of present and future players who, for a season, meet the three requirements of (1) ranking in the top 40 in AVG, (2) having an IPBA of at least .400, and (3) having enough plate appearances to qualify for a batting title, can we conclude for this population that the *true* yearly SOA > .100?

Step 2: Two hypotheses: H_0: $\mu \leq .100$; H_A: $\mu >. 100$ (claim).

Step 3: Level of significance: $\alpha = 0.05$.

Step 4: Sample data statistics: $n = 56$, $\bar{x} = .199$, $s = .058$. There are no assumptions for the X-population.

Step 5: Convert the test statistic $\bar{x}$ to the standard test statistic z. Since $n \geq 30$, by Theorem 15.1(c) we have

$$
\begin{aligned}
z &= (\bar{x} - \mu)/(s/\sqrt{n}) \\
&= (.199 - .100)/(.058/\sqrt{56}) \\
&= .099/.0078 \\
&= 12.77.
\end{aligned}
$$

Step 6: The acceptance and rejection regions are given in Figure 17.42; cv is the critical value and sts is the standard test statistic.

Step 7: Statistical decision: Since 12.77 lies under the rejection region, we reject H_0.

Step 8: Answer the research question: We conclude, for the population of current and future players who meet the three requirements, that their *true* yearly SOA is greater than .100. It would be a rare occurrence to find an elite hitter with an SOA less than or equal to .100.

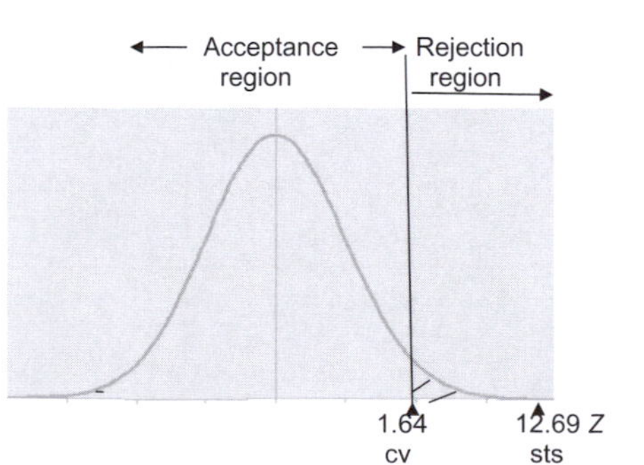

Figure 17.42 Acceptance and rejection regions for the hypothesis test for SOA > .100

Exercise 17.35. Verify that for the sample of 56 players in Table 17.16 the sample mean SOA is .199 and the sample standard deviation SOA is .058.

Why did we choose .100 for the SOA? The choice of .100 was influenced by the trend of graphs in Figures 17.10 and 17.11. These two graphs show that the league SOA was consistently above .100.

From what we said before, we would need an IPBA greater than .444, for an SOA greater than .100, to obtain an AVG of .400. Later, in the problems, you will be asked to test whether for these elite present and future players we can conclude that their true SOA is greater than .150.

Exercise 17.36. Can we conclude for the theoretical population of present and future players who meet the three requirements above for a season that their *true* yearly IPBA $<.444$? (Hint: You must calculate the sample mean and sample standard deviation for the IPBA for the 56 players.)

Exercise 17.37. Can we conclude for the population above that the *true* yearly SOA is greater than .150?

Exercise 17.38. What is the minimum IPBA needed for an SOA $>.150$ to obtain an AVG of .400?

We are now ready to conjecture on how a player in this and future eras can have a batting average of .400 for a season, based on a minimum of 502 plate appearances.

We present *six properties* we believe are keys if a player is to bat .400 for a season. The more properties a player satisfies, the better are his chances of accomplishing this fabulous hitting feat.

Property One: Be a Power Hitter. By definition of the IPBA, a player must achieve at least a .400 IPBA for a minimum of 502 plate appearances for one season. This happens rarely, for between 2000 and 2007 only 56 players (not all different) accomplished this feat, and only 10 of these players did it more than once. These 10 players are listed in Table 17.20. Of the 56 players (not all different) who had at least a .400 IPBA, 40 had at least 30 home runs. We can say that 71% of the players were power hitters. So, property one is to be a power hitter or at least a hard line drive hitter.

Property Two: Have a Home Stadium Favorable for Hitters. Of the top 28 averages in Table 17.16, 25% of the players were members of the Colorado Rockies.

Property Three: Have a Low SOA. Figure 17.37 shows a strong negative correlation between SOA and AVG. The regression equation for the 56 players in Table 17.16 is $AVG=-0.3014*SOA+.3924$. This means that an increase in the SOA of .010 (10 points) corresponds to a drop in AVG of about .003 (3 points). By today's standards, an SOA of less than .100 would be considered sensational. Of our 56 players, only *two* accomplished such a low SOA. An SOA of .100 reduces the AVG by approximately 30 points. This means that the player would have to start with an IPBA of at least .430.

From Table 17.17, only nine players had an IPBA $>.430$. Those nine players had a mean SOA of .270. Figure 17.43 below shows a moderate positive correlation between SOA and IPBA. Drawing a horizontal line through y=IPBA=.430, followed by a vertical line, from the intersection of the horizontal line and the regression line to the x-axis, we see that the intersection of the vertical line and the x-axis is approximately x=SOA=.280.

This shows that, for current players, the typical batter who had a .430 IPBA also had an SOA above .250. With an SOA of .250, a player would need an IPBA of over .530 to hit .400. The highest IPBA of the players in Table 17.16 is .478, over 50 points less than .530. The top three IPBA hitters were Manny Ramirez, Ryan Howard, and Sammy Sosa. All three sluggers are known for their hard swings and their frequency of strikeouts.

Figures 17.43 and 17.44 show the difference in SOA between the 56 current players in Table 17.16 (with an IPBA of at least .400) and the 14 players in Table 17.10 (the 14 players who either batted .400 since 1913 or came close to batting .400 since 1941). A typical player with a .430 IPBA in Table 17.11 is expected to have an SOA of .075, whereas a typical player in Table 17.16 is expected to have an SOA of .270.

The power hitters of today strike out too many times. Ted Williams in 1941 had an SOA of .059 while hitting 37 home runs. In fact, for 13 of his 17 years, Williams's SOA was below .100. Even though we have not looked at the influence of base on balls on AVG, there probably is a relationship. Williams in 1941 had 147 walks. This shows that Williams did not swing at bad pitches and probably was ahead in the count most of the time. Another advantage of getting many walks is the reduction in the number of at bats.

Property Four: Have Fewer At Bats for a Season. Many baseball people believe that having fewer at bats increases the chances of a player having a higher AVG. Statisticians use the concept of regressing to the mean to explain this. If we accept that no player was ever a *true* .400 hitter for a year, the

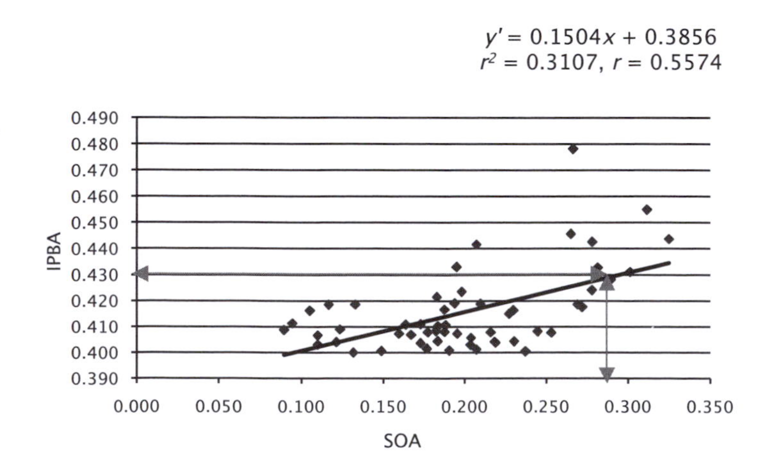

Figure 17.43 Regression line, equation, and correlation coefficient for X=SOA, Y=IPBA (56 players in Table 17.16)

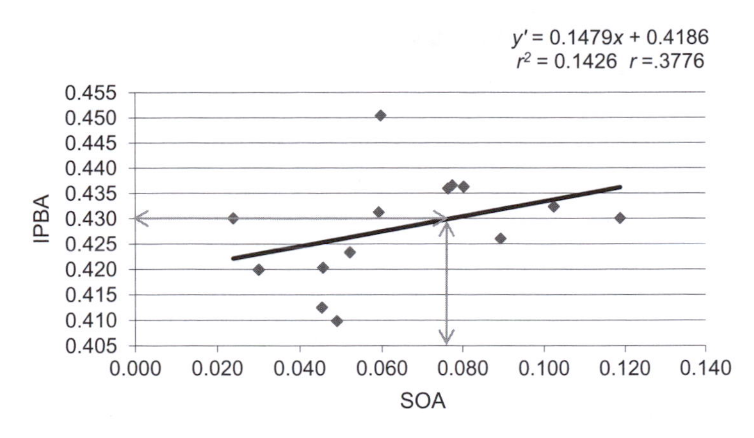

Figure 17.44 Regression line, equation, and correlation coefficient for X = SOA, Y = IPBA (14 players in Table 17.10)

more at bats a player has in a season, the more his AVG will approach his *true* AVG. The two players who came closest to batting .400 for a season illustrate how having fewer at bats helps a player obtain a higher AVG.

- Tony Gwynn, in the strike-shortened year of 1994, batted .394 in only 419 at bats (Gwynn played in only 110 games). Gwynn's strikeout average for that year was .045. Unfortunately, Gwynn, not being a power hitter, had an IPBA of only .413.
- George Brett, in 1980, had an injury-plagued year. He had only 449 at bats in 117 games, but he qualified for the batting title because of his 58 walks. Brett batted .390 with an IPBA of .410. Brett also had a very low SOA of .049. Unfortunately, Brett's IPBA was too low; he was not considered a power hitter.

Property Five: Bat from the Left-Hand Side. More than half the players in Table 17.16 meet this requirement. The two advantages of batting left-handed are that you are closer to first base and for the most part you face right-handed pitchers.

Property Six: Be Fast. It helps to be swift enough to beat out some infield hits. (Unlike Pete "Charlie Hustle" Rose, though, you don't need to sprint when you get a base on balls.)

Outside of his physical skills, we must mention a player's psychological makeup. In today's game, unlike the 1920s, a player is exposed to all sorts of media analysis. There is sports on radio, sports on satellite radio, sports on TV, sports on cable, and, of course, sports on the Internet.

So, besides the six items mentioned above, a player must be able to ignore all of these media intrusions. And then there are fans. Pedro Martinez will never be able to pitch at Yankee Stadium without hearing choruses of "Who's your Daddy."

If a clone of Ted Williams should appear, we may someday have another .400 hitter. Of course, the clone will face a different type of pitcher today. A starting pitcher is expected to pitch only through the sixth or seventh inning. There is a specialist for the seventh, eighth, and ninth innings–think Mariano Rivera. Not only would the clone face fresh arms, but there is a wider variety of pitches today: the splitter has arrived, along with two- and four-seam fastballs. The pool of pitchers is larger now, drawing talented hurlers from Japan, China, the Dominican Republic, and other countries. And, lest we forget, African American pitchers were excluded from the Major Leagues until 1947.

Of the 36 different players in Table 17.16, the two players that I feel are most like Ted Williams, as hitters, are Barry Bonds and Albert Pujols. A comparison of Barry Bonds (2002, 2004), Albert Pujols (2003), and Ted Williams (1941, 1957) with respect to the statistics studied in this chapter appears in Table 17.22. The numbers in parentheses correspond to how their statistic would rank with respect to the players in Table 17.16. Remember, BBA stands for base-on-balls average. The ranks for the BBAs are left as an exercise.

Exercise 17.39. Compute the BBA for each of the players in Table 17.16. Then sort the table in descending order based on BBA. Remember, BBA = #BB/(#AB + #BB + #SF + #HBP) = #BB/#PA. You can use either www.baseball-reference.com or www.mlb.com to compute the BBA for each player.

Exercise 17.40. Find the BBA rank for each player in Table 17.22 if they were inserted in Table 17.16.

Exercise 17.41. Chipper Jones made a run at .400 in 2008. Compute his AVG, IPBA, SOA, IPHR, and BBA for his 2008 season. Find the rank for each of these statistics if they were inserted into Table 17.16.

Exercise 17.42. Prepare a frequency distribution table for the players in Table 17.16, based on the team they played for. What conclusions can you draw from this table?

Table 17.22 **Comparison of Barry Bonds (2002, 2004), Albert Pujols (2003), and Ted Williams (1941, 1957), with respect to the statistics studied in this chapter**

Player	Year	AVG	IPBA	SOA	IPHRA	BBA
Williams	1941	.406 (1)	.431 (9)	.059 (1)	.086 (9)	.242
Williams	1957	.388 (1)	.432 (9)	.102 (3)	.101 (7)	.217
Bonds	2002	.370 (4)	.419 (18)	.117 (6)	.114 (4)	.324
Bonds	2004	.362 (6)	.407 (38)	.110 (4)	.121 (2)	.376
Pujols	2003	.359 (7)	.403 (49)	.110 (4)	.073 (5)	.115

Exercise 17.43. From what was discussed in this chapter, answer the question posed at the beginning of the chapter: *Will we ever have another .400 hitter?* Write a paragraph using descriptive and inferential statistics to justify your argument. Bring into your discussion the characteristics you feel a player would have to possess. Also, identify any current players who you feel could hit .400 for a year.

Chapter Summary

This chapter analyzed which statistics play the greatest role in determining whether a player will have a batting average greater than or equal to .400 for a season, and what it would take for a current player to bat .400 for a season.

We only considered players from 1913 to 2007. The reason for starting in 1913 was because before that year accurate records for strikeouts were not kept. Of the nine times a player batted at least .400, eight occurred between 1920 and 1930. The only exception was Ted Williams in 1941.

Formulas (17.1) and (17.2) show that AVG is completely determined by the two statistics IPBA and SOA. Since AVG is always less than or equal to IPBA, to attain an AVG of at least .400, a player must have an IPBA of at least .400.

$$\text{AVG} = \text{IPBA} * (1 - \text{SOA}), \tag{17.1}$$

$$\text{IPBA} - \text{AVG} = \text{IPBA} * \text{SOA}. \tag{17.2}$$

The four statistics studied in this chapter were AVG (batting average), IPBA (in-play batting average), SOA (strikeout average), and IPHR (in-play home run average). These were applied to the following samples:

- The years from 1913 to 2007.
- The two eras of 1920–1930 and 1994–2007.
- The career years of Ted Williams.
- The 14 players who either batted .400 for a season since 1913 or batted close to .400 in a season after 1941.
- Those players, from 2000 to 2007, who met the three requirements for a season of having (1) an IPBA of at least .400, (2) an AVG in the top 40, and (3) at least 502 plate appearances.

For the sample of years from 1913 to 2007, the four statistics were similar when comparing the American League with the National League. The differences appear for different time periods (eras). The era 1920–1930 had high batting averages. The era 1994–2007 featured home runs and strikeouts. Here are some other observations:

- Of the top 25 team leaders in home runs for a single season, 24 of the teams played between the years 1996 and 2007. The only exception was the 1961 Yankees, who featured Roger Maris and Mickey Mantle. The top team was the Seattle Mariners of 1997, with 264 home runs. Of the 24 teams, only 5 were National League teams. Clearly, this is due to the designated hitter rule.
- All 25 of the top team leaders in strikeouts for a single season played between 1994 and 2007. The team with the most strikeouts for a season (1399) was the Milwaukee Brewers of 2001. Of the 25 teams, only 4 were American League teams. Again, this is probably due to the designated hitter rule.
- Of the top 25 team leaders in batting average for a single season, 24 of the teams played between 1920 and 1930. The one exception was the '36 Cleveland Indians. Of the 24 teams, 8 of the teams played in 1930, the same year that Bill Terry hit over .400. Of the 25 teams, 13 belonged to the National League and 12 belonged to the American League.

The data in Table 17.16 show that it is rare for a player to have an IPBA of at least .400. To achieve this, a player must possess the batting ability to hit the ball where the fielders are not located. How do you hit the ball into a space not occupied by a fielder? Hit hard ground balls or line drives instead of high fly balls, hit the ball where it is pitched, hit in favorable counts, and hit either left- or right-handed pitchers with consistency.

Since half of a player's games are in his home stadium, a favorable home park helps a player's AVG. Of the top 28 AVGs in Table 17.15, 7 belong to the Colorado Rockies. It is well documented how balls fly out of Coors Field.

Besides ability, chance also plays a role in several potential hits becoming outs and several outs becoming hits.

The importance of the SOA is clear from the two formulas. Both of these show that the lower the SOA, the closer a player's AVG will be to his IPBA. They reveal that the secret of being a .400 hitter is to have a high IPBA and a very low SOA. If a player can go an entire season without striking out (this will likely never happen for a minimum of 502 plate appearances), the AVG will equal the IPBA.

Using the statistics of the 14 players in Table 17.10 who came close to or exceeded an AVG of at least .400 for a season, we concluded that the ideal conditions for a batter to hit .400 were to have an IPBA of at least .427 and an SOA of less than .066. Unfortunately, of those players from 2000 to 2007 that achieved an IPBA of at least .400, no player had an SOA less than .066. Only two current players had an SOA less than .100. To have an AVG of .400 with an SOA of .100 requires an IPBA of at least .444. Of the four current players with an IPBA that high, the lowest SOA was .265. The highest AVG for those four players was .351.

A large part of the chapter used regression and correlation analysis to look for relationships between certain pairs of variables, chosen from AVG, IPBA, SOA, IPHR, and (IPBA – AVG). These pairs of variables were used with five different samples: the 95 years from 1913 to 2007 (Table 17.2), the 11 years from 1920 to 1930, the 14 years from 1994 to 2007, the 14 players whose season AVG was at least .400 or who came closest to hitting .400 (Table 17.10), and the players from 2000 to 2007 who had an IPBA of at least .400 and ranked in the top 40 in AVG for a season (Table 17.16).

The 95 years from 1913 to 2007 consist of different baseball eras. It was important not only to look at the 95 years but more specifically to look at various eras within them. The two eras chosen for comparison were from 1920 to 1930 and from 1994 to 2007. The reason for the choice of these two eras is that eight out of the nine players who had an AVG of .400 did it between 1920 and 1930 and the era 1994–2007 describes baseball in its current state. The next .400 AVG, if we ever have one, will come from baseball as it is played today or in the future. In comparing the era 1920–1930 with the era 1994–2007, we observed the following:

- The mean yearly SOA was over 100 points higher and the mean yearly IPBA was about 20 points higher for the era 1994–2007.
- Although the mean yearly IPBA was 20 points higher between 1994 and 2007, the fact that the mean yearly SOA was also 100 points higher for this era caused the mean yearly AVG for the era 1994–2007 to be about 25 points lower than the mean yearly AVG for the era 1920–1930.

For the years 1920–1930, there was *no* correlation between SOA and AVG (Fig. 17.14a), so the AVG was almost completely determined by the IPBA (Fig. 17.15a). For the era 1994–2007, even though IPBA plays a major role in determining AVG (Fig. 17.15b), the SOA also played a moderate role in determining AVG (Fig. 17.14b).

The players from Table 17.10 consist of the 14 players (not all different) who had an AVG of either at least .400 or very close to .400 for a season. Of these 14, 8 of the AVGs occurred in the era 1920–1930. Figures 17.29 and 17.30 showed, for these 14 players, that the SOA had a moderate negative correlation with AVG while IPBA had a moderate positive correlation with AVG.

Table 17.16 consists of the players who had an IPBA of at least .400 for a year between 2000 and 2007. Figures 17.37 and 17.38 showed, for these players, that their AVG is almost completely determined by the SOA. Figure 17.37 shows a strong negative correlation between SOA and AVG. Figure 17.38 shows no correlation for this group between IPBA and AVG.

The increase in the SOA for current players will make it difficult for a player today to have an AVG of at least .400 for a season.

An entire section was devoted to Ted Williams, the only player who hit .400 outside the era 1920–1930. We compared his AVG, IPBA, SOA, and IPHR with the same four statistics for the American League for each of the years in Ted's career. These showed his greatness as a hitter relative to the time period he played in.

Two methods were introduced to compare two or more players with respect to certain baseball statistics. Rogers Hornsby was the only player, since 1913, to have a .400 AVG three times. Using these methods, Williams (1941) came out ahead of Hornsby (1924) with respect to AVG and with respect to the basket of statistics consisting of AVG, IPBA, SOA, and IPHR.

A player's *true* AVG was defined through the technique of *hypothesis testing for one population proportion*. A player's *true* AVG for a year differs from his observed (sample) AVG for a year. A *true* AVG is based on a theoretical infinite population of plate appearances for that year. The sample AVG is based on a player's actual plate appearances for that year. The *true* AVG will never be known, and the sample AVG is always known. We equate a player's *true* AVG to his ability and his sample AVG to his performance. The *true* AVG can be estimated through the use of confidence intervals.

Our definition of a *true* .400 hitter (Definition 17.1) requires the use of *hypothesis testing for one population proportion*. Rogers Hornsby's 1924 year, in which his yearly AVG was .424, was tested. The conclusion was that, at a level of significance of 0.05, we could not conclude that Rogers Hornsby was a *true* .400 hitter that year. He was chosen because, of the .400 hitters since 1913, he had the highest yearly AVG of .424.

For the theoretical population of players who both had a .400 IPBA and ranked in the top 40 in AVG for a year between 2000 and 2007, we looked at their *true* yearly SOA. Using *hypothesis testing for one mean*, based on a sample consisting of those players from 2000 to 2007 in Table 17.16, we showed that their *true* SOA was more than .100. As an exercise, the student was asked to show that their *true* IPBA was less than .444.

The chapter concluded by addressing the original question, will a future player have an AVG of at least .400 for a year? We looked at the current players who, for one season between the years 2000 and 2007, both had an IPBA of at least .400 and were in the top 40 in AVG. We compared them with the 14 players since 1941 that either had a .400 AVG (nine players) or had an AVG very close to .400 (five players). For the current players, there was *no* correlation between IPBA and AVG; however, there was a strong negative correlation between SOA and AVG. The correlation coefficient was $r=-.87$ (Fig. 17.37). From the regression equation (Fig. 17.37), an increase of 100 points in the SOA results in a decrease of 30 points in AVG.

The high SOA of a player in the current era of baseball would make it very difficult for such a player to hit .400 today.

It was conjectured that the next .400 batter should satisfy many of the following properties:

- Be a power hitter, who can generate at least a .430 IPBA.
- Be a player with a very low SOA (under .070).
- Play in a home park favorable to hitters.
- Have a high BBA (base-on-balls average), indicating a great batting eye.
- Have plate appearances close to the minimum number (502) needed for a batting title.
- Be a switch-hitter or bat from the left side.
- Have enough speed to beat out some infield hits.
- Have the mental toughness not to be influenced by outside pressures.

Apart from statistics, many other reasons can be given for why this feat may never again occur. Some of these are the ability of scouts to find batting weaknesses, the very difficult travel schedule, the use of relief specialists, the better defensive skills of players (including better and larger baseball gloves), the increase in the number of scheduled games, and the fact that most of the games are night games. Finally, baseball has become an international game with great pitchers coming from Japan (Red Sox Daisuke Matsuzaka), Korea (Phillies reliever Chan Ho Park), Cuba (World Series game winner Orlando "El Duque" Hernandez), and many other countries–including Britain, if you believe the story of Sidd Finch!

No player has batted .400 since the Major Leagues were integrated in 1947, which leads to the following questions: If baseball was integrated in 1913, how many .400 hitters would we have had? Would we have had less white .400 batters? Would we have had African American .400 hitters? The answers will never be known.

African American players competed against white Major Leaguers in barnstorming tours from the early 1900s through the 1940s. These tours took place after the World Series as a way for players to make extra money. These barnstorming tours show that African American and white players were at the same level. My feeling is that the improved pitching in an integrated baseball league probably would have reduced batting averages. Many of the great African American pitchers (Bullet Joe Rogan, Smoky Joe Williams, and Satchel Paige) had impressive winning records against white Major Leaguers. Therefore, some of the white players who hit .400 would probably have had points shaved off their batting averages had baseball been integrated. On the other hand, great African American positional players such as Pop Lloyd (1920s), Mule Suttles (1920s), Turkey Stearnes (1930s), and Josh Gibson (1930s–1940s) might well have batted .400.

CHAPTER PROBLEMS

We chose the highest five yearly AVGs for each of four great players. These players are Babe Ruth, Ted Williams, Barry Bonds, and Alex Rodriguez (A-Rod). Babe Ruth could have been the greatest hitter in the 1920s, Ted Williams could have been the greatest hitter from the 1940s to the 1950s, Barry Bonds could have been the greatest hitter from the 1990s to the early 2000s, and Alex Rodriguez could be the greatest hitter from 2000 to 2007. Problems 1–12 will concentrate on the five statistics AVG, IPBA, SOA, IPHR, and BBA for these players. Complete the following problems based on Table 17.23. Feel free to use Excel as a tool.

1. Construct a line graph similar to Figure 17.27.
2. Construct a side-by-side bar graph similar to Figure 17.28.
3. Use the Sum Method to compare the four players with respect to the five statistics. First apply the Sum Method for each year and then total the results for the five years for each player. Using these totals, rank the four players. What observations can you make?

Table 17.23 **Statistics for Babe Ruth, Ted Williams, Barry Bonds, and Alex Rodriguez (A-Rod)**

Year	Player	G	AB	H	HR	BB	SO	AVG	IPBA	SOA	IPHR	BBA
1923	Ruth	152	522	205	41	170	93	0.393	0.478	0.178	0.096	0.246
1924	Ruth	153	529	200	46	142	81	0.378	0.446	0.153	0.103	0.212
1921	Ruth	152	540	204	59	145	81	0.378	0.444	0.150	0.129	0.212
1920	Ruth	142	458	172	54	150	80	0.376	0.455	0.175	0.143	0.247
1931	Ruth	145	534	199	46	128	51	0.373	0.412	0.096	0.095	0.193
1941	Williams	143	456	185	37	147	27	0.406	0.431	0.059	0.086	0.244
1957	Williams	132	420	163	38	119	43	0.388	0.432	0.102	0.101	0.221
1948	Williams	137	509	188	25	162	47	0.369	0.402	0.081	0.053	0.198
1942	Williams	150	522	186	36	145	44	0.356	0.395	0.098	0.076	0.217
1955	Williams	98	320	114	28	91	24	0.356	0.385	0.075	0.095	0.221
2002	Bonds	143	403	149	46	198	47	0.370	0.419	0.117	0.129	0.329
2004	Bonds	147	373	135	45	232	41	0.362	0.407	0.110	0.136	0.383
2003	Bonds	130	390	133	45	148	58	0.341	0.401	0.149	0.136	0.275
1993	Bonds	159	539	181	46	126	69	0.336	0.393	0.147	0.100	0.189
2001	Bonds	153	476	156	73	177	93	0.328	0.407	0.195	0.191	0.271
1996	A-Rod	146	601	215	36	59	104	0.358	0.433	0.173	0.072	0.041
2005	A-Rod	162	605	194	48	91	139	0.321	0.416	0.230	0.103	0.131
2001	A-Rod	162	632	201	52	75	131	0.318	0.401	0.207	0.104	0.106
2000	A-Rod	148	554	175	41	100	121	0.316	0.404	0.218	0.095	0.153
2007	A-Rod	158	583	183	54	95	120	0.314	0.395	0.206	0.117	0.140

4. Use the Adjustment Method to find the adjusted AVG for each player for each year in Table 17.23.
5. Rank (1–20) the AVGs in Table 17.23 and the adjusted AVGs calculated in problem 4. What observations can you make? Based on the results, rank the four players.
6. Repeat problems 4 and 5 with the statistics BBA and SOA.
7. Let X=IPBA and Y=AVG. Construct the sample regression line for each player using for the sample his five years. Find the r and r^2 for each regression line. What conclusions can you make?
8. Let X=SOA and Y=AVG. Construct the sample regression line for each player using for the sample his five years. Find the r and r^2 for each regression line. What conclusions can you make?
9. Let X=IPBA and Y=AVG. Construct the sample regression line for each player using for the sample his five years. Find the r and r^2 for each regression line. What conclusions can you make?
10. Construct a 95% confidence interval estimate for the *true* AVG for each player's year in which they had their highest yearly AVG. What observations can you make?
11. Construct a 95% confidence interval estimate for the *true* SOA for each player's year in which they had their highest yearly AVG. What observations can you make?
12. Construct a 95% confidence interval estimate for the *true* IPBA for each player's year in which they had their highest yearly AVG. What observations can you make?

The next set of problems deals with a comparison of your two players. Choose the highest five yearly AVGs for your two players. Construct a table similar to Table 17.23.

13. Construct a line graph similar to Figure 17.27.
14. Construct a side-by-side bar graph similar to Figure 17.28.
15. Use the Sum Method to compare the two players with respect to the five statistics. First apply the Sum Method for each year and then total the results for the five years for each player. Using these totals, rank the two players. What observations can you make?
16. Use the Adjustment Method to find the adjusted AVG for each player for each year in your table.
17. Rank (1–10) the AVGs in the table and the adjusted AVGs calculated in problem 16. What observations can you make? Based on the results, rank the two players.
18. Based on a minimum of 2500 at bats, can you find any player with a *career* IPBA greater than .400?

19. Based on a minimum of 2500 at bats, which active player has the highest career IPBA?
20. Based on a minimum of 2500 at bats, which active player has the lowest career SOA?
21. Based on a minimum of 2500 at bats, which active player has the highest career AVG?
22. Find the career SOA and career IPBA for the player in problem 21. Is it possible that this player could be a .400 hitter in some future year? Explain.

Chapter 18

Postseason

This chapter not only looks in the rearview mirror at what was covered in the prior chapters but also serves as a window for future baseball research. There are many questions debated about players, teams, leagues, and seasons which were not mentioned in this book.

The first 15 chapters presented the descriptive and inferential techniques used to analyze data. Chapter 16 explored the probability of duplicating various hitting streaks (the most famous being Joe DiMaggio's 56-game hitting streak), and Chapter 17 used statistics to discover what it will take for a future player to be a .400 hitter for a season.

In this final chapter, we explore some of the baseball questions that are often debated. You, of course, are encouraged to give your opinion on each of these topics. Two such questions are as follows: What are the greatest batting accomplishments of all time? Who are the greatest hitters of all time?

The chapter problems at the end of the chapter come from the previous chapters. Some of the problems guide the student to the specific chapter that contains the information needed to solve the problem.

The chapter and book close with a tribute to the immortal Henry Louis Gehrig, the "Iron Horse."

18.1. Eight of the Greatest Batting Feats

Eight of the greatest batting accomplishments (unranked) are as follows:

1. Joe DiMaggio's 56-game hitting streak in 1941.
2. Ted Williams's 1941 season, when he became the last .400 hitter in the Major Leagues (based on 3.1 plate appearances per scheduled game).

3. Pete Rose's record of 4256 hits.
4. Barry Bonds's single-season record of 73 home runs in 2001.
5. Winning the Triple Crown.
6. Hack Wilson's 191 RBIs in the 1930 season.
7. Ty Cobb's career batting average of .366 (based on a minimum of 4000 at bats in a career).
8. Rogers Hornsby's five-year run from 1921 to 1925 in which his cumulative batting average for those five years was .402.

Remember, we are only looking at the batting feats. Of course, Nolan Ryan's seven no-hitters, Cy Young's 511 victories, and Bob Feller's three no-hitters and 12 one-hitters would qualify under the category of great pitching feats.

18.1.1. Ranking the Eight Hitting Feats

The ranking is from 1 to 8, with rank 1, the highest rank, being given to the batting feat that is the least likely to be duplicated, rank 2 given to the next hardest feat to duplicate, and so on until rank 8 is reached.

I consider Joe DiMaggio's 56-game hitting streak and Rogers Hornsby's batting average of over .400 based on five consecutive seasons (1921–1925) as the least likely feats ever to be duplicated. They tie for rank 1.

First, on DiMaggio's 56-game hitting streak, my principal argument is that a player must be successful for 56 consecutive games–he cannot make up for a single failure. A player can go hitless in a game for many reasons. He can face a Hall of Fame–caliber pitcher, or have multiple walks in a game, or sustain an injury. A player feels the pressure during each game of the streak, and there will be times when he must succeed in his last at bat. A feared hitter in today's game, or a good player on a weak team, will receive many walks, which will reduce their chance of getting a hit. A disciplined hitter who does not swing at a pitch outside the strike zone would have great difficulty repeating or breaking this streak.

In Table 16.1, the eight batters with the longest hitting streaks in the modern-day Major Leagues are listed. The streaks range from Tommy Holmes's 37 games to Joe DiMaggio's 56 games. Of the eight players listed, five are in the Hall of Fame, and one (Pete Rose) has Hall of Fame numbers. The two players who are not in the Hall of Fame are Bill Dahlen and Tommy Holmes. Dahlen had a 42-game hitting streak in 1894 when he batted .357 for that year. He played 21 years in the Major Leagues with a career batting average of .272. Tommy Holmes had a 37-game hitting streak in 1945 when he batted .352. Holmes had a career batting average of .302 and is considered a borderline Hall of Famer. The point I am making is that these eight players were elite hitters.

In Table 16.12, the 17 players with the longest hitting streaks in the Minor Leagues are listed. The streaks range from 38 games to 69 games. Of these 17 players, 8 never made an appearance in the Major Leagues; 2 players achieved their streak in 2007 (it is too soon to determine whether they will make a Major League team); and 5 had short, nonproductive careers in the Major Leagues. One player, Jack Lelivelt, batted .301 in 384 games in the Major Leagues for the years 1909–1914. In researching Lelivelt, I found that he batted .351 in 1912 for Rochester. Finally, Joe DiMaggio had a 61-game hitting streak for San Francisco in the Minor Leagues. The fact that Joe DiMaggio is the only player to appear on both lists indicates that he was a very special type of hitter. Even in 1941 when Ted Williams batted .406, his longest consecutive hitting streak was a mere 23 games.

The above discussion leads to the following question: Since it took an elite Major League player to accomplish a long hitting streak in the Major Leagues, why, except for Joe DiMaggio, did the Minor League players with long hitting streaks either never appear in the Major Leagues or have mediocre careers in the Major Leagues? I leave the answer to this question to you.

Tied with Joe for first place in difficulty is Rogers Hornsby's feat of averaging over .400 using his cumulative at bats for five consecutive seasons. From 1921 through 1925 Hornsby had 2679 at bats with 1078 hits, giving him a BA equal to .402. The last player to hit over .400 for one season was Ted Williams in 1941. That says it all.

We turn our attention to the other six great hitting feats. My third and fourth choices are having a career batting average of .366 (Ty Cobb) and batting .400 for a season. Both of these rely on having high batting averages. As pointed out before, all .400 hitters occurred before 1931, except for Ted Williams batting .406 in 1941. Most of the .400 hitters were in the lively ball era of 1920–1930. Ty Cobb, who played before 1930, had 14 years when he batted over .360 and three years when he batted over .400. Of the top 100 highest season batting averages, only five occurred after 1941. One of the five was Ted Williams in 1957, and a second was Tony Gwynn in the strike-shortened year of 1994. The other three players were Rod Carew in 1977, George Brett in 1980, and Larry Walker in 1999.

A very close third on my list for the hardest to duplicate goes to having a career batting average of .366. Except for Ted Williams (career AVG .344, rank 7), all the other members of the top 10 club for career AVG ended their careers before 1940. The career BAs for Tony Gwynn, Rod Carew, and Larry Walker were .338, .328, and .313, respectively. The three active players, as of July 24, 2009, with the highest career batting averages are Albert Pujols (.334), Ichiro Suzuki (.332), and Todd Helton (.328).

The fourth-most difficult to duplicate is hitting .400 for a season. The major reason for this selection is my belief that we will never have a *true*

.400 hitter. A *true* .400 hitter was defined in Chapter 17. Since a player is required to have 502 plate appearances to qualify and his batting average will regress to his *true* batting average, by the end of the season his batting average will be below .400. Unlike the 56-game hitting streak, there is room for an off-day. A player can get zero hits in a game and still bat .400 for a season. The last .400 hitter was Ted Williams in 1941, the same year Joe DiMaggio accomplished his 56-game hitting streak. Ted in 1941 actually had 29 hitless games. Remember, Ted drew 145 walks that year.

My fifth choice is winning the Triple Crown: a player is required to have the power to lead his league in home runs, the skill to hit for a high batting average, and the timeliness to get hits with men on base. Usually hitting with power causes more strikeouts, which reduces a player's batting average. The last Triple Crown winner was Carl Yastrzemski in 1967. Of the current active players as of 2009, I believe that Albert Pujols has the best chance of achieving this feat.

Next comes hitting 73 home runs in one season. It would take a powerful batter to accomplish this feat. With the strong enforcement of the new drug laws in baseball, a player can no longer get help from the use of human growth hormones or steroids. Since 2000, the top five home run hitters were Barry Bonds (73 in 2001), Sammy Sosa (64 in 2001), Ryan Howard (58 in 2006), Luis Gonzales (57 in 2001), and Alex Rodriguez (57 in 2002). Since 1998, six players have hit over 60 home runs. They are Barry Bonds (once), Mark McGwire (twice), and Sammy Sosa (three times). It has been alleged that all three were involved with steroid use.

Seventh place goes to having 191 RBIs in one season. A player would have to average about 1.2 RBIs per game. Even in 2001 when Bonds hit his 73 home runs, he only had 137 RBIs. Since 2000, the top five RBI seasons were Sammy Sosa (160 in 2001), Alex Rodriguez (156 in 2007), Miguel Tejada (150 in 2004), Ryan Howard (149 in 2006), and David Ortiz (148 in 2005). Between 1941 and 1999, Manny Ramirez had the highest RBI season of 165 in 1999.

My last choice is Pete Rose's career hit record. This feat requires a long career needed for the compiling of hits. A player has to average about 213 hits for 20 years to reach this record. Our example is Ichiro Suzuki. He had over 200 hits in each of his seasons from 2001 through 2009. Since his records in Japan do not count, he probably will not come close to the record. If he had started his career in this country, he would have had a very good chance of breaking Rose's record. Derek Jeter, based on current numbers, could snap the record in seven or eight years if he stays healthy and does not let advancing age reduce his skills.

18.2. The Greatest Hitters in Each of the Six Eras

Position players are evaluated by looking at the *five tools*, hitting, hitting with power, fielding, throwing, and running. My evaluation of players is restricted to hitting and hitting with power.

I list the players I feel are the greatest hitters for each of the six eras defined in Chapter 17 and give some reasons for my choices. Some players had careers that extended into two eras. For those players, I put them in the era in which they were most productive. If a player played on multiple teams, I only list his primary team.

18.2.1. Criteria for Selection

Before listing the nominees by era for the greatest hitters of all time, it is important to understand some of the criteria used for their selection.

- The statistics given the greatest importance were AVG, OBP, SLG, OPS, BRA, RC/27, HRA, #H, #HR. Other statistics such as runs scored, TB, and RBIs will also be considered.
- The groupings of baseball statistics, introduced in Chapter 4, were used to evaluate a player's ability to get on base, a player's ability to hit for power, and a player's ability to produce runs for his team.
- Players were given credit for winning the MVP Award or being selected as an All-Star.
- Players were given credit for ranking in the top 10 for their careers for the statistics AVG, OBP, SLG, OPS, BRA, RC/27, HRA, #H, and #HR.
- Players were required to complete at least 10 Major League seasons. For players active as of the 2008 season, that total is reduced to eight Major League seasons.
- Their Black-Ink Test score (discussed in Chap. 10) was used. Remember, the Black-Ink Test awards four points for leading the league in home runs, runs batted in, and batting average. Three points are awarded for leading the league in runs scored, hits, and slugging percentage. Two points are awarded for leading the league in doubles, walks, and stolen bases. One point is awarded for leading the league in games, at bats, and triples.
- The Gray-Ink Test score was also used. This test awards the same points as the Black-Ink Test, except points are awarded if the player finished in the top 10 in each of the statistics.
- Special consideration is given to a player who achieved either a Triple Crown Award or a Career Triple Crown Award, who batted .380 or above for a season, who had a consecutive hitting streak of at least 40 games, who finished his career with at least 3000 hits, who hit at least 50 home runs in a season, or who had at least 500 career home runs.

A *Career Triple Crown* winner is a player who in some year, not necessarily in the same year, won a batting title and led his league in both RBIs and home runs. The requirement of achieving these three feats in the same year is removed. Surprisingly, excluding the 11 players (Williams and Hornsby two times each) who won the Triple Crown, only nine players accomplished these three feats but not in the same year. Can you name the nine players?

Some of the above criteria favor players in certain eras. The first MVP Award was in 1931. The first All-Star Game was played in 1933. The two ink tests favor the players who played before 1961, when there were only eight teams in each league. Players who are still active are penalized by the two ink tests. Players who served in World War II or the Korean War lost prime years in their careers. All of these situations were considered in making my player selections. For each player some stats are provided, including the team with which they spent the most years.

18.2.2. Nominees for 1900–1919

18.2.2.1. Willie Keeler (Shortstop)

- Height: 5' 4"
- Weight: 140 pounds
- Threw and batted left-handed
- Career: 1882–1909
- New York Highlanders (AL)

Willie batted better than .350 for eight consecutive seasons and had a 44-game hitting streak (rank 2). His career batting average of .341 ranks 12th, and his career total of 2932 hits ranks 32nd. His Black-Ink Test score was 21 points (rank 101). His Gray-Ink Test score was 169 (rank 63).

18.2.2.2. Nap Lajoie (Second Baseman)

- Height: 6' 1"
- Weight: 191 pounds
- Threw and batted right-handed
- Career: 1896–1916
- Cleveland Naps (AL)

The Indians infielder batted at least .300 in 17 out of 21 seasons, and he batted over .350 for 10 seasons. He won the Triple Crown Award in 1901. His career batting average of .338 ranks 17th, his career total of 1599 RBIs ranks 31st, and his career total of 3242 hits ranks 12th. His Black-Ink Test score was 76 (rank 8). His Gray-Ink Test score was 266 (rank 20).

18.2.2.3. Honus Wagner (Shortstop)

- Height: 5' 11"
- Weight: 200 pounds
- Threw and batted right-handed
- Career: 1897–1917
- Pittsburgh Pirates (NL)

Wagner batted better than .300 for 17 consecutive seasons, winning eight NL batting titles. His career batting average of .327 ranks 35th, his career total of 1732 RBIs ranks 19th, and his career total of 3415 hits ranks seventh. His Black-Ink Test score was 109 (rank 6). His Gray-Ink Test score was 363 (rank 4).

18.2.2.4. Joe Jackson (Right Fielder)

- Height: 6' 1"
- Weight: 200 pounds
- Threw right-handed and batted left-handed
- Career: 1908–1920
- Chicago White Sox (AL)

"Shoeless" Joe batted better than .300 for 11 seasons and batted over .350 seven times. He batted .408 in 1911. His career batting average of .356 ranks third. His Black-Ink Test score was 14 (rank 175). His Gray-Ink Test score was less than 107 (rank > 200).

18.2.2.5. Tris Speaker (Center Fielder)

- Height: 5' 11"
- Weight: 193 pounds
- Threw and batted left-handed
- Career: 1907–1928
- Cleveland Indians (AL)

Speaker's career batting average of .345 ranks sixth, his career total of 3315 hits ranks fifth, his career total of 1529 RBIs ranks 45th, and his career .500 SLG ranks 86th. He holds the career record of 793 doubles. His Black-Ink Test score was 34 (rank 48). His Gray-Ink Test score was 346 (rank 6).

18.2.2.6. Ty Cobb (Center Fielder)

- Height: 6' 1"
- Weight: 173 pounds

- Threw right-handed and batted left-handed
- Career: 1905–1928
- Detroit Tigers (AL)

The Tigers' center fielder batted better than .300 for 23 consecutive seasons, winning 12 AL batting titles. He batted over .400 three times. His career batting average was .366 (rank 1). His career total of 4191 hits ranks second, his career total of 1937 RBIs ranks sixth, and his 5854 career total bases rank fifth. He won the Triple Crown Award in 1909. He had a 40-game hitting streak. His Black-Ink Test score was 150 (rank 2). His Gray-Ink Test score was 417 (rank 1).

My choice for best hitter in the era of 1900–1919 is Ty Cobb. Honorable mention: Honus Wagner, Joe Jackson, and Nap Lajoie.

18.2.3. Nominees for 1920–1930

18.2.3.1. Babe Ruth (Right Fielder)

- Height: 6' 2"
- Weight: 215 pounds
- Threw and batted left-handed
- Career: 1914–1935
- New York Yankees (AL)

The Sultan of Swat retired with 714 home runs, at a time when only two other players reached 300 home runs. His 2213 career RBIs rank second, his career .690 SLG ranks first, his 5793 career total bases rank fifth, his career AB/HR ratio of 11.76 ranks second, his career AVG of .342 ranks 10th, and his career 2062 walks rank third. He had eight consecutive years of hitting over .300, driving in over 100 runs, and hitting at least 30 home runs. He accomplished the Career Triple Crown by leading the league in AVG one time, leading the league in HRs 12 times, and leading the league in RBIs six times. He also led the league in SLG 13 times and in OBP 10 times. His Black-Ink Test score was 161 (rank 1). His Gray-Ink Test score was 340 (rank 7).

18.2.3.2. Rogers Hornsby (Second Baseman)

- Height: 5' 11"
- Weight: 200 pounds
- Threw and batted right-handed
- Career: 1915–1937
- St. Louis Cardinals (NL)

Hornsby hit for average and power, winning two Triple Crown Awards. He batted .400 three times. His .424 AVG in 1924 remains the highest in the Major Leagues since 1901. In 1935, he was one of three players to have 300 career home runs. His .358 career AVG ranks second all time, his career .577 SLG ranks 12th, his career total of 2930 hits ranks 34th, and his career RBI total of 1584 ranks 35th. His Black-Ink Test score was 125 (rank 3). His Gray-Ink Test score was 329 (rank 10).

18.2.3.3. George Sisler (First Baseman)

- Height: 5' 11"
- Weight: 170 pounds
- Threw and batted left-handed
- Career: 1915–1930
- St. Louis Browns (AL)

Sisler won two batting titles, batting over .400 both times. In 1920, he had 257 hits. He compiled a career .340 AVG, which ranks 15th; his career total of 2812 hits ranks 44th. He had a 41-game hitting streak in 1922. His Black-Ink Test score was 29 (rank 63). His Gray-Ink Test score was 198 (rank 42).

18.2.3.4. Bill Terry (First Baseman)

- Height: 6' 2"
- Weight: 200 pounds
- Threw and batted left-handed
- Career: 1923–1936
- New York Giants (NL)

New York Giant Terry was the last player, before Ted Williams, to bat over .400. Bill Terry had 254 hits when he batted .401 in 1930. His career AVG of .341 ranks 13th, and his .506 career SLG ranks 76th. His Black-Ink Test score was 12 (rank 197). His Gray-Ink Test score was less than 107 (rank > 200).

My choice for best hitter in the era of 1920–1930 is Babe Ruth. Honorable mention: Rogers Hornsby.

The next two eras involve some players who served in World War II between 1942 and 1946. The three years missed, usually in their prime, reduced all their totals and consequently their scores on the Black-Ink and Gray-Ink Tests. In choosing my players, I definitely considered this.

18.2.4. Nominees for 1931–1945

18.2.4.1. Lou Gehrig (First Baseman)

- Height: 6' 1"
- Weight: 212 pounds
- Threw and batted left-handed

- Career: 1923–1939
- New York Yankees (AL)

Gehrig had eight consecutive years of hitting over .300, driving in over 100 runs, and hitting at least 30 home runs. He posted 13 consecutive seasons of 100 runs and 100 RBIs. He holds the career grand slam record of 23. His consecutive games played streak of 2130 games is second overall. His career AVG of .340 ranks 15th, his career SLG of .632 ranks third, his career total of 1995 RBIs ranks fourth, his 5060 career total bases rank 16th, his career total of 2721 hits ranks 52nd, his career total of 493 home runs ranks 26th. He won the Triple Crown Award in 1934. He won the MVP Award in 1936. His Black-Ink Test score was 75 (rank 10). His Gray-Ink Test score was 315 (rank 13).

18.2.4.2. Jimmie Foxx (First Baseman)

- Height: 6'
- Weight: 190 pounds
- Threw and batted right-handed
- Career: 1925–1945
- Philadelphia Athletics (AL)

Foxx was the second player to reach 500 home runs. He slugged 30 or more home runs in 12 consecutive seasons and drove in over 100 runs in 13 consecutive seasons. He had 175 RBIs in 1938, just 16 behind the all-time record. His career AVG of .325 ranks 32nd, his career total of 1922 RBIs ranks seventh, his 534 career home runs rank 16th, his 1452 walks rank 20th, and his career total of 2646 hits ranks 68th. He won the Triple Crown Award in 1933 while smacking 48 home runs. He won the MVP Award three times, in 1932, 1933, and 1938. In 1932, he tagged 58 home runs. His Black-Ink Test score was 69 (rank 18). His Gray-Ink Test score was 267 (rank 22).

18.2.4.3. Hank Greenberg (First Baseman)

- Height: 6' 4"
- Weight: 218 pounds
- Threw and batted right-handed
- Career: 1931–1947
- Detroit Tigers (AL)

Hank was regarded as the first great Jewish player. He missed three years because of World War II (1942–1944). He was one of three Hall of Famers to win the MVP Award at two different positions (first base and in the outfield). He slugged 58 home runs in 1938 with Detroit. His .313 career AVG ranks 67th, his career SLG of .605 ranks seventh, and his career total of 331 home runs ranks 90th. In 1934, he won the MVP Award. His Black-Ink Test score was 46 (rank 30). His Gray-Ink Test score was 171 (rank 61).

18.2.4.4. Joe DiMaggio (Center Fielder)

- Height: 6' 2"
- Weight: 193 pounds
- Threw and batted right-handed
- Career: 1936–1951
- New York Yankees (AL)

DiMaggio missed 1943–1945 because of World War II. He is the owner of the most celebrated hitting streak of 56 consecutive games in 1941. He was an All-Star every season of his 13-year career. He won the MVP Award three times, in 1939, 1941, and 1947. His career AVG of .325 ranks 32nd, his 361 home runs rank 69th, his 1537 career RBI total ranks 43rd, and his career SLG of .579 ranks 10th. He won the Career Triple Crown by leading the league in AVG two times, leading the league in HRs two times, and leading the league in RBIs two times. His Black-Ink Test score was 34 (rank 38). His Gray-Ink Test score was 226 (rank 29).

My choice for best hitter in the era of 1931–1945 is a tie between Lou Gehrig and Jimmie Foxx. Honorable mention: Joe DiMaggio.

18.2.5. Nominees for 1946–1968

18.2.5.1. Ted Williams (Left Fielder)

- Height: 6' 4"
- Weight: 198 pounds
- Threw and batted left-handed
- Career: 1939–1960
- Boston Red Sox (AL)

Ted missed five years because of two wars, World War II and Korea. He was the last player to bat over .400, in 1941. He won two Triple Crown Awards, in 1942 and 1947. He won the MVP Award two times, in 1946 and 1949. He set the record of 84 consecutive games getting on base in 1949. He retired with a career AVG of .344, which ranks eighth all time. He ranks second all time in SLG with a career average of .634. His career RBI total of 1839 ranks 13th, his career walk total of 2021 ranks fourth, his career home run total of 521 ranks 19th, his career RBI total of 1839 ranks 13th, and his 4884 career total bases rank 20th. His Black-Ink Test score was 122 (rank 4). His Gray-Ink Test score was 326 (rank 11).

18.2.5.2. Stan Musial (Left Fielder)

- Height: 6'
- Weight: 180 pounds
- Threw and batted left-handed
- Career: 1941–1963
- St. Louis Cardinals (NL)

He batted over .300 for 17 consecutive seasons, winning seven NL batting titles. He played in 24 All-Star Games. He ranks fourth all time in total hits with 3630, second all time in total bases with 6134, and fifth all time in RBIs with 1951; his career AVG of .331 ranks 25th, his career total of 475 home runs ranks 28th, and his career total of 1599 walks ranks 12th. He won the MVP Award three times, in 1943, 1946, and 1948. His Black-Ink Test score was 116 (rank 5). His Gray-Ink Test score was 390 (rank 3).

18.2.5.3. Mickey Mantle (Center Fielder)

- Height: 5' 11"
- Weight: 198 pounds
- Threw right-handed and was a switch-hitter
- Career: 1951–1968
- New York Yankees (AL)

Mantle finished his 18-year career with 536 home runs and a lifetime AVG of .298. He ranks 22nd in career SLG, 40th with 4511 career total bases, 48th with 1509 career RBIs, and 15th with 536 career home runs. He won the Triple Crown Award in 1956. He won three MVP Awards, in 1956, 1957, and 1962. He was known for hitting long home runs. The expression "tape measure shot" was first introduced to describe his long home runs. His Black-Ink Test score was 65 (rank 14). His Gray-Ink Test score was 262 (rank 17).

18.2.5.4. Roberto Clemente (Right Fielder)

- Height: 5' 11"
- Weight: 175 pounds
- Threw and batted right-handed
- Career: 1955–1972
- Pittsburgh Pirates (NL)

Clemente won four batting titles and batted over .300 13 times in his 18-year career. He won the MVP Award in 1966. His career .317 AVG ranks 57th, his 4492 career total bases rank 41st, his 3000 career hits rank 26th, and his career 1305 RBIs rank 95th. His Black-Ink Test score was 23 (rank 93). His Gray-Ink Test score was 154 (rank 83).

18.2.5.5. Willie Mays (Center Fielder)

- Height: 5' 11"
- Weight: 180 pounds
- Threw and batted right-handed
- Career: 1958–1973
- New York / San Francisco Giants (NL)

Mays ranks third all time with 6066 total bases, 9th in career RBIs with 1903, 10th in career hits with 3283, 4th in career home runs with 660, and 21st in career SLG with an average of .557. He played in a record-tying 24 All-Star Games. He won two MVP Awards, in 1954 and 1965. His Black-Ink Test score was 57 (rank 20). His Gray-Ink Test score was 337 (rank 8).

18.2.5.6. Duke Snider (Center Fielder)

- Height: 6'
- Weight: 190 pounds
- Threw right-handed and batted left-handed
- Career: 1947–1964
- Brooklyn Dodgers (NL)

From 1950 to 1959, Snider led all players in both home runs (326) and RBIs (1031). He slugged over 40 homers in five consecutive seasons. His career AVG of .295 is excellent for a power hitter. His career total of 1333 RBIs ranks 82nd, and his career .540 SLG ranks 35th.

18.2.5.7. Henry Aaron (Right Fielder)

- Height: 6'
- Weight: 180 pounds
- Threw and batted right-handed
- Career: 1954–1976
- Milwaukee / Atlanta Braves (NL)

Aaron's career statistics blow you away: 755 home runs (rank 2), 6856 total bases (rank 1), 2174 runs (rank 4), 2297 RBIs (rank 1), 3771 hits (rank 3), and an SLG of .555 (rank 25). He had a Career Triple Crown by leading the league in AVG two times, leading the league in HRs four times, and leading the league in RBIs four times. He won the MVP Award in 1957. His Black-Ink Test score was 76 (rank 9). His Gray-Ink Test score was 408 (rank 2).

18.2.5.8. Frank Robinson (Right Fielder)

- Height: 6' 1"
- Weight: 194 pounds

- Threw and batted right-handed
- Career: 1956–1976
- Cincinnati Reds (NL)

Robinson concluded his career with 586 home runs (rank 7), 2943 hits (rank 30), and 1812 RBIs (rank 17). He won the Triple Crown in 1966. He won two MVP Awards, one in each league; he won the award with Cincinnati in 1961 and for Baltimore in 1966. His Black-Ink Test score was 35 (rank 45). His Gray-Ink Test score was 320 (rank 12).

My choice for best hitter in the era of 1946–1968 is a three-way tie between Ted Williams, Stan Musial, and Hank Aaron. Honorable mention: Mickey Mantle, Willie Mays, and Frank Robinson.

18.2.6. Nominees for 1969–1993

18.2.6.1. Pete Rose (Second Baseman)

- Height: 6'
- Weight: 180 pounds
- Threw right-handed and was a switch-hitter
- Career: 1963–1986
- Cincinnati Reds (NL)

Rose is the career leader in total hits with 4256. His career total of 2165 runs scored ranks fifth, his career RBI total of 1314 ranks 88th, and his 5752 career total bases rank seventh. He won the MVP Award in 1973. He had a 44-game hitting streak (second all time). His Black-Ink Test score was 64 (rank 15). His Gray-Ink Test score was 239 (rank 24).

18.2.6.2. Tony Gwynn (Right Fielder)

- Height: 5' 11"
- Weight: 225 pounds
- Threw and batted left-handed
- Career: 1982–2001
- San Diego Padres (NL)

Gwynn's career total of 3141 hits ranks 17th, and his career .338 AVG ranks 17th. He won eight batting crowns. He batted .394 in 1994. He was a 15-time All-Star. His Black-Ink Test score was 57 (rank 21). His Gray-Ink Test score was 155 (rank 80).

18.2.6.3. George Brett (Third Baseman)

- Height: 6'
- Weight: 200 pounds
- Threw right and batted left-handed

- Career: 1973–1993
- Kansas City Royals (AL)

Brett batted over .300 11 times. He was the first player to win a batting title in three different decades: 1976 (.333), 1980 (.390), and 1990 (.329). His career total of 3154 hits ranks 14th, and his career total of 1096 walks ranks 72nd. He won the MVP Award in 1980. He was a 13-time All-Star. His Black-Ink Test score was 39 (rank 37). His Gray-Ink Test score was 159 (rank 74).

18.2.6.4. Wade Boggs (Third Baseman)

- Height: 6' 1"
- Weight: 194 pounds
- Threw right-handed and batted left-handed
- Career: 1982–1999
- Boston Red Sox (AL)

Boggs was a lifetime .328 hitter. He won the batting title five times, including four in a row. He had seven consecutive 200-hit seasons. He led the league in OBP six times. His 3010 career hits rank 24th, his 4064 career total bases rank 71st, and his career AVG of .328 ranks 33rd. His Black-Ink Test score was 37 (rank 41). His Gray-Ink Test score was 138 (rank 117).

18.2.6.5. Rod Carew (Second Baseman)

- Height: 6'
- Weight: 182 pounds
- Threw right-handed and batted left-handed
- Career: 1967–1985
- Minnesota Twins (AL)

In 19 seasons, Carew won seven AL batting titles, hitting .300 or higher for 15 consecutive seasons while compiling a .328 lifetime batting average. He was selected 18 times to the All-Star Team. In 1977, when he batted .388, he won the MVP Award. His Black-Ink Test score was 42 (rank 34). His Gray-Ink Test score was 148 (rank 90).

18.2.6.6. Mike Schmidt (Third Baseman)

- Height: 6' 2"
- Weight: 195 pounds
- Threw and batted right-handed
- Career: 1972–1989
- Philadelphia Phillies (NL)

Schmidt won three MVP Awards, in 1980, 1981, and 1986. He was a 12-time All-Star. His .527 career SLG ranks 47th, his 4404 career total bases rank 46th, his 1595 career RBI total ranks 32nd, and his 548 career home runs rank 13th. His Black-Ink Test score was 74 (rank 11). His Gray-Ink Test score was 224 (rank 30).

My choice for best hitter in the era of 1969–1993 is a tie between Pete Rose and Tony Gwynn. Honorable mention: tie between George Brett and Mike Schmidt.

*18.2.7. Nominees for 1994–2008**

18.2.7.1. Barry Bonds (Left Fielder)

- Height: 6' 1"
- Weight: 194 pounds
- Threw and batted left-handed
- Career: 1986–2007
- San Francisco Giants (NL)

Bonds ranks first in career home runs with 762. He holds the record for the highest OBP for a season at .609 in 2004. That same year he drew a record 232 walks. His 1996 RBIs rank third all time. His career total of 2558 walks is number 1 all time. His AB/HR ratio of 12.94 is third all time. His 2227 runs scored are third all time. His 73 home runs in 2001 is the all-time record for a season. He achieved the Career Triple Crown by leading the league in hitting two times, leading the league in home runs two times, and leading the league in RBIs one time. His career SLG of .607 ranks sixth, and his 5976 career total bases rank fourth. He won the MVP Award seven times. His Black-Ink Test score was 69 (rank 12). His Gray-Ink Test score was 289 (rank 14).

18.2.7.2. Mike Piazza (Catcher)

- Height: 6' 3"
- Weight: 215 pounds
- Threw and batted right-handed
- Career: 1993–2007
- New York Mets (NL)

Piazza is considered by many baseball people as the greatest offensive catcher of all time. His career 427 home runs rank him number 1 for all catchers in baseball history. His .362 average in 1997 was the best all time for a season for any catcher. For his first eight years, Mike had 277 home runs, 874 RBIs, a .328 AVG, and an SLG of .582. To show how great

* The statistics for active players are through the end of the 2008 season.

these numbers were, we compare them with Willie Mays's numbers for his first eight years: 295 home runs, 844 RBIs, a .323 AVG, and an SLG of .602. Mike's career AVG was .308. He was an All-Star selection for 13 straight years. Since a catcher gets beaten up during a season with foul balls and has to work so hard during a game, these numbers are remarkable. His career total of 1335 RBIs ranks 81st, and his career home run total of 427 ranks 39th. His Black-Ink and Gray-Ink Test scores rank greater than 200.

18.2.7.3. Frank Thomas (First Baseman / Designated Hitter)

- Height: 6' 5"
- Weight: 257 pounds
- Threw and batted right-handed
- Career: 1990–2008
- Chicago White Sox (AL)

Thomas won the MVP Award twice, in 1993 and 1994. His career total of 521 home runs ranks 18th, his career total of 1704 RBIs ranks 21st, and his career SLG of .555 ranks 24th. His Black-Ink Test score was 21 (rank 99). His Gray-Ink Test score was 200 (rank 40).

18.2.7.4. Sammy Sosa (Right Fielder)

- Height: 6'
- Weight: 180 pounds
- Threw and batted right-handed
- Career: 1989–2007
- Chicago Cubs (NL)

In 1998 Sosa won an MVP Award. His career .534 SLG ranks 39th, his 609 career home runs rank sixth, and his 1667 career RBIs rank 24th. He is one of two players (the other being Mark McGwire) with four consecutive seasons of at least 50 home runs. He is also the only player to hit more than 60 home runs three times: 66 in 1998, 63 in 1999, and 64 in 2001. His Black-Ink Test score was 28 (rank 65). His Gray-Ink Test score was 138 (rank 113).

18.2.7.5. Mark McGwire (First Baseman)

- Height: 6' 5"
- Weight: 225 pounds
- Threw and batted right-handed
- Career: 1986–2001
- Oakland Athletics (AL)

He was the first player to hit 70 home runs in a season in 1998. He followed that feat by hitting 65 home runs in 1999. His career SLG of .588 ranks ninth, his career total of 583 home runs ranks eighth, his 1414 career RBIs rank 65th, and his 1317 career walks rank 35th. His Black-Ink Test score was 36 (rank 42). His Gray-Ink Test score rank was 110 (rank 196).

We conclude the selection of players by including those players still active as of the 2009 season. For each of these players I will give some of their hitting achievements. I will only consider their results through 2008.

18.2.7.6. Manny Ramirez (Left Fielder)

- Height: 6'
- Weight: 200 pounds
- Throws and bats right-handed
- Career: 1993–
- Boston Red Sox (AL)

Ramirez achieved a Career Triple Crown Award by winning one batting title, one home run title, and one RBI title. His career SLG of .593 ranks eighth, his career .314 AVG ranks 68th, his 4516 career total bases rank 39th, his career RBI total of 1725 ranks 20th, and his 527 career home runs rank 17th. His Black-Ink Test score was 31 (rank 100). His Gray-Ink Test score was 154 (rank 82).

18.2.7.7. Ken Griffey Jr. (All Outfield Positions)

- Height: 6' 3"
- Weight: 230 pounds
- Throws and bats left-handed
- Career: 1989–
- Seattle Mariners (AL)

Griffey Jr. won the MVP Award in 1997. His career total of 611 home runs ranks 5th, his career .547 SLG ranks 29th, his 5060 career total bases rank 15th, his 2679 career hits rank 58th, his 1240 career walks rank 47th, and his 1772 career RBIs rank 18th. His Black-Ink Test score was 26 (rank 74). His Gray-Ink Test score was 154 (rank 82).

18.2.7.8. Alex Rodriguez (Shortstop, Third Baseman)

- Height: 6' 3"
- Weight: 190 pounds
- Throws and bats right-handed

- Career: 1996–
- New York Yankees (AL)

Rodriguez won two MVP Awards, in 2003 and 2007. In 2007 he clobbered 54 home runs. His career .578 SLG ranks 11th, his 553 career home runs rank 12th, his 1606 career RBIs rank 30th, and his 4543 career total bases rank 37th. He has a Career Triple Crown by leading the league once in AVG, five times in home runs, and once in RBIs. His Black-Ink Test score was 68 (rank 13). His Gray-Ink Test score was 204 (rank 37).

18.2.7.9. Ichiro Suzuki (Right Fielder)

- Height: 5' 9"
- Weight: 160 pounds
- Throws right-handed and bats left-handed
- Career: 2001–
- Seattle Mariners (AL)

For his first eight years, he had over 200 hits and his AVG was over .300. He won the MVP Award in 2001. His career .331 AVG ranks 29th. His Black-Ink Test score was 36 (rank 42). His Gray-Ink Test score was 116 (rank 174).

18.2.7.10. Albert Pujols (First Baseman)

- Height: 6' 3"
- Weight: 210 pounds
- Throws and bats right-handed
- Career: 2001–
- St. Louis Cardinals (NL)

He won two MVP Awards, in 2005 and 2008. He was one of only three players to bat over .300, hit over 30 home runs, and drive in over 100 runs in eight consecutive years, the other two being Ruth and Gehrig. His career AVG of .334 ranks 20th, his career SLG of .634 ranks fourth, and his career total of 319 home runs ranks 98th. His Black-Ink Test score was 26 (rank 74). His Gray-Ink Test score was 180 (rank 55).

18.2.7.11. Todd Helton (First Baseman)

- Height: 6' 2"
- Weight: 195 pounds
- Throws right-handed and bats left-handed
- Career: 1997–
- Colorado Rockies (NL)

In the 2000 season, Helton batted .372 with 42 home runs and 147 RBIs. His career .328 AVG ranks 32nd, and his career SLG of .574 ranks 14th. His Black-Ink Test score was 16 (rank 152). His Gray-Ink Test score was 141 (rank 112).

My choice for best hitter in the era of 1994–2008 is Barry Bonds. Honorable mention: Alex Rodriguez, Albert Pujols, Manny Ramirez, Mark McGwire, and Sammy Sosa.

18.3. Who Are My Top Ten Hitters in Baseball?

My final list of candidates for my top 10 hitters comes from my choices for best hitter and honorable mention in each of the six eras.

18.3.1. Scoring System for the Selection Process

The baseball statistics that I have chosen to include are #H, #HR, AVG, OBP, SLG, OPS, BRA, HRA, and RC/27. These statistics, based on career plate appearances, will be calculated for each player in my final list.

The choices of #H, #HR, AVG, OBP, and SLG are the traditional measurements that strongly influence whether a player is elected into the Hall of Fame. For each of the nine statistics AVG, OBP, SLG, OPS, BRA, RC/27, HRA, #H, and #HR, the players with the five highest numbers for each of the statistics are assigned the ranks 1–5, with 5 being the highest. For example, for the career AVG the ranks were as follows: rank 5—Ty Cobb (.366), rank 4—Rogers Hornsby (.358), rank 3—Joe Jackson (.356), rank 2—Ted Williams (.344), and rank 1—Babe Ruth (.342). Five points are given to a player for rank 5, four points are given for rank 4, and so on. Table 18.1 displays each player's statistics. Table 18.2 assigns the five ranks for each statistic to the winning player. The first two categories in Table 18.2 are labeled TC and CA TC, which stand for Triple Crown Award and Career Triple Crown Award, respectively. In the TC column, the number represents the number of Triple Crowns the player won multiplied by 5. A number 3 in the CA TC column means that the player had a Career Triple Crown Award. If a player had a Triple Crown Award, he cannot also receive credit for a Career Triple Crown. The column next to CA TC labeled "BL/GR" represents the Black-Ink Test and Gray-Ink Test. The number 5 in that column means that the player was ranked in the top 10 in either one of the two tests or both tests.

The statistics OPS and BRA are included because these two statistics combine a player's ability to get on base with the player's ability to hit with power and advance runners. Since OPS = OBP + SLG and SLG numbers tend

Table 18.1 **The nine batting statistics for the final list of candidates for my top 10 batters**

Player	AB	BB	H	TB	HR	AVG	OBP	SLG	OPS	BRA	RC/27	HRA
Babe Ruth	8399	2062	2873	5795	714	0.342	0.474	0.690	1.164	0.327	12.68	0.0850
Ted Williams	7706	2021	2654	4886	521	0.344	0.482	0.634	1.116	0.306	11.89	0.0676
Lou Gehrig	8001	1508	2721	5057	493	0.340	0.447	0.632	1.079	0.283	10.92	0.0616
Rogers Hornsby	8173	1038	2930	4716	301	0.358	0.434	0.577	1.011	0.250	9.95	0.0368
Ty Cobb	11434	1249	4189	5854	117	0.366	0.433	0.512	0.945	0.222	8.92	0.0102
Barry Bonds	9847	2558	2935	5977	762	0.298	0.444	0.607	1.051	0.270	9.79	0.0774
Henry Aaron	12364	1402	3771	6862	755	0.305	0.374	0.555	0.929	0.208	7.62	0.0611
Nap Lajoie	9589	516	3242	4478	83	0.338	0.380	0.467	0.847	0.177	6.84	0.0087
Willie Mays	10881	1464	3283	6061	660	0.302	0.384	0.557	0.941	0.214	7.81	0.0607
Albert Pujols	4578	696	1531	2857	319	0.334	0.425	0.624	1.049	0.265	10.16	0.0697
Stan Musial	10972	1599	3630	6133	475	0.331	0.417	0.559	0.976	0.233	8.88	0.0433
Honus Wagner	10430	963	3415	4860	101	0.327	0.391	0.466	0.857	0.182	6.91	0.0097
Mickey Mantle	8102	1733	2415	4513	536	0.298	0.421	0.557	0.978	0.234	8.52	0.0662
Mark McGwire	6187	1317	1626	3638	583	0.263	0.394	0.588	0.982	0.232	8.01	0.0942
Alex Rodriguez	7860	980	2404	4543	553	0.306	0.389	0.578	0.967	0.225	8.26	0.0704
Frank Robinson	10006	1420	2943	5373	586	0.294	0.389	0.537	0.926	0.209	7.55	0.0586
Pete Rose	14053	1566	4256	5748	160	0.303	0.375	0.409	0.784	0.153	5.61	0.0114
Joe DiMaggio	6821	790	2214	3949	361	0.325	0.398	0.579	0.977	0.230	8.70	0.0529
Joe Jackson	4981	519	1772	2575	54	0.356	0.423	0.517	0.940	0.219	8.66	0.0108
Manny Ramirez	7610	1212	2392	4513	527	0.314	0.411	0.593	1.004	0.244	9.06	0.0693
Jimmie Foxx	8134	1452	2646	4954	534	0.325	0.428	0.609	1.037	0.261	9.85	0.0657
Ken Griffey Jr.	9313	1240	2679	5094	611	0.288	0.373	0.547	0.920	0.204	7.30	0.0656
George Brett	10349	1096	3154	5040	317	0.305	0.369	0.487	0.856	0.180	6.59	0.0306
Tony Gwynn	9288	790	3141	4263	135	0.338	0.388	0.459	0.847	0.178	6.86	0.0145
Mike Schmidt	8352	1507	2234	4402	548	0.267	0.380	0.527	0.907	0.200	6.97	0.0656
Sammy Sosa	8813	929	2408	4706	609	0.273	0.344	0.534	0.878	0.184	6.45	0.0691

to be larger than OBP numbers, more weight is given to the SLG. On the other hand, since BRA = OBP * SLG, more weight is given to the OBP. This is due to the fact that the product of two numbers between 0 and 1 is less than the smaller of the two numbers.

The HRA (home run average = #HR/#AB) statistic is included because in order to get a true reading of a player's home run hitting ability, the number of at bats a player has must be taken into consideration.

The RC statistic measures a player's ability to produce runs for his team. RC/27 measures a player's run production per game if he theoretically occupied each position in the batting order. The formula being used for RC/27 is the approximation formula given by RC/27 = (25.5 * OBP * TB)/(#AB – #H).

All the above baseball statistics were defined and described in Chapters 3 and 4.

Table 18.2 **Total points for the top 10 hitters**

Player	TC	CA TC	BL/ GR	RK AVG	RK OBP	RK SLG	RK OPS	RK BRA	RK RC/27	RK HRA	RK H	RK HR	Tot Pts
Ruth		3	5	1	4	5	5	5	5	4		3	40
Williams	10		5	2	5	4	4	4	4				38
Gehrig	5		5		3	3	3	3	3				25
Hornsby	10		5	4	1				1				21
Cobb	5		5	5							4		19
Bonds		3			2		2	2		3		5	17
Aaron		3	5								3	4	15
Lajoie	5		5										10
Mays			5									2	7
Pujols*						2	1	1	2	1			7
Musial			5								2		7
Wagner			5								1		6
Mantle	5												5
McGwire										5			5
Rodriguez*		3								2			5
Robinson	5												5
Rose											5		5
DiMaggio		3											3
Jackson				3									3
Ramirez*		3											3
Foxx						1							1
Griffey Jr.												1	1
Brett													0
Gwynn													0
Schmidt													0
Sosa													0

*Player active as of 2009.

Table 18.2 gives the total points for the top 10 players, displayed in Table 18.3.

18.3.2. Explaining the Selections

It was nice to see that my scoring system substantiated my belief that Babe Ruth and Ted Williams are the top two hitters in baseball history. Of course, my scoring system leads to a full-scale debate. You may have chosen more or different statistics than I chose. Some other important statistics that have been used in the past to rate players are career runs scored, career runs batted in, career ratio of home runs to strikeouts, and career ratio of strikeouts to base on balls, among others. Another difference could be that one may rank on the basis of the 10 best in each category instead of the five best in each category. In the Chapter Problems, you will be given the opportunity to produce your own scoring system, based on your choice of statistics.

Players who are still active can increase their point total as they play additional seasons. Clearly, all their career totals will increase as well as their awards. For example, many baseball people believe that Albert Pujols has the tools needed to win one or more Triple Crown Awards.

My scoring system shows that, of all active players as of 2009, Albert Pujols is the best hitter from the year 2000 onward. Albert Pujols began his career in 2001, and yet in eight years, Pujols is in a tie with Willie Mays and Stan Musial for rank 9 in my scoring system, which says it all.

Table 18.3 **Total points and rank for the top 10 players**

Player	Total points	Rank
Babe Ruth	40	1
Ted Williams	38	2
Lou Gehrig	25	3
Rogers Hornsby	21	4
Ty Cobb	19	5
Barry Bonds	17	6
Hank Aaron	15	7
Nap Lajoie	10	8
Willie Mays	7	9
Stan Musial	7	9
Albert Pujols	7	9

Besides the choice of statistics, another argument made against my scoring system is that I did not adjust my statistics for the era or the home stadium of the player. Players whose careers were mainly in the dead ball era of 1900–1919 are penalized. You will be asked in the Chapter Problems to create tables similar to Tables 18.1, 18.2, and 18.3 with adjusted statistics.

Exercise 18.1. Suppose that we added to the existing criteria the Hall of Fame Standards Test, awarding five points for the highest scores 1–5, four points for the highest scores 6–10, three points for the highest scores 11–15, two points for the highest scores 16–20, and one point for the highest scores 21–25. With this new criterion added, recalculate the player totals in Table 18.2. The information needed can be found at www.baseball-reference.com.

18.4. Future Research Questions

Here are some questions to research:

- Who are the top 10 all-time hitters at each position?
- What are the top 10 all-time batting seasons for a team?
- What is the relationship between pitch count and batting average?
- Does the designated hitter rule give the American League an advantage in the World Series?
- Is the sacrifice bunt an effective strategy in a game?
- Is pinch hitting for a positional player an effective strategy?
- Is attempting to steal a base an effective strategy?
- Which is hardest to duplicate, winning the Triple Crown or batting .400 for a season?

- Which pitching statistic is most similar to Joe DiMaggio's 56-game hitting streak? Which of the two would be hardest to duplicate?

Of course, we limited our study in this book solely to only two of the five tools of a positional player. We also ignored studying the statistics for pitchers. If we include all five tools for our positional players, the first question one might ask is, who are the top 10 positional players of all time? Of course, this task brings an extra degree of difficulty. Which of the players in our top 10 would remain? What different players might replace them?

We close the chapter with an American icon.

18.5. Lou Gehrig–An American Hero

Lou Gehrig was born on June 19, 1903, the son of German immigrants. In 1921, Lou entered Columbia University on a football scholarship to pursue a degree in engineering. He also pitched and played first base for Columbia. He was signed by the Yankees in 1923 with a $1500 bonus. He was called up to the Yankees in September of 1923. He became a Yankee for good in 1925. Replacing an injured Wally Pipp, he played every game for over 13 years.

I have already talked about Lou Gehrig's hitting accomplishments, so I will not repeat them again. I ranked Lou Gehrig number 3 on my all-time greatest hitters list, behind Babe Ruth and Ted Williams.

In 1938, Lou's batting average fell below .300 for the first time in his career since 1925. It was clear something was wrong. Lou played the first eight games of the 1939 season. On May 2, 1939, Lou removed himself from the starting lineup. The game announcer spoke these words, "Ladies and gentlemen, Lou Gehrig's consecutive game streak of 2130 has ended." Doctors at the Mayo Clinic diagnosed Gehrig with a very rare degenerate disease (ALS).

New York sportswriter Paul Gallico suggested that the team have a special day to honor Lou Gehrig on July 4, 1939. Surrounded by former and current teammates, holding back his tears, he began his famous words giving thanks to his fans and teammates. The amazing thing about his speech is that it was not prepared in advance. Baseball historians sometimes refer to Lou Gehrig's speech as the "Gettysburg Address" for baseball. Here are Gehrig's immortal words:

> Fans, for the past two weeks you have been reading about a bad break I got. Yet today I consider myself the luckiest man on the face of the earth.
>
> I have been in ballparks for seventeen years and have never received anything but kindness and encouragement from you fans. Look at these grand men. Which of you wouldn't consider it the highlight of his career just to associate with them for even one day?

Sure I'm lucky.

Who wouldn't consider it an honor to have known Jacob Ruppert? Also, the builder of baseball's greatest empire, Ed Barrow? To have spent six years with that wonderful little fellow, Miller Huggins? Then to have spent the next nine years with that outstanding leader, that smart student of psychology, the best manager in baseball today, Joe McCarthy?

Sure I'm lucky.

When the New York Giants, a team you would give your right arm to beat, and vice versa, sends you a gift–that's something. When everybody down to the groundskeepers and those boys in white coats remember you with trophies–that's something.

When you have a wonderful mother-in-law who takes sides with you in squabbles with her own daughter–that's something.

When you have a father and a mother who work all their lives so you can have an education and build your body–it's a blessing.

When you have a wife who has been a tower of strength and shown more courage than you dreamed existed–that's the finest I know.

So, I close in saying that I might have been given a bad break, but I've got an awful lot to live for.

Every time I hear or read these words, chills come over me. Lou Gehrig was elected to the Hall of Fame that December. With his health failing, he began to work in his community. He was asked to join the Parole Board in New York City so that he could help troubled children. He agreed and was sworn in for a 10-year term in 1940. His health continued to decline until he died on June 2, 1941.

His wife Eleanor described her husband as "just a square honest guy." Sportswriter Jim Murray said, "Lou Gehrig was a dedicated athlete, a caring son and husband, an honest man and an American hero."

Chapter Summary

The final chapter attempts to provide ideas for research in the field of sabermetrics. Two research questions are posed at the beginning of the chapter. What are the greatest batting feats of all time? Who are the best hitters in Major League history? Also, other possible research projects not studied in this book are presented.

Eight batting feats are presented and ranked. Joe DiMaggio's 56-game hitting streak and Rogers Hornsby's feat of averaging over .400 for his combined statistics from 1921 through 1925 tied for first place.

Candidates for the greatest hitter of all time were pulled from six eras. Candidates for each era were selected based on specific batting statistics discussed throughout the book. Also, credit was given for achieving certain batting milestones such as winning the Triple Crown, winning the MVP Award, or being selected to an All-Star Team. The 26 finalists came from the best hitters from each of the six eras. A specific scoring system was then established for the 26 finalists. The scoring system was based on awarding points for being in the top 5 (among the 26 finalists) for the baseball statistics AVG, OBP, SLG, OPS, BRA, RC/27, HRA, #H, #HR. Also, points were awarded for being in the top 10 (all time) for the Black-Ink Test / Gray-Ink Test and winning a Triple Crown or Career Triple Crown. In the chapter problems that follow, you are asked to develop your own scoring system.

The chapter ends with a special tribute to Lou Gehrig. His entire farewell speech is presented.

The chapter problems act as a review for many of the chapters studied in this book. Some of the problems refer you to the chapter where the topic in the problem is covered. Good luck in tackling these problems.

SEASON ENDING CHAPTER PROBLEMS

1. Produce your own scoring system for the players in Table 18.1 by choosing the baseball statistics you feel should be used in determining the best hitter in baseball history. Use your selected statistics to create tables similar to Tables 18.1, 18.2, and 18.3.
2. Use both your scoring system and my scoring system to rank the best five individual regular seasons. The candidates are Ruth (1920, 1921, 1923, 1927), Williams (1941, 1942, 1947, 1949), Bonds (2001, 2002, 2003), Musial (1948), Hornsby (1924 and 1925), Foxx (1932, 1933, 1938), DiMaggio (1939, 1941), Frank Robinson (1966), Cobb (1909, 1911, 1917), Mays (1955), Speaker (1912), Mantle (1956), Gehrig (1927, 1934, 1936), Brett (1980), McGwire (1980), Sisler (1920, 1922), and Alex Rodriguez (1996).
3. Let X = the team payroll on opening day of the 2008 season for all 30 Major League teams. Let Y = the winning percentage for each team at the end of the 2008 regular season. You can find this information on the Internet.
   a. Sketch the scatter-plot graph.
   b. Find the equation for the regression line.
   c. Find the correlation coefficient.
   d. Using the regression line equation, find the expected winning percentage for each team using the X-values.

   If you need to, you can review these concepts in Chapter 5.

If you need to, you can review the concepts in the next three questions in Chapters 7 and 11.

4. Suppose that it is known that a certain pitcher has the probability of 1/2 of throwing a strike. Assume the following: the batter was a pitcher who was told not to swing at any pitches, the pitches were independent events, and the probability of a pitch being a strike is the same for each pitch.
   a. Construct a tree diagram showing all the possible outcomes of the at bat.
   b. List all the possible outcomes of the at bat.
   c. Find the probability that the batter draws a walk.
   d. Find the probability that the batter strikes out.
   e. Given that the first three pitches were balls, find the probability that the fourth pitch is a ball.
   f. Given that the first three pitches were balls, find the probability that he strikes out.
   g. Find the probability that he strikes out before he walks.
   h. Repeat (c) and (d) using 7/10 for the probability that the pitcher throws a strike.
5. In how many different orders can a player hit for the cycle in one game?
6. A certain player with a career batting average of .280 has five at bats in a game. Assuming that he is a *true* .280 hitter and the five at bats are independent events, find the following:
   a. The probability that he goes hitless.
   b. The probability that he has five hits.
   c. The probability that he has at least one hit.
   d. The probability that he has at most one hit.
   e. The probability that this player has a three-game hitting streak.
   f. Starting a new season, the probability that this player doesn't get his first hit until his sixth at bat.
7. Using Ty Cobb's career at bats as your sample, find a 95% confidence interval estimate for Ty Cobb's *true* career AVG. What is the baseball statistic used to measure Cobb's batting performance? What is the baseball parameter being estimated?

   If you need to, you can review this information in Chapter 14.
8. Using Ty Cobb's career at bats as your sample, can we conclude at a level of significance $\alpha=0.05$ that Ty Cobb was a *true* career .300 hitter? Find the p-value. Was Ty Cobb a *true* career .350 hitter for $\alpha=0.05$? Find the p-value. If you need to, you can review this information in Chapter 15.
9. Can we conclude that World Series Champions beginning in 2000 and going forward in the future have a winning percentage higher than .500 against their opponents that had a .500 win percentage or higher during the regular season? Use $\alpha=0.05$ and the sample data in Table 18.4. State any assumptions needed. Find the p-value. If you need to, you can review this information in Chapter 15.

Table 18.4 **Record of World Series champions (2000–2008) against +.500 teams**

World Series champion	Record against +.500 teams	Win percentage
2000 Yankees	42 W–43 L	.494
2001 Diamondbacks	42 W–43 L	.494
2002 Angels	38 W–42 L	.475
2003 Marlins	53 W–48 L	.525
2004 Red Sox	42 W–31 L	.575
2005 White Sox	39 W–33 L	.542
2006 Cardinals	21 W–26 L	.447
2007 Red Sox	44 W–40 L	.524
2008 Phillies	43 W–46 L	.483

If you need to, you can review the concepts in the next four questions in Chapter 16.

10. Using the statistics of Tony Gwynn (1994), George Brett (1980), and Ichiro Suzuki (2004), find the probability of each one of these players duplicating the 56-game hitting streak in that season. Repeat for an 84-game on-base streak.
11. Find the probability of Ty Cobb having a two-game hitting streak somewhere during four consecutive games. Use three different methods to solve this problem. Use Ty Cobb's career average of .366 for p and assume exactly four at bats in each game.
12. Choose a current player (active as of 2008) who you feel has the best chance to duplicate the 115-game streak of not striking out. Find the probability that this player can duplicate this streak.
13. Find a consecutive game hitting streak not mentioned in this book. Identify the player or players who accomplished the streak. Find the probability of those players duplicating the streak in the year they accomplished their streak.

It is now time for you to do your own observational research study. Choose a ballgame to watch at random from the first pitch to the last pitch. For each ball used in the game, record the number of pitches before the ball either (a) winds up in the stands, (b) is discarded by a player or umpire, or (c) results in the third out of a half-inning. Assume that a new ball begins each half inning. Make a table with three columns. The three headings for the columns are the ball number, number of pitches, what happened to the ball (a, b, or c). The population for this study will be all balls used in a typical game.

The two variables in this study are X = the number of pitches for a ball and Y = what happens to the ball (a, b, or c). The balls used in your one game will be the sample for this study.

14.

a. What sampling method are you using?
b. What measurement scale would you assign to the variables X and Y?
c. Give the five-number summary for the data from the X-variable. Identify all outliers.
d. Sketch the box-and-whisker graph for the data from the X-variable.
e. Sketch the stem-and-leaf graph for the data from the X-variable.
f. What shape would you assign to the sample data from the X-variable?
g. Sketch a frequency bar graph for the data from the Y-variable.
h. Find the sample mean, the sample median, the sample standard deviation, and the sample interquartile range for the sample data from the X-variable.

From the sample data answer these questions:

i. The probability that an x-value is less than 4 is _______. (Express as a proportion.)
j. The proportion of the x-values less than 4 is _______. (Express as a fraction.)
k. The probability that an x-value is less than or equal to 4 is _______. (Express as a decimal to four decimal places.)
l. Construct a discrete probability distribution for the variable X.
m. Find a 95% confidence interval for the population mean number of pitches of a ball in a typical game.
n. Find a 90% confidence interval for the population proportion for the number of balls used for less than four pitches in a typical game.
o. Can we conclude with a level of significance of $\alpha = 0.05$ that the population mean number of pitches for a ball in a typical game is different from 4?
p. Find the p-value for the test statistic in (o).

15. We return to your decision about the player you chose as a possible candidate for the Hall of Fame. In Chapter 10, you gave a decision about the player based on the descriptive statistics covered in the first 10 chapters. It is now time for your final decision. Do you want to change your mind about your player? Use the information in the final eight chapters to either stay with your original decision or change your mind.

I HEAR THE FAT LADY SINGING.

Appendix A

Hypothesis Testing for Two Population Proportions

A.1. Introduction

The population proportion p_1 for one population is compared with the population proportion p_2 for the second population. There are many times when we want to compare two players with respect to such descriptive measures as AVG, OBP, SOA, BBA, IPBA, and HRA. Remember, "true" often replaces "population." For example, in place of the *population* proportion we can use the *true* proportion.

Table A.1 shows the career performances of Joe DiMaggio and Ted Williams with respect to their career plate appearances. Considering these as two representative samples, what can we say about comparing their *true* career abilities with respect to such parameters as their true AVG, OBP, SOA, BBA, and HRA? Remember, their respective populations are theoretical and based on each of them having an infinite number of plate appearances. In particular, DiMaggio's sample career AVG (career $\widehat{\text{AVG}}$) is equal to 2214/6821 = .325, while Williams's is equal to 2654/7706 = .344. Can we conclude, with a level of significance of 0.05, that Williams's *true* career AVG is different from DiMaggio's *true* career AVG? Questions like this will be studied.

Besides comparing two players, there are times when we want to compare two teams, two leagues, or two years with respect to certain quantitative variables.

A.1.1. Selection of the Two Samples from the Two Populations

Unlike *one-population* hypothesis testing (which relied on the selection of one random sample from one population), *two-population* hypothesis testing relies on the selection of two random samples, one from each population. The two samples selected must be *independent samples.*

Table A.1 Career batting statistics for Joe DiMaggio and Ted Williams

Player	Years	G	AB	R	H	2B	3B	HR	RBI	BB	SO	SH	SF	HBP
Joe DiMaggio	13	1736	6821	1390	2214	389	131	361	1537	790	369	14	—	46
Ted Williams	19	2292	7706	1798	2654	525	71	521	1839	2021	709	5	20	39

Two samples are *independent* if the selection of the first sample from one of the populations is not related to the selection of the second sample from the other population.

A.2. Hypothesis Testing for Two Population Proportions

In Chapter 15, we learned how to perform hypothesis testing for one population proportion. We needed to have a sampling distribution of $\hat{p}$ that was normally distributed, with its mean equal to the population proportion, p. For hypothesis testing for two population proportions, we need a sampling distribution of $\hat{p}_1 - \hat{p}_2$ that is normally distributed with its mean equal to $p_1 - p_2$.

A.2.1. The Sampling Distribution for the Difference between Two Sample Proportions

To form the sampling distribution of $\hat{p}_1 - \hat{p}_2$, we perform the following steps:

Step 1: Collect all independent random samples from each of the two populations, X_1 and X_2. The samples selected from X_1 have a fixed size n_1, and the samples from X_2 have a fixed size n_2 (it is not necessary for the two sample sizes to be equal).

Step 2: For each sample from each population we evaluate its sample proportion ($\hat{p}_1$ for the X_1-population and $\hat{p}_2$ for the X_2-population).

Step 3: We look at all possible differences $\hat{p}_1 - \hat{p}_2$. The statistic $(\hat{p}_1 - \hat{p}_2)$ is an *unbiased estimator* for $p_1 - p_2$ (the mean for the sampling distribution for the statistic $(\hat{p}_1 - \hat{p}_2)$ is equal to $p_1 - p_2$).

The next theorem provides the assumptions and conclusions needed to use the sampling distribution for $(\hat{p}_1 - \hat{p}_2)$ in hypothesis testing.

Theorem A.1. Given that steps 1–3 above are applied, let p_1 and p_2 be the population proportions for the populations X_1 and X_2, respectively. If $n_1 p_1 > 5$, $n_1(1 - p_1) > 5$, $n_2 p_2 > 5$, and $n_2(1 - p_2) > 5$, then the sampling distribution of $(\hat{p}_1 - \hat{p}_2)$ satisfies the following properties:

a. The sampling distribution of $(\hat{p}_1 - \hat{p}_2)$ is normally distributed.
b. The mean of the sampling distribution is equal to $p_1 - p_2 (\mu_{\hat{p}_1 - \hat{p}_2} = p_1 - p_2)$..

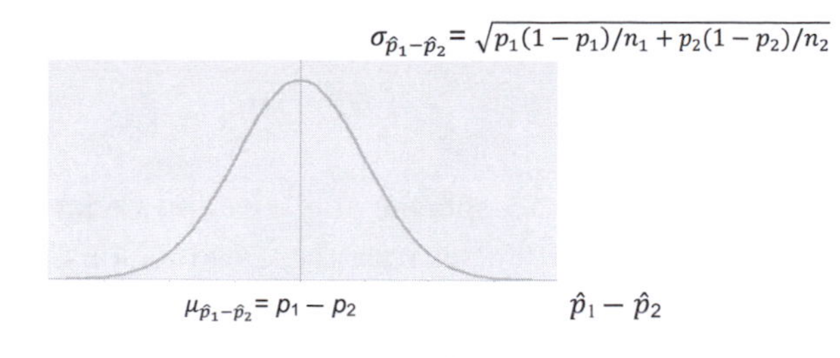

$\sigma_{\hat{p}_1-\hat{p}_2} = \sqrt{p_1(1-p_1)/n_1 + p_2(1-p_2)/n_2}$
$\mu_{\hat{p}_1-\hat{p}_2} = p_1 - p_2$
$\hat{p}_1 - \hat{p}_2$

Figure A.1 Sampling distribution for the difference for two sample proportions, $\hat{p}_1 - \hat{p}_2$

c. The standard error (SE) of the sampling distribution of $\hat{p}_1 - \hat{p}_2$ is equal to. $\sqrt{p_1(1-p_1)/n_1 + p_2(1-p_2)/n_2}$. Remember, the standard error SE is the standard deviation, $\sigma_{\hat{p}_1-\hat{p}_2}$, for the sampling distribution.

The sampling distribution for the difference for two sample proportions, $\hat{p}_1 - \hat{p}_2$, is shown in Figure A.1.

A.2.2. Classical Hypothesis Testing for $p_1 - p_2$

Once we see that the properties of the sampling distribution of $\hat{p}_1 - \hat{p}_2$ are satisfied, we are able to follow the eight steps for classical hypothesis testing. The test statistic used to estimate $p_1 - p_2$ is $\hat{p}_1 - \hat{p}_2$. The eight steps are given next.

Step 1: State the research question.
Step 2: State the two hypotheses (choose one of the three pairs below):

$H_0: p_1 - p_2 = 0$ $\quad$ $H_0: p_1 - p_2 \leq 0$ $\quad$ $H_0: p_1 - p_2 \geq 0$
$H_A: p_1 - p_2 \neq 0$ $\quad$ $H_A: p_1 - p_2 > 0$ $\quad$ $H_A: p_1 - p_2 < 0$.

(The subscripts 1 and 2 are often replaced by letters reflecting the populations.)

Step 3: State the level of confidence α.
Step 4: Data and assumptions concerning the underlying population: $\hat{p}_1 = x_1/n_1$ and $\hat{p}_2 = x_2/n_2$, where x_1 and x_2 are the number of successes and n_1 and n_2 are the sample sizes.
Step 5: Convert the test statistic $\hat{p}_1 - \hat{p}_2$ to the standard test statistic, z:

$$z = (\text{test statistic} - \text{parameter in } H_0)/\text{SE}$$
$$= [(\hat{p}_1 - \hat{p}_2) - (p_1 - p_2)]/\sqrt{p_1(1-p_1)/n_1 + p_2(1-p_2)/n_2}.$$

Regardless of which hypothesis we choose for H_0, we assume $p_1 = p_2$. This allows us to replace $(p_1 - p_2)$ with 0. Since both p_1 and p_2 are unknown, we form $\bar{p} = (x_1 + x_2)/(n_1 + n_2)$ and use $\sqrt{\bar{p}(1-\bar{p})/n_1 + \bar{p}(1-\bar{p})/n_2}$ as a *weighted* estimate for SE. Also, we require $n_1 * \bar{p} > 5$, $n_1 * (1-\bar{p}) > 5$, $n_2 * \bar{p} > 5$, and $n_2 * (1-\bar{p}) > 5$. Using these assumptions and substitutions, we continue the calculation for the standard test statistic, z:

$$z = [(\hat{p}_1 - \hat{p}_2) - 0] / \sqrt{\bar{p}(1-\bar{p})/n_1 + \bar{p}(1-\bar{p})/n_2}$$
$$= (\hat{p}_1 - \hat{p}_2) / \sqrt{\bar{p}(1-\bar{p})/n_1 + \bar{p}(1-\bar{p})/n_2}.$$

Step 6: Set up the acceptance and rejection regions by finding the critical value or values that separate these two regions. Finally, locate the result of step 5.

Step 7: Make the statistical decision about the null hypothesis. Accept or reject it.

Step 8: Answer the research question.

Example A.1. We return to the question raised in Section A.1. We are interpreting AVG as a proportion.

Step 1: Can we conclude, with a level of significance of $\alpha = 0.05$, that Williams's population career AVG is different from DiMaggio's population career AVG?

Step 2: We choose for the two hypotheses the following pair: H_0: $p_W - p_D = 0$; H_A: $p_W - p_D \neq 0$ (claim) (p_W = population AVG for Williams, p_D = population AVG for DiMaggio).

Step 3: $\alpha = 0.05$.

Step 4: $n_W = 7706$, $\hat{p}w = x_W/n_W$ = #H/#AB = 2654/7706 = .344; $n_D = 6821$, $\hat{p}_D = x_D/n_D$ = #H/#AB = 2214/6821 = .325.

Step 5: $\bar{p} = (x_W + x_D)/(n_W + n_D) = (2654 + 2214)/(7706 + 6821) = .335$. Clearly, $n_W * \bar{p} > 5$, $n_W * (1 - \bar{p}) > 5$, $n_D * \bar{p} > 5$, $n_D * (1 - \bar{p}) > 5$, and the samples are independent. With these assumptions and substitutions, we calculate the standard test statistic, z:

$$z = (\hat{p}_W - \hat{p}_D) / \sqrt{\bar{p}(1-\bar{p})/n_W + \bar{p}(1-\bar{p})/n_D}$$
$$= (.344 - .325) / \sqrt{.335(1-.335)/7706 + .335(1-.335)/6821}$$
$$= .019/.0078 = 2.44.$$

Step 6: Set up the acceptance and rejection regions. Figure A.2 shows the acceptance and rejection regions.

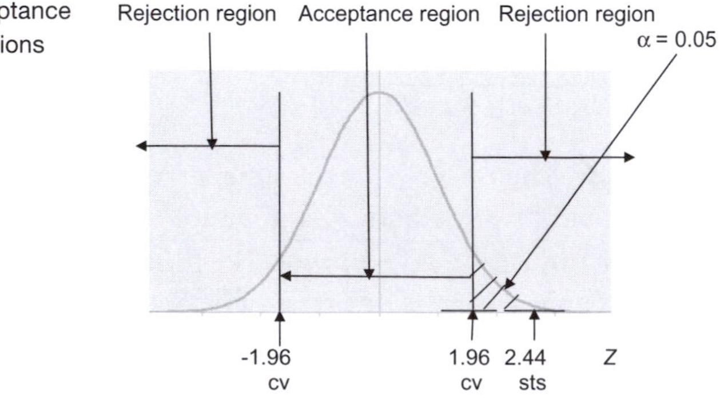

Figure A.2 Acceptance and rejection regions for Example A.1

Step 7: Reject H_0 (accept H_A).

Step 8: We can conclude that Ted Williams's population career AVG *is* different from Joe DiMaggio's population career AVG.

The Red Sox and Yankees talked about a trade involving Ted Williams for Joe DiMaggio. One reason for this was that Fenway Park with its Green Monster is more favorable to right-handed hitters and Yankee Stadium is more favorable to left-handed hitters.

A general manager might view this possible trade like this: A great defensive center fielder is more valuable than a right fielder. So, for this trade to be fair, the Yankees should be convinced that Williams has more ability as a batter than DiMaggio. The *true* AVG can be used as a measure of a player's batting ability. Can a GM conclude, using batting average as a measure of batting ability, that Williams's *true* (population) AVG is greater than DiMaggio's *true* (population) AVG? Because of the "greater than," a one-tail hypothesis test should be used. Of course, this trade never happened: the Red Sox wanted Yogi Berra to be included in the deal, so the Yankee Clipper stayed in the Bronx.

Exercise A.1. For this exercise refer to Tables 3.6 and 3.9. Use for your two samples the baseball data for both players for the 1941 season. Can we conclude, with a level of significance $\alpha=0.05$, that Williams's *true* AVG was greater than DiMaggio's *true* AVG for the year 1941?

a. Use classical hypothesis testing (eight steps)
b. Find the p-value (seven steps).

Recall that in Section 14.4 we compared the career batting averages and career on-base percentages of Aaron and Bonds by using two one-population proportion confidence intervals. We remarked at the end of that section that there was a flaw in using this approach. The flaw is that the original level of confidence is reduced during the process. When we use a two-population hypothesis test, this flaw no longer exists because the level of significance does not change during the process. We now use classical hypothesis testing to answer the following research question: Can we conclude, with a level of significance of $\alpha=0.05$, that Bonds's population career OBP is greater than Aaron's population career OBP?

Example A.2.

Step 1: Can we conclude, with a level of significance of $\alpha=0.05$, that Bonds's population career OBP is greater than Aaron's population career OBP?

Step 2: We choose for the two hypotheses the following pair: H_0: $p_B-p_A\leq 0$; H_A: $p_B-p_A>0$ (claim) (p_B = population career OBP for Bonds, p_A = career population OBP for Aaron).

Step 3: $\alpha=0.05$.

Step 4: Data: $n_B=12441$, $\hat{p}_B=x_B/n_B=$#OB/#PA$=5527/12441=.444$; $n_A=13919$, $\hat{p}_A=x_A/n_A=$#OB/#PA$=5205/13919=.374$ (for Bonds the data include up to his first 81 games of 2007).

Step 5: $\bar{p}=(x_B+x_A)/(n_B+n_A)=(5527+5205)/(12441+13919)=.407$. Clearly, $n_B*\bar{p}>5$, $n_B*(1-\bar{p})>5$, $n_A*\bar{p}>5$, and $n_A*(1-\bar{p})>5$. With these assumptions and substitutions we calculate the standard test statistic, z:

$$z=(\hat{p}_B-\hat{p}_A)/\sqrt{\bar{p}(1-\bar{p})/n_B+\bar{p}(1-\bar{p})/n_A}$$
$$=(.444-.374)/\sqrt{.407(1-.407)/12441+.407(1-.407)/13919}$$
$$=.070/.0061=11.48.$$

Step 6: Set up the acceptance and rejection regions. Figure A.3 shows the acceptance and rejection regions.

Step 7: Reject H_0 (accept H_A).

Step 8: Yes, we can conclude that the population career OBP for Bonds is greater than the population career OBP for Aaron.

Exercise A.2. Redo Example A.1 with $\alpha=0.01$.

Exercise A.3. BBA is the base-on-balls average (see Tables 3.6 and 3.9). Can we conclude, with a level of significance of $\alpha=0.10$, that Williams's population (*true*) career BBA is higher than DiMaggio's population (*true*) career BBA?

Example A.3. We now show the p-value method. The seven-step p-value approach to hypothesis testing is applied to the information in Example A.1.

Step 1: Can we conclude that Williams's population career AVG is greater than DiMaggio's population career AVG?

Step 2: We choose for the two hypotheses the following pair: H_0: $p_W-p_D\le 0$; H_A: $p_W-p_D>0$ (claim) (W for Williams, D for DiMaggio).

Step 3: $n_W=7706$, $\bar{p}_W=x_W/n_W=2654/7706=.344$; $n_D=6821$, $\bar{p}_D=x_D/n_D=2214/6821=.325$.

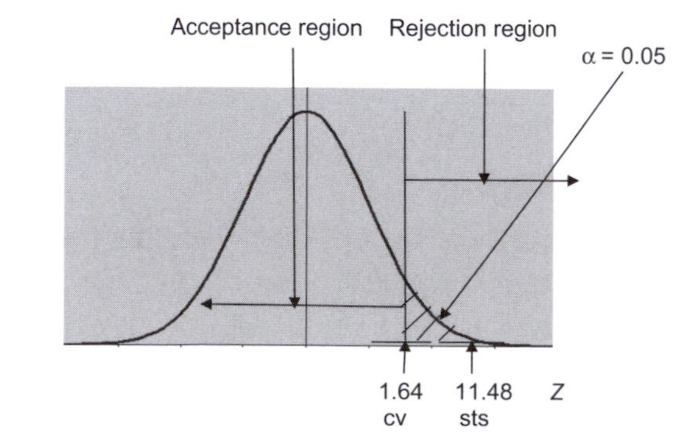

Figure A.3 Acceptance and rejection regions for Example A.2

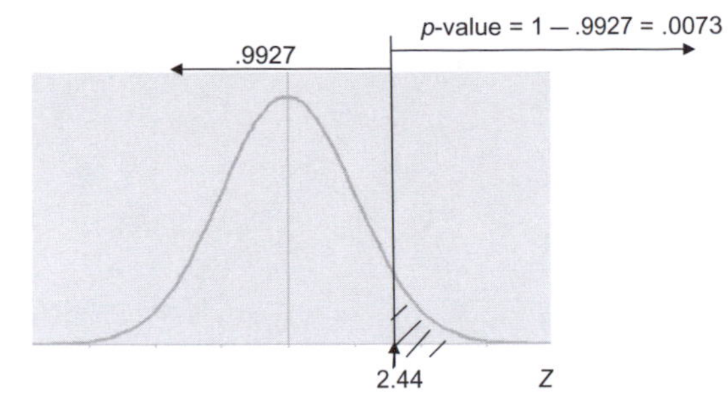

Figure A.4 Graph showing the p-value for Example A.3

Step 4: $\bar{p} = (x_W + x_D)/(n_W + n_D) = (2654 + 2214)/(7706 + 6821) = .335$. Clearly, $n_W * \bar{p} > 5$, $n_W * (1 - \bar{p}) > 5$, $n_D * \bar{p} > 5$, and $n_D * (1 - \bar{p}) > 5$. With these assumptions and substitutions we calculate the standard test statistic, z:

$$z = (\hat{p}_W - \hat{p}_D)/\sqrt{\bar{p}(1-\bar{p})/n_W + \bar{p}(1-\bar{p})/n_D}$$
$$= (.344 - .325)/\sqrt{.335(1-.335)/7706 + .335(1-.335)/6821}$$
$$= .019/.0078 = 2.44.$$

Step 5: Determine the p-value: the p-value $= .0073$. Figure A.4 displays the p-value.

Step 6: The p-value is highly significant.

Step 7: We can conclude that Ted Williams's population career AVG *is* greater than Joe DiMaggio's population career AVG.

Exercise A.4. Using the eight steps for classical hypothesis testing, can we conclude, at a level of significance of $\alpha = 0.01$, that Ted Williams's population career OBP is greater than Joe DiMaggio's population career OBP?

Exercise A.5. Using the seven steps for the p-value method of hypothesis testing, can we conclude that Ted Williams's population career OBP is greater than Joe DiMaggio's population career OBP?

Exercise A.6. Using the seven steps for the p-value method of hypothesis testing, can we conclude that Ted Williams's population career SOA is different from Joe DiMaggio's population career SOA?

Appendix B

The Chi-Square Distribution

B.1. Introduction

The chi-square distribution (χ^2-distribution), like the t-distribution, is a family of curves where each curve is determined by a specific number of degrees of freedom. Unlike the z-distribution and the t-distribution, which are symmetric around their means, the chi-square curves are skewed to the right (see Fig. B.1). The skewness of this distribution changes with the degrees of freedom (df). As the df increases, the curves become more symmetric. The values of the chi-square variable are nonnegative. Figure B.1 shows a few of the chi-square curves.

The chi-square distribution has many applications in inferential statistics. Some of these applications are finding a confidence interval for the population standard deviation, hypothesis testing for more than two population proportions, and using a goodness-of-fit test to see if observed sample data fit a particular mathematical model such as the binomial model (Chap. 11) or the normal model (Chap. 12). Many of these applications compare a set of observed data with a set of expected data. There are many statistics books that cover all of these applications.

We consider only one of the major applications of the chi-square distribution. The application covered tests whether there is a relationship between two qualitative variables. This application is called the test for independence. For this application, one random sample is selected from one population.

The two sets of frequencies considered are the *observed frequencies* and *expected frequencies*. The *observed* frequency for an event (category) is the number of subjects or objects in the sample that satisfy the event. The *expected* frequency is the number of subjects or objects in the population we would expect to satisfy the event, under the assumption that the

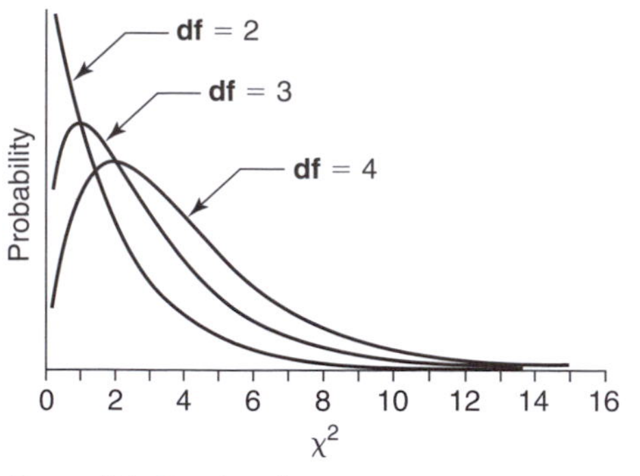

Figure B.1 Graphs of sample chi-square distributions

null hypothesis H_0 is true. The expected frequency can be calculated by multiplying the sample size times the theoretical probability of the event occurring.

The test statistic (ts) for the chi-square tests is written χ^2, where the Greek letter χ (chi) is raised to the second power. Unlike s^2, where you can take the square root to find s, χ^2 is merely a symbol. We use "Obs" or "O" for the observed frequency for each event and "Exp" or "E" for the expected frequency for each event. With these substitutions, the test statistic for chi-square tests is defined as

$$\chi^2 = \Sigma[(\text{Obs} - \text{Exp})^2/\text{Exp}] \text{ or } \Sigma[(O-E)^2/E].$$

For chi-square applications, the inferential techniques used will be either confidence intervals or hypothesis testing. Remember, hypothesis testing is based on two hypotheses. In many applications, we assume that the observed frequency distribution fits the theoretical expected distribution. For this reason, the null hypothesis (H_0) is usually that the observed frequency distribution fits the expected frequency distribution. The alternate hypothesis (H_A) is that the observed frequency distribution does *not* fit the expected frequency distribution. This type of hypothesis testing is often called *nonparametric* because the statements used in both hypotheses do *not* use parameters. (In Chap. 15, the parameters μ and p appeared in both hypotheses.)

The quantity χ^2 gives a measurement of the closeness between the observed values and the expected values. If there is perfect agreement between each observed and expected value, under the assumption that the null hypothesis H_0 is true, the test χ^2 would equal zero. When there is close agreement between the observed and expected frequencies, the χ^2 value tends to be a small positive number. A large positive number for χ^2 leads to a rejection of H_0.

For this reason, the rejection region for this test is always one tail on the right-hand side. In order to determine whether to accept or reject the null hypothesis, it is necessary to have a critical value (cv). The critical value, which separates the acceptance and rejection region, depends on the level of significance α. Knowing the level of significance, we use Table C.4 (the table for the chi-square distribution) to find the critical value. The use of this table is similar to that for the t-distribution. To select the correct critical value from the table, we must know which row and column to use. The row is determined by the degrees of freedom. The degrees of freedom df $= k -$ (number of restrictions). The number of restrictions depends on the particular application. (Later in this appendix, we provide a further discussion of degrees of freedom.) The column is determined by the level of significance α. The intersection of the row and column yields the critical value.

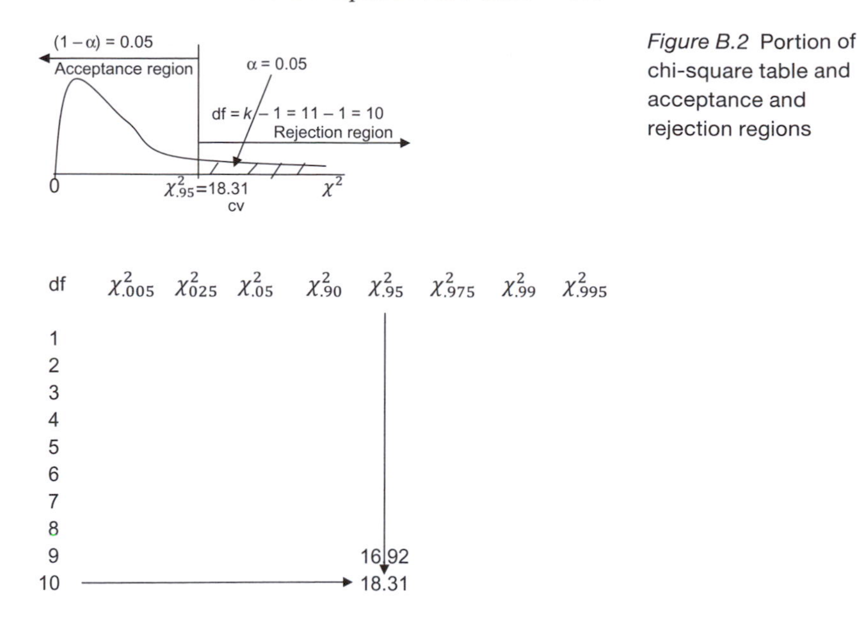

Figure B.2 Portion of chi-square table and acceptance and rejection regions

The next example demonstrates the use of this table.

Example B.1. Given 11 events and a level of significance $\alpha = 0.05$, assume df = (11 – 1) = 10. Find the critical value.

Solution: Since the rejection region is one tail to the right, the entire area of 0.05 will be contained in the right-hand tail. The area to the left of the critical value will equal $(1 - \alpha) = .95$. The degrees of freedom (10) determine which row to use. The column to use is $\chi^2_{.95}$ The subscript .95 gives the total area to the left of the critical value. The critical value for df = 10 and column = $\chi^2_{.95}$ is 18.31. The acceptance and rejection regions are shown in Figure B.2 (see also Table E).

B.2. Test for Independence between Two Qualitative Variables

In this section, we answer the question raised in Chapter 6: How can we determine if there is a relationship between two qualitative variables for a population?

In Section 6.6, we introduced contingency tables. We attempted to see if there was a relationship between the two variables of *choice of player* (Aaron or Bonds) and their *career plate appearance outcome* (1B, 2B, 3B, HR, BB, HBP, SF, O). Using side-by-side bar graphs, we gave a subjective opinion based on the sample results. The question is, can we conclude that there *is* a relationship between these two variables for the population? The population consists of all theoretically possible plate appearances for both Aaron and Bonds. The sample consists of their total career plate appearances. By looking at the various graphs, in Chapter 6, we concluded that, for a typical plate appearance, Aaron was more likely to hit a single or make an out, whereas

Bonds was more likely to draw a base on balls. These conclusions, concerning their sample of career plate appearances, were solely subjective.

In this section, we use the chi-square distribution to investigate, with respect to the theoretical population of their plate appearances, whether there is a relationship between the variables *choice of player* and *plate appearance outcome*. The statistical technique used is hypothesis testing.

For the *test for independence*, the null hypothesis H_0 will always be the statement that the two variables *are* independent, and the alternate hypothesis H_A will be the statement that there *is* a relationship between the two variables. For our example, H_0 is the statement that player choice, Aaron or Bonds, and plate appearance outcome are independent, and H_A is the statement that there is a relationship between player choice and plate appearance result.

Before proceeding, recall that the first step in a research project is to identify the population. For this population, we choose two variables (characteristics) to test for independence. Each variable is broken down into categories or events that are mutually exclusive and exhaustive (each data point of the population lies in exactly one of the categories, and all data points are used). The categories are simple events. In Section 7.8, we defined one event to be independent of another event if its occurrence had no affect on the occurrence of the other event. Using probability theory, events E and F are independent if $Pr(E \text{ and } F) = Pr(E) * Pr(F)$.

The following additional definitions are needed:

Definition B.1. Two variables, each of which is broken down into mutually exclusive categories, are independent if each of the categories of one variable is independent of each of the categories of the second variable.

Definition B.2. Contingency Table

Given:

a. Two variables W and Y defined for a certain population.
b. W is broken down into r categories, and Y is broken down into c categories (think of r for row and c for column).
c. A random sample of data points is selected from the population.

Then:

An $r \times c$ contingency table is a rectangular array consisting of r rows and c columns. The r rows represent the categories for one of the variables, and the c columns represent the categories for the other variable. The intersection of a row with a column forms a cell. Each cell contains the observed frequency for the number of subjects or objects of the sample that belong to both categories. The sum of the observed frequencies for any row (the row total) or any column

Table B.1 **Contingency table for observed frequencies for the career plate appearance results for Henry Aaron and Barry Bonds**

Player	1B	2B	3B	HR	BB	SF	HBP	Out
Aaron	2294	624	98	755	1402	121	32	8593
Bonds	1495	601	77	762	2558	91	106	6912

(the column total) is equal to the total number of subjects or objects of the sample in that category. The sum of the row totals is equal to the sum of the column totals is equal to the sample size.

Table B.1 is a 2×8 contingency table for the career plate appearances for Aaron and Bonds (for Bonds the data include his entire career).

The *test for independence* is presented through the following theorem.

Theorem B.1. Test for Independence between Two Variables

If the following conditions are satisfied:

1. The observed frequencies come from one random sample selected from one population. Two variables, each one consisting of mutually exclusive categories, are defined for the sample. Each member of the sample belongs to exactly one cell in the contingency table, constructed from the two variables.
2. The observed frequencies are the number of members of the sample in each cell of the contingency table. For each observed frequency, an expected frequency is calculated, based on the assumption that the two variables are independent of each other. The expected frequency for that cell is obtained by multiplying the sample size times the theoretical probability for that cell.
3. No more than 20% of the expected frequencies are less than 5, and no expected frequency is less than 2.

Then, we can conclude the following:

The sampling distribution for the test of independence follows the chi-square distribution with $df = (r-1) * (c-1)$, where r = the number of rows (number of categories for one of the variables) and c = the number of columns (number of categories for the other variable).The test statistic ts for the chi-square test of independence is $\chi^2 = \Sigma[(O-E)^2/E]$, where O represents the observed frequency for each cell and E represents the expected frequency for each cell.

Example B.2. We illustrate the eight steps of hypothesis testing for the test of independence for the sample consisting of the career plate appearances of Aaron and Bonds.

Table B.2 **Extended contingency table for observed frequencies for the career plate appearance results for Henry Aaron and Barry Bonds**

Player	1B	2B	3B	HR	BB	SF	HBP	Out	Total PA
Aaron	2294	624	98	755	1402	121	32	8593	13919
Bonds	1495	601	77	762	2558	91	106	6912	12602
Totals	3789	1225	175	1517	3960	212	138	15505	26521

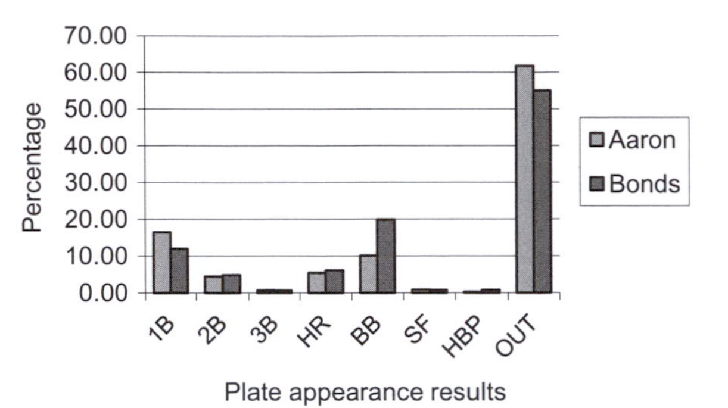

Figure B.3 Side-by-side percentage bar graph for the career plate appearances of Aaron and Bonds

Step 1: Research question: Can we conclude with a level of significance of $\alpha=0.05$ that there is a relationship between choice of player (Aaron or Bonds) and plate appearance result?

Step 2: H_0: The variables *choice of player* and *plate appearance result* are independent. H_A: There is a relationship between choice of player and plate appearance result (claim).

Step 3: The level of significance is $\alpha=0.05$.

Step 4: Data: We extend Table B.1 to include the row totals, column totals, and sample size. Table B.2 displays this extended contingency table. Figure B.3 displays their side-by-side percentage bar graph.

We illustrate the calculation for the height of the rectangles for 1B for Aaron and for 1B for Bonds in Figure B.3.

$$\begin{aligned}\text{1B for Aaron} &= \text{Observed Frequency (Aaron)/Total PAs (Aaron)}\\ &= 2294/13919 = 16.48\%,\\ \text{1B for Bonds} &= \text{Observed Frequency (Bonds)/Total PAs (Bonds)}\\ &= 1495/12602 = 11.86\%.\end{aligned}$$

For each cell, we must find the expected frequency. We label the cell with the letter E and the subscript rc, E_{rc}, where r is the row number and c is the column number. Row 1 is Aaron, and row 2 is Bonds. Column 1 is 1B, column 2 is 2B, . . . , column 8 is O (out). To illustrate the process, we compute E_{11} (the cell formed as the joint event of Aaron and 1B). Under the assumption that the two

Table B.3 **Contingency table for expected frequencies under the assumption that H_0 is true**

	Plate appearance result								
Player	1B	2B	3B	HR	BB	SF	HBP	Out	Total PA
Aaron	1988.58	642.92	91.85	796.17	2078.32	111.26	72.43	8137.48	13919
Bonds	1800.42	582.08	83.15	720.83	1881.68	100.74	65.57	7367.52	12602
Totals	3789.00	1225.00	175.00	1517.00	3960.00	212.00	138.00	15505.00	26521

variables are independent (H_0 is true), the expected probability of the joint event of Aaron and 1B occurring is the product of their probabilities:

$$Pr(\text{Aaron and 1B}) = Pr(\text{Aaron}) * Pr(\text{1B}) = (13919/26521) * (3789/26521).$$

The calculation of the expected frequency for E_{11} is

$$\begin{aligned} E_{11} &= (\text{sample size}) * Pr(\text{Aaron and 1B}) \\ &= (\text{sample size}) * Pr(\text{Aaron}) * Pr(\text{1B}) \\ &= 26521 * (13919/26521) * (3789/26521) \\ &= (13919 * 3789)/26521 \\ &= 1988.58 \\ &= (\text{row 1 total} * \text{column 1 total})/(\text{sample size}). \end{aligned}$$

The same argument could be used to find the expected frequency for any cell. This leads to the following formula for the expected frequency for a cell:

$$E_{rc} = (\text{row } r \text{ total} * \text{column } c \text{ total})/(\text{sample size}).$$

The expected frequencies should be rounded to two decimal places. Table B.3 shows the expected frequencies for each cell, under the assumption that the two variables are independent.

Figure B.4 displays the side-by-side bar graph for the expected plate appearance percentages, under the assumption of independence. We illustrate the calculation of the heights of the bars for 1B for Aaron and Bonds in Figure B.4.

$$\begin{aligned} \text{1B for Aaron} &= \text{expected frequency (Aaron)/Total PA (Aaron)} \\ &= 1988.58/13919 = 14.28\%, \\ \text{1B for Bonds} &= \text{expected frequency (Bonds)/Total PA (Bonds)} \\ &= 1800.42/12602 = 14.28\%. \end{aligned}$$

Using the expected frequencies, each plate appearance percentage is the same for both players. The side-by-side percentage bar graph in Figure B.4 demonstrates this.

Step 5: Compute the test statistic χ^2 (chi-square), using Tables B.2 and B.3. For each cell, we calculate $(O-E)^2/E$, where the O-value comes from Table B.2

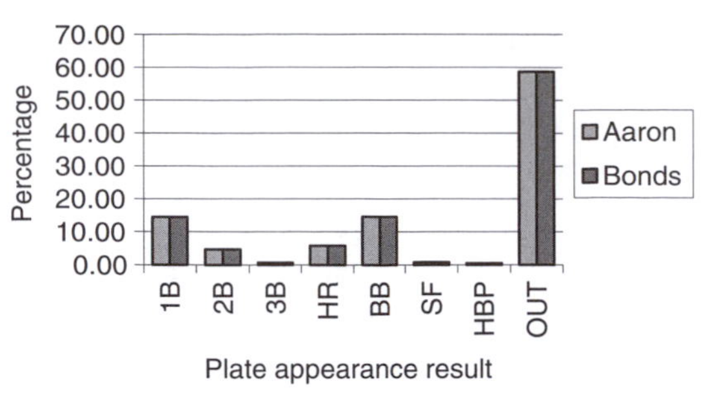

Figure B.4 Side-by-side bar graph for the expected percentages under the assumption that H_0 is true

and the *E*-value comes from Table B.3. Then, the results for these cells are totaled.

$$\chi^2 = \Sigma[(O-E)^2/E] = (2294 - 1988.58)^2/1988.58$$
$$+ (624 - 642.92)^2/642.92 + (98 - 91.85)^2/91.85$$
$$+ (755 - 796.17)^2/796.17 + (1402 - 2078.32)^2/2078.32$$
$$+ (121 - 111.26)^2/111.26 + (32 - 72.43)^2/72.43$$
$$+ (8593 - 8137.48)^2/8137.48 + (1495 - 1800.42)^2/1800.42$$
$$+ (601 - 582.08)^2/582.08 + (77 - 83.15)^2/83.15$$
$$+ (762 - 720.83)^2/720.03 + (2588 - 1881.68)^2/1881.68$$
$$+ (91 - 100.74)^2/100.74 + (106 - 65.57)^2/65.57$$
$$+ (6912 - 7367.52)^2/7367.52 = 671.36.$$

We use Table B.4 to better organize the calculations for the chi-square test statistics.

Step 6: The critical value cv = 14.07. This can be found in Table C.4 under the column $\chi^2_{.95}$ for df = (2 – 1) * (8 – 1) = 7.

Step 7: Statistical decision: Reject H_0 (see Fig. B.5).

Step 8: At the level of significance of 5%, there *is* a relationship between choice of player and plate appearance result. Yes, Aaron can be considered a different type of hitter than Bonds with respect to their plate appearance outcomes.

Table B.5 converts the observed frequencies into observed percentages.

Table B.6 converts the expected frequencies into expected percentages.

What does this really say about Aaron and Bonds as hitters? As hitters, their career statistics show that Aaron and Bonds are different. The calculations, in the table in step 5, reveal that, as hitters, Aaron and Bonds show the greatest differences in getting singles, drawing base on balls, making outs, and getting hit by pitches.

Exercise B.1. Verify the results that appear in Tables B.5 and B.6.

For the exercises below, you are free to choose the plate appearance outcomes you wish to use. Some options are as follows: combining BB with

Table B.4 **Calculations for the chi-square test statistic**

PA result	*O*	*E*	(*O*–*E*)	(*O*–*E*)²/*E*
Aaron				
1B	2294	1988.58	305.42	46.91
2B	624	642.92	–18.92	0.56
3B	98	91.85	6.15	0.41
HR	755	796.17	–41.17	2.13
BB	1402	2078.32	–676.32	220.09
SF	121	111.26	9.74	0.85
HBP	32	72.43	–40.43	22.56
Out	8593	8137.48	455.52	25.50
Bonds				
1B	1495	1800.42	–305.42	51.81
2B	601	582.08	18.92	0.61
3B	77	83.15	–6.15	0.46
HR	762	720.83	41.17	2.35
BB	2558	1881.68	676.32	243.09
SF	91	100.74	–9.74	0.94
HBP	106	65.57	40.43	24.92
Out	6912	7367.52	–455.52	28.16
Sum				671.36

HBP, combining SF with O, subdividing hits into singles and extra-base hits, and subdividing outs into strikeouts and in-play outs.

Exercise B.2. Use the test for independence to compare Ted Williams and Joe DiMaggio as hitters (use $\alpha = 0.05$).

Exercise B.3. Use the test for independence to compare Manny Ramirez and David Ortiz as hitters (use $\alpha = 0.05$).

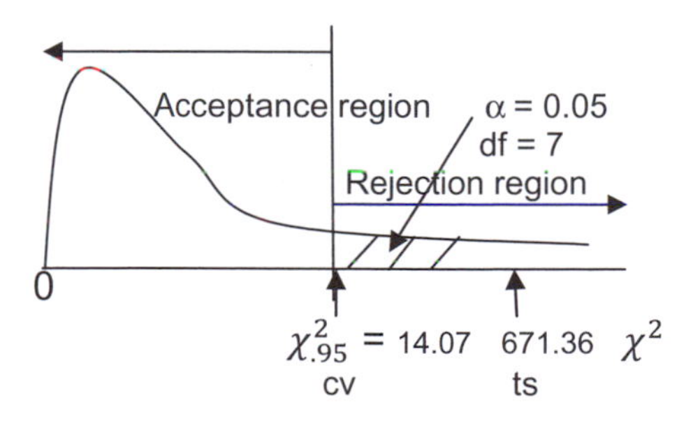

Figure B.5 Acceptance and rejection regions for the hypothesis test in Example B.2

Exercise B.4. Use the test for independence to compare your two players as hitters (use $\alpha = 0.05$).

B.3. Degrees of Freedom

For the chi-square test of independence, one random sample of size n is selected. This sample provides the observed frequency distribution. For the *test for independence*, knowing $(r-1) * (c-1)$ of the expected outcomes automatically determines all $r * c$ of the expected outcomes. This accounts for the degrees of freedom being equal to $(r-1) * (c-1)$.

Table B.5 Observed frequencies converted into observed percentages

Player	1B	2B	3B	HR	BB	SF	HBP	Out	Total PA
Aaron	16.48%	4.48%	0.70%	5.42%	10.07%	0.87%	0.23%	61.74%	13919
Bonds	11.98%	4.85%	0.61%	6.09%	19.87%	0.76%	0.80%	55.00%	12602
Totals	3697	1188	175	1463	3713	209	125	14991	25551

Table B.6 Expected frequencies converted into expected percentages

	Plate appearance result								
Player	1B	2B	3B	HR	BB	SF	HBP	Out	Total PA
Aaron	14.29%	4.62%	0.66%	5.72%	14.93%	0.80%	0.52%	58.46%	13919
Bonds	14.29%	4.62%	0.66%	5.72%	14.93%	0.80%	0.52%	58.46%	12602

Exercise B.5. Given the test of independence for a 3×4 contingency table with $E_{11}=8$, $E_{12}=18$, $E_{13}=12$, $E_{21}=9$, $E_{22}=15$, and $E_{23}=20$, the row and column totals for the observed frequencies are as follows: row 1 total = 50, row 2 total = 56, row 3 total = 44, column 1 total = 24, column 2 total = 40, column 3 total = 46, and column 4 total = 40. What is the sample size? What are the remaining expected frequencies? Observe that just knowing $(3-1)*(4-1)$ of the expected frequencies enables us to find the other six expected frequencies. Find other examples where if we know six expected values for six of the cells we will be able to find the other six expected values. Find six cells that do not work.

Appendix C

Statistical Tables

Table C.1 **Cumulative binomial probability distribution**

$n = 3$

x	$p = 0.01$	$p = 0.02$	$p = 0.03$	$p = 0.04$	$p = 0.05$	$p = 0.06$	$p = 0.07$	$p = 0.08$	$p = 0.09$
0	0.9703	0.9412	0.9127	0.8847	0.8574	0.8306	0.8044	0.7787	0.7536
1	0.9997	0.9988	0.9974	0.9953	0.9928	0.9896	0.9860	0.9818	0.9772
2	1.0000	1.0000	1.0000	0.9999	0.9999	0.9998	0.9997	0.9995	0.9993
3	1.0000	1.0000	1.0000	1.0000	1.0000	1.0000	1.0000	1.0000	1.0000

x	$p = 0.10$	$p = 0.15$	$p = 0.20$	$p = 0.25$	$p = 0.30$	$p = 0.35$	$p = 0.40$	$p = 0.45$	$p = 0.50$
0	0.7290	0.6141	0.5120	0.4219	0.3430	0.2746	0.2160	0.1664	0.1250
1	0.9720	0.9393	0.8960	0.8438	0.7840	0.7183	0.6480	0.5748	0.5000
2	0.9990	0.9966	0.9920	0.9844	0.9730	0.9571	0.9360	0.9089	0.8750
3	1.0000	1.0000	1.0000	1.0000	1.0000	1.0000	1.0000	1.0000	1.0000

$n = 4$

x	$p = 0.01$	$p = 0.02$	$p = 0.03$	$p = 0.04$	$p = 0.05$	$p = 0.06$	$p = 0.07$	$p = 0.08$	$p = 0.09$
0	0.9606	0.9224	0.8853	0.8493	0.8145	0.7807	0.7481	0.7164	0.6857
1	0.9994	0.9977	0.9948	0.9909	0.9860	0.9801	0.9733	0.9656	0.9570
2	1.0000	1.0000	0.9999	0.9998	0.9995	0.9992	0.9987	0.9981	0.9973
3	1.0000	1.0000	1.0000	1.0000	1.0000	1.0000	1.0000	1.0000	0.9999
4	1.0000	1.0000	1.0000	1.0000	1.0000	1.0000	1.0000	1.0000	1.0000

x	$p = 0.10$	$p = 0.15$	$p = 0.20$	$p = 0.25$	$p = 0.30$	$p = 0.35$	$p = 0.40$	$p = 0.45$	$p = 0.50$
0	0.6561	0.5220	0.4096	0.3164	0.2401	0.1785	0.1296	0.0915	0.0625
1	0.9477	0.8905	0.8192	0.7383	0.6517	0.5630	0.4752	0.3910	0.3125
2	0.9963	0.9880	0.9728	0.9492	0.9163	0.8735	0.8208	0.7585	0.6875
3	0.9999	0.9995	0.9984	0.9961	0.9919	0.9850	0.9744	0.9590	0.9375
4	1.0000	1.0000	1.0000	1.0000	1.0000	1.0000	1.0000	1.0000	1.0000

(continued)

$n = 5$

x	$p=0.01$	$p=0.02$	$p=0.03$	$p=0.04$	$p=0.05$	$p=0.06$	$p=0.07$	$p=0.08$	$p=0.09$
0	0.9510	0.9039	0.8587	0.8154	0.7738	0.7339	0.6957	0.6591	0.6240
1	0.9990	0.9962	0.9915	0.9852	0.9774	0.9681	0.9575	0.9456	0.9326
2	1.0000	0.9999	0.9997	0.9994	0.9988	0.9980	0.9969	0.9955	0.9937
3	1.0000	1.0000	1.0000	1.0000	1.0000	0.9999	0.9999	0.9998	0.9997
4	1.0000	1.0000	1.0000	1.0000	1.0000	1.0000	1.0000	1.0000	1.0000
5	1.0000	1.0000	1.0000	1.0000	1.0000	1.0000	1.0000	1.0000	1.0000

x	$p=0.10$	$p=0.15$	$p=0.20$	$p=0.25$	$p=0.30$	$p=0.35$	$p=0.40$	$p=0.45$	$p=0.50$
0	0.5905	0.4437	0.3277	0.2373	0.1681	0.1160	0.0778	0.0503	0.0313
1	0.9185	0.8352	0.7373	0.6328	0.5282	0.4284	0.3370	0.2562	0.1875
2	0.9914	0.9734	0.9421	0.8965	0.8369	0.7648	0.6826	0.5931	0.5000
3	0.9995	0.9978	0.9933	0.9844	0.9692	0.9460	0.9130	0.8688	0.8125
4	1.0000	0.9999	0.9997	0.9990	0.9976	0.9947	0.9898	0.9815	0.9688
5	1.0000	1.0000	1.0000	1.0000	1.0000	1.0000	1.0000	1.0000	1.0000

$n = 10$

x	$p=0.01$	$p=0.02$	$p=0.03$	$p=0.04$	$p=0.05$	$p=0.06$	$p=0.07$	$p=0.08$	$p=0.09$
0	0.9044	0.8171	0.7374	0.6648	0.5987	0.5386	0.4840	0.4344	0.3894
1	0.9957	0.9838	0.9655	0.9418	0.9139	0.8824	0.8483	0.8121	0.7746
2	0.9999	0.9991	0.9972	0.9938	0.9885	0.9812	0.9717	0.9599	0.9460
3	1.0000	1.0000	0.9999	0.9996	0.9990	0.9980	0.9964	0.9942	0.9912
4	1.0000	1.0000	1.0000	1.0000	0.9999	0.9998	0.9997	0.9994	0.9990
5	1.0000	1.0000	1.0000	1.0000	1.0000	1.0000	1.0000	1.0000	0.9999
6	1.0000	1.0000	1.0000	1.0000	1.0000	1.0000	1.0000	1.0000	1.0000
7	1.0000	1.0000	1.0000	1.0000	1.0000	1.0000	1.0000	1.0000	1.0000
8	1.0000	1.0000	1.0000	1.0000	1.0000	1.0000	1.0000	1.0000	1.0000
9	1.0000	1.0000	1.0000	1.0000	1.0000	1.0000	1.0000	1.0000	1.0000
10	1.0000	1.0000	1.0000	1.0000	1.0000	1.0000	1.0000	1.0000	1.0000

x	$p=0.10$	$p=0.15$	$p=0.20$	$p=0.25$	$p=0.30$	$p=0.35$	$p=0.40$	$p=0.45$	$p=0.50$
0	0.3487	0.1969	0.1074	0.0563	0.0282	0.0135	0.0060	0.0025	0.0010
1	0.7361	0.5443	0.3758	0.2440	0.1493	0.0860	0.0464	0.0233	0.0107
2	0.9298	0.8202	0.6778	0.5256	0.3828	0.2616	0.1673	0.0996	0.0547
3	0.9872	0.9500	0.8791	0.7759	0.6496	0.5138	0.3823	0.2660	0.1719
4	0.9984	0.9901	0.9672	0.9219	0.8497	0.7515	0.6331	0.5044	0.3770
5	0.9999	0.9986	0.9936	0.9803	0.9527	0.9051	0.8338	0.7384	0.6230
6	1.0000	0.9999	0.9991	0.9965	0.9894	0.9740	0.9452	0.8980	0.8281
7	1.0000	1.0000	0.9999	0.9996	0.9984	0.9952	0.9877	0.9726	0.9453
8	1.0000	1.0000	1.0000	1.0000	0.9999	0.9995	0.9983	0.9955	0.9893
9	1.0000	1.0000	1.0000	1.0000	1.0000	1.0000	0.9999	0.9997	0.9990
10	1.0000	1.0000	1.0000	1.0000	1.0000	1.0000	1.0000	1.0000	1.0000

Table C.1 Cumulative binomial probability distribution, *continued*

$n=20$

x	$p=0.01$	$p=0.02$	$p=0.03$	$p=0.04$	$p=0.05$	$p=0.06$	$p=0.07$	$p=0.08$	$p=0.09$
0	0.8179	0.6676	0.5438	0.4420	0.3585	0.2901	0.2342	0.1887	0.1516
1	0.9831	0.9401	0.8802	0.8103	0.7358	0.6605	0.5869	0.5169	0.4516
2	0.9990	0.9929	0.9790	0.9561	0.9245	0.8850	0.8390	0.7879	0.7334
3	1.0000	0.9994	0.9973	0.9926	0.9841	0.9710	0.9529	0.9294	0.9007
4	1.0000	1.0000	0.9997	0.9990	0.9974	0.9944	0.9893	0.9817	0.9710
5	1.0000	1.0000	1.0000	0.9999	0.9997	0.9991	0.9981	0.9962	0.9932
6	1.0000	1.0000	1.0000	1.0000	1.0000	0.9999	0.9997	0.9994	0.9987
7	1.0000	1.0000	1.0000	1.0000	1.0000	1.0000	1.0000	0.9999	0.9998
8	1.0000	1.0000	1.0000	1.0000	1.0000	1.0000	1.0000	1.0000	1.0000
9	1.0000	1.0000	1.0000	1.0000	1.0000	1.0000	1.0000	1.0000	1.0000
10	1.0000	1.0000	1.0000	1.0000	1.0000	1.0000	1.0000	1.0000	1.0000
11	1.0000	1.0000	1.0000	1.0000	1.0000	1.0000	1.0000	1.0000	1.0000
12	1.0000	1.0000	1.0000	1.0000	1.0000	1.0000	1.0000	1.0000	1.0000
13	1.0000	1.0000	1.0000	1.0000	1.0000	1.0000	1.0000	1.0000	1.0000
14	1.0000	1.0000	1.0000	1.0000	1.0000	1.0000	1.0000	1.0000	1.0000
15	1.0000	1.0000	1.0000	1.0000	1.0000	1.0000	1.0000	1.0000	1.0000
16	1.0000	1.0000	1.0000	1.0000	1.0000	1.0000	1.0000	1.0000	1.0000
17	1.0000	1.0000	1.0000	1.0000	1.0000	1.0000	1.0000	1.0000	1.0000
18	1.0000	1.0000	1.0000	1.0000	1.0000	1.0000	1.0000	1.0000	1.0000
19	1.0000	1.0000	1.0000	1.0000	1.0000	1.0000	1.0000	1.0000	1.0000
20	1.0000	1.0000	1.0000	1.0000	1.0000	1.0000	1.0000	1.0000	1.0000

x	$p=0.10$	$p=0.15$	$p=0.20$	$p=0.25$	$p=0.30$	$p=0.35$	$p=0.40$	$p=0.45$	$p=0.50$
0	0.1216	0.0388	0.0115	0.0032	0.0008	0.0002	0.0000	0.0000	0.0000
1	0.3917	0.1756	0.0692	0.0243	0.0076	0.0021	0.0005	0.0001	0.0000
2	0.6769	0.4049	0.2061	0.0913	0.0355	0.0121	0.0036	0.0009	0.0002
3	0.8670	0.6477	0.4114	0.2252	0.1071	0.0444	0.0160	0.0049	0.0013
4	0.9568	0.8298	0.6296	0.4148	0.2375	0.1182	0.0510	0.0189	0.0059
5	0.9887	0.9327	0.8042	0.6172	0.4164	0.2454	0.1256	0.0553	0.0207
6	0.9976	0.9781	0.9133	0.7858	0.6080	0.4166	0.2500	0.1299	0.0577
7	0.9996	0.9941	0.9679	0.8982	0.7723	0.6010	0.4159	0.2520	0.1316
8	0.9999	0.9987	0.9900	0.9591	0.8867	0.7624	0.5956	0.4143	0.2517
9	1.0000	0.9998	0.9974	0.9861	0.9520	0.8782	0.7553	0.5914	0.4119
10	1.0000	1.0000	0.9994	0.9961	0.9829	0.9468	0.8725	0.7507	0.5881
11	1.0000	1.0000	0.9999	0.9991	0.9949	0.9804	0.9435	0.8692	0.7483
12	1.0000	1.0000	1.0000	0.9998	0.9987	0.9940	0.9790	0.9420	0.8684
13	1.0000	1.0000	1.0000	1.0000	0.9997	0.9985	0.9935	0.9786	0.9423
14	1.0000	1.0000	1.0000	1.0000	1.0000	0.9997	0.9984	0.9936	0.9793
15	1.0000	1.0000	1.0000	1.0000	1.0000	1.0000	0.9997	0.9985	0.9941
16	1.0000	1.0000	1.0000	1.0000	1.0000	1.0000	1.0000	0.9997	0.9987
17	1.0000	1.0000	1.0000	1.0000	1.0000	1.0000	1.0000	1.0000	0.9998
18	1.0000	1.0000	1.0000	1.0000	1.0000	1.0000	1.0000	1.0000	1.0000
19	1.0000	1.0000	1.0000	1.0000	1.0000	1.0000	1.0000	1.0000	1.0000
20	1.0000	1.0000	1.0000	1.0000	1.0000	1.0000	1.0000	1.0000	1.0000

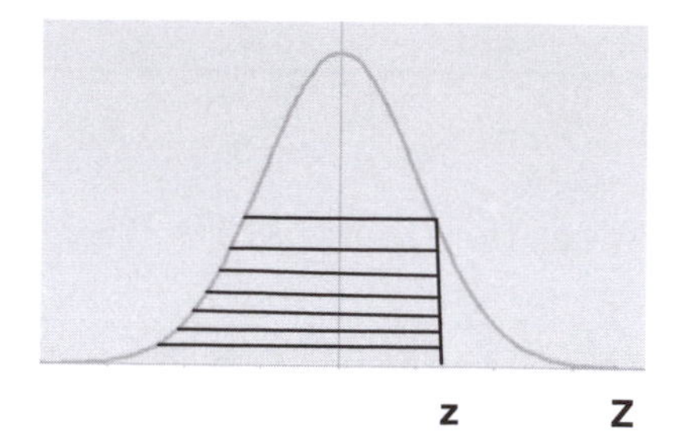

Table C.2 Normal curve area Pr($Z \leq z$)

Z	−0.09	−0.08	−0.07	0.06	−0.05	−0.04	−0.03	−0.02	−0.01	0.00	z
−3.80	.0001	.0001	.0001	.0001	.0001	.0001	.0001	.0001	.0001	.0001	−3.80
−3.70	.0001	.0001	.0001	.0001	.0001	.0001	.0001	.0001	.0001	.0001	−3.70
−3.60	.0001	.0001	.0001	.0001	.0001	.0001	.0001	.0001	.0002	.0002	−3.60
−3.50	.0002	.0002	.0002	.0002	.0002	.0002	.0002	.0002	.0002	.0002	−3.50
−3.40	.0002	.0003	.0003	.0003	.0003	.0003	.0003	.0003	.0003	.0003	−3.40
−3.30	.0003	.0004	.0004	.0004	.0004	.0004	.0004	.0005	.0005	.0005	−3.30
−3.20	.0005	.0005	.0005	.0006	.0006	.0006	.0006	.0006	.0007	.0007	−3.20
−3.10	.0007	.0007	.0008	.0008	.0008	.0008	.0009	.0009	.0009	.0010	−3.10
−3.00	.0010	.0010	.0011	.0011	.0011	.0012	.0012	.0013	.0013	.0013	−3.00
−2.90	.0014	.0014	.0015	.0015	.0016	.0016	.0017	.0018	.0018	.0019	−2.90
−2.80	.0019	.0020	.0021	.0021	.0022	.0023	.0023	.0024	.0025	.0026	−2.80
−2.70	.0026	.0027	.0028	.0029	.0030	.0031	.0032	.0033	.0034	.0035	−2.70
−2.60	.0036	.0037	.0038	.0039	.0040	.0041	.0043	.0044	.0045	.0047	−2.60
−2.50	.0048	.0049	.0051	.0052	.0054	.0055	.0057	.0059	.0060	.0062	−2.50
−2.40	.0064	.0066	.0068	.0069	.0071	.0073	.0075	.0078	.0080	.0082	−2.40
−2.30	.0084	.0087	.0089	.0091	.0094	.0096	.0099	.0102	.0104	.0107	−2.30
−2.20	.0110	.0113	.0116	.0119	.0122	.0125	.0129	.0132	.0136	.0139	−2.20
−2.10	.0143	.0146	.0150	.0154	.0158	.0162	.0166	.0170	.0174	.0179	−2.10
−2.00	.0183	.0188	.0192	.0197	.0202	.0207	.0212	.0217	.0222	.0228	−2.00
−1.90	.0233	.0239	.0244	.0250	.0256	.0262	.0268	.0274	.0281	.0287	−1.90
−1.80	.0294	.0301	.0307	.0314	.0322	.0329	.0336	.0344	.0351	.0359	−1.80
−1.70	.0367	.0375	.0384	.0392	.0401	.0409	.0418	.0427	.0436	.0446	−1.70
−1.60	.0455	.0465	.0475	.0485	.0495	.0505	.0516	.0526	.0537	.0548	−1.60
−1.50	.0559	.0571	.0582	.0594	.0606	.0618	.0630	.0643	.0655	.0668	−1.50
−1.40	.0681	.0694	.0708	.0721	.0735	.0749	.0764	.0778	.0793	.0808	−1.40
−1.30	.0823	.0838	.0853	.0869	.0885	.0901	.0918	.0934	.0951	.0968	−1.30
−1.20	.0985	.1003	.1020	.1038	.1056	.1075	.1093	.1112	.1131	.1151	−1.20
−1.10	.1170	.1190	.1210	.1230	.1251	.1271	.1292	.1314	.1335	.1357	−1.10
−1.00	.1379	.1401	.1423	.1446	.1469	.1492	.1515	.1539	.1562	.1587	−1.00
−0.90	.1611	.1635	.1660	.1685	.1711	.1736	.1762	.1788	.1814	.1841	−0.90
−0.80	.1867	.1894	.1922	.1949	.1977	.2005	.2033	.2061	.2090	.2119	−0.80
−0.70	.2148	.2177	.2206	.2236	.2266	.2296	.2327	.2358	.2389	.2420	−0.70
−0.60	.2451	.2483	.2514	.2546	.2578	.2611	.2643	.2676	.2709	.2743	−0.60
−0.50	.2776	.2810	.2843	.2877	.2912	.2946	.2981	.3015	.3050	.3085	−0.50
−0.40	.3121	.3156	.3192	.3228	.3264	.3300	.3336	.3372	.3409	.3446	−0.40

Table C.2 Normal curve area Pr($Z \leq z$), *continued*

Z	−0.09	−0.08	−0.07	0.06	−0.05	−0.04	−0.03	−0.02	−0.01	0.00	z
−0.30	.3483	.3520	.3557	.3594	.3632	.3669	.3707	.3745	.3783	.3821	−0.30
−0.20	.3859	.3897	.3936	.3974	.4013	.4052	.4090	.4129	.4168	.4207	−0.20
−0.10	.4247	.4286	.4325	.4364	.4404	.4443	.4483	.4522	.4562	.4602	−0.10
0.00	.4641	.4681	.4721	.4761	.4801	.4840	.4880	.4920	.4960	.5000	0.00

z	0.00	0.01	0.02	0.03	0.04	0.05	0.06	0.07	0.08	0.09	z
0.00	.5000	.5040	.5080	.5120	.5160	.5199	.5239	.5279	.5319	.5359	0.00
0.10	.5398	.5438	.5478	.5517	.5557	.5596	.5636	.5675	.5714	.5753	0.10
0.20	.5793	.5832	.5871	.5910	.5948	.5987	.6026	.6064	.6103	.6141	0.20
0.30	.6179	.6217	.6255	.6293	.6331	.6368	.6406	.6443	.6480	.6517	0.30
0.40	.6554	.6591	.6628	.6664	.6700	.6736	.6772	.6808	.6844	.6879	0.40
0.50	.6915	.6950	.6985	.7019	.7054	.7088	.7123	.7157	.7190	.7224	0.50
0.60	.7257	.7291	.7324	.7357	.7389	.7422	.7454	.7486	.7517	.7549	0.60
0.70	.7580	.7611	.7642	.7673	.7704	.7734	.7764	.7794	.7823	.7852	0.70
0.80	.7881	.7910	.7939	.7967	.7995	.8023	.8051	.8078	.8106	.8133	0.80
0.90	.8159	.8186	.8212	.8238	.8264	.8289	.8315	.8340	.8365	.8389	0.90
1.00	.8413	.8438	.8461	.8485	.8508	.8531	.8554	.8577	.8599	.8621	1.00
1.10	.8643	.8665	.8686	.8708	.8729	.8749	.8770	.8790	.8810	.8830	1.10
1.20	.8849	.8869	.8888	.8907	.8925	.8944	.8962	.8980	.8997	.9015	1.20
1.30	.9032	.9049	.9066	.9082	.9099	.9115	.9131	.9147	.9162	.9177	1.30
1.40	.9192	.9207	.9222	.9236	.9251	.9265	.9279	.9292	.9306	.9319	1.40
1.50	.9332	.9345	.9357	.9370	.9382	.9394	.9406	.9418	.9429	.9441	1.50
1.60	.9452	.9463	.9474	.9484	.9495	.9505	.9515	.9525	.9535	.9545	1.60
1.70	.9554	.9564	.9573	.9582	.9591	.9599	.9608	.9616	.9625	.9633	1.70
1.80	.9641	.9649	.9656	.9664	.9671	.9678	.9686	.9693	.9699	.9706	1.80
1.90	.9713	.9719	.9726	.9732	.9738	.9744	.9750	.9756	.9761	.9767	1.90
2.00	.9772	.9778	.9783	.9788	.9793	.9798	.9803	.9808	.9812	.9817	2.00
2.10	.9821	.9826	.9830	.9834	.9838	.9842	.9846	.9850	.9854	.9857	2.10
2.20	.9861	.9864	.9868	.9871	.9875	.9878	.9881	.9884	.9887	.9890	2.20
2.30	.9893	.9896	.9898	.9901	.9904	.9906	.9909	.9911	.9913	.9916	2.30
2.40	.9918	.9920	.9922	.9925	.9927	.9929	.9931	.9932	.9934	.9936	2.40
2.50	.9938	.9940	.9941	.9943	.9945	.9946	.9948	.9949	.9951	.9952	2.50
2.60	.9953	.9955	.9956	.9957	.9959	.9960	.9961	.9962	.9963	.9964	2.60
2.70	.9965	.9966	.9967	.9968	.9969	.9970	.9971	.9972	.9973	.9974	2.70
2.80	.9974	.9975	.9976	.9977	.9977	.9978	.9979	.9979	.9980	.9981	2.80
2.90	.9981	.9982	.9982	.9983	.9984	.9984	.9985	.9985	.9986	.9986	2.90
3.00	.9987	.9987	.9987	.9988	.9988	.9989	.9989	.9989	.9990	.9990	3.00
3.10	.9990	.9991	.9991	.9991	.9992	.9992	.9992	.9992	.9993	.9993	3.10
3.20	.9993	.9993	.9994	.9994	.9994	.9994	.9994	.9995	.9995	.9995	3.20
3.30	.9995	.9995	.9995	.9996	.9996	.9996	.9996	.9996	.9996	.9997	3.30
3.40	.9997	.9997	.9997	.9997	.9997	.9997	.9997	.9997	.9997	.9998	3.40
3.50	.9998	.9998	.9998	.9998	.9998	.9998	.9998	.9998	.9998	.9998	3.50
3.60	.9998	.9998	.9999	.9999	.9999	.9999	.9999	.9999	.9999	.9999	3.60
3.70	.9999	.9999	.9999	.9999	.9999	.9999	.9999	.9999	.9999	.9999	3.70
3.80	.9999	.9999	.9999	.9999	.9999	.9999	.9999	.9999	.9999	.9999	3.80

Note: Entries in the body of the table are areas between $-\infty$ and z.

Table C.3 *t*-distribution

d.f.	$t'_{.90}$	$t'_{.95}$	$t'_{.975}$	$t'_{.99}$	$t'_{.995}$
1	3.078	6.3138	12.706	31.821	63.657
2	1.886	2.9200	4.3027	6.965	9.9248
3	1.638	2.3534	3.1825	4.541	5.8409
4	1.533	2.1318	2.7764	3.747	4.6041
5	1.476	2.0150	2.5706	3.365	4.0321
6	1.440	1.9432	2.4469	3.143	3.7074
7	1.415	1.8946	2.3646	2.998	3.4995
8	1.397	1.8595	2.3060	2.896	3.3554
9	1.383	1.8331	2.2622	2.821	3.2498
10	1.372	1.8125	2.2281	2.764	3.1693
11	1.363	1.7959	2.2010	2.718	3.1058
12	1.356	1.7823	2.1788	2.681	3.0545
13	1.350	1.7709	2.1604	2.650	3.0123
14	1.345	1.7613	2.1448	2.624	2.9768
15	1.341	1.7530	2.1315	2.602	2.9467
16	1.337	1.7459	2.1199	2.583	2.9208
17	1.333	1.7396	2.1098	2.567	2.8982
18	1.330	1.7341	2.1009	2.552	2.8784
19	1.328	1.7291	2.0930	2.539	2.8609
20	1.325	1.7247	2.0860	2.528	2.8453
21	1.323	1.7207	2.0796	2.518	2.8314
22	1.321	1.7171	2.0739	2.508	2.8188
23	1.319	1.7139	2.0687	2.500	2.8073
24	1.3i8	1.7109	2.0639	2.492	2.7969
25	1.316	1.7081	2.0595	2.485	2.7874
26	1.315	1.7056	2.0555	2.479	2.7787
27	1.314	1.7033	2.0518	2.473	2.7707
28	1.313	1.7011	2.0484	2.467	2.7633
29	1.311	1.6991	2.0452	2.462	2.7564
30	1.310	1.6973	2.0423	2.457	2.7500
35	1.3062	1.6896	2.0301	2.438	2.7239
40	1.3031	1.6839	2.0211	21423	2.7045
45	1.3007	1.6794	2.0141	2.412	2.6896
50	1.2987	1.6759	2.0086	2.403	2.6778
60	1.2959	1.6707	2.0003	2.390	2.6603
70	1.2938	1.6669	1.9945	2.381	2.6480
80	1.2922	1.6641	1.9901	2.374	2.6388
∞	1.282	1.645	1.96	2.326	2.576

Notes: Each row represents a different *t*-distribution. Example: d.f. = 4, $t'_{.95}$ yields the value $t = 2.1318$. This means that the area from $-\infty$ to $t = 2.1318$ is .9500.

Table C.4 Chi-square table

d.f.	$\chi^2_{.005}$	$\chi^2_{.025}$	$\chi^2_{.05}$	$\chi^2_{.90}$	$\chi^2_{.95}$	$\chi^2_{.975}$	$\chi^2_{.99}$	$\chi^2_{.995}$
1	.0000393	.000982	.00393	2.706	3.841	5.024	6.635	7.879
2	.0100	.0506	.103	4.605	5.991	7.378	9.210	10.597
3	.0717	.216	.352	6.251	7.815	9.348	11.345	12.838
4	.207	.484	.711	7.779	9.488	11.143	13.277	14.860
5	.412	.831	1.145	9.236	11.070	12.832	15.086	16.750
6	.676	1.237	1.635	10.645	12.592	14.449	16.812	18.548
7	.989	1.690	2.167	12.017	14.067	16.013	18.475	20.278
8	1.344	2.180	2.733	13.362	15.507	17.535	20.090	21.955
9	1.735	2.700	3.325	14.684	16.919	19.023	21.666	23.589
10	2.156	3.247	3.940	15.987	18.307	20.483	23.209	25.188
11	2.603	3.816	4.575	17.275	19.675	21.920	24.725	26.757
12	3.074	4.404	5.226	18.549	21.026	23.336	26.217	28.300
13	3.565	5.009	5.892	19.812	22.362	24.736	27.688	29.819
14	4.075	5.629	6.571	21.064	23.685	26.119	29.141	31.319
15	4.601	6.262	7.261	22.307	24.996	27.488	30.578	32.801
16	5.142	6.908	7.962	23.542	26.296	28.845	32.000	34.267
17	5.697	7.564	8.672	24.769	27.587	30.191	33.409	35.718
18	6.265	8.231	9.390	25.989	28.869	31.526	34.805	37.156
19	6.844	8.907	10.117	27.204	30.144	32.852	36.191	38.582
20	7.434	9.591	10.851	28.412	31.410	34.170	37.566	39.997
21	8.034	10.283	11.591	29.615	32.671	35.479	38.932	41.401
22	8.643	10.982	12.338	30.813	33.924	36.781	40.289	42.796
23	9.260	11.688	13.091	32.007	35.172	38.076	41.638	44.181
24	9.886	12.401	13.848	33.196	36.415	39.364	42.980	45.558
25	10.520	13.120	14.611	34.382	37.652	40.646	44.314	46.928
26	11.160	13.844	15.379	35.563	38.885	41.923	45.642	48.290
27	11.808	14.573	16.151	36.741	40.113	43.194	46.963	49.645
28	12.461	15.308	16.928	37.916	41.337	44.461	48.278	50.993
29	13.121	16.047	17.708	39.087	42.557	45.722	49.588	52.336
30	13.787	16.791	18.493	40.256	43.773	46.979	50.892	53.672
35	17.192	20.569	22.465	46.059	49.802	53.203	57.342	60.275
40	20.707	24.433	26.509	51.805	55.758	59.342	63.691	66.766
45	24.311	28.366	30.612	57.505	61.656	65.410	69.957	73.166
50	27.991	32.357	34.764	63.167	67.505	71.420	76.154	79.490
60	35.535	40.482	43.188	74.397	79.082	83.298	88.379	91.952
70	43.275	48.758	51.739	85.527	90.531	95.023	100.425	104.215
80	51.172	57.153	60.391	96.578	101.879	106.629	112.329	116.321
90	59.196	65.647	69.126	107.565	113.145	118.136	124.116	128.299
100	67.328	74.222	77.929	118.498	124.342	129.561	135.807	140.169

Notes: Each row represents a different chi-square distribution. Example: d.f. = 4, $\chi^2_{.95}$ yields the value $\chi^2 = 9.488$. This means that the area from 0 to $\chi^2 = 9.488$ is .9500.

Index